Letter or Symbol	*Primary Use*
α (alpha)	Direction of slip line, or a constant
β (beta)	Direction of slip line, or a constant
γ (gamma)	Unit weight (density times acceleration of gravity)
Γ (cap. gamma)	Shear-stress function
δ (delta)	Increment of some quantity, or slope angle
Δ (cap. delta)	Increment of some quantity
∇ (del)	Partial differential operator
E (Epsilon)	Component of strain tensor
ϵ (epsilon)	Strain
$\dot{\epsilon}$	Rate of strain
η (eta)	Coefficient of viscosity
θ (theta)	Angle of polar coordinate
λ (lambda)	Elastic constant or wave number
ν (nu)	Poisson's ratio of elasticity
ξ (xi)	A constant
π (pi)	A constant (3.14159 . . .)
ρ (rho)	Radius of curvature
σ (sigma)	Stress
Φ (cap. phi)	Airy's stress function
ϕ (phi)	Angle of internal friction
ψ (psi)	Angle between *x*-direction and principal stress trajectory
Ω (cap. omega)	Angular velocity
ω (omega)	Angular displacement

Physical Processes in Geology

Physical Processes in Geology

A method for interpretation of natural phenomena — intrusions in igneous rocks, fractures and folds, flow of debris and ice

ARVID M. JOHNSON
Stanford University

Freeman, Cooper & Company
1736 Stockton Street, San Francisco, California 94133

This book is dedicated to two searchers

Alfred M. Johnson, my father
of Dee, Oregon

Eugene G. Williams, Professor of Geology
The Pennsylvania State University
State College, Pennsylvania

Printed in the United States of America

Library of Congress Catalogue Card Number 70–119373
ISBN 0–87735–319–0

Preface

While I was a student, I began to sense the need for an introductory course and textbook, designed primarily for geology students, that would describe theories of the mechanical behavior of materials and apply the theories to geological problems. There are plenty of courses on mechanics in departments of engineering and physics, but geology students find these courses inadequate for solving geological problems. In fact, many geology students believe that, in order to eke out ten grams of concepts they can use in analyzing geological processes, they must sort through a kilogram of unusable information and hypothetical problems.

Recognition of the need for an introductory course on methods of applying mechanics to the solution of geological problems has led me to teach such a course at Stanford University. This book is a product of that course.

The student will find that once he has gained some experience in the application of mechanics to specific geological problems, he will be able to apply mechanics to other problems in geology; and once he begins to apply the fundamental concepts of mechanics in his search for understanding of geological processes, he will recognize the pertinence of the analytical techniques developed by engineers and physicists that he once believed to be inapplicable to geology. He will also be able to comprehend and evaluate research papers on mechanical analyses of geological problems.

This book presents the rheological approach to mechanical processes that is fundamental to our understanding of the behavior of earth materials. Briefly, the rheological approach consists of deciding whether a real material is predominantly elastic, plastic, viscous, plastico-viscous, or something else. Its method is to compare the behavior of real materials with the response, under controlled conditions, of theoretical or experimental materials whose behaviors can be described mathematically. Through the rheological approach we can get a fairly good idea of how earth materials have been deformed during the formation of many geologic features. After we have decided how earth materials responded to natural forces, we are in a position to search for answers to the question of why the materials behaved the way they did.

The primary purposes of this book are to emphasize the value of understanding physical processes in studies of diverse geologic features such as folds, dikes, and glacial deposits; to provide the geologist with a background in mechanics that will allow him to study more advanced texts on mechanical behavior of materials; to stimulate interest in certain geological problems; and to illustrate what I consider to be a fruitful approach to their solution.

The first chapter states and discusses briefly the fundamental concepts of physical processes of earth materials—concepts of equilibrium, rheology, and boundary conditions.

Part I, *Flexures*, introduces the reader to certain mathematical techniques that have been used to solve some problems of the formation of flexures. These mathematical techniques, collectively called *beam and plate theory*, are widely used by engineers, who have developed them to a high degree of usefulness. The theory, however, is not limited to use in engineering and in the analysis of folding; it is useful in an analysis of any physical process that emphasizes the layered nature of rock. For example, students at Stanford have applied beam and plate theory to understand deformations measured in the ground surface in the vicinity of a pumping well (*16*)* and to understand the behavior of strata through which salt domes pierce. In Part I, concepts of equilibrium, force, bending moment, stress, strain, and three-dimensional representations of the behavior of elastic and viscous substances are discussed where they are first required in the mathematical analysis of processes of flexure. The fundamentals of the theory of elasticity are developed, and similarities between elastic and viscous behaviors are emphasized. All of these concepts are introduced within a context familiar to geologists—flexures.

Part II, *Patterns of faults, joints, and dikes*, presents analyses that serve six purposes: to develop the meaning of Mohr's Circle—that is, the correspondence between physical space and stress space; to introduce polar coordinates; to discuss briefly some criteria of fracture; to illustrate the method of calculating patterns of stresses within an elastic body; to introduce the important concept of initial stress; and to indicate how little we know about the processes of formation of joints and faults, particularly emphasizing the fact that currently there is no logical basis for the almost universal assumption that criteria of failure predict orientations of large-scale fractures.

In Part III, *Flow of ice, lava, and debris*, we examine the marked similarities among patterns of flow and among depositional products of ice, lava, and mud. Then we show that these similarities can be explained in terms of the mechanical behavior of the three substances: ice, lava, and mud. An important goal of this part of the book

* See references at end of Chapter 1.

is to demonstrate that a study of processes, of flow for example, can greatly aid our understanding of geomorphological features. Part III also examines the Navier-Stokes equations of viscous flow, the Fourier series, and combinations of rheological equations.

I have made no attempt to compile all the available knowledge of the behavior of earth materials, nor have I tried to replace comprehensive texts on elasticity, plasticity, viscosity, and fracturing. My aim is, rather, to provide a background and a perspective that will allow the student to gain greater profit from a study of more comprehensive texts on these subjects.

The mathematical background required is differential and integral calculus; other mathematical techniques needed are developed in the text. The reader should search for further illustrations of some of these mathematical techniques, and for these he is referred to the following texts on applied mathematics:

Kreyszig, Erwin, 1962, *Advanced Engineering Mathematics*: John Wiley and Sons, Inc., N.Y.

Sokolnikoff, I. S., and Redheffer, R. M., 1958, *Mathematics of Physics and Modern Engineering*: McGraw-Hill Book Co., Inc., N.Y.

Wylie, C. R., Jr., 1960, *Advanced Engineering Mathematics*: McGraw-Hill Book Co., Inc., N.Y.

The book by Sokolnikoff and Redheffer treats more subjects than do the books by Wylie or Kreyszig, but its language is more abstract.

I have found that, of the many mathematics and mechanics books in print, all the preceding books on applied mathematics, as well as the following books on applied mechanics, are the most easily understood.

ELASTICITY:

Timoshenko, S., and Goodier, J. N., 1951, *Theory of Elasticity*: McGraw-Hill Book Co., Inc., N.Y.

Frocht, M. M., 1941, *Photoelasticity*: John Wiley and Sons, Inc., N.Y. 2 vol.

Love, A. E. H., 1944, *A Treatise on the Mathematical Theory of Elasticity*: Dover Publications, N.Y.

Biot, M. A., 1965, *Mechanics of Incremental Deformations*: John Wiley and Sons, Inc., N.Y.

Biot is rather difficult to follow, because he introduces many new concepts, but the theories Biot develops will be very useful to geologists concerned with the behavior of initially stressed rock. The book by Timoshenko and Goodier is exceptionally lucid.

PLASTICITY:

Prager, W., and Hodge, P. G., 1951, *Perfectly Plastic Solids*: John Wiley and Sons, Inc., N.Y.

Hill, R., 1950, *The Mathematical Theory of Plasticity*: Oxford at the Clarendon Press, London.

Johnson, W., and Mellor, P. B., 1962, *Plasticity for Mechanical Engineers*: D. Van Nostrand Co., Ltd., London.

VISCOSITY:

Rouse, Hunter, 1961, *Fluid Mechanics for Hydraulic Engineers*: Dover Publications, Inc., N.Y.

Langlois, W., 1964, *Slow Viscous Flow*: Macmillan Co., N.Y.

Lamb, Sir Horace, 1945, *Hydrodynamics*: Dover Publications, Inc., N.Y.

For an introduction to the theory of viscosity the book by Rouse is superior to the other two.

FRACTURE:

Tetleman, A. S., and McEvily, A. J., Jr., 1967, *Fracture of Structural Materials*: John Wiley and Sons, Inc., N.Y.

Jaeger, J. C., and Cook, N. G. W., 1969, *Fundamentals of Rock Mechanics*: Methuen and Co., Ltd., London.

STRAIN:

Ramsay, J. G., 1967, *Folding and Fracturing of Rocks*: McGraw-Hill Book Co., Inc., N.Y.

Ramsay presents the best treatment of the theory of strain that I have seen anywhere.

By studying this book, the reader can become familiar with fundamental concepts that are necessary for the solution of problems in mechanics—the concepts of rheological equations, boundary conditions, and equations of equilibrium—but without practice he cannot become proficient in their use. I recommend, therefore, that while he is still reading the book he begin to formulate and solve a mechanical problem in geology. For example, after studying Part I, he might formulate realistic rheological equations and boundary conditions that would explain the rate of deflection of the earth's surface at Hoover Dam (*13*)* or at Lake Bonneville (*2*)*. Other problems will be suggested by those studied in the text.

In addition, I have found that a detailed, step-by-step analysis of the classic paper on overthrust faulting by M. K. Hubbert and W. W. Rubey (*7*)† is especially instructive for geology students. The paper contains an excellent analysis of the concept of effective stresses in porous media, a concept introduced by Karl Terzaghi

* See references at end of Chapter 1.

† See Ref 7: p. 117–123 and 149–162, Statement of problem; p. 129–142, Solution of problem. Also see discussions by Laubscher, Moore, and Birch (refs. 8, 9, and 10).

(*19*)‡ and clarified by Hubbert and Rubey. An analysis of the paper also provides an introduction to vector notation.|| I have found that students learn a great deal by writing papers and then holding a seminar to exchange ideas on effective stress, including a clear explanation of each step in the mathematical derivations by Hubbert and Rubey. The papers may be written as though a student were explaining the analyses of Hubbert and Rubey to a reader who is unfamiliar with vector notation.

By making such an application of the fundamental concepts that are becoming familiar to him, the student should attain sufficient mastery of the concepts to apply them to other problems in geology. By the time he finishes this book, he should be able to make use of the principles of physical processes, and he should be prepared to study texts dealing with most subjects in mechanics.

‡ See references at end of Chapter 1.

|| The book by Gibbs(*5*) is a good treatment of vector notation and analysis.

Acknowledgements

During the years of developing my theme and writing this textbook, I have become indebted to many persons. Such debts I acknowledge while absolving these persons from any shortcomings the book may have.

Professors Eugene G. Williams and the late Paul D. Krynine, both of Pennsylvania State University, and Dean Richard H. Jahns, now of Stanford University, introduced me to the method that is developed here as the framework of the book. Professor Jahns encouraged my writing of this analysis of the method, not without many a constructive criticism. Similarly, David Gold and Mark Meier were helpfully critical and encouraging.

I am especially grateful to the editors, William H. and Margaret C. Freeman, whose collaborative efforts are reflected throughout the published work. And the illustrative equipment, in its extent and quality, reflects my debt to the artist, Perfecto Mary.

And to my students, my thanks: David Pollard, as a graduate student at Stanford, has been unusually helpful. He read with the critical eye of a student most of the manuscript, making numerous valuable suggestions. In addition he, Stephen Ellen, and I spent many searching hours discussing the philosophy I present as well as developing among us the method for studying physical processes at work in the earth. Chapter 2 is based largely on research by David Pollard, and Chapter 8 on work by Stephen Ellen. Certainly the greatest contributions have been made by the collective efforts of graduate students who have taken my course and who offered criticism. Especially helpful have been Monty Hampton, Robert Lawrence, Bruce Clark, Robert Wesson, Elmont Honea, Charles Price, George Pflafker, S. D. Peng, Stephen Ellen, and David Pollard.

Stanford University
Summer, 1970

ARVID M. JOHNSON

Contents

Page

Topics in Mechanics xii
1. Method of Solving Mechanical Problems in Geology . . . 2

Flexures

2. Laccoliths of Henry Mountains of Utah 30
3. Fold Trains in Appalachian Mountains 74
4. Folding of Single Members 132
5. Theoretical Interlude: Stress, Strain, and Elastic Constants . . 174
6. Development of Folds in Carmel Formation, Near Moab, Utah . 218
7. Theoretical Interlude: Introduction to Theories of Elasticity and Viscosity 248
8. Folding of Interbedded Chert and Shale of Franciscan Formation, San Francisco, California 292

Patterns of Faults, Joints and Dikes

9. Fault Pattern at Timber Mountain Caldera, Nevada 334
10. Formation of Sheet Structure in Granite 356
11. Dike Patterns at Spanish Peaks, Colorado 400

Flow of Ice, Lava, and Debris

12. Formation of Debris-Flow Deposits 432
13. Transport of Boulders and Blocks by Debris Flow 460
14. Rheological Properties of Debris, Ice, and Lava 494
15. U-Shape of Valleys 536

Closing Comments 572
Index 574

Topics in Mechanics

	Chapter
Equilibrium of Forces and Bending Moments	1, 2
Rheological Models	1
Meaning of Boundary Conditions	1
Flexure of Unconfined Members	
Forces and Moments within Members	2
Deflection as a Result of Uniform Transverse Load per Unit of Length	2
Large Deflection — Elastica	4
Deflection of Material with Nonlinear Properties	8
Flexure of Members Confined Laterally	
Single Members	
Deflection under Axial Loads	3
Influence of End Loading on Deflection	3
Influence of Initial Deflection on Wavelength	3, 4
Effect of Gravity on Wavelength	4
Multiple Members	
Layers with Frictionless Contacts	4
Interbedded Soft and Stiff Layers Bonded together	6, 8
Stress and Strain	
Engineering Stress and Strain	5
Initial Stress	10
Plane Stress and Plane Strain	7
Strain Rate	7
Mohr Circle	5, 9
Theory of Elasticity	
Elastic Constants	

	Chapter
Isotropic Material	5
Orthotropic (Layered) Material	6
Airy Stress Function	
x-y Coordinates	7
Polar Coordinates	11
Superposition of Solutions	7, 11
Method of Images	7, 11
Sinusoidal Deflection of One Surface of Thick Body	7
Stresses around Circular Hole in Elastic Plate	9, 10, 11
Stresses around Elliptic Hole in Elastic Plate	10
Theory of Viscosity	
Generalized Rheological Equations	7
Sinusoidal Deflection of One Surface of Thick Body	7
Buckling of Viscous Member in Viscous Medium	7
Flow in Circular and Wide Channels	14
Flow in Rectangular Channels	15
Theory of Plasticity	
Bending of Plastic Member	8
Slip Lines	13
Velocity Relations along Slip Lines	13
Penetration of Surface by Flat Punch	13
Theories of Fracture	
Coulomb Theory	9
Griffith Theory	10
Theory of Bingham Substance	
Generalization of Rheological Equations	15
Flow in Circular and Wide Channels	14
Flow in Rectangular and Triangular Channels	15

PART I

FLEXURES

1

Chapter Sections

Introduction
Analytical Tools Needed for Deciphering Physical Processes
Mechanics
  Equilibrium of Moments and Forces
  Force
  Newton's Laws
  Stress and Strain
Rheology
  Linear Rheological Models — Elasticity, Plasticity, and Viscosity
  Combinations of Linear Rheological Models
    Bingham Model
    Prandtl Model
    Kelvin and Maxwell Models
  Higher-Order Rheological Models
Boundary Conditions
References Cited in Chapter 1

1

Method of Solving Mechanical Problems in Geology

Introduction

The study of physical processes of geology is a study of a method: a method introduced by G. K. Gilbert in the latter part of the 1800's and clearly illustrated in his professional papers. His method was quite different from that of his predecessors and, indeed, is quite different from that of most of his successors. Many geologists, when they examine an outcrop, see certain features that they recognize, such as drag folds next to a fault or a dike transecting bedding in the outcrop, and they name and "explain" them: "These are drag folds. They indicate that the rock on this side has moved upward along the fault with respect to the rock on the other side." And, "There is a dike of basalt, solidified in the earth's plumbing system. It probably fed surface flows."

When I read Gilbert's papers, however, I get the distinct impression that he did not operate this way at all. Indeed, he performed the normal operations of gathering and classifying field data, mapping and describing the shapes of structures. But he had an additional goal—he tried to *understand* what he saw in each outcrop. He was not satisfied, necessarily, with classical explanations of features he observed. And he refused to interpret natural phenomena without following sets of mechanically sound rules or without imagining analogies between the geologic process and a process with which he was familiar.

Thus, Gilbert would examine in detail the contorted beds next to the fault and try to guess how the strata behaved at the time the contortions formed. He would draw upon his experience and intuition in order to imagine an analogous behavior in familiar materials. He probably would recognize that different shapes of the flexures are related to details of the path of the fault. This relation would suggest that the loads to which each stratum was subjected depended upon the orientation of the fault as it cut across the strata. Thus, he would imagine a thin strip of metal held firmly at one end and loaded on the other end. If the free end is loaded along the axis of the strip, the strip buckles laterally, forming a long arch (Fig. 1.1A). If, however, one end of the strip is pushed laterally, it simply drags upward (Fig. 1.1B). If there is a combination of axial and lateral loading, the form of the deflection is a combination of these two modes of deflection (Fig. 1.1C). He would reason, therefore, that details of the shapes of the drag features can be understood in terms of the types of loading, which, in turn, are related to details of the path of the fault.

When Gilbert looked at the outcrop containing the basalt dike, he would see more than a vision of part of the earth's plumbing system. First, he would note that the dike does not extend across the entire outcrop but that it terminates. Further, he would note that the termination of the dike is flat. Some geologists probably would not even notice that the dike terminates within the outcrop, but, if they did, they

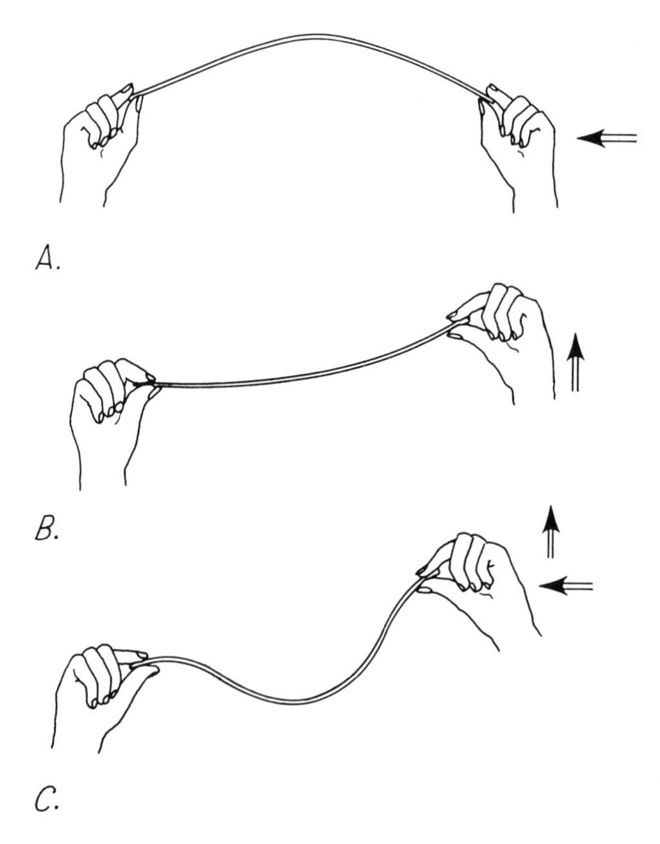

Figure 1.1. Reaction of a thin strip of metal to various types of loads applied to its ends.

probably would draw it on their maps as a tabular body thinning laterally, finally terminating in a wedge shape, because that is the way they learned to draw dike terminations.

Gilbert may not have understood the cause of the blunt terminations of dikes and sills, but he would have realized immediately that they are related to the properties of the molten magma and the country rock and that he could learn something by contemplating the forms.

It seems to me that there are two fundamental differences between the approach of many geologists and the approach of Gilbert. One is the ability to create sound mechanical analogies and the other is a matter of attitude. Many geologists operate with a dearth of quantitative rules or laws. They learn a few rules in geology courses such as structural geology and geomorphology, but, unfortunately, many of the rules cannot weather careful scrutiny—they are full of half-truths or even absolute misconceptions. Gilbert, however, did not suffer from a lack of quantitative rules. Most of the rules he applied arose from analogies. A phenomenon, such as a drag flexure, reminded him of another, more familiar phenomenon which he felt he could understand, such as the bending of a strip of metal under various loading conditions. One of Gilbert's unique qualities was his ability to create scores of analogies, almost at the drop of a hat. These analogies represented the rules with which he examined geologic features.

Much of this book is concerned with introducing various rules and analogies of the behavior of materials. Most of the rules have been invented by engineers and physicists but we can use them where they seem to apply to geological problems. Further, by studying and learning the rules and laws that engineers and physicists have created, we can learn to create others that are required only to solve geological problems. By studying the way engineers and physicists have developed their rules we can learn to develop our own rules.

The other fundamental difference between Gilbert's approach and the approach of many geologists is one of attitude. We geologists tend to emphasize the complexity of our endeavor to decipher the geologic record. We excuse our apparent inability to quantify our science, as physicists and chemists have quantified theirs, by stating and eventually believing that all geological problems are much more complex than physical and chemical problems. Besides, we sometimes proudly proclaim, how can we understand, with theory or experiment, phenomena that it has required millions of years to create? Gilbert, it seems, took the opposite point of view: We can understand nearly any geologic process if we work at it, and the essential features of most processes have analogies.

This confidence in ability to solve geological problems is a personal matter that cannot be transmitted directly by a book or teacher. Perhaps, however, careful study of the examples discussed in following pages and elsewhere in the published literature will encourage students to solve a geological problem involving physical processes.

One seems, by successfully solving a problem, to develop the confidence that he can solve other problems.

In order to probe more deeply into Gilbert's method, let us examine some peculiar folds in the Franciscan Formation of central California (Fig. 1.2). One limb of the syncline is about 60 cm and the other is 30 cm in length. The resistant beds are chert and the soft beds are shale. The thicknesses of the chert beds range from a few millimeters to about eight centimeters within the area shown by the photograph. The chert beds characteristically are highly fractured, but slippage along the fractures is negligible. A peculiar feature of the folds shown in Fig. 1.2 is lack of thickening or thinning of chert beds in their crests or troughs. However, the thicknesses of shale interbeds are different in different places along the folds.

Some questions that immediately come to mind about the process that produced

Figure 1.2. Folds in Franciscan Formation. The lack of thickening in crests and troughs of folds is peculiar.

these folds are: Why do the folds have relatively straight limbs, with most of the bending during folding apparently having been concentrated in the crests of folds? How did the materials behave during deformation; that is, were their behaviors predominantly elastic or plastic or viscous? If we tested samples of the chert and shale would their present mechanical behaviors give clues to their behaviors when the folds formed? What were the conditions at the time of folding?

To answer these questions we would use several tools. First we would study the folds in the field, making detailed maps of outcrops. One of the most important functions of our mapping is to relate the folds to each other and to adjacent, undeformed rocks. Once we know how adjacent rocks behaved during folding we can place limits on the types of deformation that could have led to the folds. Thus we can determine the conditions of folding.

While the field mapping is in progress we begin to develop or adopt rules of behavior of theoretical materials which seem to fulfill conditions we deduce from field relations. By comparing the behaviors of theoretical materials with field observations, we should be able to determine what theoretical material most closely behaves as the rock did at the time it folded. This is the rheological method, to the understanding of which many of the following pages are devoted.

When we have decided which theoretical material most closely corresponds with the rock during folding, we can select experimental materials whose behavior reproduces approximately its behavior. With the experimental materials so selected we can construct physical models to study effects of conditions more complex than those we can study analytically with theoretical methods.

In addition, once we have decided how the materials behaved, we are in a position to understand the mechanisms of deformation. For example, we might investigate the internal structure of silica jel, so that we can better understand the deformation of chert in a highly fluid form. Perhaps we would make thin sections and x-rays of cherts and shales to study their internal structures. We can get important clues about the behavior of the chert and shale beds by these methods. For example, microscopic evidence suggests that the cherts had a very low strength during the deformation; even though there had to be intense deformation in the fold hinges in order to produce the appression observed, remains of radiolaria, consisting of very delicate, siliceous tests of microscopic organisms, generally were undeformed. The only places in the folds where radiolaria within the chert beds are markedly sheared are along the edges of the chert beds, within about one millimeter of the chert-shale contacts. Thus the cherts must have been soft when they folded, not hard and brittle as they are today.

This outline of possible steps in an analysis of folds in the Franciscan Formation illustrates four facets of an approach that should lead to solutions of a host of problems in physical geology:

1. Analysis of geologic features in the field. Any problem in physical geology must begin in the field situation and any solution must fit what one sees in the field. This is where a hypothesis is truly tested.

2. Theoretical approach. In a theoretical, or mathematical, analysis of geologic features we idealize earth materials by simple models—elasticity, plasticity, viscosity, and strength, or combinations of these—and apply different conditions to the idealized materials to reproduce forms seen in the field. The power of this tool of the geologist has not been exploited. A function of this book is to allow him to use this tool.

3. Experimentation with physical models. Experimentation is useful in at least three ways: It can check predictions based upon theory; it is helpful in an investigation of processes in more complex, yet still idealized, systems than those that can be treated readily by mathematical manipulation; and it stimulates the imagination of the investigator.

4. Investigation of mechanisms of deformation. A knowledge of mechanisms of deformation commonly is necessary to unify experimental and field observations. For example, by working with different rheological models one might discover that certain types of folds in quartzite could have formed in a rigid-plastic substance. We should be able to explain why the quartzite behaves as a rigid-plastic substance by investigating the ways quartz sand and quartz crystals deform internally.

This is what we mean by saying that the study of physical processes of geology is a study of a method. None of these four aspects of the method, alone, is adequate to solve most problems in physical geology. All four of them, executed at various times during an investigation, will provide information essential to the solution of the problems. At least two and in some cases all four of the facets of the method are needed to solve the problems I have selected for this book, most of which originated in the field.

Analytical tools needed for deciphering physical processes

Our major thrust will be toward a working knowledge of one facet of the method of physical processes—toward an understanding of the theoretical approach. The elements of the theoretical approach are few, perhaps surprisingly few. They are: mechanics, rheology, and boundary conditions. Mechanics and rheology are the rules. Mechanics is the collection of rules that tell us how forces and motions of bodies are related. Equations describing motions of gravitating bodies and definitions of stresses are examples of mechanical equations. Rheology is a collection of rules that tell us how certain imaginary materials respond to stresses and strains. Boundary conditions place restrictions on types of possible behaviors or configurations of bodies. For example, if the position of one end of an inflexible rod is fixed in space, we know that the other end is somewhere on an imaginary sphere that has a radius equal to the length of the rod. The boundary condition, the position of one end of the rod, places a limit on where we would search for the other end of the rod.

Mechanics

Mechanics is a branch of physical science that deals with effects of forces upon the motions of bodies. The subject of classical mechanics usually is divided into statics and dynamics. If the forces that act on a body are balanced so that there is no change in its motion, a state of equilibrium exists and the problem is placed under the division of *statics*. Most of the problems we will study fall under this category. For example, we will assume that the rocks over a laccolith are in static equilibrium with the pressure exerted by the magma in the laccolith, in order to relate magma pressure to deflection of the overburden.

Dynamics, the second subdivision of mechanics, concerns the motion of bodies. It commonly is further subdivided into two categories: *kinematics*, which is the study of the motion of bodies without reference to forces that cause the motion, and *kinetics*, which relates the forces and the resulting motion. An analysis of strain in rocks is an example of a study of kinematics; strain is the kinematics of deformation. Our analysis of the flow of mud in a channel is a study of kinetics, because we are concerned with the relation between the driving and resisting forces and the velocity distribution within the channel.

Equilibrium of Moments and Forces.—Most of the fundamental concepts of mechanics that we will use in solving problems are in the category of statics. They are remarkably few. They are simply those of *equilibrium* or balance of moments and forces, first introduced by Archimedes in about 250 B.C. and clarified by Stevinus in about 1600; Newton's definition of *force* and his *second law* relating force and acceleration, published in 1687; and Cauchy's definition of *stress*, first published in 1823 (*1,20,14*). The other concepts of mechanics we will use are derived from these.

The Greeks, notably Archimedes (287–212 B.C.), developed statics. Building constructors of Archimedes' time and even before recognized intuitively that a small force applied at one end of a lever will balance a larger force at the other end of the lever if the fulcrum of the lever is nearer the larger force. But it was left to Archimedes to discover the rule relating the forces and their distances from the fulcrum. He began his analysis by assuming that equal weights will balance if they are placed at equal distances from the point of support. With geometrical arguments he then proposed the principle of the lever: For equilibrium, the weights must be inversely proportional to the distances. In the problems we investigate, Archimedes' principle takes the following form: For a body to be in equilibrium the sum of torques about any point in the body is zero.

Symbolically:

$$\sum M = 0 \tag{1.1}$$

where $\sum$ (sigma) indicates summation. Each torque or moment, M, can be represented by a force or a weight on a lever, times the distance between the point of application of the force and the point of reference, for example the fulcrum (*1*).

Stevinus (1548–1620) devised a different proof for Archimedes' principle of the lever and recognized that Archimedes' principle readily can be extended to a concept of balance of forces (*1*). We will repeatedly use the principle of equilibrium of forces proposed by Stevinus. The principle is that the sum of all the forces on a body in equilibrium is zero. Symbolically,

$$\sum F = 0. \tag{1.2}$$

Force.—Newton's formulation, in 1687, of the concept of force was an outgrowth of his study of one of the fundamental problems of his day: explaining Kepler's laws of planetary motion. Almost incidentally to this work, he developed the fundamental principles of Dynamics and stated the Laws of Motion in a form which, with the aid of mathematical analysis, is sufficient for solving most mechanical problems (*1*).

When we push or pull an object we say that we apply a force to it. Newton generalized this intuitive concept of force so that the concept would include all kinds of forces. Thus, in a series of eight definitions in the introductory chapter of his book (*15*) *Mathematical Principles of Natural Philosophy*, first published in 1685, he defined three kinds of forces: one that we now call inertia, another that we now call centripetal force, and a third that he called an *impressed force*. He defined an impressed force as an action exerted upon a body in order to change its state, either of rest or of uniform motion in a straight line (*11*).

Newton's Laws.—After defining the various kinds of forces and explaining them, Newton stated his famous Laws of Motion (*15*).

I. Every body continues in its state of rest or of uniform motion in a straight line unless it is compelled to change that state by forces impressed upon it.

II. Rate of change of momentum is proportional to the impressed force, and is in the direction in which the force acts. Or, the acceleration of a particle is proportional to the resultant force acting on it and is in the direction of this force.

Newton's *second law* forms the basis for most analyses in classical *dynamics*. This law can be written in several forms, three of which are

$$\vec{F} = \frac{\vec{mv}}{t},$$

$$\vec{F} = \vec{ma},$$

$$\vec{F} = \frac{d(\vec{mv})}{dt}$$

where $\vec{mv}$ is momentum, t is time, and $\vec{a}$ is acceleration. The arrows over the terms indicate that the relationship is vectorial. For example, the direction of the force is the same as the direction of the acceleration. In addition, the magnitudes of the force and the mass times its acceleration are equal.

III. To every action there is always opposed an equal reaction. Or, the forces of action and reaction between contacting bodies are equal in magnitude, opposite in direction, and colinear.

The *third law* of Newton is basic to an understanding of forces. It states that forces always occur in pairs that are equal in magnitude but opposite in direction. When we analyze the equilibrium or motion of a body under the action of forces, it is absolutely necessary to be clear about which of the pair of forces is being considered. First we isolate the body of interest and then we consider only the forces that act *on* the body.

Stress and Strain.—The concept of stress is intimately related to the development of the mathematical theory of elasticity. Navier, in 1821, advanced a theory of molecular forces to explain elastic behavior of solids (*14*). He assumed that there are two systems of forces acting on the particles of an elastic solid. One system consists of forces that bond the molecules and that balance each other. The other system of forces is related to changes in distances between molecules. They balance external forces that cause changes of positions of particles, that is, that cause deformation of the solid. He assumed that elastic reactions arise from variations in the distances between molecules which, in turn, cause variations in the forces between molecules. Thus, he assumed that if the distance between two particles is decreased, the forces between the particles increase proportionately (*14,20*).

After reading Navier's paper on the theory of elasticity, Cauchy, in about 1825, became interested in the theory and began work on it himself. Instead of considering forces between particles in an elastic substance, Cauchy applied the notion of pressure on various imaginary planes passing through the substance. Thus Cauchy chose a larger fundamental element of an elastic solid than did Navier. He had become familiar with the concept of pressure in his studies of hydrodynamics. To Cauchy, stress was visualized as a type of pressure that is not necessarily normal to the plane on which it acts. The magnitude of pressure near a point is the same in all directions from that point, but the magnitudes of Cauchy's stresses near a point can be different in different directions. Cauchy defined the stresses on the surfaces of a very small element of a material as the magnitudes of the forces acting across the surfaces, divided by the areas of the surfaces. He showed, by geometrical considerations similar to those that we will develop in subsequent pages, that the stresses on the element are expressible by means of nine components. Later Cauchy redefined stress so that it could be related to the structure of matter. He defined the total force on a small plane (that is large relative to the size of molecules) as the resultant of all the forces exerted by molecules on one side of the plane upon the molecules on the other side of the plane. Then, by dividing the total force by the area of the plane, he obtained the magnitude of the stress (*14*). Symbolically we can define a stress as the limit of the part of a force acting on an area divided by that area, as the area approaches zero. Thus, if

$$\sigma = \text{stress},$$

and if

$$\Delta\vec{F} = \text{the part of the force acting on}$$

$$\Delta A = \text{the small element of area},$$

then

$$\sigma = \lim_{\Delta A \to 0} \frac{\Delta\vec{F}}{\Delta A}.$$

Now the part of the force, $\Delta\vec{F}$, acting on the element is a vector quantity. That is, it is defined entirely by its magnitude and the direction in which it acts. The stress (sigma), however, is a tensor quantity, which means that it is defined by a magnitude, a direction, and a plane on which it acts. Thus a force is defined by two quantities and a stress by three.

Stress has the dimensions of force per unit of area, but it is defined near a point. Therefore, the stress exerted on the bottom of a pogo stick by the rider is known if and only if the force is uniformly distributed over an end of the pogo stick. Otherwise only the *average* stress is known; it is the force divided by the area.

Cauchy also expressed the state of strain near a point in a body in terms of nine components. Following Cauchy we can define *strain* as the limit of the change in length of any line element within a body, divided by the original length of the line element, as the length of the line element approaches zero. Symbolically, if

$$\epsilon = \text{strain},$$

Δl = change in length of a line element,

ΔL = original length of the same line element,

$$\epsilon = \lim_{\Delta L \to 0} \frac{\Delta l}{\Delta L}.$$

Strain, therefore, is similar to stress in that it is defined near a point within a body. Strain is dimensionless, it is a length divided by a length.

Stress and strain are analyzed more fully in later pages.

Rheology

The flow of matter, or rheology, is today a subject of intensive study. The phenomena of flow impinge upon nearly every aspect of life. Civil engineers attempt to predict how soil beneath buildings will respond to long-term loading conditions. They also try to explain and to avoid failure of slopes and earth-fill dams when the slopes and dams are under short-term cyclic loading induced by earthquakes. Medical doctors want to know how blood flows in arteries. Forest-product technologists need to know how suspended fibers can be most efficiently transported in pipes. Geologists want to understand how and why rocks deform under rather low but extremely long-term loads. Thus a wide range of materials flow, and the subject of flow is of interest to people with a wide range of interests.

Rheology is not restricted to study of materials that deform permanently, however. It is a branch of physics that is concerned with the behavior of all materials under stress. To illustrate what rheology is, let us perform the following mental experiment. Drop onto a concrete floor four bodies, a gum eraser, a cube of halite, a ball of soft clay, and a cubic centimeter of honey. Mechanics tells us no more about the behavior of these materials during their fall than is contained in Newton's second law, $\vec{F} = m\vec{g}$. The force acting on each body is the same. It is proportional to the mass of each body times the acceleration of gravity. Also, Newton's law tells us that the direction of fall of each body corresponds with the direction of the gravitational field where the experiment is conducted. The four different materials, therefore, behave essentially alike during the fall. Their differences are not evident until they strike the concrete floor. Then the gum eraser rebounds and bounces around the floor for a few seconds, the soft clay sticks to the floor, the halite fractures and fragments scatter about the floor, and the honey slowly spreads out on the floor. Of these entirely different behaviors mechanics gives us no account. This is where rheology takes over (*17*).

The behaviors of these four substances, rubber, clay, honey, and halite, illustrate the basic rheological behaviors—elastic, plastic, viscous, and fracture. The behaviors of most real substances can be represented adequately in terms of the four basic ones.

The basic substances have fundamentally different properties: In response to stress an *elastic* body strains and all the strain is recoverable; a viscous body flows, and a plastic body deforms permanently. Ideal plastic and viscous substances are closely related in the sense that their deformations are not recovered when the stress causing the deformation is removed. But they are significantly different in a very important respect. For a *viscous* substance there is no threshold strength; that is, a shear stress of any magnitude other than zero causes a viscous substance to flow. Its rate of flow depends upon the magnitude of the stress. For a *rigid-plastic* substance, in contrast, there is no deformation unless the stress is equal to the threshold strength; the threshold strength cannot be exceeded. Neither the strain nor the strain rate of a plastic substance is related to stress, which differentiates the plastic substance from viscous and elastic substances.

Strength is a twofold phenomenon. If the strain in a material is increased indefinitely two things might happen. With increasing strain, a soft material will yield at its yield strength. However, if the material is brittle, it will *fracture* at a certain stress that is called its fracture strength. This twofold phenomenon involving both plastic yielding and fracture is called strength.

Linear Rheological Models.—All the models of rheology are mathematical, but some aspects of their behaviors can be visualized with the aid of simple mechanical contrivances (*18*). The linear elasticity, or Hookean, model can be represented in one dimension by the equation

$$\sigma_s = G\epsilon_s \qquad \text{(Hookean model)} \tag{1.3}$$

where σ_s is shear stress, ϵ_s is shear strain, and G is the modulus of rigidity. In terms of normal stress and normal strain, Hooke's law is

$$\sigma_n = E\epsilon_n \qquad \text{(Hookean model)} \tag{1.4}$$

where E is Young's modulus, σ_n is normal stress, and ϵ_n is normal strain. A normal stress applies a compression or a tension to the body. A shear stress applies a couple or a torque. Thus when an elastic body is twisted or compressed its deformation is proportional to the applied load. Strain is proportional to stress in the Hookean model. The fundamental aspect of the Hookean model can be represented by a helical spring rigidly supported at one end (Fig. 1.3A). When a force is applied to the free end the spring extends. When the force is removed the spring recoils.

The linear viscosity, or Newtonian, model can be represented in one dimension by an equation of the form

$$\sigma_s = \eta \frac{d\epsilon_s}{dt} \qquad \text{(Newtonian model)} \tag{1.5}$$

where η is the coefficient of viscosity, ϵ_s is strain, and t is time. Thus, according to the Newtonian model, stress and rate of strain are proportional. The Newtonian

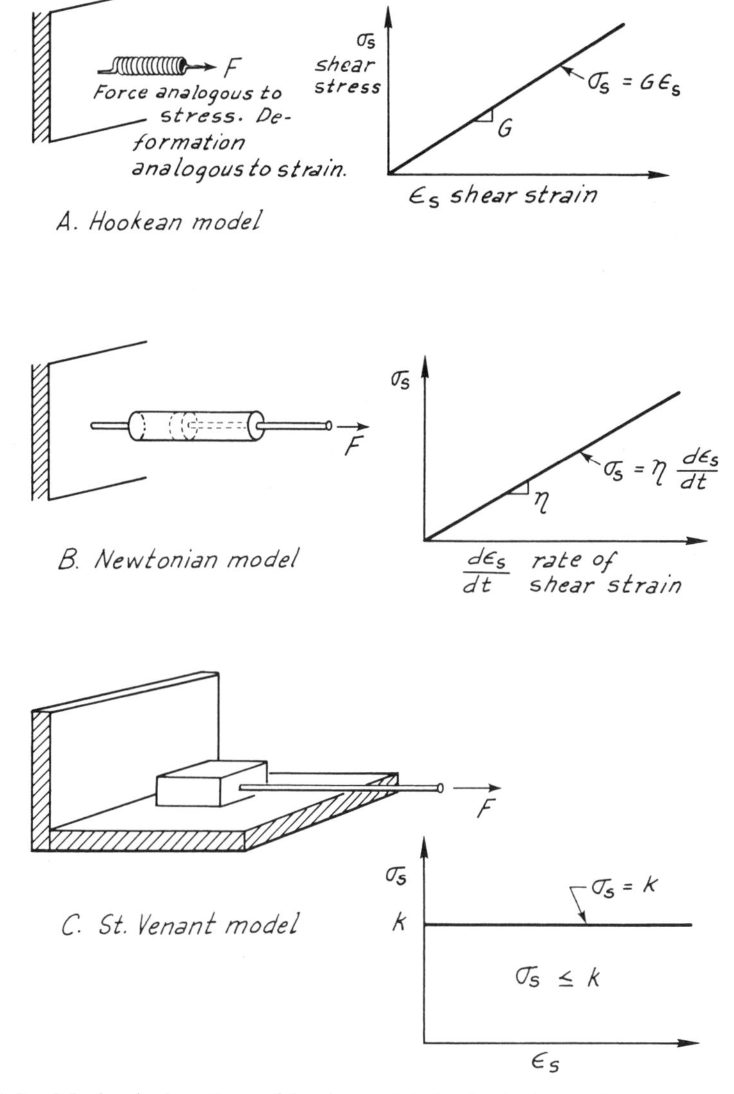

Figure 1.3. Mechanical analogs of fundamental rheological equations.

A. Hookean model; a helical spring rigidly supported at one end.

B. Newtonian model; a tube filled with oil in which a loosely fitting piston can be pushed back and forth.

C. Saint Venant model; a weight on a flat, rough surface.

substance commonly is represented by a dashpot, a tube filled with oil, in which a loosely fitting piston can be pushed back and forth (Fig. 1.3B). Pushing or pulling the piston causes oil to move from one end to the other of the cylinder by flowing past the edges of the piston. The rate at which the oil moves past the piston is proportional to the force applied to the piston. When the force is removed the oil stops flowing and the piston stays where it is.

The behavior of the plastic, or Saint Venant, substance can be described in one dimension with the equation

$$\sigma_s \leq k \qquad \text{(Saint Venant model)} \tag{1.6}$$

which states that the shear stress is less than or equal to the yield strength. If the shear stress is equal to the yield strength the Saint Venant body deforms; if it is less than the yield strength the Saint Venant body is rigid. This type of behavior can be modeled as is shown in Fig. 1.3C; the model shown consists of a weight on a flat surface. In order for the weight to move, the friction between it and the underlying surface must be overcome. A force too small to overcome the friction does not move the weight. If the force is sufficiently large the weight moves. As in the case of a viscous substance, but unlike that of the elastic substance, if the force is removed the weight stops moving; deformation is permanent.

Combinations of Linear Rheological Models.—Perhaps the earliest recorded attempt to analyze mathematically the resistance of a solid to fracture was made by Galileo in about 1630 (*14*). Galileo considered the problem of the resistance of a cantilever beam when the tendency for it to fail is due to its own weight or to an applied load (*3*). Thus Galileo laid the foundation for present-day beam theory. A few years after Galileo made his analysis, Hooke, in 1660, postulated an ideal elastic substance, in which the strains are proportional to the stresses. Until about 1820 most of the work with Hooke's law was devoted to the beam problem posed by Galileo, and to related problems. In 1820 Navier formulated the general differential equations of theoretical elasticity. And by 1822 Cauchy had discovered most of the remainder of the fundamentals of the modern theory of elasticity (*14*).

Newton, in 1670, was the first to consider the resistance of fluids to motion. He postulated that resistance to motion of a fluid is due to lack of slipperiness, that is, to viscosity. The equations of viscous flow were derived by Navier and Stokes in the early 1800's (*17*).

The plasticity model was invented in 1773 by Coulomb so that he could calculate earth pressures on retaining walls. But the model was not intensively developed until it became of interest to metallurgists. In about 1860 Tresca showed that lead at high pressures could be forced to flow through tubes. On the basis of Tresca's experiments, Saint Venant developed a theory of plastic flow of materials.

Thus, by about 1660 two of the fundamental rheological models, elasticity and viscosity, had been erected. By the time the plasticity model had been invented, about 100 years later, the theories of elasticity and viscosity had reached a high level of development.

Until about 1920 only three rheological models were recognized. Materials were classified as solid, either elastic or elastic-plastic, and as viscous liquids. This classification seemed to be adequate until 1919, when Bingham considered the properties of ordinary house paint. Good paint is capable of being spread very thinly on a wall, but once it has been spread, it does not flow off the wall. Paints that satisfy the first condition, Bingham reasoned, have a low viscosity. But to satisfy the second condition of staying on the wall, it would seem that the same paints must have a high viscosity, because they flow very little by the time they have dried. Bingham resolved this inconsistency by assuming that paint has a low yield strength, and that if the yield strength is exceeded the paint flows viscously. Its strength is so low that the slight pressure of the brush stroke can easily make it flow, but it is high enough to cause a thin film of paint to stay on the wall until it dries (*17*).

Bingham model.—The discovery by Bingham, that combinations of the basic rheological models can predict quite well the behavior of many substances, initiated wide interest in what is now called rheology. Thus Bingham and his colleagues at Lafayette College in Pennsylvania and, in 1922, Markus Reiner and his colleagues in Palestine began working on the problem of developing the modern theory of rheology (*17*).

In his analysis of the flow of paint, Bingham combined the plasticity and viscosity models in such a way that the behavior of the combination is similar to that of a dashpot and friction block in parallel (Fig. 1.4). The block, or plasticity component, represents approximately the behavior of paint below its yield point.

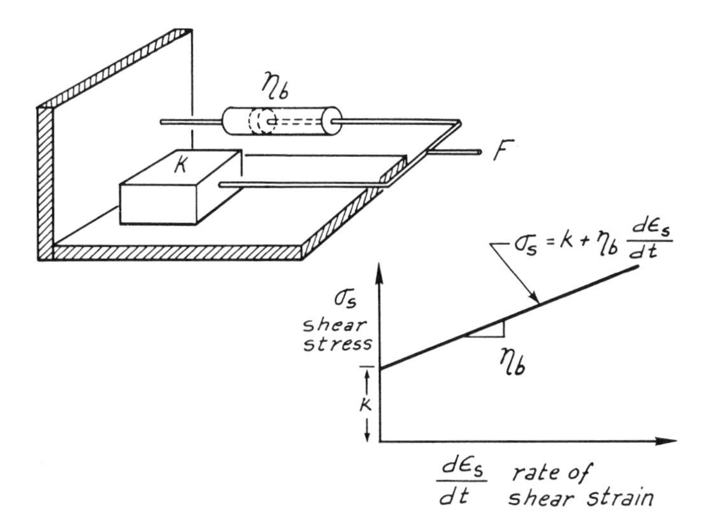

Figure 1.4. Analog of Bingham model. Force is analogous to shear stress and velocity is analogous to rate of shear strain.

As soon as the block begins to slide, that is, at the yield point, resistance is a combination of viscosity and strength. The combination of the dashpot and the block in parallel and the mathematical expressions

$$\left.\begin{array}{ll}\sigma_s = k + \eta_b \dot{\epsilon}_s; & \sigma_s > k \\ \dot{\epsilon}_s = 0; & \sigma_s \leq k\end{array}\right\} \text{ (Bingham model)} \tag{1.7}$$

are called the Bingham model.

Prandtl model.—Different combinations of the fundamental models can be formulated to suit a variety of applications. For example, if we were concerned with the mechanical behavior of a bar of aluminum we might use the Prandtl model (Fig. 1.5),

$$\sigma_s = G\epsilon_s; \qquad \sigma_s < k_p \qquad \text{(Prandtl model)}, \tag{1.8}$$

where G is the modulus of rigidity and k_p is the yield strength. According to this model the aluminum bar will snap back again if it is deformed and released. But if the bar is bent enough, it will begin to deform permanently and only part of the deformation is recovered when the load is released (Fig. 1.5).

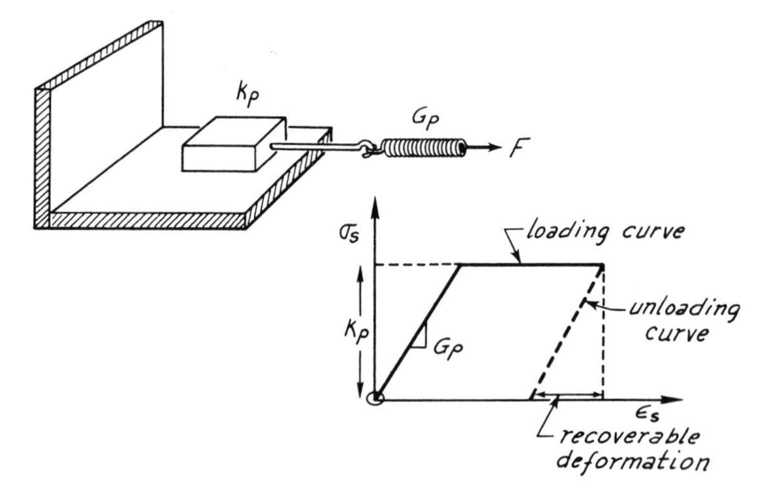

Figure 1.5. Analog of Prandtl model.

Kelvin and Maxwell models.—Many other combinations of the fundamental models have been named, being proposed as adequate descriptions of the behavior of some real materials. For example, the Kelvin model can be visualized as a spring in parallel with a dashpot (Fig. 1.6), so that

$$\sigma_s = \eta_k \dot{\epsilon}_s + G_k \epsilon_s \qquad \text{(Kelvin model)}. \tag{1.9}$$

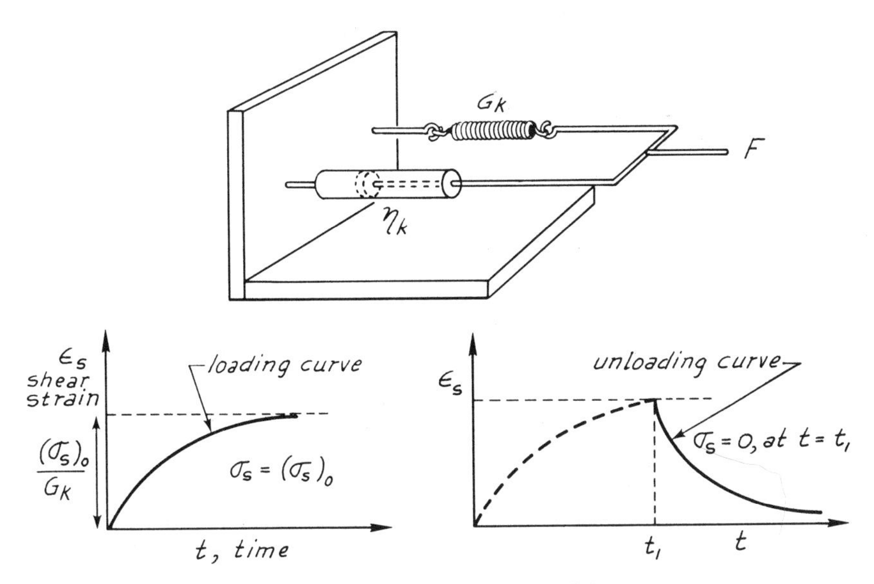

Figure 1.6. Analog of Kelvin model.

According to this model, if the shear stress is constant, the shear strain, ϵ_s, increases rapidly at first, then progressively slower with increasing time. After a very long time the strain approaches the strain that the spring alone would have attained immediately had the dashpot not hindered it.

A Maxwell body behaves similarly to a spring and dashpot in series (Fig. 1.7), so that

$$\sigma_s = \eta_m \dot{\epsilon}_s = G_m \epsilon_s \qquad \text{(Maxwell model).} \tag{1.10}$$

The elastic strains are instantaneous in this substance, but if the stress is maintained for a long time strain increases indefinitely because of the viscous component of strain (Fig. 1.7). Depending on the magnitude of the viscosity coefficient, it may be seconds, years, or even millions of years before the fluid properties of a Maxwell substance become apparent.

Higher-Order Rheological Models.—Although the rheological approach described thus far has proved to be a powerful tool of engineers and geologists, it is not the last word in our understanding of the behavior of materials. The ideal materials we have considered are all incomplete for at least two reasons. First, all the materials were assumed to behave linearly. That is, we assumed that effects are directly proportional to causes. In most of the mathematical analyses of geologic phenomena in following pages we will maintain the assumption of linearity between stress and strain or strain rate. But it is important to bear in mind that some aspects of many natural processes may be explained better if nonlinear models are assumed

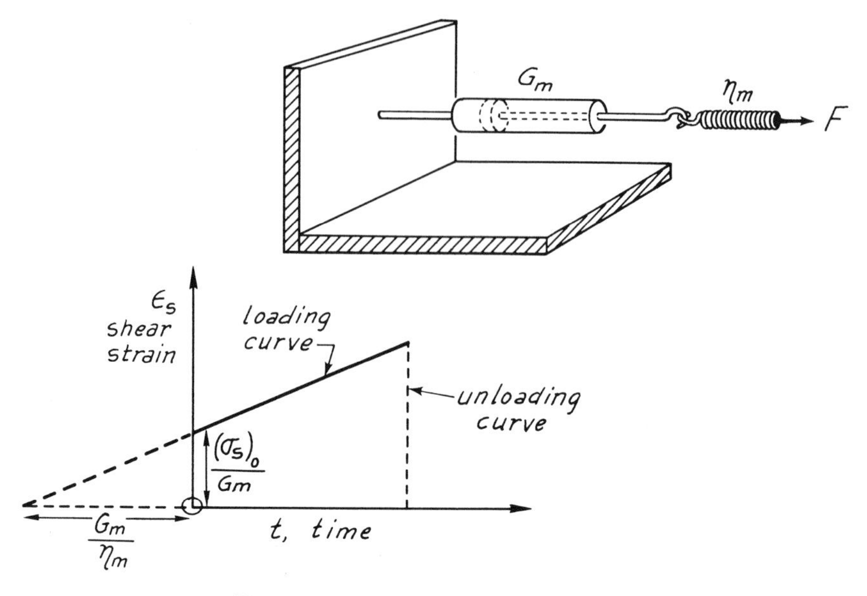

Figure 1.7. Analog of Maxwell model.

for earth materials. Second, none of the rheological models directly relates to the structure of matter.

An example of a real material that seems to respond nonlinearly is thixotropic clay. When such a clay is sheared, for example, we find that doubling the shearing stress on the clay may triple or quadruple the rate of flow. Thus its viscosity appears to decrease with increasing shear strain, rather than to stay constant as it would if clay were linearly viscous. This behavior might be due to the rearrangement of the clay particles so that the clay flows more easily at higher rates of shear strain. Regardless of the mechanism, however, the stress and strain rate of a thixotropic clay are not linearly related. Some aspects of the flow of thixotropic clay might be studied with the simple viscous or plastic-viscous models. But whether we select a linear or a nonlinear model depends upon the magnitude of the nonlinearity between stress and strain rate and the type of problem we are trying to solve.

The first proposed theory of nonlinear viscosity of fluids appeared in a paper by Reiner in 1945 (*21*). He predicted nonlinear effects in the shearing of fluids such that for an adequate description of even a simple shearing flow, usually assumed to be a typical one-dimensional situation, all three spatial dimensions are needed. A practical consequence of the rather strange effects predicted by Reiner was discovered by Weissenberg in 1947. Weissenberg found that when a rotating rod, something like the shaft of the beater in a milkshake maker, is lowered into a cylinder of some liquids a very peculiar thing happens. Although the rotational motion of a liquid should cause radial acceleration, and thereby cause the surface of the liquid to rise up on the outer cylinder, the experiments show that the liquids climb up the rotating stirring rod (Figs. 1.8A and 1.8B). This peculiar phenomenon is called the Weissenberg effect (*17*).

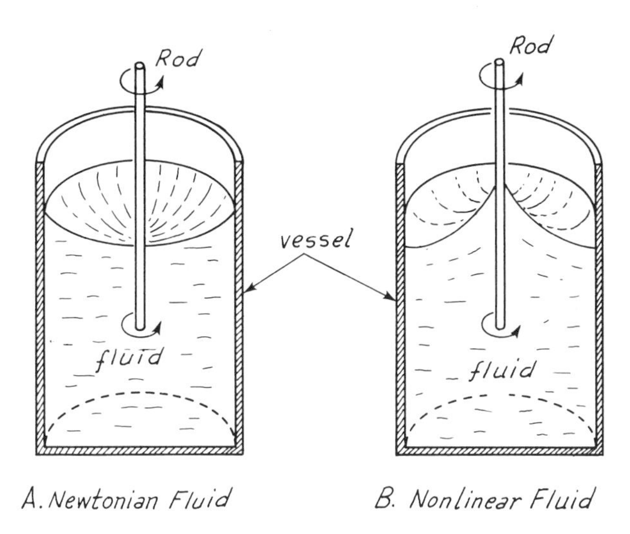

Figure 1.8. Weissenberg effect.

A. A Newtonian fluid reacts to the rotation of the rod by moving down at the center of the fixed cylindrical vessel and by moving up near the walls.

B. Some fluids, in which strain rate and stress are nonlinearly related, react to stirring by climbing the stirring rod. When sheared, each lamina of fluid behaves like a stretched rubber band, contracting and driving fluid inward and up the stirring rod.

Reiner (*17*) explained this peculiar phenomenon approximately as follows. A liquid set into circular motion by a rotating rod is sheared, because cylindrical layers or laminae nearer the rod rotate faster than the ones farther away. Thus the shear decreases outward from the stirring rod. Now if the liquid is elastic-viscous and if its rate of shear is nonlinearly related to the shearing stresses, there will be a tension in the same direction as the shear. Thus each cylindrical layer of shearing liquid acts like a stretched rubber sleeve, tending to shrink and to force the liquid inward and up the rod.

This phenomenon of a fluid climbing a rotating rod apparently convinced many investigators that deformations of most materials, earlier thought to be well explained by linear rheology, might be more subtle than they seemed. A rather extreme example of this view was expressed by Truesdell (*21*) in a paper presented at a symposium on second-order effects in the mechanics of materials. He stated, "I wish I could tell you that the one-dimensional concept of rheology had been shot dead [by the above-mentioned work by Reiner and Weissenberg]; all I can say is that apparently this illusion, like some noxious weeds and insects, is so sturdy as to survive all the efforts of science."

Nonlinear rheological equations have been solved for several problems, including the flow of a fluid in a noncircular pipe and the flow of dilute suspensions. Green and Rivlin (*6*) applied nonlinear rheological models to the flow of a visco-elastic fluid through a straight, noncircular pipe. They showed that high-order terms in the general visco-elastic model of a fluid flowing slowly should cause the fluid to circulate in noncircular pipes rather than flow in straight lines parallel to the boundaries of the pipes. Giesekus (*4*) derived nonlinear rheological equations that will describe the Weissenberg effect, shown in Fig. 1.8, and proposed that dilute suspensions of rigid particles flow according to his equations. For other references see Truesdell (*21*).

Boundary conditions

Two of the fundamentals of the physical processes of earth materials have been introduced so far: We have discussed rheological equations, which tell us how a material responds to loads, and mechanics, which tells us how several loads acting on the material are interrelated. The third fundamental is called *boundary condition*. Boundary conditions limit the types of possible behaviors of bodies and they specify initial conditions of bodies.

An example of a boundary condition is a known pressure in a field of varying pressures. The change in pressure in a static liquid within the earth's gravitational field is equal to the density of the liquid times the acceleration of gravity. Denoting the density by ρ (rho), we can express the relation as

$$\frac{dp}{dy} = \rho g, \qquad \text{differential equation,} \tag{1.11}$$

where p is pressure and y is the spatial coordinate, values of which increase downward. The general solution of eq. (1.11) is

$$p = \rho g y + C_1, \qquad \text{general solution,} \tag{1.12}$$

where C_1 is an arbitrary constant of integration. This result is not surprising, for we learned it in introductory physics. In its present form, however, the equation is not particularly valuable. For example, it will not tell us what the pressure is in a bucket of water, or in the earth's mantle. To answer these questions our mathematical description of the physical situations must be more complete. The description is made more complete by introducing *boundary conditions*. To calculate the pressure at any point in a bucket of water, for example, we specify a pressure at one point within the water. The specified pressure is the boundary condition.

To illustrate further what we mean by boundary conditions, let us examine the pressure distribution within a bucket of water the top of which we have sealed with a rubber membrane. Suppose that the rubber membrane is inflated by air pressurized at p_0 (Fig. 1.9). The latter supposition imposes the condition that the pressure in the water is equal to the pressure in the air at the air-water interface. Thus, according to notation shown in Fig. 1.9,

$$p = p_0 \quad \text{at} \quad y = 0 \qquad \text{boundary condition.} \tag{1.13}$$

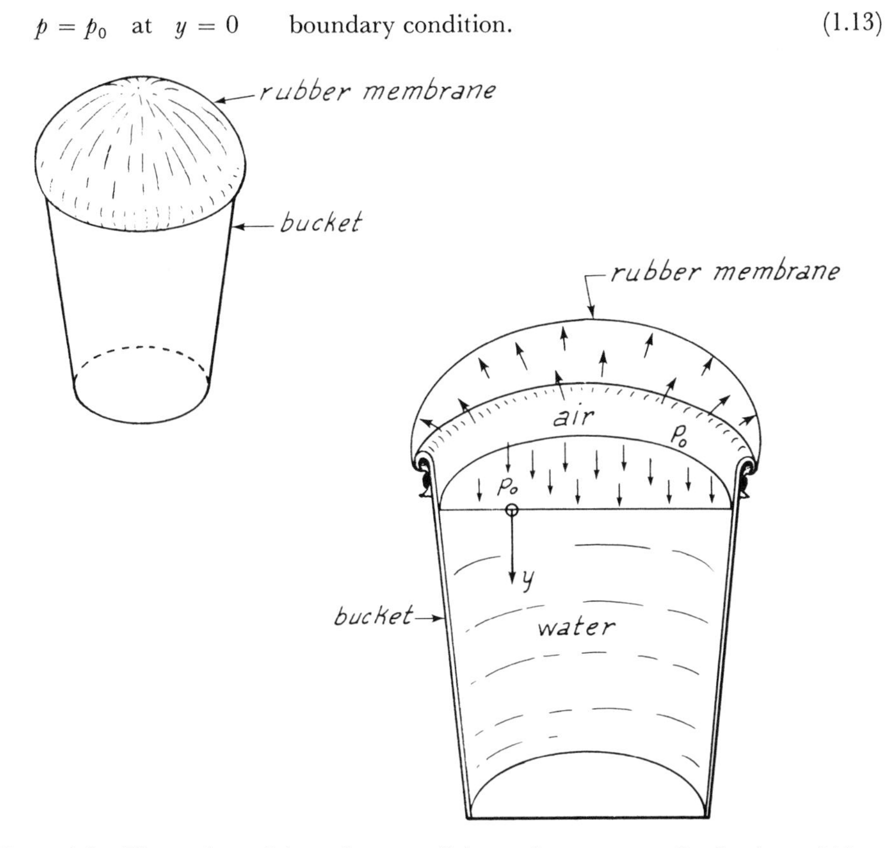

Figure 1.9. Illustration of boundary conditions; the pressure distribution within a bucket of water. Top of bucket is sealed with rubber membrane inflated at a pressure, p_0. Thus, pressure at water-air interface is p_0.

For simplicity we have chosen our spatial coordinates so that the value of y is zero at the air-water interface. With the boundary condition, eq. (1.13), we can determine the value of the constant, C_1, in the general solution, eq. (1.12), of the differential equation of pressure, eq. (1.11). Thus,

$$C_1 = p_0,$$

and eq. (1.12) becomes

$$p = \rho g y + p_0 \qquad \text{particular solution.} \tag{1.14}$$

We see that imposition of *a boundary condition removes the arbitrariness of the general solution*. This is a characteristic of boundary conditions in general. They introduce concreteness into a solution of a problem. In the problem of the pressure distribution within a bucket of water, a boundary condition simply allowed us to evaluate an arbitrary constant of integration. Boundary conditions in the most general sense, however, are of the utmost importance in the solution of mechanical problems in geology, physics, and engineering. They impose special restrictions on the way a problem is formulated and they allow us to specify values of arbitrary constants.

According to Lindsay and Margenau (*12*, p. 49), "a *specific* boundary condition is simply a postulated event in space and time expressed by the statement that a symbol . . . shall have a definite value or set of values throughout a specified region of space and within a specified interval of time." In calculating the pressure within a bucket of water we were able to evaluate the constant of integration because we knew the pressure at the air-water interface.

Other types of boundary conditions become important in the solution of problems not quite as familiar as that of the pressure distribution within a bucket of water. These are *general boundary conditions*. They . . . "impose fundamental restrictions on the type of activity possible for the system considered" (*12*, p. 53). For example, our knowledge of general boundary conditions leads us to consider rock to be a layered substance when we solve such problems as the forms of laccoliths and the wave lengths of concentric folds. Also, the radial dike pattern at Spanish Peaks, Colorado, led Odé to formulate the problem of the origin of the dike pattern by writing the appropriate differential equations in polar coordinates. Stratification and radial form are examples of general boundary conditions. The general conditions are used to formulate a problem. Specific conditions are used to solve an equation related to a problem.

General boundary conditions are even more fundamental than postulates about the behaviors of materials. This is illustrated in our study in Chapter 2 of the laccoliths of the Henry Mountains of Utah. G. K. Gilbert, who first studied the laccoliths, recognized that the most important control of the form and growth of the laccoliths is the layered nature of the rocks into which the laccolith magmas were intruded. We will start with the same general boundary conditions recognized by Gilbert, but we will use a different rheological model.

I think that the most important task of the geologist-mechanic is to recognize general and specific boundary conditions. They can be read only from field relations. Until we recognize the pertinent boundary conditions there is little hope of solving important problems, such as how slaty cleavage forms, how flow folds form, or why joints and faults display distinct patterns. The purpose of field mapping by the geologist-mechanic is to discover the boundary conditions of a process.*

> In summing up the whole matter we may say that closely associated with every physical law are to be found important boundary conditions, of both a general and a specific nature, forming a very significant part of the physical meaning of the law as well as its practical application. The specific conditions show us how to use the law to predict physical events, and ignorance of them forces us back on probability considerations; the general conditions fix the possible types of laws or the possible kinds of functions which enter into them. The mind can conceive countless forms of differential laws. The general boundary conditions serve to pick out the useful ones. Our search should forever turn in the direction of these conditions, the discovery and clear statement of which represent much of the progress of mathematical physics (*12*, p. 55).

We could justifiably substitute *physical geology* for "mathematical physics" in the last sentence of the statement by Lindsay and Margenau.

Let us now return to the folds in the Franciscan Formation, shown in Fig. 1.2, and apply in a general way what we have learned about the fundamentals of mechanics of earth materials, particularly the fundamental concept of boundary conditions.

The fundamentals are mechanics, which gives us relations amongst forces, stresses, or strains on the beds at the time of folding; rheology, which gives us equations and concepts of responses of materials to stresses and strains; and finally, boundary conditions, which apparently give us everything else.

Questions that arise immediately as we introduce more and more fundamentals are, how do we apply these fundamentals toward improving our understanding of the folds and what can we deduce from field study about their processes of formation? First we try mechanics. We cannot see stresses or forces in the field, or anywhere else for that matter. We can estimate the strain within the hard chert beds and soft shale interbeds by making certain assumptions about the original configurations of the layers. Such an analysis merely tells us that most of the strain has been concentrated in the fold hinges, and that the limbs of the folds could reach their present attitudes by rotation, without appreciable internal deformation or strain.

Thus thwarted, we put mechanics back into our briefcases and pull out rheology to see what it can do. It also disappoints us somewhat. We might select a sample of the chert, perhaps even one of the shale, and test it in the laboratory to determine its physical properties. But even a few moments of reflection reveal that the beds could not have had their present physical characteristics at the time of folding. The cherts are extremely brittle—very small strains are sufficient to fracture the chert beds in their present form—yet to form the sharp hinges they were intensely strained.

Apparently, therefore, we must deduce indirectly the rheological behavior of the rocks as well as the mechanics of the process that folded the rocks, because we cannot see or directly observe either the mechanics of the folds or the rheological behavior of the rocks at the time of folding. In these ways the study of physical processes of geology is quite different from the study of engineering mechanics or of the mechanics of theoretical physics. Whereas in physics and engineering mechanics investigators are usually able to impose certain forces on bodies whose rheological behaviors can be measured experimentally, in the mechanics of earth materials we are usually unable to measure the forces or the rheological behaviors of the bodies of interest. Usually the geologic experiment has already been made by nature and we can see only the final products, sometimes probably only the results of the dying phase of a deformation. There are minor exceptions, of course, such as processes of stream flow, glacier flow, mudflow, and landsliding, all of which are taking place today.

Frustrated once more, then, we throw rheology away and pull the concept of boundary condition from our briefcases. Now we are getting somewhere. The general boundary conditions can be deduced by study of outcrops. For example, we note that the folded beds consist of interbedded cherts and shales, which are of about equal thickness; the shale beds probably were thicker at the time of folding. Other general boundary conditions are concentration of bending in hinges of folds, only minor bending within limbs of folds, gross shortening and thickening by folding of the chert-shale sequence, lack of thickening or thinning of individual chert beds in hinges of folds, lack of deformation in delicate tests of radiolaria within the cherts, and probably many others.

We note that about 50 feet stratigraphically above the folds shown in Fig. 1.2, that is, beyond the upper left corner of that figure, the beds are relatively planar and inclined toward the left. Also, about 100 feet stratigraphically below the beds in the folds, that is, beyond the lower right corner of Fig. 1.2, there is a relatively planar stratum of greenstone, which seems to be tilted without major bending. A tentative specific boundary condition is, therefore, that the folds formed in a sequence of interbedded cherts and shale that was bounded above and below by units that were undeformed during folding. We could tentatively assign zero displacement to chert beds along these boundaries. This boundary condition is necessarily tentative because we have no way of determining the actual displacements along the boundaries.

I think that the following conclusion of this brief analysis is applicable to most problems in physical geology: We can deduce, even with considerable effort sometimes, only the general boundary conditions of a product, such as a fold, formed by a physical process, such as folding. We use the field observations to guide us in our selection of rheological models and of specific boundary conditions, by trying various rheological behaviors and specific boundary conditions until we are able to describe quite well the field relations. The field relations also help us to select appropriate forms of the equations of mechanics. For example, the chert and shale beds in the Franciscan Formation probably were plate-like prior to folding so that we could probably use the mechanical equations of plate theory in our analysis of the folds. Thus, by trial and error, we grope for combinations of specific boundary conditions, mechanical equations, and rheological behaviors that will describe the final products we see in nature. The more experience we have with interactions of various combinations of these three concepts, the more rapidly we can select suitable combinations to solve problems.

Of the three fundamentals of the mechanics of earth materials, mechanics, rheology, and boundary conditions, therefore, we can usually directly deduce only the general boundary conditions from the field relations. We must select the rheological behavior of the materials and the appropriate forms of the mechanical equations by analyzing the general boundary conditions. The general conditions can be deduced only in the field. For this reason the quality of the field work associated with a study plays an important, perhaps controlling, role in determining the quality of our applications.

References cited in Chapter 1

1. Cox, John, 1909, *Mechanics*: Cambridge University Press, Cambridge.
2. Crittenden, M. D., Jr., 1963, *New Data on the Isostatic Deformation of Lake Bonneville*, U.S. Geological Survey Prof. Paper 454-E.
3. Galilei, Galileo, 1914, *Two New Sciences*: transl. by H. Crew and A. DeSalvo, Dover Publications, Inc., N.Y.
4. Giesekus, Hanswalter, 1964, *Statistical Rheology of Suspensions and Solutions with Special Reference to Normal Stress Effects*: International Union of Theor. and Applied Mech., Symposium, Haifa, Israel, 1962; Macmillan Co., N.Y.
5. Gibbs, J. W., 1960, *Vector Analysis*: Dover Publications, Inc., N.Y.
6. Green, A. E., and Rivlin, R. S., 1956, "Steady Flow on Non-Newtonian Fluids Through Tubes": *Quart. Appl. Math.*, V. 14, p. 299–308.
7. Hubbert, M. K., and Rubey, W. W., 1959, "Role of Fluid Pressure in Mechanics of Overthrust Faulting; Part I: Mechanics of Fluid-Filled Porous Solids and Its Application to Overthrust Faulting": *Geol. Soc. America Bull.*, V. 70, p. 115–206.
8. ———, 1960, Reply to discussion of above paper by Laubscher: *Geol. Soc. America Bull.*, V. 71, p. 611–628.
9. ———, 1961, Reply to discussion of above paper by Moore: *Geol. Soc. America Bull.*, V. 72, p. 1581–1594.
10. ———, 1961, Reply to discussion of above paper by Birch: *Geol. Soc. America Bull.*, V. 72, p. 1441–1452.
11. Jammer, Max, 1957, *Concepts of Force*: Harper and Brothers, N.Y.
12. Lindsay, R. B., and Margenau, H., 1957, *Foundations of Physics*: Dover Publications, Inc., N.Y.
13. Longwell, C. R., 1960, "Interpretation of the Leveling Data," in Smith, W. O., et al., 1960, *Comprehensive Survey of Sedimentation in Lake Mead*, 1948–49, U.S. Geological Survey Prof. Paper p. 33–38.

14. Love, A. E. H., 1944, *A Treatise on the Mathematical Theory of Elasticity*: Dover Publications, Inc., N.Y.
15. Newton, Isaac, 1943, *Mathematical Principles of Natural Philosophy*, transl. by Andrew Motte in 1729, revised by Florian Cajori, 1931: University of California Press.
16. Peterson, F. L., 1967, "Short-term Surface Responses in the Vicinity of Flowing Wells": Ph.D. thesis, Stanford University.
17. Reiner, Markus, 1959, "The Flow of Matter": *Scientific American.*
18. ———, 1960, *Deformation, Strain, and Flow—an Elementary Introduction to Rheology*: Interscience Publishers, N.Y.
19. Terzaghi, Karl, 1960, *From Theory to Practice in Soil Mechanics*: John Wiley and Sons, New York.
20. Timoshenko, S. P., 1953, *History of Strength of Materials*: McGraw-Hill Book Co., Inc., N.Y.
21. Truesdell, C., 1964, *Second-Order Effects in the Mechanics of Materials*, in International Union of Theor. and Applied Mech., Symposium, Haifa, Israel, 1962, Macmillan Co., N.Y.

2

Chapter Sections

Introduction
The Henry Mountains
  Regional Setting
  Gilbert's Concept of the Ideal Laccolith
Gilbert's Mechanical Analysis of the Sizes of Laccoliths
Bending of Strata over Laccoliths
Analysis Based on Beam and Plate Theory
  Internal Forces and Bending Moments
    Sign Conventions
    Meaning of Bending Moment
    Free-Body Diagram
  Forces and Moments Induced within Beams Carrying Distributed Loads
  Approximate Relation between Bending Moment and Curvature of Elastic Members
  Mathematical Expression for Form of Overburden Deflected by Magma of Laccolith
Reason for Lack of Small Laccoliths in Henry Mountains
Explanation of the Dependency of Size of Laccoliths on Stratigraphic Position in the Henry Mountains
References Cited in Chapter 2

2

Laccoliths of Henry Mountains of Utah

Introduction

The report by G. K. Gilbert on the geomorphology and structure of the Henry Mountains of southeastern Utah is a classic of geological literature. In this remarkable report, Gilbert (*3*) recorded his impressions of an intrusive of a form that had never before been recognized—the laccolith. After examining Table Mountain (Fig. 2.1), formed by an intrusive body on the north flank of Mount Ellen, he visualized the ideal laccolith as being mushroom-shaped, having steep sides and a gently rounded upper surface (*4*). Recognition of the laccolithic form of intrusion was a significant contribution to geology, but Gilbert went even further. He explained, via a plausible mechanical analysis, why the sizes of the intrusive bodies increase with increasing depths of emplacement.

Prior to Gilbert's study of the Henry Mountains, they had been visited only by explorers, Indians, scouts, rabbits, and lizards. In 1869, John Wesley Powell, in charge of the U.S. Geographical and Geological Survey of the Rocky Mountain Region, passed near the foot of the Henry Mountains while descending the Colorado River on his famous boat ride. Powell recognized the fact that the Henry Mountains are different from most volcanic mountains, and therefore he assigned Gilbert to make a special study of them. Gilbert spent a week in the mountains during 1875 and returned for about two months of field work there during 1876 (*4*). "Report on the Geology of the Henry Mountains" was published in 1877 (*3*).

In this chapter we are going to begin to develop the theory of beams and plates which has found wide application in engineering and geology. We will continue developing this theory in four subsequent chapters of the book. This chapter introduces concepts of force and moment equilibrium, which are fundamental to all

Figure 2.1. Table Mountain, a laccolith at the northern edge of Mount Ellen. View southwest. After examining Table Mountain, Gilbert visualized the ideal laccolith as being mushroom-shaped, having steep sides and a gently rounded top (photo by David Pollard).

of mechanics. Beam and plate theory will allow us to explain several features of the laccoliths in the Henry Mountains.

The Henry Mountains

Regional Setting.—To Gilbert (*3*), "the Henry Mountains are not a range, and they have no trend; they are simply a group of five individual mountains, separated by low passes and arranged without discernible system." The Mountains are within an area of generally flat-lying sedimentary rocks that form a major structural depression in the Colorado Plateau. Fifteen miles to the west is the Waterpocket flexure, a prominent monocline. The San Rafael Swell is thirty miles to the north.

Each of the Henry Mountains (Fig. 2.2) is a structural dome. The four southern domes are 6 to 8 miles in diameter and the northern dome, Mount Ellen, is twice as large. All the domes except Mount Ellsworth have superimposed many small anticlinal masses and domes produced by individual laccoliths or other intrusions.

Ninety-five percent of the intrusives in the Henry Mountains are diorite porphyry,

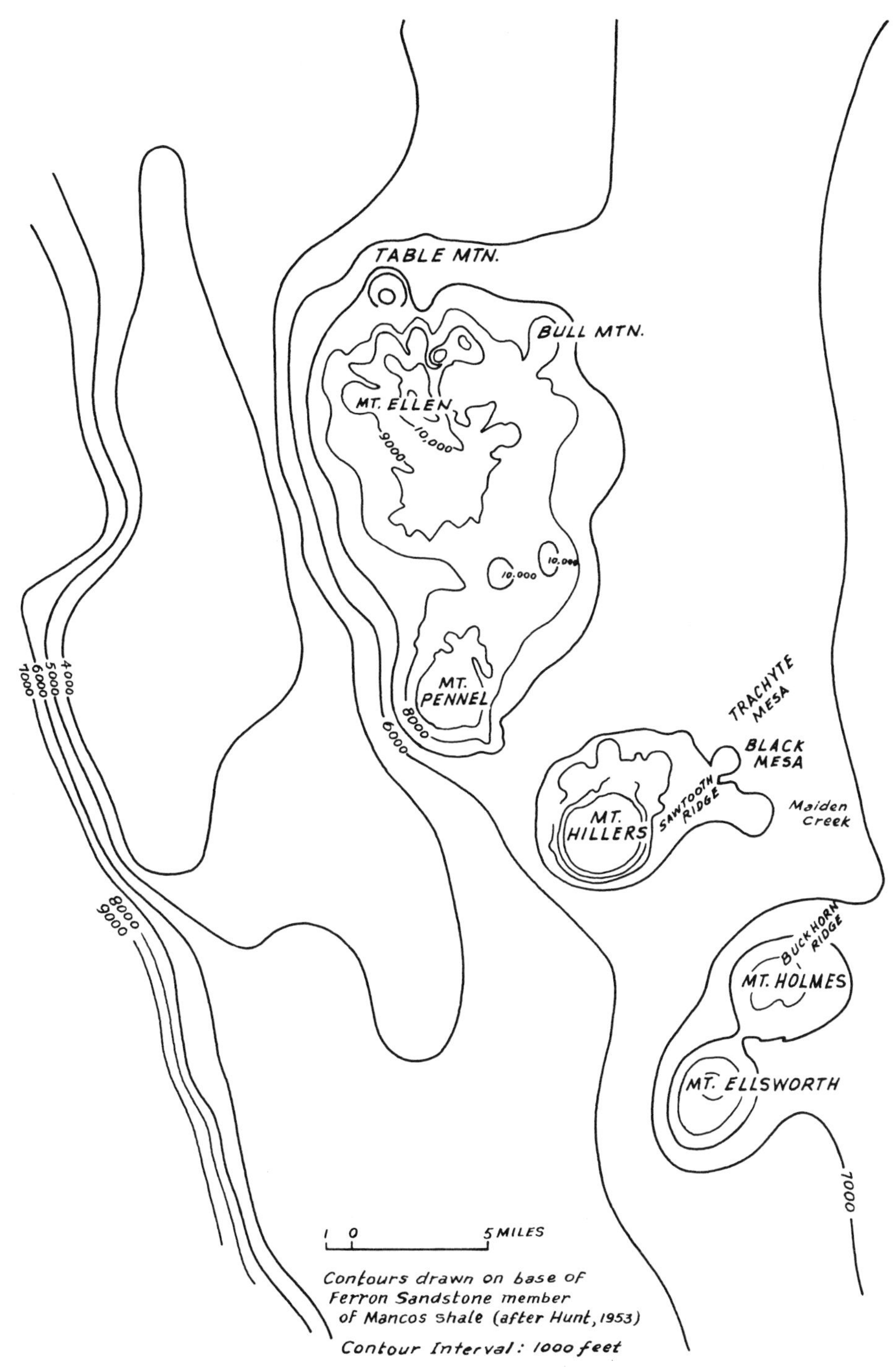

Figure 2.2. The Henry Mountains. Forms of mountains shown with structural contours drawn on base of Ferron sandstone member of Mancos Shale (after Hunt, ref. *4*).

but there are also monzonite porphyry, aplite, and basalt (*4*). The intrusives invaded rocks ranging in age from Jurassic through Cretaceous and in rock type from conglomerate through shale. The formations intruded are the Summerville, Morrison, Dakota, and Mancos. The laccoliths usually are concentrated in shaly units and are absent in the parts of the stratigraphic section that are predominantly sandstone.

Gilbert's Concept of the Ideal Laccolith.—Gilbert's laccolith has the shape of a toadstool. The stem of the toadstool corresponds to the conduit, which feeds magma to the carapace above (Fig. 2.3A). According to Gilbert, the bases of laccoliths generally are flat. The plan view is a short oval. The ratios of the diameters of the oval usually are less than three to two, so that the type plan approximates a circle and the type form probably is half of an oblate spheroid. "The height is never more than one-third of the width, but is frequently much less, and the average ratio of all the measurements I am able to combine is one to seven" (*3*).

As far as I know, none of the conduits that fed the laccoliths has been definitely identified anywhere in the Henry Mountains. Gilbert was unable to find a feeder: "Dikes are represented [in Figs. 2.3A and 2.3B] beneath as well as above the lacco-

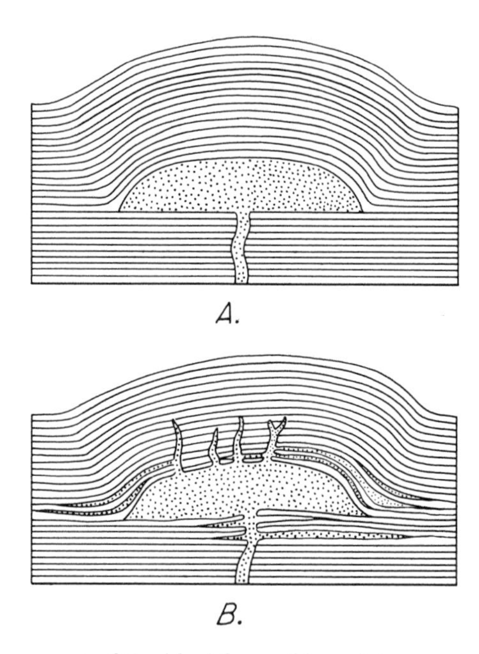

Figure 2.3. Gilbert's concept of the ideal form of laccoliths.

A. The shape of the idealized laccolith is like that of a toadstool. The stem of the toadstool corresponds to the conduit that feeds magma to the carapace above.

B. Dikes and sills are represented above and below the ideal laccolith. The dikes are largest and most common above the center.

liths. These are purely hypothetical, since they have not been seen. In a general way, the molten rock must have come from below, but the channel by which it came has in no instance been determined by observation" (*3*, p. 20). Hunt (*4*) believed that most of the laccoliths were injected as tongue-like projections from a central stock near the center of each of the Henry Mountains. However, as far as I know, Hunt did not actually observe any of the feeders. David Pollard and I believe that we have recognized a feeder dike beneath the Maiden Creek laccolith.

Gilbert's mechanical analysis of the sizes of laccoliths

Gilbert imagined that the laccoliths in the Henry Mountains formed as molten magma, rising in a conduit, which "stopped at a . . . horizon, insinuated itself between two strata, and opened for itself a chamber by lifting all the superior beds. In this chamber it congealed, forming a massive body of trap. For this body the name *Laccolite* . . . will be used" (*3*, p. 19). The name was later changed to laccolith.

The lifting process Gilbert visualized is clearly recognizable in outcrops at the east end of the Sawtooth Ridge laccolith, on the eastern side of Mount Hillers (Fig. 2.4). The diorite porphyry of the laccolith intruded the Morrison Formation and it has a blunt front that terminates against truncated sandstone and shale beds (*5*). The contact is nearly vertical, and vertical fractures between blocks of sandstone near the contact are slickensided in the vertical direction.

Sills extend beyond the terminations of the Sawtooth Ridge laccolith in some places, as is shown in Fig. 2.4A. Also, the sandstone beds in front of the termination shown in Fig. 2.4A appear to be relatively undisturbed a few feet from the termination; if the magma had advanced laterally with the thick termination visible today, the beds in front of it would have been severely crumpled and otherwise distorted. Thus, David Pollard (*5*), who recently studied several of the laccoliths, suggested that the magma spread out laterally as a thin sheet or sill until the magma gained sufficient leverage to fracture the overburden near its periphery. Then the overburden started to lift and the intrusion thickened to form the laccolith we see today at Sawtooth Ridge.

Gilbert's analysis of the driving and resisting forces acting on the overburden of a laccolith assumes conditions similar to those we can visualize in the Sawtooth Ridge laccolith. According to Gilbert, the dome of strata that overlies a laccolith has a definite diameter; within the dome the strata are bent and outside the dome the strata are undisturbed (Fig. 2.5A). The lifted rock of the dome is bounded, approximately, by a cylindrical boundary between deformed and undeformed strata (Fig. 2.5B). The diameter of the cylinder of rock is about the same as the diameter of the laccolith and the cylinder extends to the earth's surface.

Gilbert idealized the surface of the cylinder of uplifted rock by a cylindrical fault and assumed that the only condition required to cause the cylinder of over-

Figure 2.4. East end of Sawtooth Ridge Laccolith, Henry Mountains of Utah.

A. Thick, blunt termination of laccolith. Sandstone beds partly truncated, partly upwarped by intrusion.

B. Blunt termination of sill located at *B* in Fig. 2.4A. Sill moved from right to left.

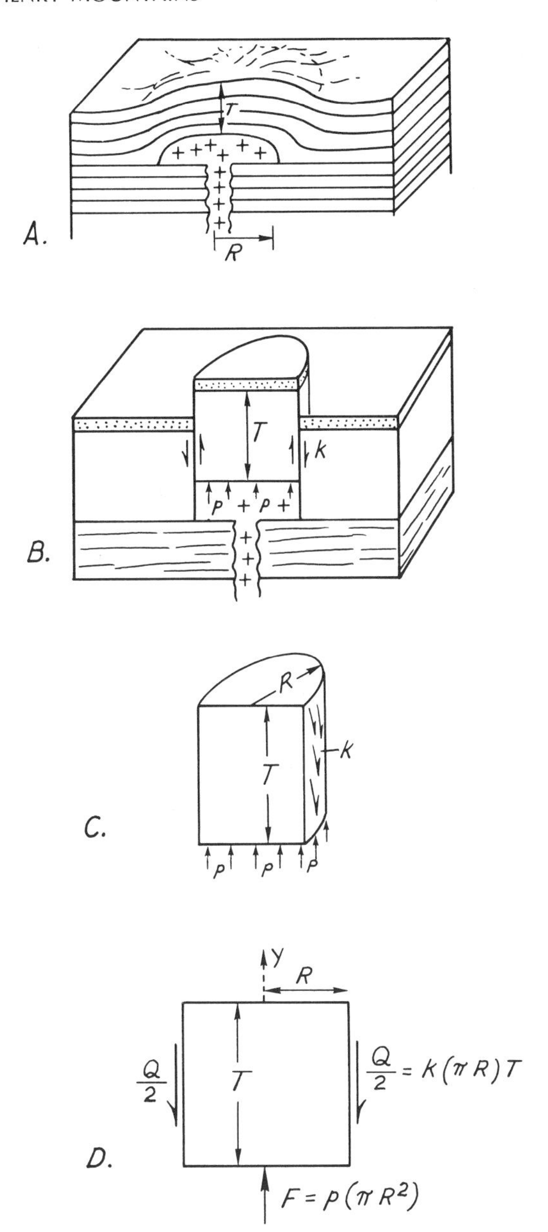

Figure 2.5. Gilbert's analysis of forces acting on overburden.

A. Gilbert's ideal laccolith.

B. Gilbert's model of ideal laccolith. Overburden is represented by cylindrical plug, bounded by vertical fault.

C. k is strength of contact between cylinder and surrounding rock.

D. Free-body diagram of plug of overburden.

burden to move upwards is that the strength of the contact between the moving cylinder and the stationary surrounding rock must be overcome. Gilbert assumed a rheological model of the form

$$\sigma_s \leq k,$$

which states that the shear stress, σ_s, acting between the cylinder and surrounding rock is less than or equal to the shear strength, k, of the contact between the cylinder and the surrounding rock (Fig. 2.5C). Accordingly, the force resisting uplift of the cylinder is the shear strength times the surface area of the wall of the cylinder.

The driving force is a function of the pressure in the magma, which pushes the cylinder upward (Fig. 2.5C). For equilibrium, according to Newton's laws of mechanics, the summation of the forces acting on the cylinder must be zero. Thus, if y is the coordinate in the vertical direction, we can write the requirements for equilibrium as

$$\left[\sum F_y = 0\right], \tag{2.1}$$

where $\sum$ is "sigma," indicating summation.

In order to use this equilibrium relation, we need to isolate the body in question, the cylinder, and determine which forces act *on* the body. In two dimensions, we can represent the downward, resisting force by Q, as is shown in Fig. 2.5D. The resisting force is divided into two parts in the figure because it acts over both sides of the cylinder—indeed, over the entire surface of the cylinder. The body is also acted upon by a driving force, F, due to the pressure in the magma.

Thus the equation of equilibrium, eq. (2.1), takes the form:

$$\left[\sum F_y = 0\right], \qquad F - Q = 0. \tag{2.2}$$

Both the pressure and the shear strength are distributed over parts of the surface area of the cylinder. In order to determine the *force* arising from the pressure, for example, we multiply the pressure by the area over which it acts. The pressure is distributed over the base of the cylinder and the area of the base is πR^2, where R is the radius of the cylinder, so that the driving force is

$$F = p\pi R^2. \tag{2.3}$$

The resisting shear strength is distributed over the surface of the cylinder, which has an area of $2\pi RT$, where T is the height of the cylinder, that is, the thickness of the overburden. If the shear strength is the same everywhere along the surface of the cylinder, the resisting force is

$$Q = k(2\pi R)T. \tag{2.4}$$

Substituting eqs. (2.3) and (2.4) into eq. (2.2), we have

$$p\pi R^2 - k(2\pi R)T = 0,$$

or

$$\underbrace{pR}_{\substack{\text{pressure}\\ \text{term}}} = \underbrace{2kT}_{\substack{\text{strength}\\ \text{term}}}. \tag{2.5}$$

Gilbert was able to explain two peculiarities of the laccoliths in the Henry Mountains by analyzing an equation similar to eq. (2.5):

1. He saw no small laccoliths in the Henry Mountains.
2. The larger laccoliths were generally intruded at lower horizons in the Henry Mountains. The mean diameter of the laccoliths in the upper zone of laccoliths is 1.2 miles, or about 1.9 km. That of the lower zone is about 2.6 miles, or about 4.2 km. The difference in altitude of the centers of the two zones is about one kilometer and the laccoliths of the lower zone are about twice as large as those in the upper zone.

Gilbert concluded that there are no small laccoliths in the Henry Mountains for the following reasons: For a given magma pressure, p, strength, k, and thickness of overburden, T, [eq. (2.5)], there is a certain radius, R, below which the maximum possible resisting force is greater than the actual driving force, that is, for which

$$R < \frac{2kT}{p},$$

where $<$ means "less than." For a larger value of R, the driving force might equal the maximum possible resisting force because the resisting force increases much more slowly than does the driving force with an increase in R [see eqs. (2.3) and (2.4)]. Thus, Gilbert reasoned, small laccoliths do not form because they are incapable of lifting their overburdens.

He explained the apparent increase in size with increase in depth of laccoliths in the Henry Mountains approximately as follows: For a laccolith to form, according to our analysis, the driving force must equal the maximum possible resisting force. Thus, according to eq. (2.5),

$$pR = 2kT.$$

Solving for R,

$$R = \frac{2kT}{p}. \tag{2.6}$$

Now, if the magma pressure, p, and rock strength, k, are essentially constant, the radius of the laccolith should be proportional to the thickness of overburden, T, according to eq. (2.6). Laccoliths that form at greater depths should be larger than those that form nearer the earth's surface. This is in accord with Gilbert's observation that laccoliths in the lower zone are larger than those in the upper zone in the Henry Mountains.

Bending of strata over laccoliths

Gilbert clearly recognized that the overburden of a laccolith is deformed initially by bending. For example, his sketch of the ideal laccolith shows bent strata (Fig. 2.5A). He was unsure, however, how to account for the influence of bending resistance on the growth of laccoliths (*3*). An exposure on one side of the Trachyte Mesa laccolith, on the northeastern side of Mount Hillers, provides an excellent field example of the shape taken by the bent strata (Fig. 2.6). The Entrada Sandstone bends up and over the edge of the diorite porphyry of the laccolith. Even the sandstone beds at the contact with the porphyry seem to be continuous, relatively unbroken by faulting. In contrast, the sandstone beds in front of the blunt Sawtooth Ridge laccolith, shown in Fig. 2.4, are sheared and truncated.

Figure 2.6. Edge of Trachyte Mesa Laccolith. The Entrada Sandstone bends up and over the edge of the diorite porphyry of the laccolith, just as is shown in Gilbert's ideal laccolith.

Analysis based on beam and plate theory

The influence of bending of strata on the form of instrusive bodies received minor attention until, in 1968, Pollard (*5*) investigated the effect of stratification on theoretical forms of laccoliths and on modes and locations of failure and diking of the overburden. He also searched for variables which control the theoretical deformations, in order to explain peculiar features of the laccoliths in the Henry Mountains. In addition, Pollard studied, by means of theory and experiments, the transition from lateral sill propagation to vertical uplift of the overburden as a laccolith forms.

We will restrict our attention to the theoretical forms of laccoliths and to Gilbert's observations that there is a conspicuous lack of small laccoliths in the Henry Mountains and that the sizes of laccoliths seem to increase with increasing depths of emplacement. (Other theoretical considerations are noted in this chapter's reference 5.)

In order to derive the mathematical form of the strata bent by Gilbert's laccolith, we will need to make certain assumptions about the properties of the intruding magma and of the intruded strata. We will represent the magma by a fluid that is static and that exerts the same pressure everywhere along the surface of the laccolith. This assumption probably does not lead our analysis far from the actual situation in nature. Even if the magma were flowing at an appreciable rate, the pressure distribution it would exert on the overburden would not greatly alter our solutions, which are based on a constant pressure. Also, whether the magma behaved as a viscous or plastic substance, which are two likely extremes of its behavior while molten, its pressure distribution would decrease essentially linearly with distance from the feeder dike or sill. Pollard (*5*) has shown that a linear pressure distribution does not markedly alter the form of the theoretical laccolith or the stresses leading to failure of the overburden.

We will assume, also, that each layer of the overburden has homogeneous and isotropic elastic properties. By elastic homogeneity we mean that a sample that is very small relative to the size of the bent layer of overburden has the same elastic properties as a larger sample, or even as the entire layer of overburden. By elastic isotropy, we mean that the elastic properties of a layer of the overburden are the same normal to bedding as they are parallel to bedding, or in any other direction within the layer. Just as a crystal is isotropic optically if indices of refraction are the same in every orientation of the crystal, a rock is isotropic elastically if the elastic indices, or elastic constants, are the same in every direction through the rock.

Our analysis is somewhat simplified if we assume that the overburden, lifted by the magma to form the laccolith, deforms into an anticlinal form. Thus our theoretical model of the overburden of a laccolith consists of an elastic plate bent around an axis similar to a fold axis (Fig. 2.7). Under this condition of deformation, the theory of

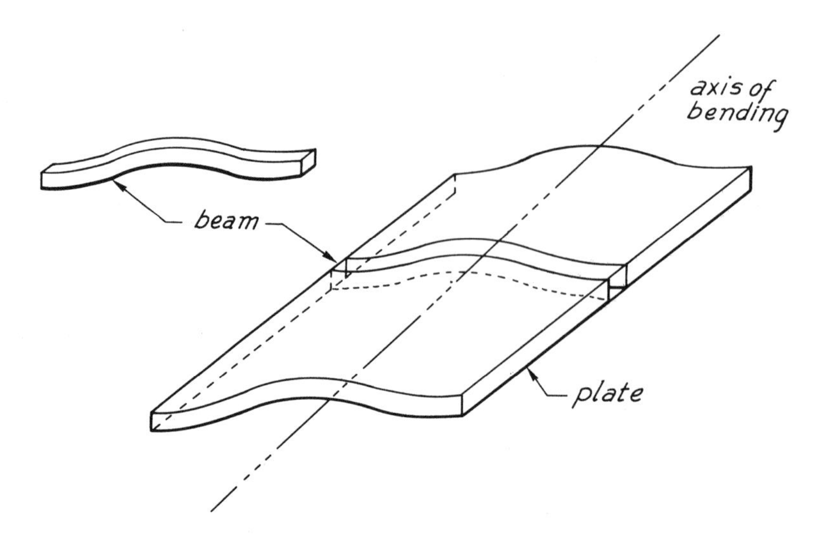

Figure 2.7. Overburden is deflected into anticlinal form. It is represented by an elastic plate bent around an axis similar to a fold axis. Under this condition of bending, plate and beam theories are essentially the same.

bending of plates is the same as that of beams (Fig. 2.7). Therefore, we can apply either beam or plate theory to our analyses of bending of single rock layers. Nearly everywhere the term *beam* is used, the term *plate* could be substituted in our analyses.

In the following pages we will study forces induced within beams by other forces and pressures applied at various places along the surfaces of the beams. Then we will assume that the beams are elastic so that we can calculate stresses within them. At that point we will return to the problems of forms and sizes of laccoliths in the Henry Mountains.

Internal Forces and Bending Moments.—The analysis of the deformation of a loaded beam consists of two steps. First, the equilibrium requirements of the beam as a whole or any part of it are considered so that the distribution of forces within the beam can be calculated. Second, relations between forces within the beam and the internal resistance that counteracts these forces are established. Thus the deformation of a beam is related to internal forces. The first analysis requires the principle of static equilibrium introduced by Archimedes almost 2200 years ago. The second analysis involves the behavior in response to stress of the material of which the beam is composed. We will return to the second analysis later. In this and the following section we will be concerned with forces induced within beams.

Consider a beam supported below by two rollers and loaded by three forces (Fig. 2.8A). The forces within the beam can be calculated by pretending that we cut the beam with a transverse section, such as the one shown in Fig. 2.8B. For the two

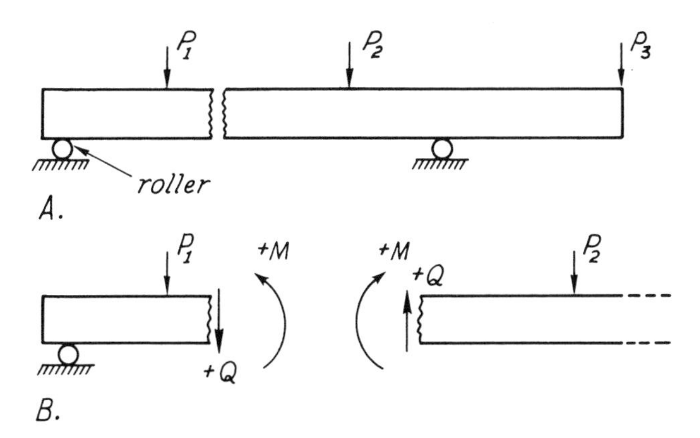

Figure 2.8. Representation of forces acting on a beam and of internal forces acting on a cross-section within a beam.

A. Beam on rollers supporting three loads.

B. To study internal forces, we examine forces acting on two parts of beam at certain cross-sections. For two parts of beam to be in equilibrium, there must be certain forces exerted by one half of the beam on the other half.

halves of the beam severed by the imaginary cut to be in equilibrium, the resultant of all forces on each side of the imaginary cut must have the same magnitude.

Only the resultants of the force distribution across a section of the beam need to be considered. The resultants can be expressed in terms of a force and a bending moment, or couple,

$$\sum F = 0, \quad \text{and} \quad \sum M = 0.$$

In accord with Newton's third law, the forces and moments acting on each side of the section must be equal in magnitude but opposite in direction.

Sign conventions.—The vertical force on each side of the section in Fig. 2.8B is the *shear force*, Q. It is a shear force because it is tangent to the section. The couple, M, is the *bending moment*, acting on each side of the section. The positive directions of Q and M are, by convention, as is shown by the arrows in Figs. 2.9A and 2.9B. Thus, for example, positive Q is upward on the right-hand part and downward on the left-hand part of the beam.

Meaning of bending moment.—The bending moment, M, can be interpreted as follows. Figure 2.10A is a section, I–II, of a beam. A bending moment on the section will be expressed by a tendency for the cross-section to rotate around point O at the center of the cross-section. Perhaps the origin of the bending moment can be visualized if section I-O-II is represented by a seesaw (Fig. 2.10B). The seesaw is an old one, so that the bearing between the board and the support is rusty and difficult to

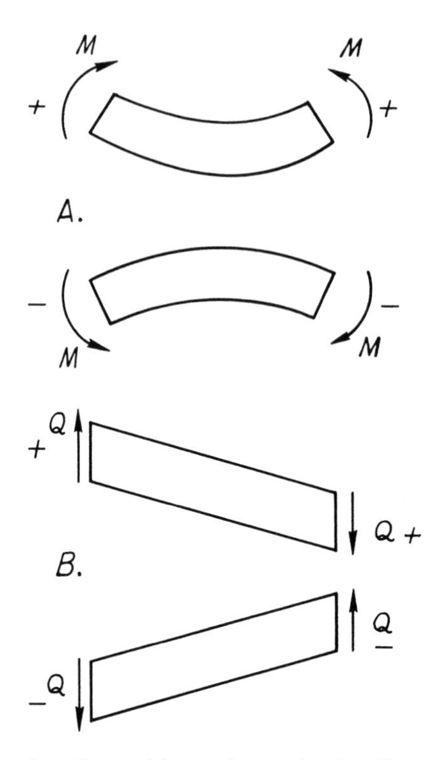

Figure 2.9. Sign conventions for positive and negative bending moments, M, and shear forces, Q.

work. If equal forces, F, are applied to the ends of the seesaw, there is no tendency for rotation. Therefore, the bending moment transferred to the support by the board is zero. But there is a net force of $2F$ on the support of the seesaw (Fig. 2.10B). Now, if forces of opposite directions are applied to the ends of the seesaw, it tends to rotate. That is, the section of the beam is subjected to a bending moment equal to the sum of each force times its lever arm (Fig. 2.10C). Thus, the bending moment is Fl. If the two forces are equal, the support carries no load. If, however, the two forces are unequal, the support is subjected to both a bending moment and a force. The most general state of force is analogous to the forces sensed by the support if the vertical forces were different and if there were, in addition, horizontal forces applied to the seesaw.

Another way to visualize the meaning of bending moment is by analyzing the bending of an "I" beam. The cross-section of the beam consists of a narrow web and heavy top and bottom flanges (Fig. 2.11). Suppose that the beam is bent by equal and opposite moments applied at its ends. Because of the bending, the upper flange is shortened and the lower one is lengthened. Thus the upper flange is in compression

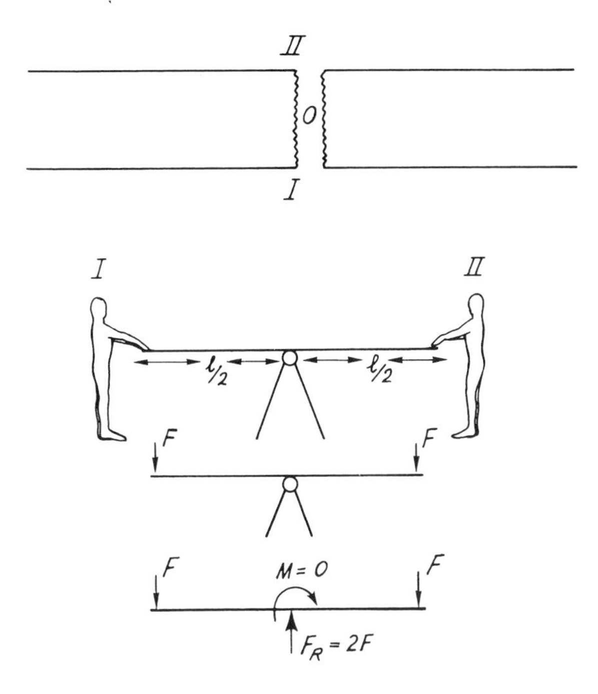

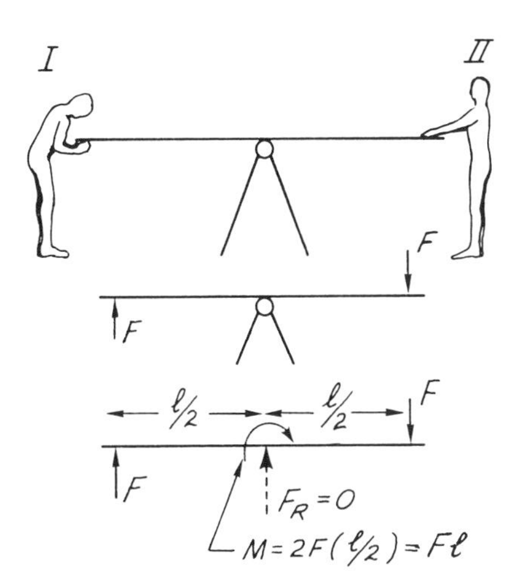

Figure 2.10. Illustration of meaning of bending moment.

A. Part of a beam severed at section I-*O*-II.

B. Section I-*O*-II represented by a seesaw. Bending moment at support is zero because forces F are equal in direction and magnitude.

C. Bending moment at support is not zero because forces F are equal in magnitude but opposite in direction.

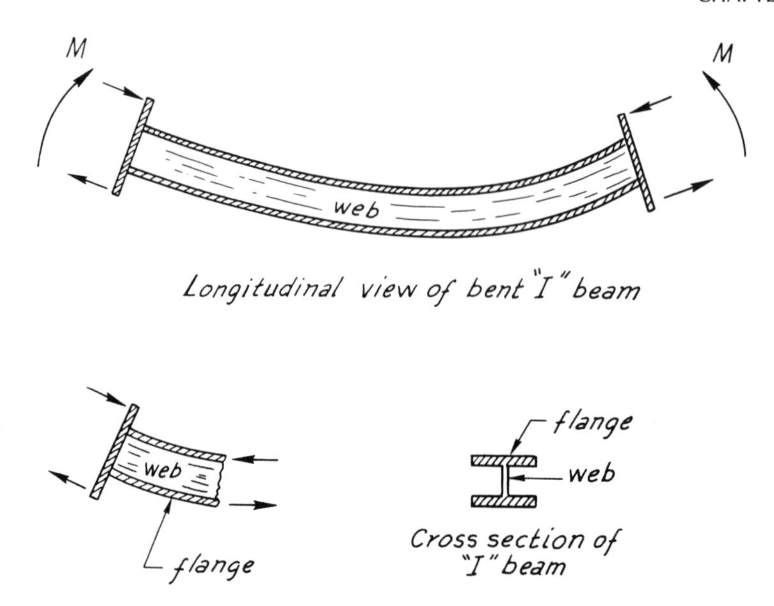

Figure 2.11. Another illustration of the meaning of bending moment. In an "I" beam loads are carried largely by the flanges. Compression in the top flange and extension in the lower flange correspond to a bending moment or torque.

and the lower one is in tension. The compression and tension of the flanges form a couple, or bending moment.

Free-body diagram.—The first step in the calculation of internal shear forces and bending moments of a beam is to establish the values of all forces acting on the exterior of the beam, usually by applying the equations of equilibrium. This operation is facilitated by drawing what is called a *free-body diagram* of the beam as a whole. In a free-body diagram, all contacting and attached bodies are removed and their effects on the beam are replaced by vectors. The vectors represent the external forces these bodies exert *on* the beam.

Figure 2.12A shows a beam supporting three vertical forces at various places. The beam itself is supported from below by two rollers. We will ignore the weight of the beam. The rollers can exert only vertical forces on the beam because they roll under the action of horizontal forces (Fig. 2.12B). Accordingly, the free-body diagram of the beam as a whole shows three forces exerted downward on top of the beam and it shows the effects of the rollers replaced by vertical forces, F_1 and F_2. These unknown forces can be calculated by means of the equilibrium equations for moment and forces,

$$\sum M = 0, \quad \sum F_y = 0, \quad \text{and} \quad \sum F_x = 0,$$

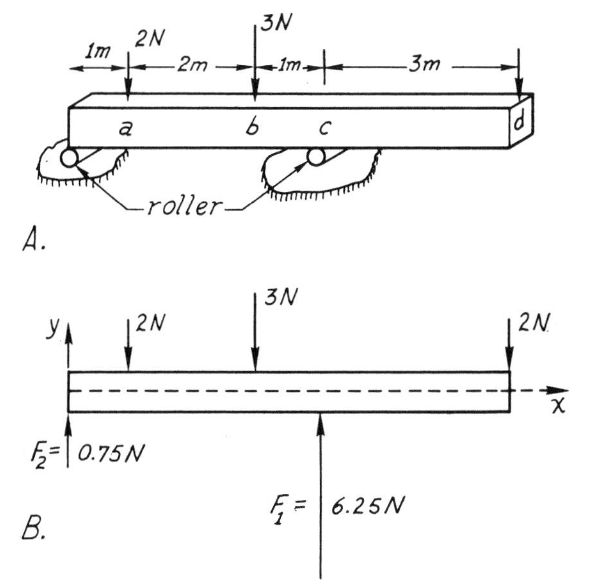

Figure 2.12. Loaded beam on rollers.
A. Positions of rollers and magnitudes of loads.
B. Free-body diagram of beam.

where $\sum$ indicates summation. These equations state that the sum of the moments about any point in the plane of the paper, the sum of the forces in the y-direction, and the sum of the forces in the x-direction are zero. The last condition is automatically satisfied for the problem posed here because there are no forces applied in the x-direction.

Taking moments about point O we eliminate the effect of the force applied at this point, F_2, because its lever arm is zero. Thus,

$$\left[\sum M_0 = 0\right],$$

$$F_1(4) - (2)(1) - (3)(3) - (2)(7) = 0,$$

or

$$F_1 = 6.25 \text{ N } [= 6.25 \times 10^5 \text{ dn}].$$

Here N is newtons and dn is dynes.

To calculate the other reaction force, F_2, sum forces in the y-direction. Thus,

$$\left[\sum F_y = 0\right],$$

$$F_2 + 6.25 - 2 - 3 - 2 = 0,$$

or

$$F_2 = 0.75 \text{ N } [= 7.5 \times 10^4 \text{ dn}].$$

The next step is to isolate, with a free-body diagram, part of the beam to the right or left of an arbitrary transverse section. For example, see Fig. 2.13B. The

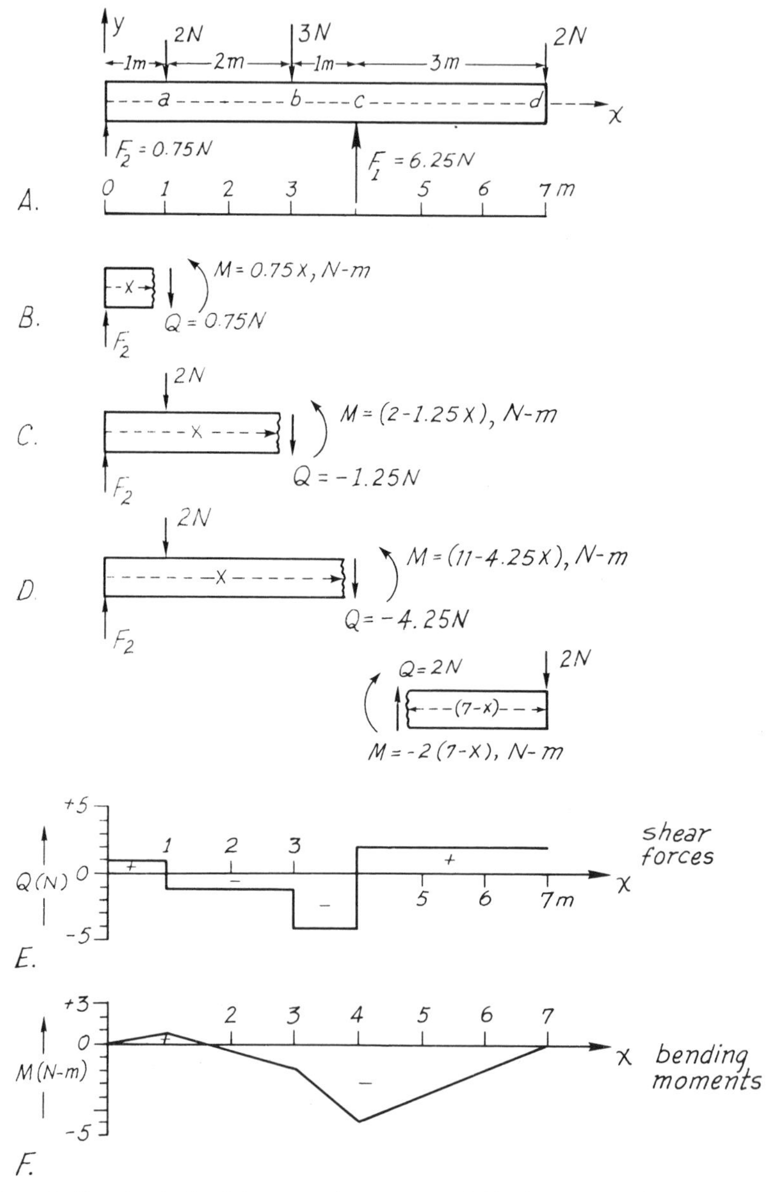

Figure 2.13. Analysis of forces and moments within beam carrying point loads.

equations of equilibrium are then applied to the isolated section of the beam. In this way the shear force, Q, and the bending moment, M, acting on each cross-section within the beam are calculated. In order to calculate the shear force and bending moment to the left of the force applied at point a in Fig. 2.13A, we construct the free-body diagram shown in Fig. 2.13B. The length of the element of beam is x, where x is less than or equal to one meter. We can calculate the shear force by summing forces in the vertical direction:

$$\left[\sum F_y = 0\right], \qquad Q - F_2 = 0,$$

or

$$Q = 0.75 \text{ N}.$$

We can calculate the bending moment, M, acting at the cross-section a distance x from the origin of coordinates by summing moments about the origin. Thus,

$$\left[\sum M_0 = 0\right], \qquad M - Q(x) = 0,$$

or

$$M = 0.75(x) \qquad \text{N}-m, \qquad \text{where } 0 \leq x \leq 1m.$$

The free-body diagram for the segment of beam up to the next force applied to the beam is shown in Fig. 2.13C. Equilibrium requires

$$\left[\sum F_y = 0\right], \qquad Q + 2 - 0.75 = 0,$$

or

$$Q = -1.25 \text{ N}.$$

$$\left[\sum M_0 = 0\right], \qquad M - Q(x) - 2(1) = 0.$$

But $Q = -1.25$, so that

$$M = 2.0 - 1.25(x) \qquad \text{N}-m, \qquad \text{where } 1m \leq x \leq 3m.$$

The shear forces and moments for the next interval of the beam are analyzed with the free-body diagram shown in Fig. 2.13D. Equilibrium requires

$$\left[\sum F_y = 0\right], \qquad Q + 3 - 2 - 0.75 = 0,$$

or

$$Q = -4.25 \text{ N}.$$

$$\left[\sum M_0 = 0\right], \qquad M - Q(x) - (3)(3) - 2(1) = 0.$$

But $Q = -4.25$, so that

$$M = 11.0 - 4.25(x) \qquad \text{N}-m, \qquad \text{where } 3m \leq x \leq 4m.$$

The final interval of the beam, the interval between points c and d in Fig. 2.13A, can be analyzed most simply by considering a free-body diagram of the right-hand end of the beam, shown in Fig. 2.13E. Note that the direction of positive Q, indicated by the direction of the arrow, is in accord with the convention established in earlier paragraphs, and derived from the principle of action and reaction, or Newton's third law. Equilibrium of the last element of the beam requires that

$$\left[\sum F_y = 0\right], \qquad Q - 2 = 0, \quad \text{or} \quad Q = 2\,\text{N}.$$

$$\left[\sum M_d = 0\right], \qquad M + Q(7x) = 0, \quad \text{or} \quad M = -2(7 - x) \qquad \text{N}-m,$$

where $7m \geq x \geq 4m$.

Thus we have computed the shear forces and bending moments on all cross-sections within the beam. To visualize the distributions, we can plot the relations on diagrams such as those shown in Figs. 2.13F and 13G. The magnitude of the maximum shear force is -4.25 N and it is between one of the roller supports and the largest applied load, that is, in the interval $3m \leq x \leq 4m$ (Fig. 2.13F). The bending moment of maximum magnitude is -6 N$-m$. It is at the right-hand roller, that is, at $x = 4m$. Thus if the beam fails in shear, failure should occur in the interval $3m \leq x \leq 4m$, and if it fails in tension, a crack should appear at the place of the largest bending moment, at $4m$. The tension crack will appear at the top of the beam, over the second roller.

This analysis of the shear forces and bending moments within a beam is valid irrespective of the type of material. Thus, if the loading conditions are the same, the shear forces and bending moments are the same within a beam whether it is elastic, viscous, plastic, or something else. The stresses and strains within the beam depend upon the type of material, but the forces do not. We used only the equations of equilibrium to calculate the forces.

For additional examples of the type of problem illustrated here, the reader is referred to any textbook on the subject of statics used in elementary courses in engineering mechanics.

Exercises

1. Draw the shear-force and bending-moment diagrams for the cantilever beam shown below. It has a length of l. A load, P, is applied at a distance of $l/4$ from the free end.

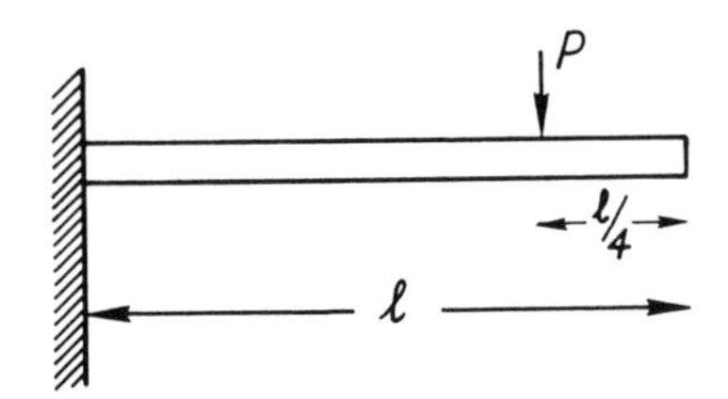

2. The following problem was first formulated and solved by Galileo in 1665. Galileo's experiments and observations of beams in buildings indicated that the moment that resists bending of a cantilever beam of uniform width, b, is proportional to the square of the depth of the beam, that is, to y^2. Thus, $M_r = ky^2$, where k is a constant, dependent upon the properties of the material of the beam. A load, P, is applied at the free end of the cantilever beam, and T is the depth of the beam at its fixed end.

   Derive an expression for the depth, y, in terms of x, T, k, and P, such that the resistance to bending is the same for all cross-sections as it is at the fixed end of the beam.

   Ans: $y = \left(T^2 - \frac{Px}{k}\right)^{1/2}$.

   The problem is to compute the form of the longitudinal profile of the ideal beam, which, if it fails, will fail simultaneously everywhere along its length. Presumably you would use a beam of the shape calculated if you were trying to minimize the amount of material used in constructing the beam. The beam is shown below.

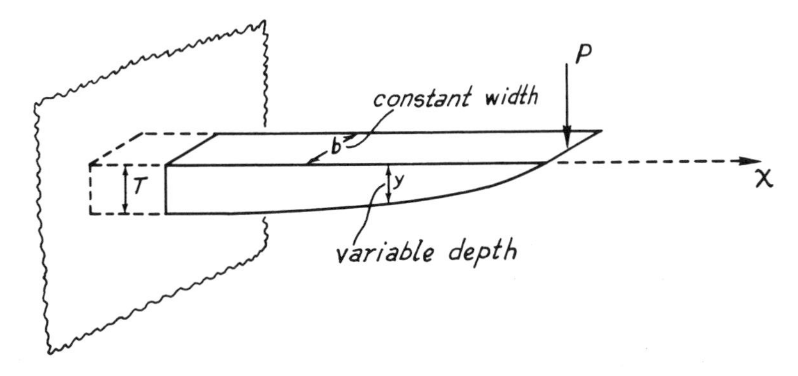

3. Draw the shear-force and bending-moment diagrams for the beam shown below. The beam is supported at each end by rollers and its upper surface is loaded by n equal forces, which are applied at points spaced a distance, $l/(n - 1)$ apart.

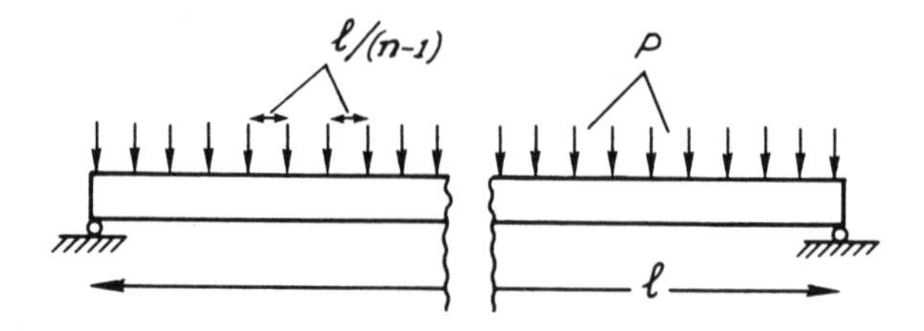

Forces and Moments Induced within Beams Carrying Distributed Loads.—The means for determining shear forces and bending moments in simply supported beams under concentrated loads has been developed adequately for our purposes. Now the analysis will be extended to beams with distributed loads, such as those applied via magmatic pressure.

In addition to the conventions for the positive directions of Q and M, there are several useful relations involving Q and M that can be established for a beam with distributed loads. Figure 2.14A is part of a beam supporting a distributed load. An

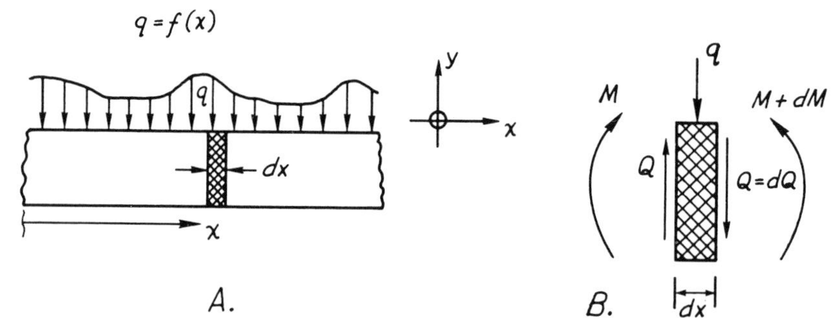

Figure 2.14. Part of beam supporting distributed load.

A. Distributed load of arbitrary form. $q = f(x)$, that is, q is some function of x.

B. Forces and moments on an element of the beam. Distance to element is x, width of element is dx. With a change of distance of dx, the shear force changes by an amount dQ and the bending moment changes by an amount dM.

element of width, dx, of the beam has been isolated (Fig. 2.14B). The distributed load, q, represents the force *per unit of length* on the beam. It can have several origins, such as weight of the beam or pressure exerted on the beam. The shear, Q, and the moment, M, acting on the element at point x are drawn in their positive directions. On the opposite side of the element, where the coordinate is $x + dx$, the shear force is $Q + dQ$ and the bending moment is $M + dM$, where dQ and dM are increments of shear force and bending moment, respectively. The applied load, q, per unit of length can be considered to be a constant over the surface of the element because the length of the element is infinitesimal. Equilibrium requires that the sum of the vertical forces be zero, so that

$$\left[\sum F_y = 0\right], \qquad Q - q(dx) - (Q + dQ) = 0,$$

or

$$q = -\frac{dQ}{dx}. \tag{2.7}$$

According to this equation, the slope of the shear diagram is equal to the negative of the distributed load applied to the beam.

Equilibrium of the element requires also that the sum of moments must be zero. Taking moments about the lower left corner of the element gives

$$M + q\,dx\left(\frac{dx}{2}\right) + (Q + dQ)\,dx - M - dM = 0.$$

We will ignore the $(dx)^2$ and $(dx\,dQ)$ terms because dx and dQ are small, so that their products are extremely small. Therefore, we derive

$$Q = \frac{dM}{dx}, \tag{2.8}$$

which states that the shear force is equal to the slope of the moment curve.

Comparing eqs. (2.7) and (2.8), it is apparent that

$$q = -\frac{d^2M}{dx^2}. \tag{2.9}$$

Approximate Relation between Bending Moment and Curvature of Elastic Members.—We have examined in some detail relations among bending moment, shear force, and externally applied point and distributed loads. Those relations are valid for beams of any composition because our analysis was based solely on the concept of equilibrium of forces and moments. But the problem that we want to solve involves the form of a beam that is subjected to pressure from below. To determine form, we need to consider the stress-strain relations of the material of the beam. Let us assume that the beam is elastic.

Consider such an elastic beam bent by a certain moment, M, applied at its ends. Figure 2.15A shows a beam that is bent sharply into a circular arc with a radius of curvature, ρ. Somewhere within the beam there is a plane, normal to the plane of the paper, above which the material is compressed and below which it is extended under the action of the moment. But along that plane there is neither extension nor compression; this is why we call it the *neutral plane*.

Examine a small rectangular element of length δx (delta x) and of height T, removed from some part of the beam before the beam is bent (Fig. 2.15B). When this element is bent, along with the remainder of the beam, the length of its neutral plane

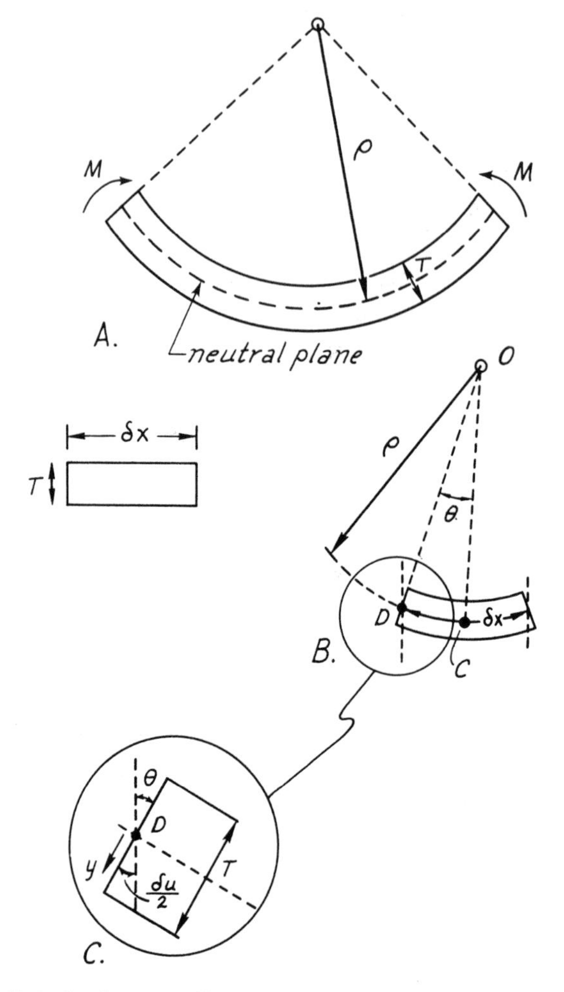

Figure 2.15. Relation between fiber strain and radius of curvature of a bent beam.
A. Beam bent into circular arc with radius of curvature ρ.
B. An element of width δx of the bent beam.
C. Enlargement of element.

is unchanged, to a first approximation. But the top of the element is definitely shortened and the bottom of the element is definitely lengthened when the element is bent, as is shown in the figure. Therefore, the ends of the element are rotated slightly during bending. Let this angle of rotation be called θ (theta).

Drop a perpendicular line down from the center of curvature to the center of the element in Fig. 2.15C. The distance, OC, is ρ, the radius of curvature. Now draw another radius of curvature, OD, to the end of the element. The angle between these two radii of curvature is equal to θ, the angle of rotation of the end of the element. The angle is equal to the arc length, $\delta x/2$, divided by the radius of curvature, ρ, so that

$$\theta = \frac{\delta x}{2\rho}. \tag{2.10}$$

Let us denote the change in length in the x-direction of the element by δu. The change in length of each half of the element, of course, is $\delta u/2$ (Fig. 2.15C). Now let us postulate that the stretching and contracting of parts of the element are functions of the distance from the neutral plane of the element. If we let this distance be y, then δu is a function of y. Assume that the relation is linear (see enlargement in Fig. 2.15C). Then, according to Fig. 2.15C, y, δu, and θ are related by the following equation,

$$\tan\theta = \frac{\delta u}{2y}.$$

For small angles, however, $\theta \approx \tan\theta$, so that

$$\theta = \frac{\delta u}{2y}. \tag{2.11}$$

Comparing eqs. (2.10) and (2.11) we see that

$$\frac{\delta u}{2y} = \frac{\delta x}{2\rho}, \tag{2.12}$$

which states that the change in length, δu, of an element with respect to distance from its neutral plane is the same as the length of the neutral plane, δx, divided by its distance, ρ, from the center of curvature of the neutral plane.

Rewriting eq. (2.12) in the form

$$\frac{\delta u}{\delta x} = \frac{y}{\rho}, \tag{2.13}$$

we derive a most useful relation. The importance of the relation becomes clear when we remember that normal strain is defined as the change in distance between two points within a body, divided by the original distance between the two points. Accordingly, if we make the element extremely small we can write eq. (2.13) in differential form,

$$\frac{du}{dx} = \epsilon_f = \frac{y}{\rho}. \tag{2.14}$$

This equation states that the normal strain, or *fiber strain*, ϵ_f, within a bent beam is, to a first approximation, proportional to the distance from the neutral plane, and inversely proportional to the radius of curvature of the beam.

For a simple elastic substance, normal stresses and strains are proportional,

$$\sigma = E\epsilon, \tag{2.15}$$

where σ (sigma) is normal stress, E is Young's elastic modulus, and ϵ (epsilon) is normal strain. Thus, eq. (2.14) can be rewritten as

$$\epsilon_f = \frac{\sigma_f}{E} = \frac{y}{\rho}, \tag{2.16}$$

where σ_f is the fiber stress.

Bending moment can be calculated by integrating the fiber stresses over the cross-section of the beam. Consider a small element, dA, of the cross-section of the beam shown in Fig. 2.16. The total force on an infinitesimal area is $(\sigma_f)(dA)$, so that at any distance, y, from the neutral plane the moment around the neutral plane is

$$\int dM = \int \sigma_f(y)(dA). \tag{2.17}$$

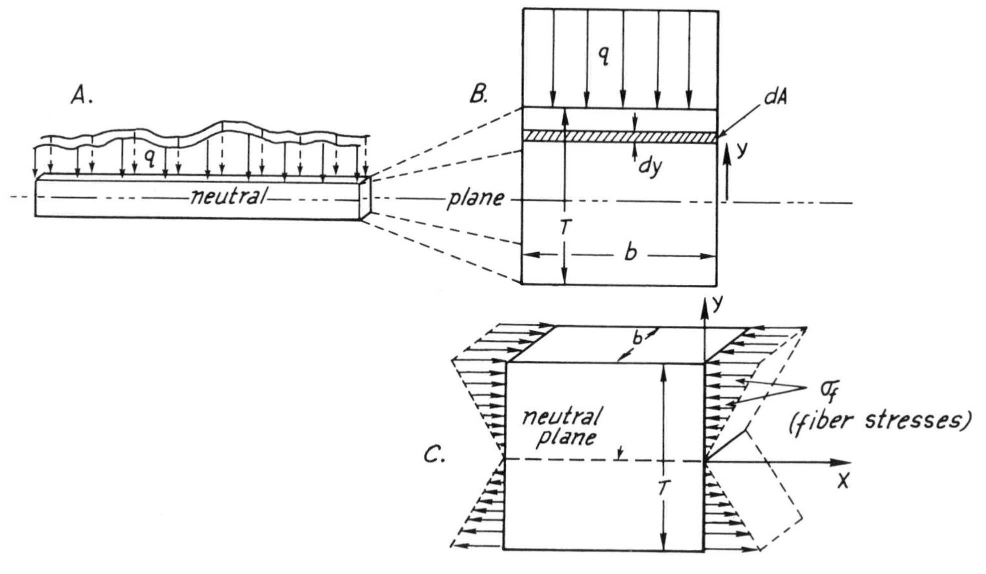

Figure 2.16. Calculation of bending moment from distribution of fiber stresses.

A. Longitudinal view of beam, showing neutral plane and a distributed load on its upper surface.

B. View of an arbitrary cross-section of the beam, showing thickness, T; width, b; distance from neutral plane, y; and distributed load, q. An infinitesimal area, dA, times the fiber stress acting on it, times the lever arm, y, is an increment of bending moment, dM.

C. Small element of beam, showing assumed distribution of fiber stresses across sections such as the one in B.

Substituting the value of σ_f from eq. (2.16) into eq. (2.17),

$$M = \int_A \frac{Ey^2\,dA}{\rho},$$

or

$$M = \frac{E}{\rho}\int_A y^2\,dA. \tag{2.18}$$

Equation (2.18) can be rewritten as

$$M = \frac{EI}{\rho}, \tag{2.19}$$

where I is the *moment of inertia* of the cross-section of a beam with respect to its neutral plane. For a beam of rectangular cross-section, $I = (bT^3)/12$, where b is its width and T is its thickness. This result can be verified readily by evaluation of the integral, $\int_A y^2\,dA$, above and below the neutral plane of a rectangular beam.

The radius of curvature can also be written in terms of the coordinates v and x, where v is the displacement of the neutral plane in the y-direction. From the calculus, the radius of curvature is

$$\rho = \frac{[1 + (dv/dx)^2]^{3/2}}{|d^2v/dx^2|}, \tag{2.20}$$

where the vertical lines indicate "absolute value of."

Equation (2.20) can be simplified if we assume that the slope, dv/dx, of the neutral plane is small. Then the $(dv/dx)^2$ term can be neglected, because it is much smaller than one, to which it is added. The relation between the slope angle and the neutral plane of a beam and the numerator of eq. (2.20) is shown in Fig. 2.17. Apparently the numerator is not significantly different from unity as long as the neutral plane of the beam has a slope of less than about 20 degrees (*5*). For slopes less than about 10 or 20 degrees, then, the radius of curvature is approximately

$$\rho \approx \frac{1}{|d^2v/dx^2|}, \tag{2.21}$$

so that eq. (2.19) can be rewritten

$$\frac{d^2v}{dx^2} = \frac{M}{EI}. \tag{2.22}$$

This differential equation, relating deflection to bending moment of a beam, was derived by approximation methods involving several simplifying assumptions. We will use the equation many times in analyzing forms of laccoliths and forms of folds, and the equation is potentially useful for many other problems. Therefore, it would

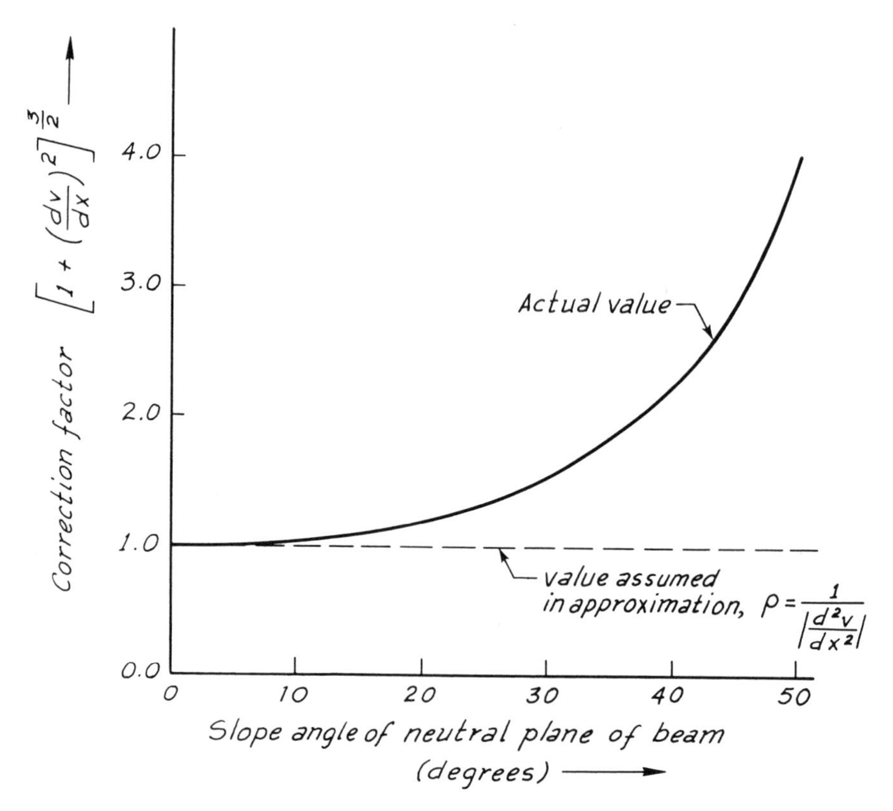

Figure 2.17. Relation between slope of neutral plane of beam and correction term in expression for radius of curvature (*5*).

be prudent to state explicitly the most important simplifying assumptions we used to derive the equation.

One assumption is that lateral deflections are small, so that the radius of curvature of the beam is unaffected by the square of the first derivative of the deflection [eq. (2.20)]. We will remove this restriction in Chapter 4 when we analyze the "elastica," the form of highly deflected, thin beams.

Another assumption is that the beam is thin, so that bending is resisted solely by fiber stresses. Biot (*1,2*) has solved the problem of buckling of a thick beam and has shown that the assumption of thinness, for that problem at least, does not lead to serious error if the thickness of the beam is less than about $\frac{1}{10}$ its length. Timoshenko and Goodier (*6*) indicate that the curvature of a uniformly loaded cantilever beam is given by the expression

$$\frac{d^2v}{dx^2} = \frac{M}{IE} + \frac{qT^2}{EI}\left(\frac{4}{5} + \frac{\nu}{2}\right), \tag{2.23}$$

where q is the load per unit of length of the beam and ν (nu) is Poisson's ratio, which ranges from 0 to 0.5 for most materials and is about 0.3 for steel. According to Timoshenko and Goodier, the second term in the exact solution, derived from elasticity theory, is largely due to internal shear stresses. It is not contained in our approximation, eq. (2.22).

Let us rewrite eq. (2.23) in equivalent form,

$$\frac{d^2v}{dx^2} = \frac{12M}{EbT^3} + \frac{12q}{EbT}\left(\frac{4}{5} + \frac{\nu}{2}\right),$$

where $I = bT^3/12$, T is thickness, and b is width of the beam. With the equation in the latter form it becomes apparent that the error caused by using the approximate expression, eq. (2.22), is small if the thickness is small, particularly if the thickness is much less than unity. As thickness gets smaller, the first term increases in magnitude much more rapidly than the second term.

Example.—A cantilever beam is deformed by a load, q, per unit of width distributed over its upper surface. Determine the expression for the deflection, v, at any

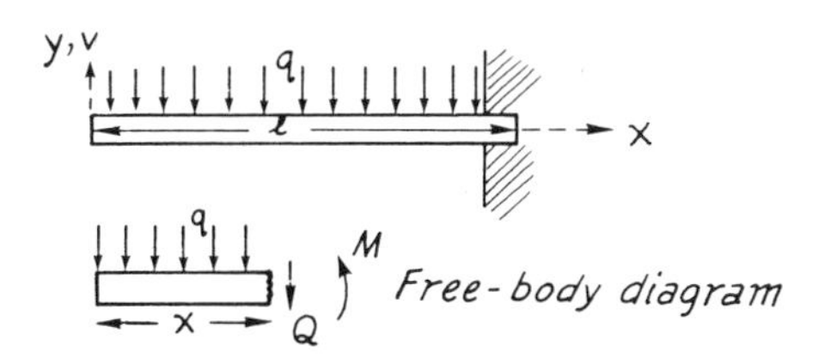

distance, x, from the free end of the beam. According to the free-body diagram,

$$\left[\sum M = 0\right], \qquad M - \int_0^x qx\, dx - Qx = 0,$$

or

$$M - \int_0^x qx\, dx + qx^2 = 0,$$

or

$$M = -\frac{qx^2}{2}.$$

Thus the equation for the deflection of the beam, eq. (2.22), becomes

$$EI\frac{d^2v}{dx^2} = -\frac{qx^2}{2}.$$

Integrating,

$$EI\frac{dv}{dx} = -\frac{qx^3}{6} + C_1 \qquad (2.24)$$

and

$$EIv = -\frac{qx^4}{24} + C_1x + C_2, \tag{2.25}$$

where C_1 and C_2 are constants. The boundary conditions are

$$\left(\frac{dv}{dx}\right)_{[x=l]} = 0, \qquad v_{[x=l]} = 0,$$

which state that the slope and the deflection of the beam are zero at the fixed end of the cantilever. Thus,

$$C_1 = \frac{ql^3}{6} \quad \text{and} \quad C_2 = -\frac{ql^4}{8}.$$

Equations (2.24) and (2.25), then, become

$$EI\frac{dv}{dx} = -\left(\frac{q}{6}\right)(x^3 - l^3)$$

and

$$EIv = -\left(\frac{q}{24}\right)(x^4 - 4xl^3 + 3l^4).$$

Exercise

Solve the preceding problem exactly, using the exact differential equation for the deflection, eq. (2.23). Calculate the percent error caused by using eq. (2.22) for deriving the deflection of the free end of the cantilever beam. Assume that

$$E = 10^{12}\,\frac{\text{dn}}{\text{cm}^2}$$

$$\nu = 0.25$$

$$q = 2 \times 10^6\,\frac{\text{dn}}{\text{cm}}$$

$$l = 10 \text{ cm}$$

$$b = 1 \text{ cm}$$

and

$$T = 1 \text{ cm}.$$

Ans: error $\approx 4\%$.

Mathematical Expression for Form of Overburden Deflected by Magma of Laccolith.—Figure 2.18A shows an idealized laccolith and feeder dike. The laccolith has been emplaced within a sequence of strata, which overlie and underlie it. To relate the form of the idealized laccolith to properties of strata overlying it and to its maximum thickness, let us represent the strata overlying the laccolith by an elastic plate. The elastic plate has a thickness, T, an elastic modulus, B, and width, b. The length of the laccolith is l. Let us choose the origin of coordinates to be at the left-hand side of the laccolith shown in Fig. 2.18B.

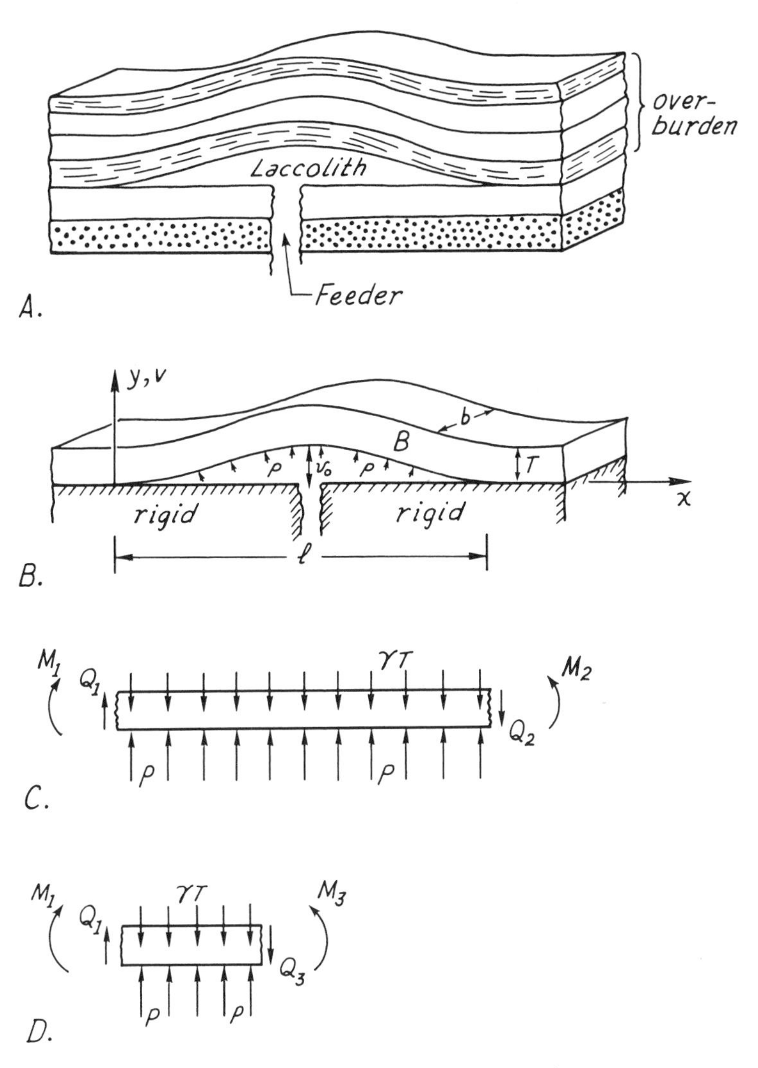

Figure 2.18. Analysis of ideal laccolithic form.
A. Idealized anticlinal laccolith.
B. Overburden replaced by plate.
C. Free-body diagram of overburden.
D. Free-body diagram of part of overburden.

We will show in later pages that if the member is an infinitely wide plate,

$$B = \frac{E}{1 - \nu^2},$$

and if it is a beam,

$$B = E,$$

where ν is a Poisson's ratio.

The pressure, p, tends to drive the plate upwards to form the laccolith. Resisting the pressure are the elastic resistance of the plate and the weight of the plate. The weight is $b\gamma T(x)$, where x is any length, b is the width, and γ is the unit weight of the plate. The elastic resistance is a combination of shearing forces, Q, and bending moments, M, at each end of any segment of the plate.

Figure 2.18C shows a free-body diagram of the part of the plate over the laccolith of length l. For the plate to be in equilibrium, the sum of forces in the y-direction must be zero. By symmetry, the shear forces, Q_1 and Q_2, at each end of the plate, must be equal. Their sum is the difference between the downward force of the weight of the plate and the upward force exerted by the pressure in the magma. Thus,

$$Q_1 - Q_2 + lb(p - \gamma T) = 0.$$

But $Q_1 = -Q_2$, so that

$$Q_1 = -Q_2 = -\frac{lb}{2}(p - \gamma T). \tag{2.26}$$

Figure 2.18D shows a free-body diagram for part of the plate over the laccolith. Summing forces in the y-direction,

$$Q_1 - Q_3 + bx(p - \gamma T) = 0.$$

But, according to eq. (2.26),

$$Q_1 = -\frac{lb}{2}(p - \gamma T),$$

so that

$$Q_3 = b(p - \gamma T)\left(x - \frac{l}{2}\right). \tag{2.27}$$

Thus, the shear stress is a linear function of the distance along the plate over the laccolith. It is maximum at each end, $x = O$ and $x = l$, and is zero at the center, $x = l/2$. If we write eq. (2.27) in dimensionless form,

$$\frac{Q_3}{lb(p - \gamma T)} = \left(\frac{x}{l} - \frac{1}{2}\right),$$

the relation can be readily plotted as in Fig. 2.19B.

Now the problem is to relate the deflection of the plate to the various driving and resisting forces. According to eq. (2.8),

$$\frac{dM}{dx} = Q,$$

so that the derivative of the moment with respect to x is, from eq. (2.27),

$$\frac{dM_3}{dx} = Q_3 = -(p - \gamma T)\left(\frac{l}{2} - x\right)b.$$

Integrating,

$$M_3 = -(l - x)\left(\frac{x}{2}\right)(p - \gamma T)b + C_1, \tag{2.28}$$

where C_1 is a constant of integration.

The curvature of the plate is related to the bending moment by eq. (2.22),

$$BI\frac{d^2v}{dx^2} = M,$$

so that

$$BI\frac{d^2v}{dx^2} = -(l - x)\left(\frac{x}{2}\right)(p - \gamma T)b + C_1. \tag{2.29}$$

Integrating twice, we calculate v:

$$BI\frac{dv}{dx} = -(p - \gamma T)\left(\frac{lx^2}{4} - \frac{x^3}{6}\right)b + C_1x + C_2, \tag{2.30}$$

and

$$BIv = -(p - \gamma T)\left(\frac{lx^3}{12} - \frac{x^4}{24}\right)b + \frac{C_1x^2}{2} + C_2x + C_3, \tag{2.31}$$

where C_2 and C_3 are constants of integration.

The boundary conditions are our means of eliminating the constants of integration. At the center of the laccolith the deflection of the plate is a maximum and the slope of the plate is zero. Thus, at $x = l/2$,

$$\frac{dv}{dx} = 0, \quad \text{and} \quad v = v_0, \tag{2.32}$$

where v_0 is the maximum deflection. This is the deflection at the center of the laccolith. At the ends of the laccolith the deflection and the slope of the plate are zero because the surface of the plate becomes parallel to the surface of the rigid base there. Thus, at $x = 0$, and at $x = l$,

$$\frac{dv}{dx} = 0, \quad \text{and} \quad v = 0. \tag{2.33}$$

Now C_2 must be zero to satisfy the condition that

$$\frac{dv}{dx} = 0, \qquad \text{at } x = 0,$$

as is seen by inspection of eq. (2.30). Also, C_3 must be zero to satisfy the condition that $v = 0$ at $x = 0$. Either of the boundary conditions (2.33) can be used to calculate C_1. From eq. (2.30),

$$C_1 l = -\left(\frac{l^3}{6} - \frac{l^3}{4}\right)(p - \gamma T)b$$

or

$$C_1 = \frac{l^2}{12}(p - \gamma T)b. \tag{2.34}$$

Substituting both boundary conditions (2.33) into eqs. (2.30) and (2.31) verifies that they both are satisfied by our solution.

Finally, we can relate the deflection, v_0, of the plate at its center to the length of the laccolith, l, by the second of boundary conditions (2.32)

$$v = v_0 \qquad \text{at } x = \frac{l}{2}$$

and eq. (2.31)

$$BIv_0 = -(p - \gamma T)b\left(\frac{l^4}{96} - \frac{l^4}{384}\right) + \frac{l^4}{96}b(p - \gamma T),$$

or

$$v_0 = \frac{(p - \gamma T)bl^4}{384BI}.$$

However, $I = bT^3/12$, so that

$$v_0 = \frac{(p - \gamma T)l^4}{32BT^3}. \tag{2.35}$$

The general equation for the deflection of the member is eq. (2.31). Making appropriate substitutions for constants, we derive

$$v = \frac{b(p - \gamma T)}{24BI}(x^4 - 2lx^3 + l^2x^2), \tag{2.36}$$

where v is the deflection at any point, x, along the plate between ends of the laccolith, $x = 0$, and $x = l$.

Substituting $I = bT^2/12$ into eq. (2.36) we derive

$$v = \frac{(p - \gamma T)}{BT^3}\left(\frac{x^4}{2} - lx^3 + \frac{l^2x^2}{2}\right). \quad (2.37)$$

To make a diagram illustrating the form of eq. (2.37) it will be convenient to put the equation in *dimensionless form*. To do this we can divide each side of eq. (2.37) by

$$\frac{(p - \gamma T)l^4}{2BT^3}.$$

Thus we have

$$v\left[\frac{2BT^3}{(p - \gamma T)l^4}\right] = \left(\frac{x}{l}\right)^4 - 2\left(\frac{x}{l}\right)^3 + \left(\frac{x}{l}\right)^2. \quad (2.38)$$

It can be verified that each term in eq. (2.38) is dimensionless. Now the equation is in a form that is particularly suited for graphical presentation.

Figure 2.19A is a graph of eq. (2.38). The equation seems to represent quite well the deflection of the strata overlying Gilbert's ideal laccolith.

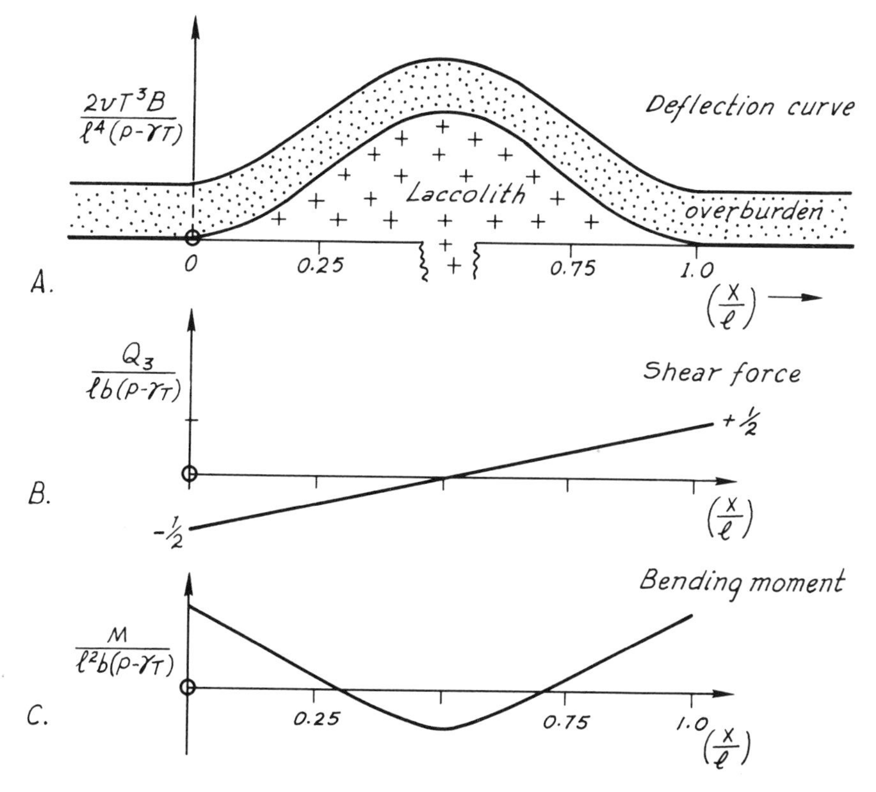

Figure 2.19. Theoretical form and distribution of shear force and bending moment in overburden.

Exercises

1. Calculate the position of greatest and least fiber stress within the stratum overlying the idealized laccolith shown in Fig. 2.19A.
2. Derive an expression for the relation between fiber stress and position along the upper and lower boundaries of the stratum over the idealized laccolith.
3. Put the expression in dimensionless form and plot the relation between a dimensionless stress variable and x/l. Indicate on your graph the positions of greatest and least fiber stresses.
4. If dikes tend to form in regions of maximum tension in the base of the stratum forming the roof of a laccolith, where in the idealized laccolith shown in Fig. 2.19A would you expect dikes to occur?

The mathematical form displayed graphically in Fig. 2.19A is strikingly similar to the form of Gilbert's ideal laccolith (Fig. 2.3A). However, there are at least two discrepancies between the concept of a laccolith analyzed mathematically and the concept of the ideal laccolith presented by Gilbert. First, laccoliths are circular or tongue-shaped in plane view, not anticlinal as was assumed in our analysis. Second, our analysis assumed that the overlying rock consists of a single stratum of thickness T, but the sedimentary rocks that overlay the laccoliths in the Henry Mountains at the time the laccoliths formed were definitely stratified.

First let us consider the problem of uplift of more than one stratum by laccoliths. If the overlying stratum of thickness T in our previous analysis is replaced by an arbitrary number, n, of strata, the equation for the bending moment, eq. (2.29),

$$\frac{bBT^3}{12}\frac{d^2v}{dx^2} = (x - l)\left(\frac{x}{2}\right)(b)(p - \gamma T) + C_1, \tag{2.29}$$

should be replaced by

$$\frac{Bb}{12}\frac{d^2v}{dx^2}(t_1^3 + t_2^3 + t_3^3 + \cdots + t_n^3) = (x - l)\left(\frac{x}{2}\right)b(p - \gamma T) + C_1, \tag{2.39}$$

where the small t's with subscripts refer to thicknesses of individual strata and where T is, now, the total thickness of the overburden above the laccolith. Equation (2.39) should replace (2.29) if there are multiple strata because the sum of the resisting moments contributed by each stratum must equal the difference between the driving moment induced by the pressure, p, and the resisting moment contributed by the weight of the overburden. In writing eq. (2.39) we have assumed that contacts between strata are frictionless, that each stratum has the same elastic modulus, B, and that the unit weight, γ, of each stratum is the same. If unit weights are different, γ should be replaced by $\bar{\gamma}$, the mean unit weight of the overburden. If moduli of different layers are different, we could simply remove B from outside the parentheses and replace it with different moduli for each stratum, such as B_1, B_2, and so forth. The restriction that contacts between strata be frictionless is more difficult to alter. However, the problem considered here can be considered as one extreme. The other extreme already was considered; if the strata are perfectly stuck together they behave as one thick stratum.

By integrating eq. (2.39) we can show that the deflection of a stack of layers over an idealized laccolith is given by the relation

$$\frac{Bv}{(p - \gamma T)} (t_1^3 + t_2^3 + t_3^3 + \cdots + t_n^3) = \frac{x^4}{2} + \frac{l^2x^2}{2} - x^3l, \tag{2.40}$$

which is different from eq. (2.37) only in that the total thickness, T, of the members is replaced by individual members of thicknesses t_1, t_2, and so forth.

The problem of the influence of the circular or near-circular plan form of laccoliths on their height and width cannot be analyzed with the theoretical background we have developed thus far. However, the background necessary to analyze this problem can be obtained readily from the excellent treatise on the theory of plates and shells by Timoshenko and Woinowsky-Krieger (*7*).

According to Timoshenko and Woinowsky-Krieger (*7*), the deflection of a circular plate with fixed edges and supporting a uniform pressure on one surface is

$$v = \frac{3(p - \gamma T)}{16BT^3}\left[\left(\frac{l}{2}\right)^2 - r^2\right]^2. \tag{2.41}$$

Here r is the distance from the center of the laccolith to some point and l is the diameter of the laccolith.

The maximum deflection of the circular plate is, at $r = o$,

$$v_0 = \left(\frac{3}{8}\right)\frac{(p - \gamma T)l^4}{32BT^3}. \tag{2.42}$$

By comparing eqs. (2.35) and (2.42) we see that the amplitude of the circular laccolith is three-eights that of a laccolith that is anticlinal in form. However, the equations are identical in form. Indeed, the equations for the cross-sectional shape of the upper surfaces of the two types of laccoliths are nearly identical also. Therefore, whether the plan form of a laccolith is treated mathematically as circular or anticlinal is immaterial for our purposes. Conclusions derived from a study of one solution will be qualitatively the same as those derived from a study of the other.

Pollard (*5*) has derived the solution for the deflection of strata over a laccolith that is elliptic in plan. An elliptical laccolith is intermediate in form between a circular and an anticlinal laccolith, so that the maximum deflection over an elliptical laccolith ranges between the deflection over a laccolith that is circular and the deflection over one that is anticlinal in form. Therefore, we can conclude that, where our theoretical model is a reasonably accurate description of actual conditions in nature, the deflection shape of the overburden should be similar, regardless of plan shape, and the magnitudes of the deflection of the overburden are essentially independent of the plan shape of the laccolith, varying only by a factor of two.

Reason for lack of small laccoliths in Henry Mountains

Gilbert noted that there are no small laccoliths in the Henry Mountains; the smallest one he saw is more than one-half mile in diameter. Thus, he concluded, laccoliths do not occur at a small scale; their size seems to have a definite lower limit (*3*).

As was stated in earlier pages, Gilbert explained the lack of small laccoliths approximately as follows: He represented the overburden by a cylinder of rock that is pushed upwards by the magma of the laccolith. The driving force caused by the pressure is proportional to the area of the end of the cylinder, so that it is proportional to the square of the radius of the laccolith. The resisting force, however, is proportional to the surface area of the sides of the cylinder, so that it is proportional to the radius. As a result, the driving force decreases more rapidly than does the resisting force, with a decrease of radius of the laccolith. For a given magmatic pressure, therefore, there is a limiting radius of laccolith above which the magma can lift its overburden and below which the magma cannot. Thus, Gilbert reasoned, small laccoliths do not form (*3*).

We can derive an alternative explanation for the lack of small laccoliths in the Henry Mountains by detailed analysis of the expression for the amplitude, that is, the maximum magma thickness, of a laccolith [eq. (2.35)]. We can write the expression in the following form,

$$v_0 = \frac{k_1(p - \gamma T)l^4}{T^3}, \tag{2.43}$$

where v_0 is the magma thickness at the center of the laccolith, l is the laccolith diameter, T is the effective thickness of strata overlying the laccolith, p is pressure in the magma at the time the laccolith formed, γ is the unit weight of the intruded strata, and k_1 is a constant. For a laccolith of circular plan,

$$k_1 = \left(\frac{3}{8}\right)\left(\frac{1}{32B}\right).$$

For a laccolith of anticlinal form,

$$k_1 = \frac{1}{32B}.$$

We can simplify eq. (2.43) if the effective driving pressure, $(p - \gamma T)$, is approximately constant. Imagine two laccoliths forming at two different stratigraphic horizons but being fed by the same source of magma. Then, *assuming* that the pressure, p, in the column of magma is hydrostatic, the only difference of pressure in the two laccoliths is that due to a difference in altitude, that is, a difference in head. However, the difference in effective driving pressure, $(p - \gamma T)$, in the two laccoliths probably is minimal because the pressure of the magma decreases as the resisting pressure of the overlying strata decreases. They decrease together because they are both proportional to the thickness of overburden, T. The mean specific gravity (*3*) of the intrusive rock is about 2.6, whereas that of the overlying sedimentary rock is about 2.3. Thus the change of resisting pressure, which is proportional to the specific gravity of the sedimentary rock, differs little from the change of the driving pressure, which is proportional to the specific gravity of the magma. Accordingly, eq. (2.43) can be reduced to

$$v_0 = \frac{k_2 l^4}{T^3}, \tag{2.44}$$

where k_2 is a constant.

Suppose for our analysis of the question of a minimum size of laccoliths that several laccoliths of different sizes are beginning to form at a certain depth. Then T, as well as k_2, in eq. (2.44) is constant and the equation can be written

$$v_0 = k_3 l^4, \tag{2.45}$$

where k_3 is a new constant.

Now eq. (2.45) states that the deflection at the center of a laccolith is proportional to the magnitude of the width of the laccolith to the fourth power. Because of this peculiar relation between width and amplitude, the amplitude of a laccolith is highly sensitive to its width. Thus, if the diameter is very small, the amplitude is extremely small; if the diameter is large, the amplitude is extremely large. It is probable that intrusive bodies with relatively small diameters would not be called laccoliths; rather, they would be called sheets or sills, which are abundant in the Henry Mountains. Perhaps sills are baby laccoliths, as suggested by Pollard (*5*).

Gilbert (*3*) estimated that the most common ratio of amplitude to diameter of laccoliths in the Henry Mountains is about 1/7. According to Hunt's (*4*) estimates of the plan shapes of the laccoliths, a ratio of 1/15 is representative.

Our explanation of the lack of small laccoliths is based on the rather fundamental assumption that the overburden being lifted by the magma is flexed elastically. Our solution probably is not completely applicable to the actual situation, because real rocks can withstand very little flexing, or strain, before they fail. According to Pollard (*5*), the overburden of a laccolith fails by shear along a nearly vertical, cylindrical fault, or by diking near the periphery of the laccolith. Pollard shows that the maximum ratio of the deflection amplitude to overburden thickness is on the order of 1/50 or 1/100 at the time of some type of overburden failure if the properties of the overburden are similar to properties of rocks tested experimentally. Thus, he concludes, an intrusive body that we would definitely call a laccolith rather than a sill has nearly always deformed its overburden inelastically. His study indicates that our explanation of the lack of small laccoliths in the Henry Mountains is incomplete.

Explanation of the dependency of size of laccoliths on stratigraphic position in the Henry Mountains

The range in altitudes at which laccoliths have been found in the Henry Mountains is at least 1500 meters. As we stated, Gilbert (*3*) recognized two zones of laccoliths there. The mean diameter of the laccoliths in his upper zone is about 1.9 km and the mean diameter of the laccoliths in his lower zone is about 4.2 km. Thus, laccoliths of the lower zone are about twice as large as those in the upper zone. The difference in altitude of the centers of the two zones is about one kilometer.

According to our solution, based on elastic plate theory, the relation between diameter and depth of laccoliths should be [eq. (2.43)]

$$l = \left[\frac{v_0 T^3}{k_1(p - \gamma T)}\right]^{1/4}, \tag{2.46}$$

where v_0 is the amplitude of a laccolith, l is its diameter, T is the thickness of overburden, γ is the unit weight of the strata overlying the laccolith, p is the pressure in the magma, and k_1 is a constant. Equation (2.46) shows that the diameter of laccoliths should be approximately proportional to the depth of emplacement.

The total depth of overburden that existed at the time of intrusion of the laccoliths can be estimated with eq. (2.46) and with field data supplied by Gilbert on the sizes of the laccoliths. First we will simplify eq. (2.46), then we will turn to the data. We argued in earlier paragraphs that it is reasonable to assume that the pressure term in eq. (2.46),

$$(p - \gamma T),$$

is approximately constant, independent of depth. Thus eq. (2.46) becomes

$$l^3 = \left(\frac{v_0}{l}\right) k_4 T^3.$$

Also, Gilbert stated that the mean value of the ratio of amplitude to diameter of the laccoliths is approximately 1/7, so that

$$\frac{v_0}{l} \approx \frac{1}{7},$$

and eq. (2.46) further simplifies to

$$l^3 \approx k_5 T^3,$$

or

$$l \approx (k_5)^{-1/3} T. \tag{2.47}$$

Now we will examine the data. Remember that, according to Gilbert, there are two zones of laccoliths of different diameters in the Henry Mountains. The laccoliths in the upper zone have a mean diameter of 1.9 km and those in the lower zone have a mean diameter of about 4.2 km. The centers of the two zones are separated by about one kilometer. Thus, the change in diameter, δl, of about 2.3 km corresponds with a change of depth, δT, of about 1.0 km, or

$$\frac{\delta l}{\delta T} \approx 2.3.$$

But according to eq. (2.47),

$$\frac{dl}{dT} = (k_5)^{-1/3},$$

so that

$$\frac{dl}{dT} = (k_5)^{-1/3} \approx \frac{\delta l}{\delta T} \approx 2.3,$$

or

$$k_5 \approx (2.3)^3.$$

Thus we have crudely determined the constant of proportionality between thicknesses of overburden and diameters of laccoliths in the Henry Mountains. Substituting the calculated value of k_5 and a diameter of 1.9 km into eq. (2.47), we estimate that the overburden of the laccoliths in Gilbert's upper zone was approximately 800 meters thick.

This estimate of overburden thickness correlates remarkably, and probably fortuitously, with the thickness of the stratigraphic section, above the laccoliths, preserved in the Henry Mountains region today. That thickness, involving the Mesaverde Formation and the Masuk, Emergy, Blue Gate, and Ferron members of the Mancos Shale, is roughly 800 meters (*4*). All Tertiary rocks that may have been in the area at the time of intrusion of the laccoliths have since been removed and the age of the intrusions is unknown, so that there is little basis for an independent estimate of the total thickness of overburden. Hunt (*4*) states that about 3000 meters is a reasonable estimate of the maximum thickness of Cretaceous, Paleocene, and Eocene strata that have been removed from above the Mesaverde Formation in the Henry Mountains region. However, we do not know whether any of this rock rested on the Mesaverde at the time the laccoliths were intruded.

I have used part of Hunt's data of the sizes and stratigraphic positions of laccoliths to estimate the thickness of overburden by a method similar to that outlined above. With his data, I estimated a thickness of about 750 $\pm$ 250 meters of overburden, which estimate is close to that of 800 meters calculated with Gilbert's data. We will not investigate Hunt's data in detail, because the data are highly interpretive, as are Gilbert's.

References cited in Chapter 2

1. Biot, M. A., 1963, "Exact Theory of Buckling of a Thick Slab": *Applied Scientific Research*, Ser. A, V. 12, p. 182–198.
2. Biot, M. A., 1965, *Mechanics of Incremental Deformations*: McGraw-Hill Book Co., Inc., N.Y.
3. Gilbert, G. K., 1877, "Report on the Geology of the Henry Mountains": U.S. Geog. and Geol. Survey of Rocky Mountain Region.
4. Hunt, C. B., 1953, "Geology and Geography of the Henry Mountains Region, Utah": U.S. Geological Survey Professional Paper 228.
5. Pollard, D. D., 1968, "Deformation of Host Rocks during Sill and Laccolith Formation": Ph.D. Dissertation, School of Earth Sciences, Stanford University, Stanford, California.
6. Timoshenko, S. P., and Goodier, N. J., 1951, *Theory of Elasticity*: McGraw-Hill Book Co., Inc., N.Y.
7. Timoshenko, S. P., and Woinowsky-Krieger, S., 1959, *Theory of Plates and Shells*: McGraw-Hill Book Co., Inc., N.Y.

3

Chapter Sections

Introduction
Folds of the Appalachians
The Problems
Willis' Theoretical Analysis of the Appalachian Folds
Mathematical Expression for the Form of the Fold Trains
  Unconfined Member Loaded Axially
  Confined Member Loaded Axially
Effects of Shear Force and Initial Deflection on Buckling
  Bailey's Folds
  Influence of Shear Force on Bending
  Influence on Initial Dips on Folding
    Experimental
    Initial Dips in the Appalachians
    Simple Initial Deflections
    Crooked Members
    Interpretation of Theory
References Cited in Chapter 3
Other References on Folding
Recapitulation

3

Fold Trains in Appalachian Mountains

Introduction

Bailey Willis contributed in similar measure to our understanding of folding as G. K. Gilbert did to our understanding of laccoliths. For a good reason: Gilbert urged Willis to investigate, not only the forms of the folds in the Appalachians, but also the processes that produced the folds (*9*).

In the early 1890's, Willis studied parts of the belt of folded Paleozoic rocks that forms the Appalachian Mountains between New York and Alabama. The folds of the Appalachians are on a vast scale and their forms have been etched into the landscape by erosion, so that the folds are clearly outlined by valleys in soft strata and ridges in hard strata (Fig. 3.1). Consequently, the gross forms of the folds were recognized by early geologists and the forms became known as a type example: "Appalachian Structure" (*8*). The term "Appalachian Structure" in those days, much as it does today, implied strata compressed into a bundle of long, narrow folds, usually parallel to each other, and sometimes overturned or overthrusted (*8*).

The strata were *compressed* to form the folds, according to early investigators. Apparently Sir James Hall, in 1788, was one of the first geologists to recognize that lateral compression of stratified rocks is a major cause of folds. According to Hall, the idea occurred to him as he was examining folds in the cliffs of the coast of Berwickshire, Scotland (*3*):

> I reckoned sixteen bendings in the course of about six miles each of the largest size and reaching from top to bottom of the cliffs, their curvatures being alternatively concave and convex upwards.... These strata seem to have been originally deposited in a position nearly horizontal.... In the year 1788 it occurred to me that this peculiar conformation might be accounted for by supposing that these strata, originally lying flat...had been urged when in a soft but tough and ductile state by a powerful force acting

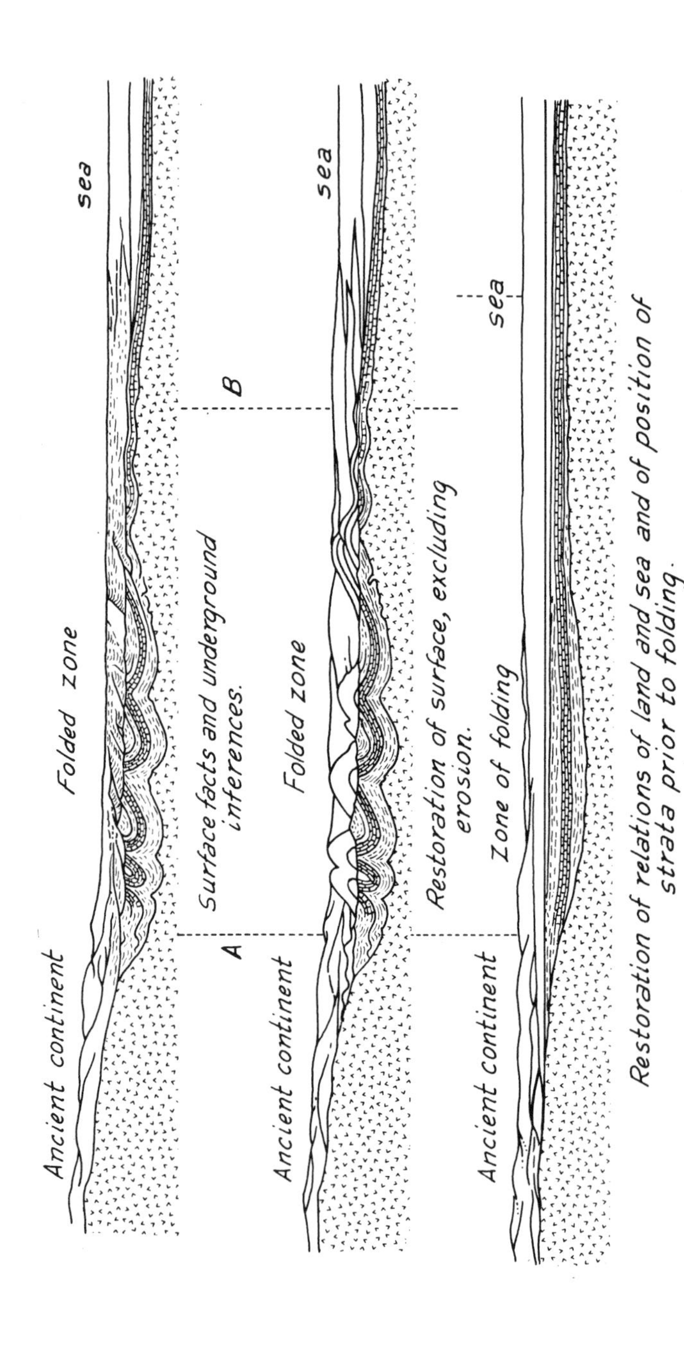

Figure 3.1. Willis' concept of the events that led to the folds in the Appalachian Mountains.

> horizontally.... I conceived that, two opposite extremities of each bed being made to approach, the intervening substance could only dispose of itself in a succession of folds, which might assume a set of parallel curves alternatively convex and concave towards the center of the earth....

Hall demonstrated experimentally his concept of the formation of folds (*3*):

> ...Several pieces of cloth, some linen, some woolen, were spread upon a table, one above the other, each piece representing a single stratum; a door (which happened to be off the hinges) was then laid above the mass, and being loaded with weights, confined it under considerable pressure; two boards being next applied vertically to the ends of the stratified mass were forced towards each other by repeated blows of a mallet applied horizontally. The consequence was that the extremities were brought nearer to each other, the heavy door was gradually raised, and the strata were constrained to assume folds bent up and down....

Thus, prior to the time of Willis, the concept of fold implied flexures and contortions of layers and other datum planes in rocks, usually caused by lateral compression. Also, "Appalachian Structure" implies a belt of anticlines and synclines and a few faults, all caused by lateral compression. The amplitudes of the folds of the typical belt of folds diminish from one edge of the belt to the other, presumably diminishing with increasing distance from the source of the push that produced the folds (*8*).

Willis demonstrated, however, that this simple picture of the structures in the Appalachians disregards many of the important features of the structures. By means of a research program combining experimentation, field work, and analysis, he discovered significant relations between details of the stratigraphy of the folded Paleozoic rocks and the folds themselves. First, he described the types of structures in the Appalachians and he related the structures and the stratigraphy. Then, he experimentally buckled various materials in the laboratory and proposed laws that seemed to govern the experiments and to provide important insights into the natural processes that formed the folds in the Appalachians. We will follow certain parts of his investigation and we will extend his theoretical analysis by using beam and plate theory.

Folds of the Appalachians (*8*)

The Appalachian Mountains are about 900 miles long and 50 to 125 miles wide. On the eastern side of the mountains are largely crystalline rocks, composed of metamorphic and plutonic rocks (Fig. 3.1). Toward the west of the present mountains was the continent, which was covered by a shallow sea. Between the continental margin and the crystalline rocks was the geosyncline into which clastic, chemical, and organic sediments accumulated in great thicknesses. The sediments in the geosyncline ultimately were strongly compressed to produce the "Appalachian Structure" (*8*).

Willis separated the Appalachian province into four districts according to structural types: open folds, closed folds, folds and faults, and folds and schistosity. Open and closed folds are characterized in Fig. 3.2. Anticlines and synclines of the

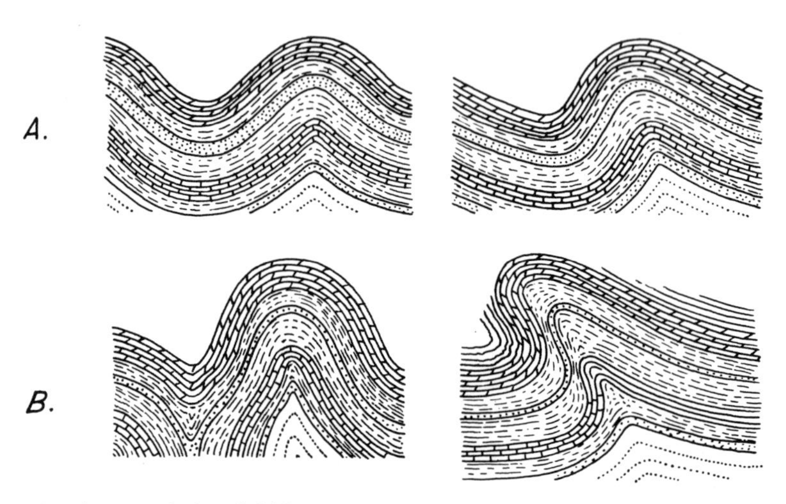

Figure 3.2. Open and closed folds.
A. Anticlines of the open variety are wide open so that further shortening would be possible by further bending, without the necessity of thickening or thinning of the strata.
B. Closed folds cannot be further shortened without squeezing of the strata. (After Willis, ref. *8*.)

open variety are wide open so that further shortening would be possible by further bending, without the necessity of thickening or thinning of the strata. Those of the closed variety cannot become further shortened without squeezing of the strata.

The district of open folding in the Appalachian Mountains extends from southern New York State to the southern part of West Virginia and from Blue Ridge on the east to the Allegheny front on the west. According to Willis, the typical fold in the district is the open anticline, tens of miles long, a mile or two wide, sweeping in a gentle curve from north to south. The exposed limbs of the open folds are approximately parallel, but they diverge in some places and converge in others, so that the fold axes rise out of the earth in some places and plunge into the earth in others. They are much like waves on the ocean: Some waves are considerably larger than their neighbors and smaller waves are superimposed on large ones (Fig. 3.3).

The early concept of the folds in the Appalachians was that the larger folds are

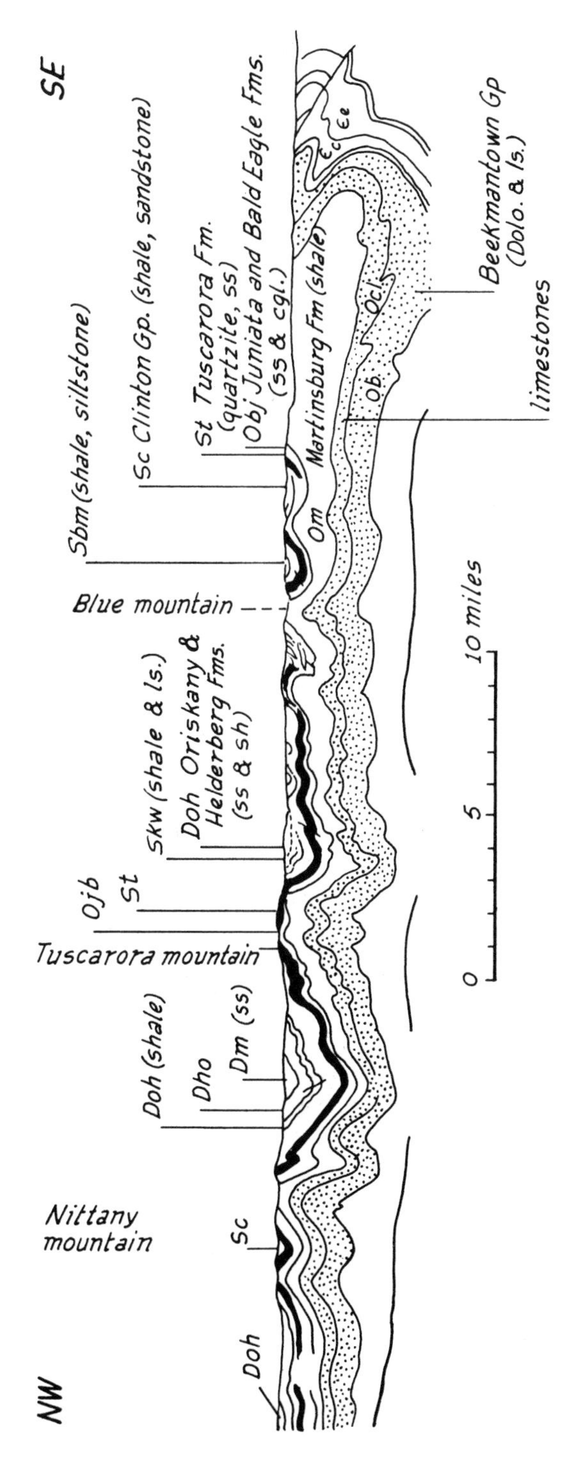

Figure 3.3. Folds in the central Appalachian Mountains of Pennsylvania (from geologic map of Pennsylvania by Carlyle Gray et al., 1960, Pennsylvania Geological Survey).

in the east and the smaller ones are in the west, so that, presumably, the strata were pushed from east to west. Willis noted, however, that perhaps the largest fold in Pennsylvania is the Nittany arch, on the western edge of the zone of pronounced open folds, near the Allegheny front. Thus the folds do not decrease in size in an orderly manner from east to west. Rather, they are more appressed at their extreme eastern limit than at their western limit.

The thickness of strata that folded in the district of open folds ranged from about 18,000 to 27,000 feet, according to Willis (*8*). At the base of the section are limestones and dolomites and above are sandstones and shales (Fig. 3.3). Structurally, stated Willis, the rocks fall into two major categories: The carbonates form one unit and the shale and sandstone form the other. The carbonates are massive, and they behaved as a rigid unit during folding. The sandstones and shales are relatively thinly bedded. Even though individual beds might have been thick and hard, the entire unit of sandstones and shales resisted folding as a mass of flexible beds. Thus Willis seems to have introduced the concept of the *strut member*: Competent beds or groups of beds in a folded sequence control the type of size of folds.

The problems

Some questions arise from this brief description of the structures and the stratigraphy of the belt of open folds in the Appalachians: Folds in the belt of open folds lie side by side and their appression generally decreases from east to west. That is, the folds in the west are more open than those in the east. Why? Can this behavior be explained in terms of a driving force that was applied at the eastern edge of the stack of layers and that was transferred toward the west partly through the relatively competent members, partly through the relatively incompetent substrata?

Why is one fold, distant from the supposed origin of the applied load in the southeast, larger than its neighbors? Such is the relation of Nittany arch to smaller folds southeast of it. What condition caused giant Nittany arch to form where it did?

According to Willis, over the entire Appalachian belt there is gross similarity of fold patterns, which implies unity of cause. Yet, the effects are varied in detail, which implies dissimilar conditions. The conditions that set the stage for folding were the result of sedimentation and downwarping of the mass of sediment. The hypothesis that Willis presented and tested is that "...circumstances of sedimentation determined conditions which afterward controlled the place and type of deformation and influenced the size and relations of individual structures" (*8*, p. 230). The circumstances of sedimentation that he refers to are weight of the overburden and initial dips of the strata, that is, incipient arches formed during sedimentation and compaction.

In the following pages we will examine these three questions in detail, by studying Willis' experiments and analyses and by further developing beam and plate theory.

Willis' theoretical analysis of the Appalachian folds

Early students of Appalachian structure recognized that the folds in the Appalachian belt generally diminish in magnitude from southeast to northwest. They believed that the folds were formed by a force applied at the southeast edge of the strata within the Appalachian geosyncline, and that the force transmitted through the beds diminished in magnitude toward the northwest. The term applied to their concept was "unbalanced forces," which implied that the force applied in the southeast was not balanced by an equal and opposite force applied in the northwest. This misnomer, unfortunately, has been critized so severely that I, and perhaps other students, had the impression that the concept itself is invalid. We will show, however, that a train of waves whose amplitudes diminish away from an applied load can, in fact, be predicted with slight modification of the theory that we have developed already.

Willis envisioned the regional conditions that gave rise to the Appalachian structure approximately as follows (*8*, p. 238–241). The behavior of the earth's crust is similar to that of a thin shell, a few miles thick, resting upon and grading downward into another, softer shell. Thus, the crust is relatively rigid and the underlying shell is relatively soft.

The part of the crust that folded in the Appalachians is stratified, which greatly aided folding. A massive layer is much more difficult to bend than a stack of thin layers with the same thickness, just as a thick beam is more difficult to bend than the same wood sawed into boards (*8*).

The strata that folded in the Appalachians ranged in thickness from about five miles in the southeast to about two miles in the northwest (*8*). These thicknesses are great, indeed, but they are small relative to the width of the folded belt, which is about 75 miles wide in central Pennsylvania (see Fig. 3.1). The thickness of the pile of folded strata was less than $\frac{1}{15}$ of the width of the folded belt, so that the strata might be modeled with a plate 15 inches wide and tapering in thickness from one inch to about 0.4 inch. But the pile of strata was much more flexible than our model envisions. The pile was divided into thousands of relatively rigid layers separated by relatively weaker layers. The pile, therefore, bent not as a simple stratum but individually and irregularly. In resisting bending, the pile of strata was similar, not to a thick piece of cardboard, but to a tablet of paper.

According to Willis, the deforming force which folded the strata in the Appalachians was compressive and tangential to the earth's circumference. The result was shortening of the crust. He suggested that the deforming force may have been applied by buttresses of crystalline rock at each edge of the stack of layers that moved toward each other, or perhaps one buttress stayed put and the other approached it. In either case, the loads were transmitted from one buttress to the other through the layers and the substratum. It is the mode of this transmission that controlled the gross type of deformation. The mode was determined by the boundary conditions of the stack of layers, and it is these conditions that we can investigate experimentally and theoretically.

The significance of different boundary conditions can be illustrated with experiments. Figure 3.4A shows a thin metal strip that has been bent by equal loads applied at its ends. A push at one end is transmitted to the other end. If a strip is placed on the surface of a layer of easily deformed gelatin, however, a push at one end of the strip is not transmitted to the other end. Instead, the load is distributed throughout the gelatin and is ultimately balanced by the reaction of the container of the gelatin. This latter behavior is, presumably, what earlier investigators referred to as "unbalanced forces."

This is about as far as Willis carried his theoretical analysis of the transfer of

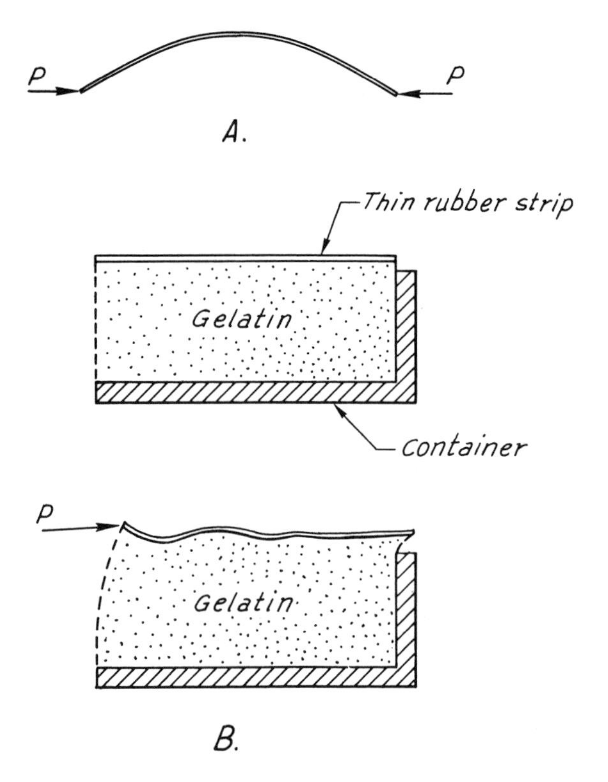

Figure 3.4. Deflection of thin, elastic strips under different conditions of confinement.

A. Thin, elastic strip buckles into a half-wave when equal and opposite loads are applied to its ends.

B. Thin rubber strip on soft, elastic, gelatin substratum. A push at one end of the strip is not transmitted to the other end. Instead, the load is distributed throughout the gelatin and is ultimately balanced by the reaction of the container.

forces through a layered medium resting on a soft substratum. It provides the general boundary conditions with which we can formulate a theory of the gross form of the folds in the Appalachians. Figure 3.5A shows an idealized fold train such as the one we see in the Appalachians. The strata rest upon a massive, soft substratum, in accord with Willis' concept of the situation in the Appalachians. We will further simplify Willis' concept by a fold train in an elastic plate, with modulus B, which

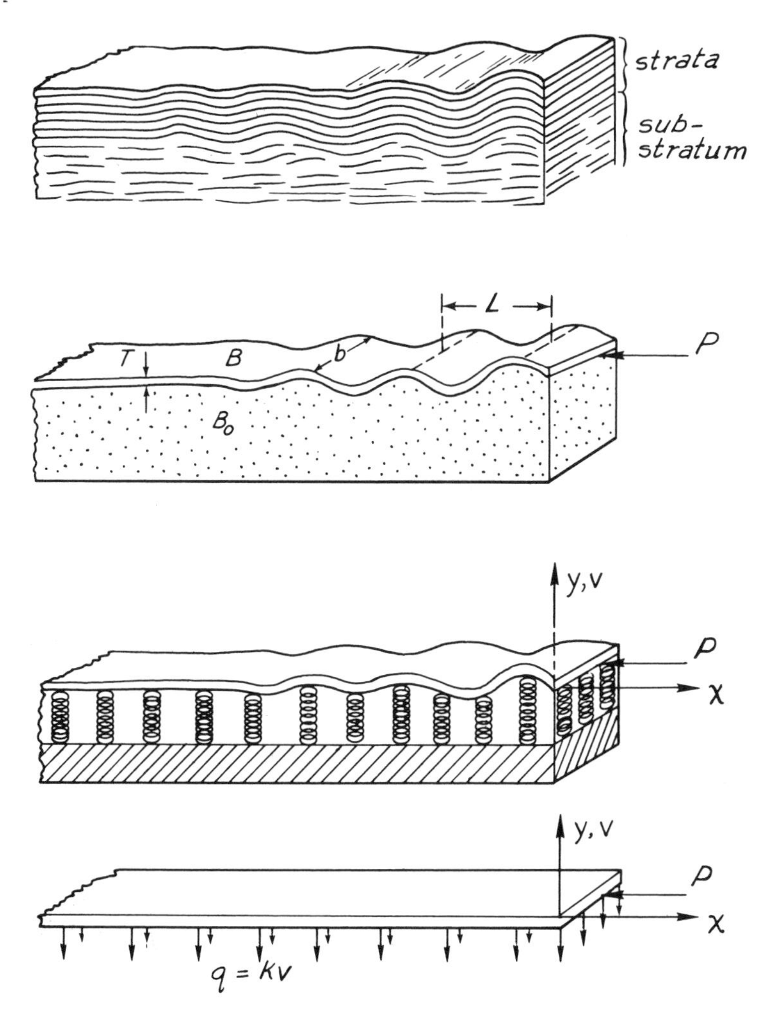

Figure 3.5. Fold train. Stages of abstraction leading to free-body diagram of member.
A. Idealization of fold train such as we see in Appalachians.
B. Single member on elastic substratum.
C. Single member on battery of springs.
D. Free-body diagram of member.

rests on an elastic substratum, with modulus B_0 (Fig. 3.5B). As we have shown already, when we studied forms of beds arched by laccoliths, the single elastic plate could be replaced by a multilayer of several elastic plates, representing several strata, but the extra work involved in analyzing the bending of multilayered sequences does not yield a proportionate amount of new information. Finally, for our mathematical analysis, we will replace the elastic substratum with a battery of identical springs of stiffness k so that the amount of deflection, v, is proportional to the load per unit of length, q, applied to the springs by the plate:

$$q = kv.$$

Mathematical expression for the form of the fold trains

There are two quantities we can derive from our analysis. We can determine the magnitude of the axial load, P, required to initiate buckling of the member and we can determine the form of the buckles.

The essential difference between the present problem and the problem of describing the form of a laccolith is that the axial load introduces a third term in the basic differential equation for the deflection of the member.

Unconfined Member Loaded Axially.—The differential equation will be somewhat complicated by the additional term, so we will begin by solving the differential equation for the deflection of an unconfined elastic member subjected to an axial load (Fig. 3.6A).

The moment that resists bending of a member is, according to eq. (2.22),

$$M_R = BI\left(\frac{d^2v}{dx^2}\right).$$

Here v is the deflection of the neutral plane of the member in the y-direction, and B is the elastic modulus, which is E for a beam or, as will be shown subsequently, $B = E/(1 - \nu^2)$ for a plate, where ν (nu) is Poisson's ratio. The bending moment given by eq. (2.22) is the member's elastic resistance to bending.

The driving moment, M_D, which the resisting moment balances is

$$M_D = -Pv,$$

as we now proceed to show. We assume that P, the axial load, acts in the x-direction, even when the member deflects (Fig. 3.6A). Figure 3.6B shows a free-body diagram of part of the buckled member. To be in equilibrium, the forces and moments on any section of the member must satisfy the equations of equilibrium. Therefore,

$$\left[\sum F_x = 0\right] \qquad P - N = 0, \quad \text{or} \quad N = +P,$$

where N is the force normal to the section of the beam. Also,

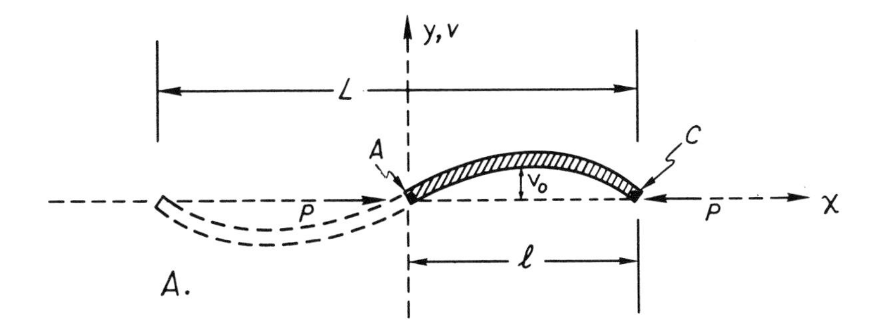

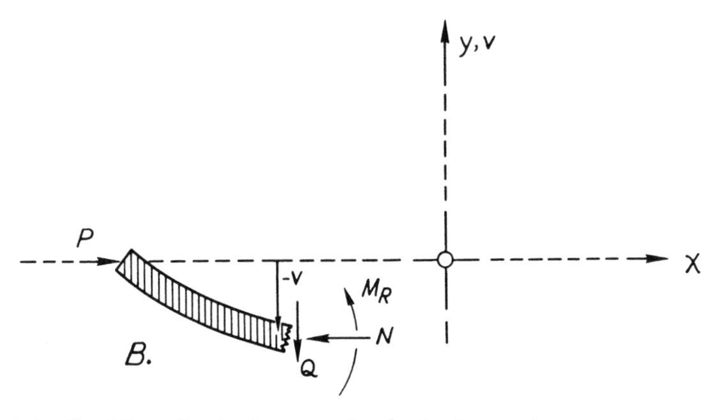

Figure 3.6. Buckling of a single, unconfined, elastic member.

A. P is the axial load, assumed to be parallel to the x-axis even when the member deflects. Here v_0 is the amplitude, l is the length, and L is the wavelength of the buckled member.

B. Free-body diagram of part of buckled member.

$$\left[\sum F_y = 0\right] \qquad Q = 0.$$

$$\left[\sum M_{\text{D}} = 0\right] \qquad M_{\text{D}} + Nv - Q(0) = 0, \quad \text{or} \quad M_{\text{D}} = -Nv = -Pv.$$

Thus we have derived the expression for the driving moment. Note that the shear force on any section cut parallel to the y-axis is zero. This is because the applied load, P, is in the x-direction.

The driving and resisting moments must be equal for equilibrium, so that $M_{\text{R}} - M_{\text{D}} = 0$,

$$\frac{d^2v}{dx^2} + \frac{Pv}{BI} = 0. \tag{3.1}$$

Equation (3.1) is called a homogeneous, second-order, differential equation with constant coefficients. It is *homogeneous* because it contains no constant terms or terms involving x, other than derivatives. Examples of nonhomogeneous equations are:

$$\frac{d^2v}{dx^2} + \frac{Pv}{BI} = k$$

and

$$\frac{d^2v}{dx^2} + \frac{Pv}{BI} = k + ne^x,$$

where k and n are constants. Equation (3.1) is of *second order* because the largest derivative is a second derivative. Finally, it has *constant coefficients* because the quantities preceding all terms containing v and derivatives of v are constants.

This type of equation can be solved by substituting the trial solution (*6*)

$$v = e^{mx}, \tag{3.2}$$

where m is some constant, to be determined by substituting the trial solution into the differential equation, eq. (3.1). Thus,

$$\frac{d^2v}{dx^2} = m^2e^{mx},$$

and eq. (3.1) becomes

$$m^2e^{mx} + \frac{Pe^{mx}}{BI} = 0.$$

Dividing through by e^{mx} and solving for m,

$$m = \pm\left(\frac{-P}{BI}\right)^{1/2}. \tag{3.3}$$

Thus if m is given by eq. (3.3), eq. (3.2) is a solution (*6*) to the differential equation. Substituting eq. (3.3) into eq. (3.2), the solution of eq. (3.1) is

$$v = \exp\left[ix\left(\frac{P}{BI}\right)^{1/2} + C_5\right] + \exp\left[-ix\left(\frac{P}{BI}\right)^{1/2} + C_6\right],$$

where $i = (-1)^{1/2}$, C_5 and C_6 are arbitrary constants, and $\exp(x) = e^x$.

The solution of eq. (3.1) can be rewritten

$$v = C_1 \exp\left[ix\left(\frac{P}{BI}\right)^{1/2}\right] + C_2 \exp\left[-ix\left(\frac{P}{BI}\right)^{1/2}\right], \tag{3.4}$$

because $\exp(x + C_5) = C_1 \exp(x)$, where C_5 and C_1 are arbitrary constants.

This solution satisfies the differential equation, as can be verified by substituting it into the differential equation.

Now trigonometric functions can be written in terms of complex functions:

$$\sin x = \frac{1}{2i}(e^{ix} - e^{-ix}),$$

$$\cos x = \tfrac{1}{2}(e^{ix} + e^{-ix}),$$

$$e^{ix} = \cos x + i \sin x,$$

and

$$e^{-ix} = \cos x - i \sin x,$$

where $i = (-1)^{1/2}$.

Thus, eq. (3.4) can be written

$$v = C_1[\cos(rx) + i \sin(rx)] + C_2[\cos(rx) - i \sin(rx)],$$

as the general solution, where $r^2 = (P/BI)$, or

$$v = (C_1 + C_2)\cos(rx) + (C_1 - C_2)i \sin(rx),$$

or

$$v = C_3 \cos\left(x\left[\frac{P}{BI}\right]^{1/2}\right) + C_4 \sin\left(x\left[\frac{P}{BI}\right]^{1/2}\right), \tag{3.5}$$

where $C_3 = C_1 + C_2$ and $C_4 = (C_1 - C_2)i$ are constants.

Equation (3.5) expresses the form of an unconfined, buckled member; the member will buckle into a sinusoidal curve, and its exact form can be determined by evaluating the arbitrary constants in eq. (3.5).

Exercises

Find the general solution to the following differential equation:

$$\frac{d^2v}{dx^2} + a\frac{dv}{dx} + bv = 0:$$

1. Where $b = a^2/4$.

   Ans: $v = [C_1 + C_2x]\exp\left(-\frac{ax}{2}\right)$.

2. Where $a^2 = b$.

   Ans. $v = e^{-(ax/2)}\left(C_1 \cos\left[x\left(\frac{3b}{4}\right)^{1/2}\right] + C_2 \sin\left[x\left(\frac{3b}{4}\right)^{1/2}\right]\right)$.

To evaluate the arbitrary constants in the general solution, eq. (3.5), assume that the origin of coordinates is at the left-hand end of the bent member, point A in

Fig. 3.6A. Then C_3 in eq. (3.5) must be zero, because $v = 0$ at $x = 0$ for this coordinate system. Thus,

$$v = C_3 \cos(xr) + C_4 \sin(xr),$$

where $r^2 = P/BI$. At $x = 0$, $v = 0$, so that

$$v = 0 = C_3 \cos(0) + 0,$$

and C_3 must equal zero.

At the center of the member shown in Fig. 3.6A, that is, at $x = l/2$, the deflection is a maximum, v_0, so that

$$v = v_0 = C_4 \sin\left(\frac{rl}{2}\right).$$

But $(rl/2)$, according to Fig. 3.6A, must be $\pi/2$, because at $x = l$, $(rl) = \pi$ and $\sin(\pi) = 0 = v$. Therefore,

$$v_0 = C_4 \sin\left(\frac{\pi}{2}\right) = C_4,$$

and the equation for the deflection curve of the member shown in Fig. 3.6A is

$$v = v_0 \sin(rx). \tag{3.6}$$

We can write the equation for the displacement of the member in a form that is completely dependent upon parameters shown in Fig. 3.7A,

$$v = v_0 \sin\left(\frac{n\pi x}{l}\right), \tag{3.7}$$

where l is the length of the member and n is twice the number of wavelengths, L, in the member. That is, n is the *mode* of buckling. For the first mode, the one shown in Fig. 3.7A, $n = 1$ and

$$v = v_0 \sin\left(\frac{\pi x}{l}\right). \tag{3.8}$$

By comparing eqs. (3.6) and (3.8), it is apparent that for the first mode

$$r = \frac{\pi}{l}.$$

But we already defined r as $(P/BI)^{1/2}$, where P is the axial load. Therefore, the critical load for the first mode of buckling is

$$P_{1\ \mathrm{crit}} = BI\left(\frac{\pi}{l}\right)^2.$$

The second mode of buckling is shown in Fig. 3.7B. For the second mode

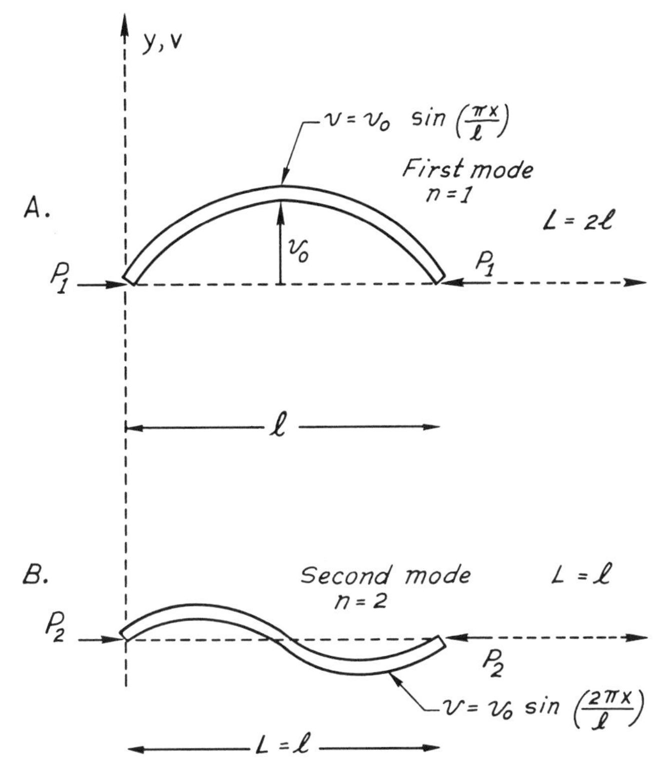

Figure 3.7. First and second buckling modes of a member.
A. First mode. The wavelength of buckling is twice the length of the member.
B. Second mode. The wavelength and length of the member are equal.

$n = 2$, so that

$$v = v_0 \sin\left(\frac{2\pi x}{l}\right),$$

where, as before, l is the length of the member. Proceeding as before,

$$r = \frac{2\pi}{l} = \left(\frac{P_2}{BI}\right)^{1/2},$$

so that the critical load for the second mode is:

$$P_{2\,\mathrm{crit}} = BI\left(\frac{2\pi}{l}\right)^2.$$

For the third mode,

$$P_{3\,\mathrm{crit}} = BI\left(\frac{3\pi}{l}\right)^2,$$

and so on for higher modes.

Now, what do these equations mean? They indicate the loads required to hold a member of length l in different modes of buckling. We have stated nothing about how the members attain their configurations and nothing about which mode is most likely. If the member is initially straight and the axial load is gradually increased from a small value to the critical value of the first mode, we would expect the member to buckle into the first mode. This follows from our analysis because, to maintain modes of higher order, greater loads are required. In fact, we have shown that the critical loads increase in proportion to the square of the order of the buckling mode.

But if we cause the member to buckle into a higher-order mode, say by applying temporary lateral constraints to the member, the load necessary to maintain that mode after the temporary constraints are removed is the critical load, either P_1 or P_2 or $\cdots P_n$, depending upon the order of the buckle (Fig. 3.8).

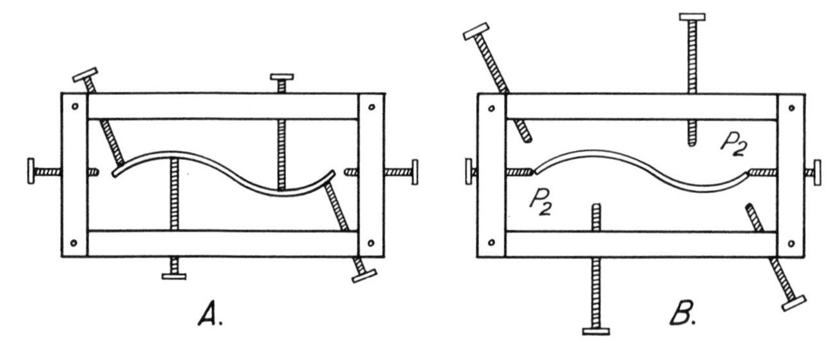

Figure 3.8. A way to produce second-order buckles in an unconfined member.

A. We can cause the member to buckle into second mode by applying temporary lateral supports with screws.

B. The load necessary to maintain the second mode by the end screws, after the temporary constraints are released, is the critical load, P_2, for the second mode.

The load necessary for buckling of a member, therefore, depends upon the dimensions and the elastic properties of the member, and upon the order of the mode of buckling. In general, if the unconfined, elastic member is subjected to only an axial load, the critical load is equal to that of the first mode, $P_{1\,\mathrm{crit}}$,

$$P_{1\,\mathrm{crit}} = BI\left(\frac{\pi}{l}\right)^2 = BI\left(\frac{2\pi}{L}\right)^2 = \frac{Bb}{3}\left(\frac{\pi}{L}\right)^2 T^3, \tag{3.9}$$

where $I = bT^3/12$, and the wavelength (Fig. 3.7) is

$$L = 2l, \tag{3.10}$$

because this mode requires the smallest axial load.

The greater the length of the member, the longer will be the wavelength and the smaller will be the load necessary to initiate buckling. If the applied load is greater than or equal to the critical load, the member will buckle. That is, it will deflect laterally, *although the amount of deflection is indeterminate.* But if the load is less than the critical one the member will shorten without buckling.

If we solve eq. (3.9) for L,

$$L_e = 2\pi\left(\frac{BI}{P_1}\right)^{1/2}, \qquad \textit{Euler Wavelength} \tag{3.11}$$

we derive the wavelength of the unconfined, elastic member that will buckle under an axial load, P_1. This wavelength is the so-called Euler wavelength, the significance of which will become apparent in following pages. For the present it is sufficient to note that the critical load, P_1, for buckling decreases as the length of the member increases.

Confined Member Loaded Axially.—We have shown that an unconfined member will buckle into a sinusoidal form if axial loads of sufficient magnitude are applied to it. Our solution describes a form that corresponds quite well to the form of the unconfined, thin, metal strip shown in Fig. 3.3. Usually a mode consisting of one-half a wave forms in the buckled, unconfined member, because that mode requires the least axial load. The unconfined member, therefore, is a singularly poor approximation to the real folding situation in the Appalachian Mountains, where there is a train of waves. Our approximation will improve with the addition of a confining medium to our model.

We can derive the differential equation which governs the deflection of a competent elastic member in an incompetent elastic medium by considering forces and moments acting on a very small element of the member. Figure 3.9A shows a member supporting axial loads, P; shear forces, Q; bending moments, M; and a distributed load, q. The distributed load is due to the elastic medium, the elastic constant of which is k, so that the magnitude of the distributed load is proportional to the deflection of the member. The positive directions of the loads and moments are indicated by the directions of the arrows in the figure.

A small element with a length δx and a thickness equal to that of the member, T, has been isolated in Fig. 3.9C. Acting on the element are equal and opposite axial loads, P, on each end. The lines of action of the two axial loads are in the x-direction at both ends, so that they are parallel. However, they are not colinear, that is, they act along different parallel lines so that they bend the member. The bending moment acting on one end of the element is M and that acting on the other is $M + \delta M$. Similarly the shear forces are Q and $Q + \delta Q$.

If we sum moments around the center of the right-hand end of the element, the moments exerted by P and $Q + \delta Q$ on the right-hand end vanish, because their lever arms are zero. Summing moments,

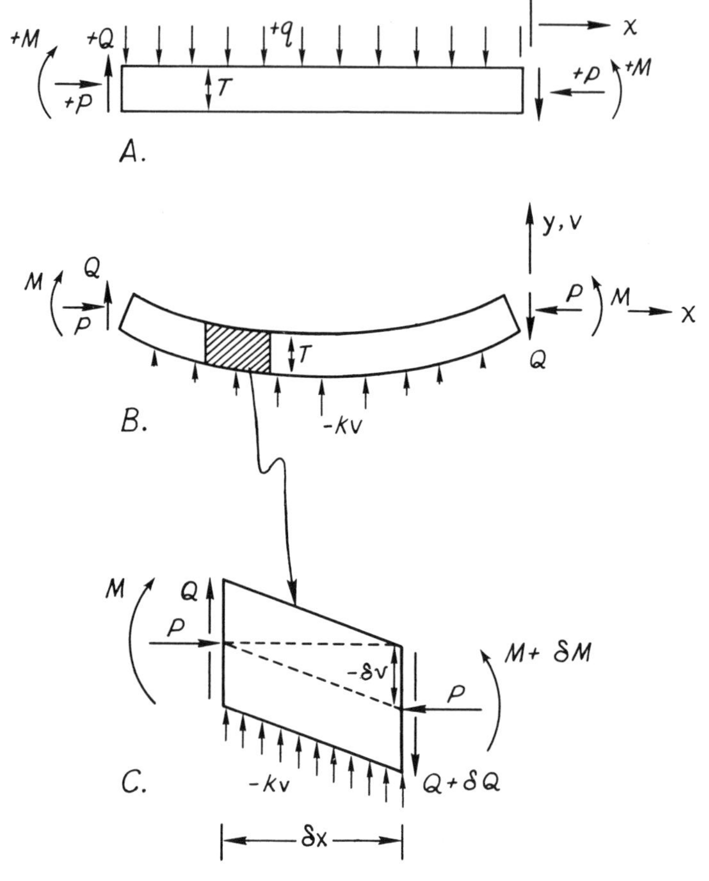

Figure 3.9. Free-body diagrams of member supporting axial loads.
A. Sign convention for loads and moments.
B. Bent member.
C. Forces and moments acting on small element of bent member.

$$\left[\sum M = 0\right], \qquad -M + (M + \delta M) - Q\,\delta x + P(\delta v) = 0,$$

or

$$\delta M - Q\,\delta x + P\,\delta v = 0. \tag{3.12}$$

If we divide by δx and take limits as δx approaches zero, we have a differential equation,

$$\frac{dM}{dx} - Q + P\frac{dv}{dx} = 0,$$

or

$$Q = P\frac{dv}{dx} + \frac{dM}{dx}. \tag{3.13}$$

This expression for Q is different from that derived earlier, eq. (2.8), because the axial load introduces a second term, $P\,dv/dx$.

By summing forces in the y-direction we find that

$$\left[\sum F_y = 0\right], \qquad Q - (Q + \delta Q) - k(v)\,\delta x = 0,$$

or

$$\frac{dQ}{dx} = -kv, \tag{3.14}$$

which is equivalent to the relation between the shear force and the distributed load derived earlier for a member that was not supporting an axial load, eq. (2.9). The axial load is in the x-direction, so that it does not affect the shear force, Q, in the y-direction.

Combining eqs. (3.13) and (3.14) we derive

$$\frac{d^2M}{dx^2} + P\frac{d^2v}{dx^2} = -kv. \tag{3.15}$$

However,

$$M = BI\frac{d^2v}{dx^2}, \tag{3.16}$$

so that the differential equation we have derived for deflection of the member is:

$$BI\left(\frac{d^4v}{dx^4}\right) + P\left(\frac{d^2v}{dx^2}\right) + kv = 0. \tag{3.17}$$

We can solve eq. (3.17) in the same way that we have solved equations of similar form, that is, by trying a solution of the form

$$v = e^{mx}, \tag{3.18}$$

where m is a constant. Substituting the trial solution into the differential equation, eq. (3.17), the latter becomes

$$(BIm^4 + Pm^2 + k)e^{mx} = 0. \tag{3.19}$$

For finite values of m and x, the terms in parentheses must sum to zero:

$$BIm^4 + Pm^2 + k = 0. \tag{3.20}$$

Let $n = m^2$. Then eq. (3.20) becomes

$$BIn^2 + Pn + k = 0,$$

or

$$n = \frac{-P \pm (P^2 - 4BIk)^{1/2}}{2BI} = -\left(\frac{P}{2BI}\right) \pm \left[\left(\frac{P}{2BI}\right)^2 - \left(\frac{k}{BI}\right)\right]^{1/2}$$

and

$$m = \pm (n)^{1/2} = \pm\left[-\left(\frac{P}{2BI}\right) \pm \left\{\left(\frac{P}{2BI}\right)^2 - \left(\frac{k}{BI}\right)\right\}^{1/2}\right]^{1/2},$$

so that the four roots of eq. (3.20) are

$$\left.\begin{aligned} m_1 &= -m_2 = \left[-\left(\frac{P}{2BI}\right) + \left\{\left(\frac{P}{2BI}\right)^2 - \left(\frac{k}{BI}\right)\right\}^{1/2}\right]^{1/2} \\ m_3 &= -m_4 = \left[-\left(\frac{P}{2BI}\right) - \left\{\left(\frac{P}{2BI}\right)^2 - \left(\frac{k}{BI}\right)\right\}^{1/2}\right]^{1/2} \end{aligned}\right\} \tag{3.21}$$

The general solution of eq. (3.17) is, therefore,

$$v = A_1 e^{m_1 x} + A_2 e^{m_2 x} + A_3 e^{m_3 x} + A_4 e^{m_4 x}, \tag{3.22}$$

where $A_1 \cdots A_4$ are arbitrary constants.

To proceed further with the solution for the deflection curve of a member confined by an elastic medium and loaded axially, we need to evaluate further the roots, $m_1, \ldots, m_4$. Equation (3.22) and the four roots defined by eqs. (3.21) are valid whether the axial load, P, is tensile or compressive. Only compressive axial loads can produce folding, however, so that P in eq. (3.21) is positive, in accord with our sign convention. The constant, k, is always positive also, so that the value of the term

$$\left[\left(\frac{P}{2BI}\right)^2 - \left(\frac{k}{BI}\right)\right]^{1/2}, \tag{3.23}$$

under the second radical sign, always will be less than the value of the term

$$\left(\frac{P}{2BI}\right),$$

under the first radical sign. Therefore, we can write eqs. (3.21) in the form

$$\left.\begin{aligned} m_1 &= -m_2 = i\left[\left(\frac{P}{2BI}\right) - \left\{\left(\frac{P}{2BI}\right)^2 - \left(\frac{k}{BI}\right)\right\}^{1/2}\right]^{1/2}, \\ m_3 &= -m_4 = i\left[\left(\frac{P}{2BI}\right) + \left\{\left(\frac{P}{2BI}\right)^2 - \left(\frac{k}{BI}\right)\right\}^{1/2}\right]^{1/2}. \end{aligned}\right\} \tag{3.24}$$

By inspection of eqs. (3.24) we can see that the deflection curve is described by a series of sine and cosine terms. We showed earlier, eqs. (3.4) and (3.5), that an exponential containing an imaginary number can be expressed as the sum of a sine and a cosine. To proceed further with the solution, we need to distinguish three cases, depending upon the magnitude of the axial load. According to eqs. (3.24), if the term under the second radical sign is negative, that is, if

$$\left(\frac{P}{2BI}\right)^2 - \left(\frac{k}{BI}\right) < 0,$$

or if

$$P < 2(kBI)^{1/2},$$

there is a second imaginary number under the first radical sign:

$$\left.\begin{aligned} m_1 &= -m_2 = i\left[\left(\frac{P}{2BI}\right) - i\left\{\left(\frac{k}{BI}\right) - \left(\frac{P}{2BI}\right)^2\right\}^{1/2}\right]^{1/2}, \\ m_3 &= -m_4 = i\left[\left(\frac{P}{2BI}\right) + i\left\{\left(\frac{k}{BI}\right) - \left(\frac{P}{2BI}\right)^2\right\}^{1/2}\right]^{1/2}. \end{aligned}\right\} \tag{3.25}$$

Equations (3.25) can be expressed in a different, more useful form by factoring (*4*). Thus, let

$$\left.\begin{aligned} \alpha &= \left[\left(\frac{k}{4BI}\right)^{1/2} + \left(\frac{P}{4BI}\right)\right]^{1/2}, \\ &\text{and} \\ \beta &= \left[\left(\frac{k}{4BI}\right)^{1/2} - \left(\frac{P}{4BI}\right)\right]^{1/2}, \end{aligned}\right\} \tag{3.26}$$

then, for example,

$$m_1 = i(\alpha - i\beta). \tag{3.27}$$

The statement that the first of eqs. (3.25) and eq. (3.27) are equivalent can be verified by squaring and then taking the square root of eq. (3.27). Substituting eqs. (3.26) into (3.27) and squaring

$$m_1 = i(\alpha - i\beta) = i\left[\left(\left\{\frac{k}{4BI}\right\}^{1/2} + \frac{P}{4BI}\right)^{1/2} - i\left(\left\{\frac{k}{4BI}\right\}^{1/2} - \frac{P}{4BI}\right)^{1/2}\right]$$

$$(m_1)^2 = -\left[\left(\frac{k}{4BI}\right)^{1/2} + \frac{P}{4BI} - 2i\left(\frac{k}{4BI} - \left(\frac{P}{4BI}\right)^2\right)^{1/2} - \left(\frac{k}{4BI}\right)^{1/2} + \frac{P}{4BI}\right],$$

or

$$(m_1)^2 = -\left[\frac{P}{2BI} - i\left(\frac{k}{BI} - \left(\frac{P}{2BI}\right)^2\right)^{1/2}\right],$$

and

$$m_1 = i\left[\frac{P}{2BI} - i\left(\frac{k}{BI} - \left(\frac{P}{2BI}\right)^2\right)^{1/2}\right]^{1/2},$$

which is identical to the first of eqs. (3.25).

Similarly (*4*),

$$\left.\begin{aligned} m_1 &= -m_2 = i(\alpha - i\beta) = (\beta + i\alpha), \\ &\text{and} \\ m_3 &= -m_4 = i(\alpha + i\beta) = (i\alpha - \beta), \end{aligned}\right\} \tag{3.28}$$

where

$$\alpha = \left[\left(\frac{k}{4BI}\right)^{1/2} + \frac{P}{4BI}\right]^{1/2},$$

and

$$\beta = \left[\left(\frac{k}{4BI}\right)^{1/2} - \frac{P}{4BI}\right]^{1/2}.$$

The four roots of the solution, therefore, are given by eqs. (3.28). With the notation indicated there, the general solution, eq. (3.22), becomes (*4*)

$$v = (C_1 e^{\beta x} + C_2 e^{-\beta x}) \cos(\alpha x) + (C_3 e^{\beta x} + C_4 e^{-\beta x}) \sin(\alpha x). \tag{3.29}$$

The arbitrary constants, $C_1, \ldots, C_4$, can be evaluated as follows: According to the coordinate system shown in Fig. 3.10A, the amplitudes of the folds are zero as x approaches negative infinity. Constants C_2 and C_4, therefore, must be zero and the equation for the deflection curve becomes:

$$v = e^{\beta x}(C_1 \cos \alpha x + C_3 \sin \alpha x).$$

The two remaining constants, C_1 and C_3, can be evaluated for particular loading conditions of the end of the member (for examples, see ref. *4*, p. 128). We have already assumed that the applied load is axial, so that the loading condition is as shown in Fig. 3.10A. If the deflection at the end of the member is zero, that is, if at

$$x = 0, \qquad v = 0,$$

then $C_1 = 0$ and the form of the deflection curve is a damped sine function,

$$v = C_3 e^{\beta x} \sin(\alpha x). \tag{3.30}$$

The sine curve, eq. (3.30), is shown in Fig. 3.10B. The wavelengths of the theoretical folds are constant but the amplitudes diminish with increasing distances from the applied load.

If the axial load is less than $2(kBI)^{1/2}$, therefore, the wave train is a damped sinusoidal function. Suppose, however, that $P = 2(kBI)^{1/2}$. According to eqs. (3.26), if $P = 2(kBI)^{1/2}$,

$$\alpha = \left[\left(\frac{k}{4BI}\right)^{1/2} + \frac{P}{4BI}\right]^{1/2} = \left(\frac{k}{BI}\right)^{1/4},$$

and $\beta = 0$, so that the expression for the wave train, eq. (3.30), becomes

$$v = C_3 \sin(\alpha x), \qquad \text{if} \quad P = 2(kBI)^{1/2}. \tag{3.31}$$

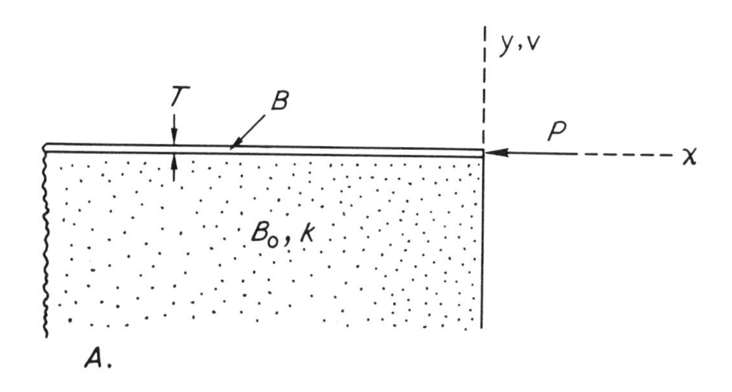

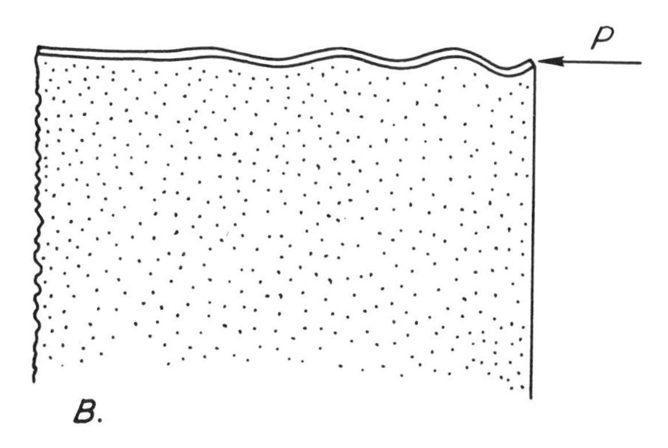

Figure 3.10. Buckling of member on elastic medium by load applied to one end of member.

A. Situation before buckling.

B. Deflection shape of buckled member. Deflection is in form of damped sine curve.

Thus the wave train appears to become a series of undamped sine waves, that is, a series of waves each having the same amplitude. Actually, though, an axial load of such magnitude cannot be supported by a member of semi-infinite length (*4*). The critical load for deflection of a semi-infinite member is one-half that value. It is $P_{crit} = (kBI)^{1/2}$. We will derive this relation in following paragraphs when we investigate the effect of vertical forces on folding.

Effects of shear force and initial deflection on buckling

The maximum load the member can support without buckling is $(kBI)^{1/2}$, apparently, and under such loading conditions eq. (3.30) should describe the deflection of the member. There are, however, serious discrepancies between the forms of the idealized folds described by eq. (3.30), and shown in Fig. 3.10B, and the actual folds in the Appalachians. Our solution predicts a wave train of folds of constant wavelength, but the amplitudes of the folds diminish away from the place where the load is applied. Neither of these features is evident in cross-sections of folds in the Appalachians. For example, the cross-section shown in Fig. 3.3 was copied from a recent geologic map of Pennsylvania. As predicted by theory, the fold with the largest amplitude is, indeed, in the southeast, near the place where most investigators have assumed the driving force to have been applied. However, the larger waves have superimposed on them numerous smaller waves. Also, there are two large waves, which form Tuscarora Mountain and Nittany Mountain, distant from the applied load, where we would expect smaller folds.

Bailey Willis (*8*) discovered a reasonable explanation for these differences between actual and theoretical behaviors.

Bailey's Folds.—Bailey Willis proceeded to design experiments in order to extend his analysis of folding in the Appalachian Mountains beyond what he considered to be the safe limits of his theoretical analysis.

Willis selected experimental materials and an apparatus to fulfill as closely as possible the general boundary conditions that he recognized during his theoretical and field studies. For example, the beds must be thin enough to be flexible, the substratum must be soft, the load must be transmitted to the strata primarily from one end, and the materials must be moderately coherent and firm.

The basic substance Willis selected for the strata of his experiments was beeswax, and the properties of the beeswax were changed by adding softeners and hardeners. Plaster of Paris was added to harden and turpentine was added to soften the beeswax. The wax was melted, other substances were stirred in, and the mixture was cast into layers of various thicknesses. Apparently part of the turpentine in the wax evaporated during tests lasting a few days, causing the properties of the material to change. In order to prevent this problem, yet make the wax flow easily, Willis added less turpentine to the wax and loaded the top of the stack of wax layers with shot (Fig. 3.11). Wax flows more easily when it is under pressure.

The apparatus Willis used for his experiments was a massive oak box provided with a piston that was advanced with a screw. The piston compressed the layers of wax edgewise (Fig. 3.11).

Figure 3.12 shows some of Bailey's folds. The original wax strata were about 39 inches long and 5 inches wide and the total thickness of the layers was $3\frac{1}{2}$ inches. The upper 13 layers were composed of a mixture of one part of wax and one part of plaster, so that they were relatively hard. Their brittleness is indicated by the fractures at the hinge of the fold. The bottom layer was soft, consisting of two parts of wax and one part of turpentine. The top layer was $\frac{1}{2}$ inch thick, layers 2 to 13 were $\frac{1}{3}$ inch

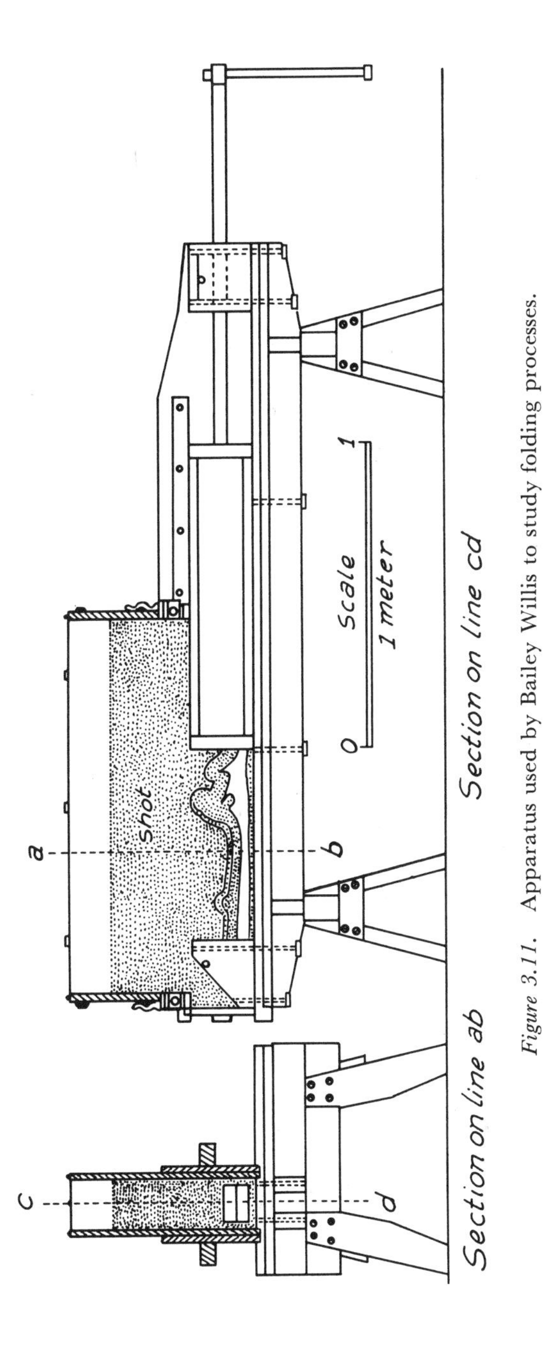

Figure 3.11. Apparatus used by Bailey Willis to study folding processes.

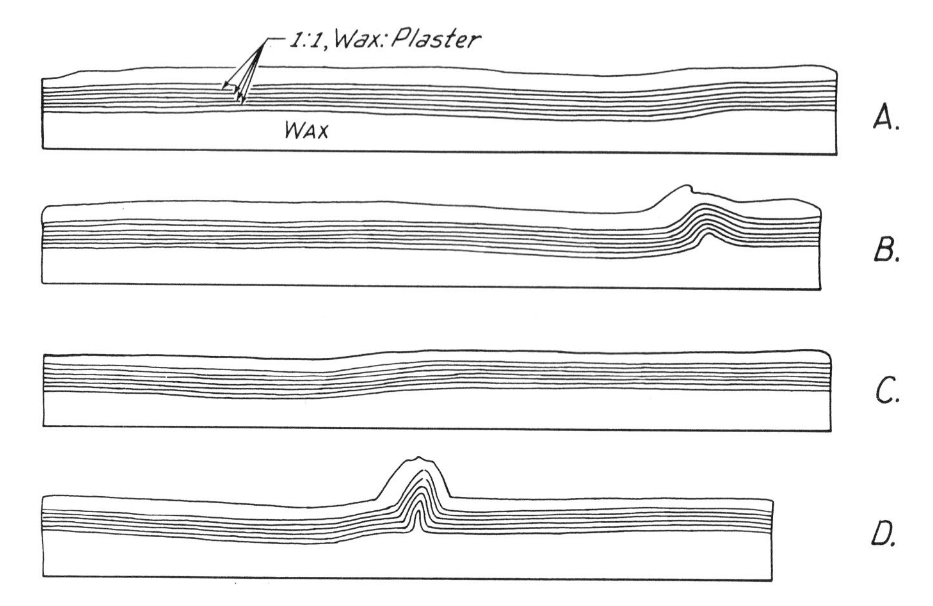

Figure 3.12. Localization of buckle by position of initial warp in experimental layers. Original length of layers was one meter, original thickness was 8.8 centimeters. Thin, hard layers consisted of one part of wax and one part of plaster. Thick, soft layer was mixture of two parts wax and one part turpentine. Top of layers was loaded by lead shot.

A. Initial warp near driving piston, at right end of layers.
B. Buckle formed at position of warp.
C. Initial warp about one-half meter from driving piston.
D. Buckle formed at position of initial warp. (Drawn from photographs by Willis, ref. *8*, pl. LXXIX.)

thick, and the bottom layer was 2½ inches thick. The softest material used in the experiments behaved much like butter. Most of the soft materials used for the base, however, were stiffer. Under the conditions of the experiments, the load imposed on the base caused it to flow relatively easily. The base usually yielded under the load of the piston by raising the overlying strata and the load imposed by the shot. The harder materials used for the thin strata were granular. They could be carved with a knife and dented by a dull blow, but they shattered under the impact of a sharp blow. The strata did not thicken readily, as did the softer base, when a thrust was applied (*8*).

The upper surface of the model shown in Fig. 3.11 was loaded with 1100 pounds of shot, roughly equivalent to 5½ pounds per square inch. Then the model was compressed in four steps, one of which is shown in Fig. 3.12B. The piston was advanced at a rate of about one inch in five minutes.

Willis experienced several difficulties during the initial experiments. The pressures developed in compressing the models proved to be so great that several boxes burst. Also, the friction between the wax layers and the sides of the box was disturbingly large. Another difficulty that he encountered during early experiments was that the layers persistently buckled immediately in front of the piston, whereas the remainder of the layer was undisturbed. In Willis' words: "I began to be embarrassed to explain the constant occurrence of an anticline at the end of the model nearest the piston. . . ." (*8*, p. 243).

When Willis began his experiments he supposed that the entire stack of hard strata would bow upward, just as a steel strap bends if loads are applied to its edges: ". . . It was thought that they would bend over the longest possible chord with the minimum curvature, and that the crown of the curve would be near the middle of the entire length . . ." (*8*, p. 247). Instead, the strata usually buckled near the piston. He first thought the position of the buckle was caused by friction between the layers and the sides of the squeezing box. However, the effect continued after the friction was removed. He then considered friction between beds, rates of loading, and a host of other possible causes of the peculiarity, but the possible explanations were rejected one by one as he tested them experimentally.

Eventually, Willis decided that the major control of the position of the first buckle in his experiments was the existence of irregularities in dips of the strata. The strata could not be cast perfectly horizontally. Also, the dips near the piston sometimes changed slightly because the layers were lifted slightly as the piston advanced. Other times the soft base thickened and lifted the strata at early stages of compression. In Willis' words (*8*): "The conclusion reached through the experiments was that, when a firm but flexible stratum transmits pressure, it tends to yield to bending along any line where there is a slight change in dip, and this deviation may be due to an initial uplift or depression; the fold is further developed by that component of the thrust which is diverted by the inclined strata."

Willis' explanation of the peculiar occurrence of buckles near the driving piston in his apparatus probably is his most important contribution to our understanding of folding processes. He discovered that initial dips of bedding can control the positions and the relative growths of folds and he recognized the marked effect a small vertical force can have on the initiation of a fold near the point of application of the force. Both of these concepts will be developed mathematically with beam theory.

Influence of Shear Force on Bending.—Willis observed two causes for shear forces being transmitted from the driving piston of his apparatus to the stiff layers of his experimental strata. One cause was minor vertical movements of the driving piston because of imperfections in his apparatus. The other cause is related to thickening of the soft substratum underlying the stiff experimental strata. When the piston was advanced, the soft substratum shortened. As the substratum shortened it necessarily thickened, in order to preserve its original volume. Thus the substratum lifted the overlying strata as the piston was advanced. The lifting produced drag between the strata and the piston, inducing a shear force. Both causes of shear forces are results of relative movement between piston and strata.

Natural analogs of the types of behavior Willis noted in his squeeze box probably are common. Probably the most frequent occurrence of a buckling condition involving both vertical and horizontal loads applied to one end of a stack of layers is along faults. A fault severs the mechanical tie between beds across it so that, for example, bending moments might not be readily transferred across it. Beds on each side do not bend the same amount; instead, a crack opens. Friction between the two sides of the fault, however, can cause high loads acting tangentially to the surface of the fault to be transferred from beds on one side of a fault to beds on the other side.

Figure 3.13 shows drag flexures on two different scales in a roadcut east of St.

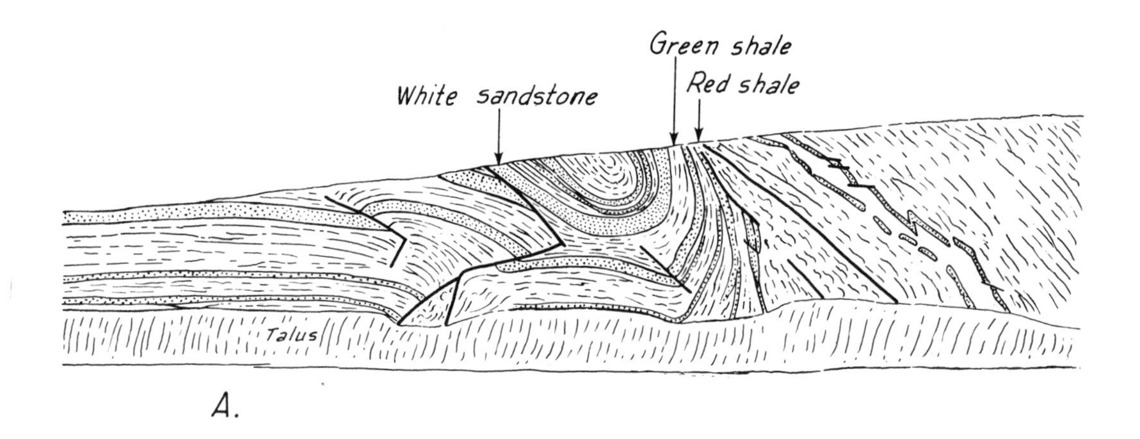

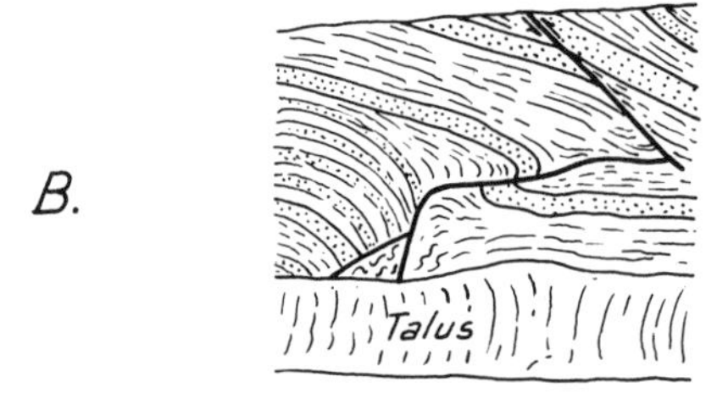

Figure 3.13. Drag flexure along high-angle thrust fault. Roadcut along south side of State Highway 17, east of St. George, Utah.

A. Beds on left are nearly flat-lying. Near center of photograph they bend downward and then sharply upward near the fault. Beds on right dip steeply toward right.

B. Traces of some of bedding planes, showing outline of major and minor drag flexures.

George, Utah, about one-half mile east of the intersection of State Highway 17 and Interstate Highway 15. The light-toned beds are sandstone and the dark-toned beds are shale and siltstone. The largest, most obvious flexure is bounded on the right by a fault, which dips steeply toward the right and which separates the flexure and the contorted beds that generally dip toward the right (Fig. 3.13A). The beds in the flexure have been turned sharply upward near the fault, presumably implying that there was drag along the fault as the rock on the left moved downward relative to the rock on the right. The flexure, however, could not have formed entirely by drag, because the bent beds swoop down below the general level of the strata on the left. Presumably there must have been an axial load on the beds as well as a load induced by drag along the fault; this is in accord with the interpretation that the fault is of the high-angle thrust variety. The combined effects of an axial load and a shear force could conceivably cause the beds to drag upwards as well as buckle.

The smaller flexures, not clearly visible in Fig. 3.13A but sketched in Fig. 3.13B, also were related to faults. They are adjacent to secondary faults that bound a wedge of rock driven toward the core of the larger flexure.

The types of flexure shown in Fig. 3.13 and those produced near the driving piston in Bailey's squeeze box can be predicted with the theory we have developed.

Let us represent the situation in its simplest form: a single competent elastic member in a soft, elastic medium (Fig. 3.14B). The problem would be the same if the soft, elastic medium were only on one side of the competent member; the resistance of the medium would simply be halved. Also, multiple beds could be considered, but, as stated earlier, the analysis would simply be slightly complicated without providing new information. The situation can be further idealized to the model shown in Fig. 3.14C, which shows an elastic member on a foundation of elastic springs with a spring constant, k. Thus the model consists of a member of semi-infinite length resting on a bed of springs and loaded by a shear force, Q_0, and an axial load, P_0, at one end.

Thus the basic problem is the same as that solved in the previous example, where we studied a train of folds. The differential equation for both problems is:

$$BI\left(\frac{d^4v}{dx^4}\right) + P\left(\frac{d^2v}{dx^2}\right) + kv = 0. \tag{3.32}$$

The general solution we derived for this equation is [eq. (3.29)]:

$$v = (C_1e^{\beta x} + C_2e^{-\beta x}) \cos \alpha x + (C_3e^{\beta x} + C_4e^{-\beta x}) \sin \alpha x, \tag{3.33}$$

in which

$$\alpha = \left(\left[\frac{k}{4BI}\right]^{1/2} + \frac{P}{4BI}\right)^{1/2}, \tag{3.34}$$

$$\beta = \left(\left[\frac{k}{4BI}\right]^{1/2} - \frac{P}{4BI}\right)^{1/2}.$$

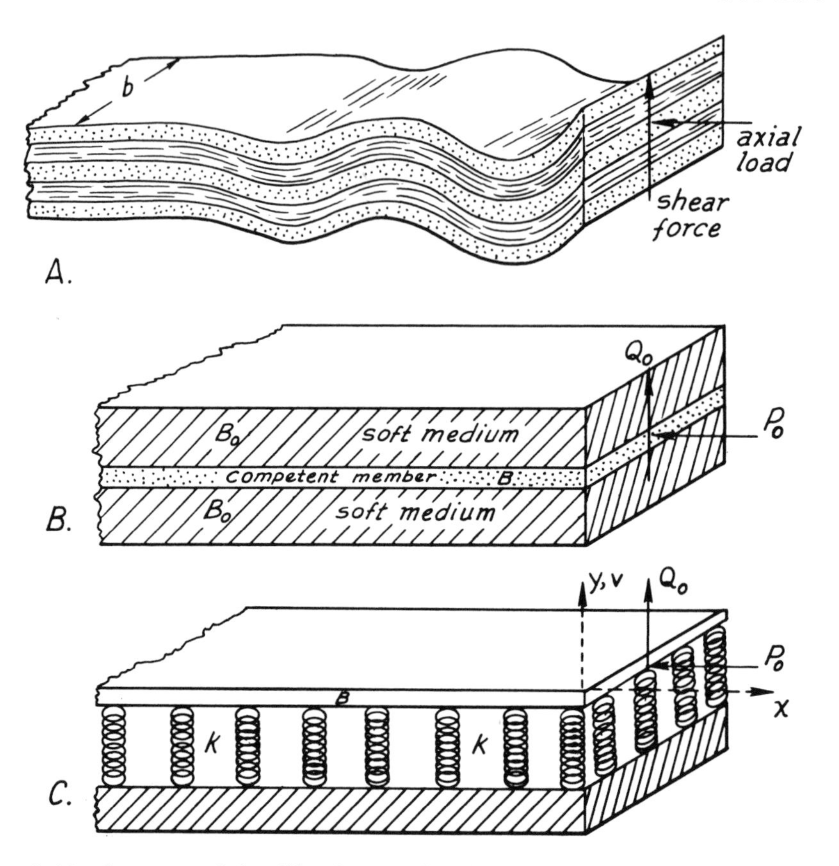

Figure 3.14. Sequence of simplifications used to derive expression for deflection of beds subjected to axial and shear forces.
A. Sequence of beds deflected into wave train.
B. Single competent member embedded in soft medium.
C. Single member on bed of springs.

The arbitrary constants, $C_1, \ldots, C_4$, can be evaluated with the boundary conditions illustrated in Fig. 3.14C:

$$\text{at} \quad x = 0, \qquad P = P_0, \quad Q = Q_0, \quad M = 0, \tag{3.35}$$

and

$$\text{as} \quad x \to -\infty, \qquad v \to 0.$$

The last of boundary conditions (3.35) is in accord with the field observation that the amplitudes of the flexures decrease with increasing distance from the fault, and it requires that constants C_2 and C_4 be zero. If they were not zero, the deflections would increase boundlessly with increasing distance from the end of the member where loads are applied.

The other two arbitrary constants, C_1 and C_3, can be evaluated with the other boundary conditions, (3.35). The bending moment and the shear force are defined as [eqs. (3.13) and (3.16)]:

$$M = BI\frac{d^2v}{dx^2}, \qquad \text{the bending moment,}$$

$$Q = BI\frac{d^3v}{dx^3} + P\frac{dv}{dx}, \qquad \text{the vertical shear force.}$$

Thus we can write four equations that relate deflection, slope, bending moment, and shear force, all involving the arbitrary constants, C_1 and C_3:

$$v = e^{\beta x}(C_1 \cos \alpha x + C_3 \sin \alpha x), \tag{3.36}$$

$$\frac{dv}{dx} = e^{\beta x}[C_1(\beta \cos \alpha x - \alpha \sin \alpha x) + C_3(\beta \sin \alpha x + \alpha \cos \alpha x)], \tag{3.37}$$

$$M = BI\frac{d^2v}{dx^2} = BIe^{\beta x}\{C_1[(\beta^2 - \alpha^2) \cos \alpha x - 2\alpha\beta \sin \alpha x] + C_3[(\beta^2 - \alpha^2) \sin \alpha x + 2\alpha\beta \cos \alpha x]\}, \tag{3.38}$$

and

$$Q = BI\frac{d^3v}{dx^3} + P\frac{dv}{dx},$$

in which

$$BI\frac{d^3v}{dx^3} = BIe^{\beta x}\{C_1[\alpha(\alpha^2 - 3\beta^2) \sin \alpha x + \beta(\beta^2 - 3\alpha^2) \cos \alpha x] + C_3\beta(\beta^2 - 3\alpha^2) \sin \alpha x - \alpha(\alpha^2 - 3\beta^2) \cos \alpha x]\}. \tag{3.39}$$

If we know any two of the quantities defined by eqs. (3.36)–(3.39) at some point in the member, the two arbitrary constants, C_1 and C_3, can be evaluated and we can determine all four of the quantities (3.36)–(3.39) throughout the member. In the present problem we know that no bending moment is applied to the end of the member, so that eq. (3.38) must be zero at $x = 0$:

$$M = 0 = C_1(\beta^2 - \alpha^2) + C_3(2\alpha\beta),$$

or

$$C_1 = -\left(\frac{2\alpha\beta}{\beta^2 - \alpha^2}\right)C_3. \tag{3.40}$$

Also, we know that at $x = 0$, the shear force on the member is the shear force applied across the fault: at $x = 0$,

$$Q = -Q_0. \tag{3.41}$$

Here Q_0 is the magnitude of the shear force that acts upwards on the right-hand end of the member. The convention we established in earlier pages was that a shear force is positive if it is acting downwards on that end. Thus the negative sign in eq. (3.41).

By substituting eq. (3.41) into eq. (3.39) we derive a second expression involving C_1 and C_3: at $x = 0$,

$$Q = -Q_0 = BI[C_1\beta(\beta^2 - 3\alpha^2) - C_3\alpha(\alpha^2 - 3\beta^2)] + P\frac{dv}{dx}$$

$$= BI[C_1\beta(\beta^2 - 3\alpha^2) - C_3\alpha(\alpha^2 - 3\beta^2)] + P(C_1\beta + C_3\alpha).$$

Dividing by $BI\alpha$,

$$\frac{-Q_0}{BI\alpha} = C_1\frac{\beta}{\alpha}(\beta^2 - 3\alpha^2) - C_3(\alpha^2 - 3\beta^2) + \frac{P}{\alpha BI}(C_1\beta + C_3\alpha). \tag{3.42}$$

However, according to eqs. (3.34),

$$\alpha^2 = \left(\frac{k}{4BI}\right)^{1/2} + \frac{P}{4BI}$$

and

$$\beta^2 = \left(\frac{k}{4BI}\right)^{1/2} - \frac{P}{4BI}$$

so that

$$\frac{P}{BI} = 2(\alpha^2 - \beta^2),$$

and eq. (3.42) becomes

$$\frac{-Q_0}{BI\alpha} = \frac{-\beta}{\alpha}C_1(\alpha^2 + \beta^2) + C_3(\alpha^2 + \beta^2). \tag{3.43}$$

To solve for C_3, we substitute eq. (3.40) for C_1 into eq. (3.43):

$$\frac{-Q_0}{BI\alpha} = C_3\left[\frac{2\beta^2}{\beta^2 - \alpha^2}(\alpha^2 + \beta^2) + (\alpha^2 + \beta^2)\right]$$

$$= C_3\frac{(\alpha^2 + \beta^2)(3\beta^2 - \alpha^2)}{\beta^2 - \alpha^2}$$

or

$$C_3 = -\frac{Q_0(\beta^2 - \alpha^2)}{BI\alpha(3\beta^2 - \alpha^2)(\alpha^2 + \beta^2)}. \tag{3.44}$$

Substituting eq. (3.44) into eq. (3.40),

$$C_1 = \frac{Q_0 2\alpha\beta}{BI\alpha(3\beta^2 - \alpha^2)(\alpha^2 + \beta^2)}. \tag{3.45}$$

The deflection of the member and the shear force and bending moment acting at any cross-section within the member can be derived by substituting eqs. (3.44) and (3.45) into eqs. (3.36)–(3.39). The deflection curve is defined by (*4*)

$$v = \frac{Q_0 e^{\beta x}}{(kBI)^{1/2}\alpha(3\beta^2 - \alpha^2)} [2\alpha\beta \cos \alpha x + (\alpha^2 - \beta^2) \sin \alpha x], \tag{3.46}$$

for the range $(x \leq 0)$, because $\alpha^2 + \beta^2 = (k/BI)^{1/2}$. According to eq. (3.46), the amount of deflection of the member is proportional to the shear force, Q_0, applied to the end of the member. Also, we could show that the slope of the center line of the member and the bending moment and the shear force acting at any section within the member are all proportional to the shear force applied at the end.

The deflection curve is of the damped sinusoidal type, in which the amount of deflection decreases very rapidly with increasing distance from the applied load (Fig. 3.15). The deflection curve in Fig. 3.15C resembles quite closely the flexure

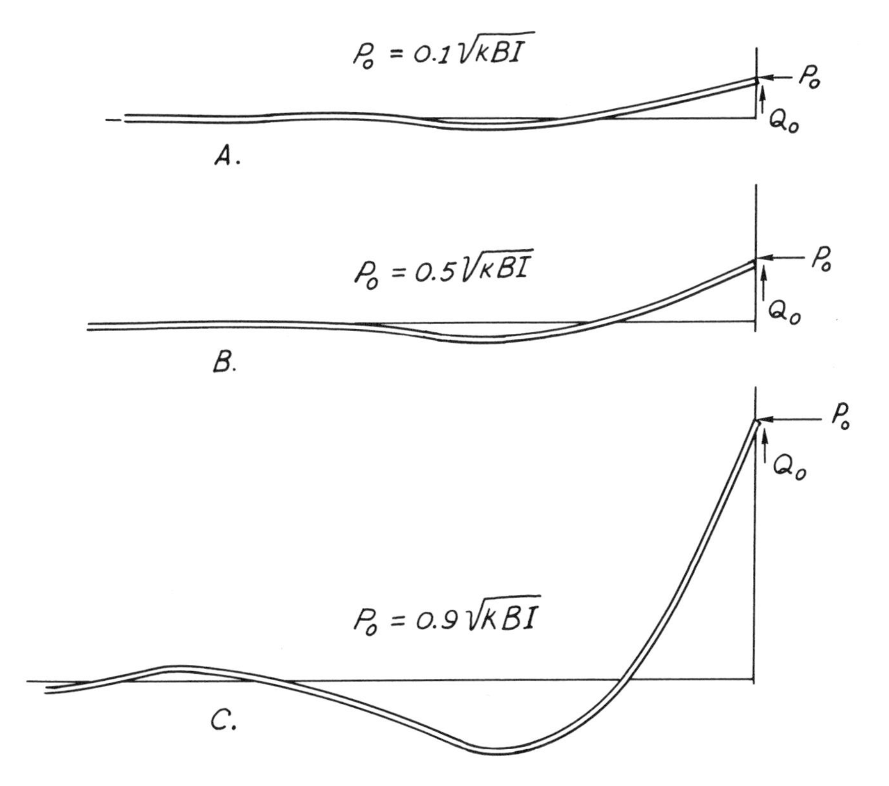

Figure 3.15. Deflection of member under action of constant shear force, but different axial loads.

next to the fault shown in Fig. 3.13. The ends of the members in the field and in the theoretical examples have been bent sharply upward and in both examples there is a distinct syncline, formed by downward bending of the member a short distance from the fault. Figure 3.15C shows the effect of a large axial load on the form of the flexure. As we will show very soon, the assumed axial load of 0.9 $(kBI)^{1/2}$ is nine-tenths of the load required to buckle the member without a shear force applied at the end. Figures 3.15A and 3.15B show flexures formed under conditions where the magnitude of the axial load is less. For one the axial load is one-tenth of the buckling load and for the other it is one-half. As the axial load approaches zero the syncline shifts slightly away from the point where the load is applied, the fault surface, and its amplitude becomes very small. Thus we can conclude that there was a relatively large horizontal load transferred across the fault in the natural situation shown in Fig. 3.13. If the horizontal load had been small, the syncline probably would have been invisible.

Now we can understand why Bailey Willis had difficulties producing folds anywhere but immediately in front of the piston of his squeezing apparatus. The slightest upward or downward force exerted by the piston during its advance against the layers of wax would cause a buckle to initiate and grow next to the piston. The contact between the piston and the wax layers behaved mechanically much like a vertical fault along which there is a small amount of vertical motion.

Equation (3.46), describing the deflection of a member by axial and shear forces applied at one of its ends, contains some very useful information that we have not tapped *(4)*: It is rewritten here for reference,

$$v = \frac{Q_0 e^{\beta x}}{(kBI)^{1/2}\alpha(3\beta^2 - \alpha^2)} [2\alpha\beta \cos \alpha x + (\alpha^2 - \beta^2) \sin \alpha x],$$

where

$$\alpha = \left(\left[\frac{k}{4BI}\right]^{1/2} + \frac{P}{4BI}\right)^{1/2},$$

and

$$\beta = \left(\left[\frac{k}{4BI}\right]^{1/2} - \frac{P}{4BI}\right)^{1/2}.$$

According to this equation, the deflection of the member appears to tend toward zero as the shear force, Q_0, becomes vanishingly small. It turns out, however, that this is not the entire story. The denominator contains a term,

$$3\beta^2 - \alpha^2,$$

which depends upon the axial load. If this term tends to zero at the same time the shear force becomes zero, the deflection becomes zero divided by zero, an indeterminate quantity. Also, if the shear force is finite, the deflection tends toward infinity as the denominator tends toward zero. Thus Hetenyi (*4*) calls the compressive axial load at which $3\beta^2 - \alpha^2$ approaches zero the *buckling load.* We can derive the buckling load by substituting the appropriate values of α and β into the term,

$$3\beta^2 - \alpha^2$$

and setting the result equal to zero. Thus,

$$3\left(\left[\frac{k}{4BI}\right]^{1/2} - \frac{P_{\text{crit}}}{4BI}\right) - \left(\left[\frac{k}{4BI}\right]^{1/2} + \frac{P_{\text{crit}}}{4BI}\right) = 0,$$

or

$$P_{\text{crit}} = (kBI)^{1/2}, \tag{3.47}$$

the buckling load for the semi-infinite member.

Thus the critical load for buckling of a perfectly straight, very long, elastic member in an elastic medium is given by eq. (3.47). If the member supports a shear force as well as an axial load, P_{crit} in eq. (3.47) is the axial load at which buckling becomes large. If there is no shear force, P_{crit} is the load at which buckling will begin, but the actual magnitude of the deflection is undefined.

If there is an appreciable bending moment transferred from one side of the fault to the other, that is, if a certain bending moment is applied to the end of the member we have been considering, our conclusions and derivations apply in a general way but there are slight differences. For example, if the ends of the beds on each side of the fault remained in contact with each other along the fault, the resulting structure would appear similar to a monocline. The drag produced by slippage along the fault would apply a bending moment to the beds so that the deflection would be approximately expressed by (*4*, p. 129):

$$v = \frac{-M_0}{2BI\alpha\beta} e^{-\beta x} \sin \alpha x \qquad (x \geq 0). \tag{3.48}$$

This equation can be derived by assuming that the boundary conditions are, at $x = 0$,

$$M = M_0,$$

$$v = 0,$$

$$Q = 0,$$

and as x becomes very large, $v \rightarrow 0$. Compressive axial loads are applied at each end of a long member.

Exercises

1. Derive eq. (3.48) by evaluating the constants in the general solution, eq. (3.33),

of the differential equation for the deflection of a member loaded axially (see ref. *4*, p. 129).

2. Show that the critical axial load for buckling of an infinitely long member in an elastic medium is twice that given by eq. (3.47).

Influence of Initial Dips on Folding.—*Experimental.*—Most of the experiments Bailey Willis made were focused on the problem of causes of initiation of buckling in his squeezing machine, but the ideas he formulated should apply in natural situations as well. As a result of his analysis of the experiments, Willis developed a concept that he named (*8*) "the law of competent structure." The concept is composed of two parts. First, initial dips in the strata determine places where folds will form. Second, for a flexure to form there must be competent beds to transmit the load, as through an arch. Thus Willis states that (*8*, p. 250): "If the thrust be not powerful enough to raise the load there will be no uplift, or if the layers be so plastic that they yield to the thrust by swelling, then the principal result of deformation is change of form other than by simple flexure, and it assumes some phase of flowing. This is incompetent structure." Our attention will be focused on the effect of initial dips in beds supporting axial loads. We will assume that the beds are sufficiently competent to buckle.

Willis made several experiments to check his ideas qualitatively. Two of them are shown in Fig. 3.12. According to his ideas, if the component of vertical shear between the piston and the beds is negligible, a flexure should begin to form where the change in dip in the beds is maximum. As is shown in Fig. 3.12, the position of the fold is determined by the position where the experimental strata bend downward sharply. Figures 3.12A and 3.12B show the initial dips in the strata. Succeeding photographs show folds that grew as the piston advanced from right to left.

Willis built several irregularities into the strata used in some of his experiments. The example shown in Fig. 3.16 shows typical results. A pronounced initial dip was preformed in the layers near the end of the apparatus, opposite to the piston. When the piston had advanced from right to left a distance of a few centimeters, a small syncline began to form near the piston, and anticlines developed near the center of the apparatus and at the place of the prominent initial dip. Thus, even though the initial dip was distant from the applied load, the initial dip influenced buckling sufficiently for a fold to form on it. The anticline in the center seems to have been localized by a slight change in dip of the layers, made visible in Fig. 3.16A by laying a straight-edge along the layers. The syncline that formed near the piston probably is a result of a shear force transferred from the piston to the layers, a shear force similar to the one we analyzed in the previous section.

The first nine layers in the model were composed of equal portions of wax and plaster mixed together. The lowest, thick layer was composed of two parts of wax and one part turpentine. The superincumbent load was 1100 pounds of shot.

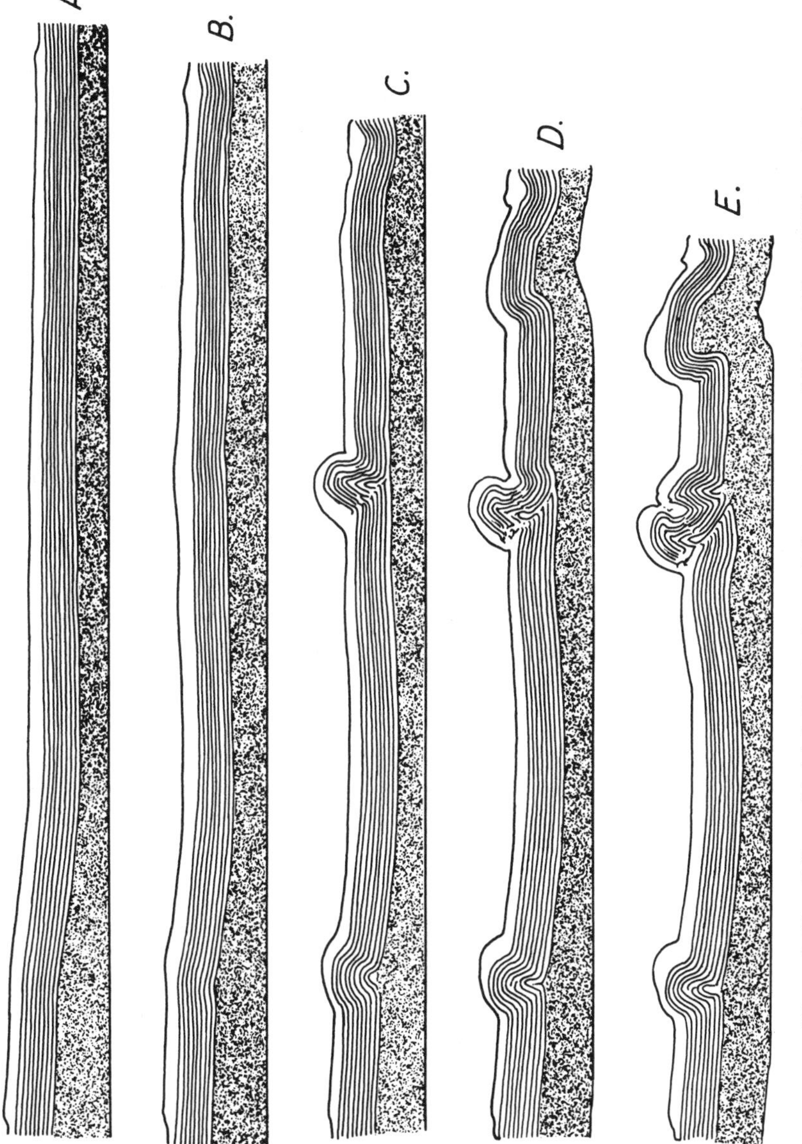

Figure 3.16. Effect of initial dips in determining positions of folds. (After Willis, ref. *8.*)

Initial dips in the Appalachians.—Cross-sections of the fold belt that forms the Appalachian Mountains, such as the one in Fig. 3.3, typically show large folds distant from the assumed source of the push from the southeast. For example, Nittany Mountain and Tuscarora Mountain in Fig. 3.6 are separated from the large recumbent syncline in the southeast by an area where folds are smaller and, therefore, the strata appear to be relatively horizontal. Such cross-sections are strikingly similar to cross-sections of some of Bailey's experimental folds, which were localized by initial dips, so that Willis searched for stratigraphic evidence of initial dips in the Appalachians. His search turned up rather convincing evidence that his concept explains the position of some of the folds in the Appalachians.

The upper diagram in Fig. 3.17 is Bailey's interpretation of initial dips in the direction normal to fold axes in central Pennsylvania. The lower diagram shows the folds. Willis noted that the gentle southeastward initial dips and one northwestward initial dip correspond with anticlinal folds and the steeper southeastward dips underlie synclinal folds. Thus the arches and troughs induced by unequal thicknesses of deposits correspond with anticlinal and synclinal folds. Bailey's "law of competent structure" seems to apply to folds in the Appalachian Mountains.

Simple initial deflections.—We will examine the bending of an initially deflected, elastic stratum to explain why a minor initial deflection can have a major effect on determining where a fold will form. The following discussion is based largely on lectures by S. N. Goodier (*2*) and a book by Timoshenko and Gere (*7*).

Thus far we have limited our theoretical analyses to members that are perfectly elastic and flawless, but all real members, including rock layers, are imperfect; for example, they all contain flaws and irregularities in thickness, and none of them is perfectly elastic or straight. The effects of irregularities in thickness have been analyzed by Timoshenko and Gere (*7*), to whom the reader is referred. Nonelastic behavior will be discussed in a subsequent chapter. Here we will consider only the effects of crookedness of members on buckling loads by assuming, first, that members contain simple deflections prior to loading, and, second, that members are extremely crooked.

If a member is submitted to the action of lateral loads only, that is, to the action of loads normal to its flat surface, a small initial curvature of the member has no appreciable effect on the bending. If there is an axial load on the member, however, the deflection will be markedly affected by an initial curvature of the member.

Suppose that the initial deflection of the neutral plane of the member is given by the equation, (Fig. 3.18),

$$v_0 = \delta_0 \sin\left(\frac{2\pi x}{L}\right). \tag{3.49}$$

The neutral plane of the member, therefore, is initially in the form of a sine curve, with a maximum amplitude, δ_0, at the mid-length of the member. If the member is

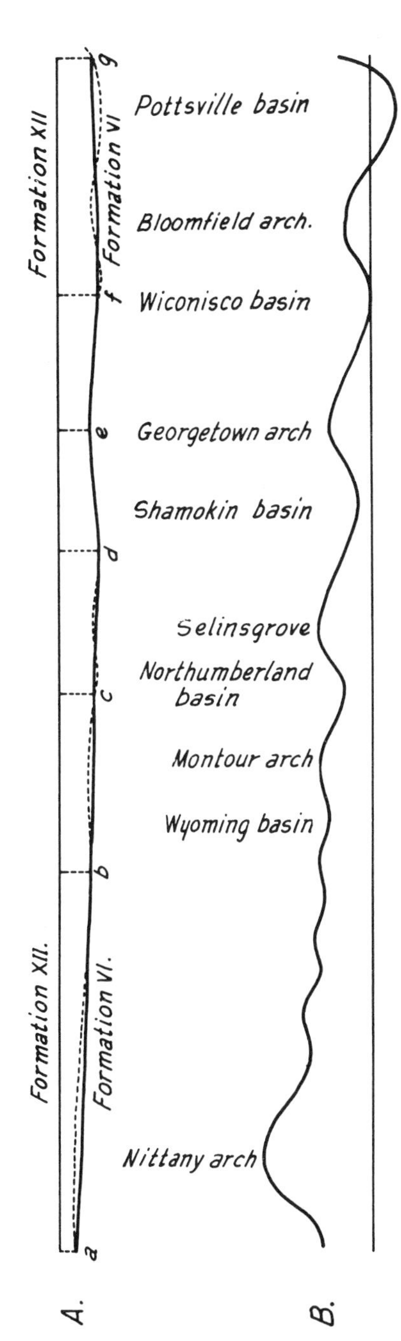

Figure 3.17. Initial dips in Appalachian Mountains (after Willis, ref. *8*).
A. Initial dips in "Formation VI," central Pennsylvania.
B. Positions of folds in "Formation VI."

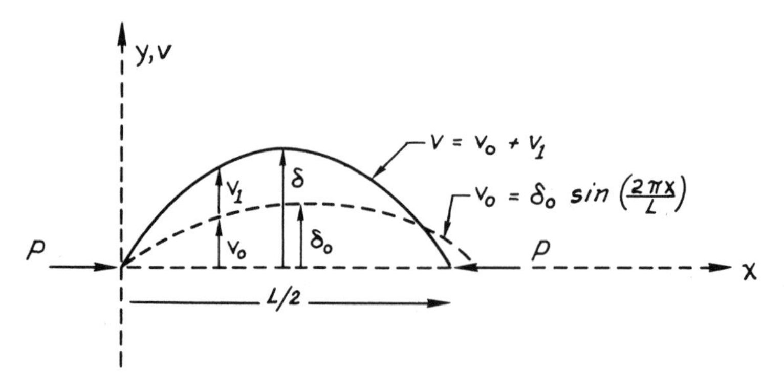

Figure 3.18. Further bending of an initially deflected member. The initial deflection, v_0, is sinusoidal, with a maximum of δ_0 at its center. Additional deflection, v_1, will result from an axial load. The total deflection v is $v_0 + v_1$. The maximum deflection, that at the center, is δ.

acted upon by an axial load, P, additional deflection, v_1, will result so that the total deflection is:

$$v = v_0 + v_1, \tag{3.50}$$

and the *driving*, bending moment at any cross-section of the member is

$$M = -P(v_0 + v_1) = -Pv.$$

The *resisting* moment is in the same form as before,

$$M_R = BI\frac{d^2v_1}{dx^2},$$

but we have replaced the total deflection, v, by the *additional* deflection, v_1, caused by the axial load. The reason for this change in variables is that we assumed that the member maintains its initial curvature when it is supporting no bending moments. Only the additional curvature,

$$\frac{d^2v_1}{dx^2},$$

is resisted by the rigidity of the member.

Thus the differential equation for the deflection of the member with initial curvature becomes

$$BI\frac{d^2v_1}{dx^2} + P(v_0 + v_1) = 0. \tag{3.51}$$

Combining eqs. (3.49) and (3.51),

$$\frac{d^2v_1}{dx^2} + \frac{P}{BI}v_1 = -\frac{P\delta_0}{BI}\sin\left(\frac{2\pi x}{L}\right). \tag{3.52}$$

Here L is the wavelength of the initial curvature.

Equation (3.52) is a nonhomogeneous, second-order differential equation with constant coefficients. It is nonhomogeneous because it contains a term that does not involve the dependent variable, v_1. We will solve it by finding a solution to the homogeneous part, that is, the part to the left of the equal sign, and then adding to this solution a particular solution that satisfies the entire equation (*6*). The homogeneous part of the differential equation, eq. (3.52), is

$$\frac{d^2v_1}{dx^2} + \frac{P}{BI}v_1 = 0,$$

the solution to which is, according to eq. (3.5),

$$v_1 = C_3\cos\left(x\left[\frac{P}{BI}\right]^{1/2}\right) + C_4\sin\left(x\left[\frac{P}{BI}\right]^{1/2}\right). \tag{3.53}$$

Now we have the general solution to the homogeneous part of differential equation (3.52). The problem is to find a *particular solution* that satisfies the nonhomogeneous part,

$$-\frac{P\delta_0}{BI}\sin\left(\frac{2\pi x}{L}\right),$$

as well as the homogeneous part. Try

$$v_1 = A\sin\left(\frac{2\pi x}{L}\right) \tag{3.54}$$

as a particular solution where A is a constant. To determine the value of A, substitute the particular solution, eq. (3.54), into the entire differential equation, eq. (3.52). Thus

$$\frac{d^2v_1}{dx^2} = -A\left(\frac{2\pi}{L}\right)^2\sin\left(\frac{2\pi x}{L}\right),$$

and

$$\frac{P}{BI}v_1 = \frac{PA}{BI}\sin\left(\frac{2\pi x}{L}\right),$$

so that the differential equation, eq. (3.52), becomes

$$-A\left(\frac{2\pi}{L}\right)^2\sin\left(\frac{2\pi x}{L}\right) + \frac{PA}{BI}\sin\left(\frac{2\pi x}{L}\right) = -\frac{P\delta_0}{BI}\sin\left(\frac{2\pi x}{L}\right),$$

or

$$A\left[\frac{P}{BI} - \left(\frac{2\pi}{L}\right)^2\right] = -\frac{P\delta_0}{BI}.$$

Dividing through by P/BI and solving for A,

$$A = \frac{\delta_0}{[(2\pi/L)^2(BI/P) - 1]}. \tag{3.55}$$

By substituting this value for A into eq. (3.54), and combining the resulting eq. (3.54) with eq. (3.53), we have the *general solution* to the nonhomogeneous differential equation, eq. (3.52):

$$v_1 = C_3 \cos\left(x\left[\frac{P}{BI}\right]^{1/2}\right) + C_4 \sin\left(x\left[\frac{P}{BI}\right]^{1/2}\right) + \frac{\delta_0}{[(2\pi/L)^2(BI/P) - 1]} \sin\left(\frac{2\pi x}{L}\right). \tag{3.56}$$

Now we will evaluate the arbitrary constants, C_3 and C_4, in terms of the boundary conditions:

$$\text{at} \quad x = 0, \qquad v_1 = 0,$$

and

$$\text{at} \quad x = \frac{L}{2}, \qquad v_1 = 0.$$

The boundary conditions indicate that the additional deflection is zero at each end of the member. That is, the ends are fixed in place. Consider, first, the conditions that $v_1 = 0$ at $x = 0$. The second and last terms in eq. (3.56) are zero if $x = 0$ but the first term is a constant, C_3. In order to satisfy the boundary condition, therefore, C_3 must be zero. Now consider the boundary condition that $v_1 = 0$ at $x = L/2$. The last term in eq. (3.56) is zero because $\sin(\pi) = 0$. The first term is zero because $C_3 = 0$. The second term, however, is not zero unless

$$\sin\left(x\left[\frac{P}{BI}\right]^{1/2}\right) = \sin\left(\frac{L}{2}\left[\frac{P}{BI}\right]^{1/2}\right) = \sin(\pi) = 0,$$

that is, $P = BI(2\pi/L)^2$, or $C_4 = 0$. This latter possibility, $C_4 = 0$, is the only feasible one because the end of the member must remain fixed at $v_1 = 0$ regardless of the values of the axial load. For nonzero axial loads, therefore, the condition that the vertical displacement is zero at $x = L/2$ requires that $C_4 = 0$.

Thus the expression for the additional displacement must be

$$v_1 = \frac{\delta_0}{[(2\pi/L)^2(BI/P) - 1]} \sin\left(\frac{2\pi x}{L}\right). \tag{3.57}$$

The total deflection is the sum of the additional and the initial deflections of the member,

$$v = v_0 + v_1,$$

so that from eqs. (3.57) and (3.49),

$$v = \delta_0 \sin\left(\frac{2\pi x}{L}\right) + \frac{\delta_0}{[(2\pi/L)^2(BI/P) - 1]} \sin\left(\frac{2\pi x}{L}\right),$$

or

$$v = \frac{(2\pi/L)^2(BI/P)}{[(2\pi/L)^2(BI/P) - 1]} \delta_0 \sin\left(\frac{2\pi x}{L}\right),$$

or, multiplying top and bottom by $(P/BI)(L/2\pi)^2$,

$$v = \frac{\delta_0}{[1 - (P/BI)(L/2\pi)^2]} \sin\left(\frac{2\pi x}{L}\right). \tag{3.58}$$

Equation (3.58) is interesting in several respects. We see that as the axial load, P, approaches the quantity $BI(2\pi/L)^2$, the denominator approaches zero, which causes the theoretical deflection to become boundlessly large, or to approach infinity. Now the quantity $BI(2\pi/L)^2$, we recall, is the critical load for buckling of a member that is initially perfectly straight. That is,

$$P_1 = BI\left(\frac{2\pi}{L}\right)^2. \tag{3.59}$$

Therefore, as the axial load, P, acting on an initially bent member approaches the critical load, P_1, the deflection becomes very large.

Figure 3.19 shows relations between axial load and deflection for a member initially deflected into one half-wave. When the axial load is zero, the total deflection is equal to the initial deflection. That is, $\delta/\delta_0 = 1$, if $P/P_1 = 0$. As the axial load

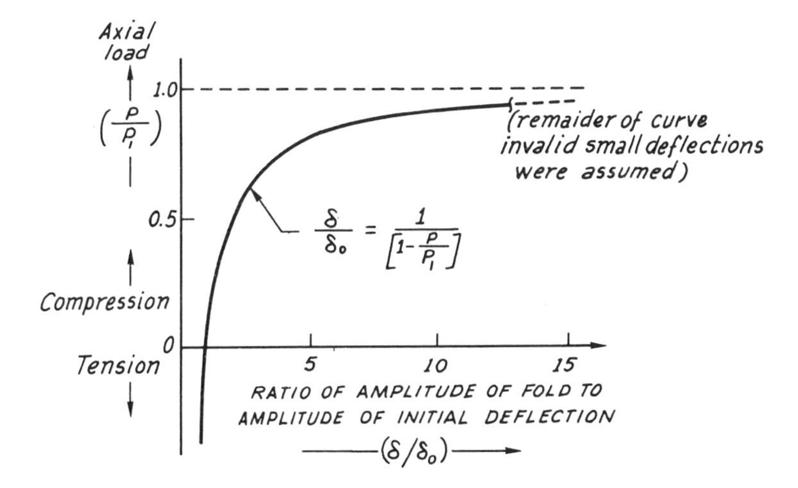

Figure 3.19. Relation between deflection at center of member and axial load. Member is initially deflected. The horizontal line near top of figure indicates critical load for buckling of initially straight members.

becomes larger, the deflection increases slowly at first, then increases more and more rapidly as the axial load approaches the critical load, that is, as $P/P_1 \rightarrow 1$.

We can understand better the meaning of the critical load for the deflection of an initially straight member by studying buckling of initially deflected members. If we let the magnitude of the initial deflection become smaller and smaller, the load required for a given amount of deflection becomes larger and larger (Fig. 3.20). As the magnitude of the initial deflection approaches zero, the load required for a given amount of deflection approaches the critical load. In the limit, the load required equals the critical load, and the deflection is zero until the axial load reaches the critical load, at which time the member buckles and the amount of its deflection is undefined (Fig. 3.20).

All real members contain initial deflections. Even members that appear to be straight are slightly deflected. Therefore, we can state that the critical buckling load is a property of ideal members and that it is the maximum load required for measurable deflections of real members.

So far we have considered relations between axial loads and deflections of members with a very special kind of deflection: an initial deflection in the form of a half sine-wave. Suppose, instead, that the members were initially deflected into four

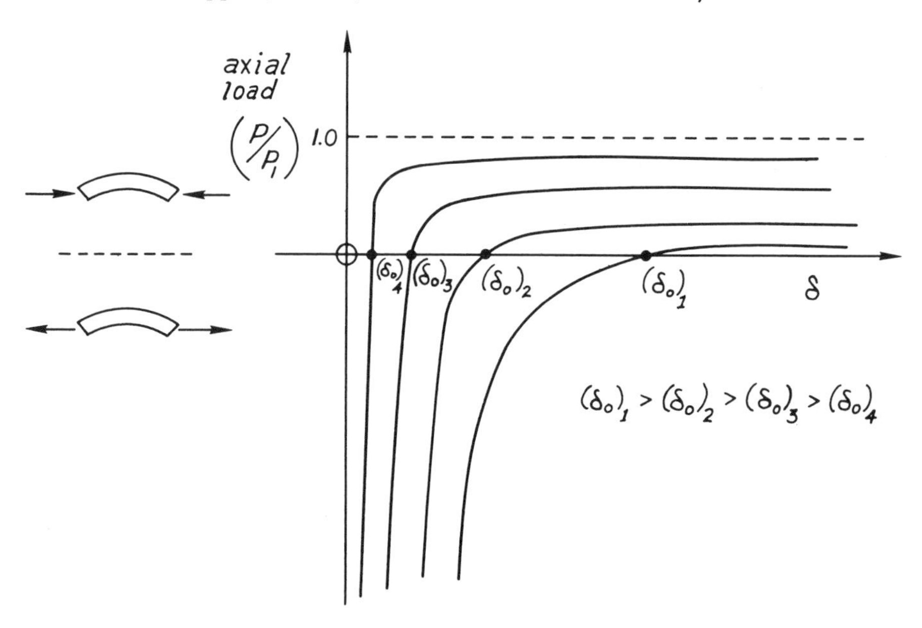

Figure 3.20. Effect of amplitude of initial deflection on relation between axial load and amplitude of fold. The increase of deflection with an increase in load depends upon the magnitude of the initial deflection. As the magnitude of the initial deflection approaches zero, the load required for appreciable deflection approaches the critical load, P_1, for the Euler wavelength.

complete waves, that is, into eighth-order waves. The expression for the initial deflection would be either

$$v_0 = \delta_0 \sin\left(\frac{2\pi x}{L}\right),$$

where L is the wavelength of the initial deflection, or

$$v_0 = \delta_0 \sin\left(\frac{8\pi x}{l}\right),$$

where l is the length of the member, which contains four initial waves.

We could derive a solution for the relation between axial load and deflection just as we did for the member that was initially deflected into a half-wave, eq. (3.49) and (3.57). If we did this, we would find that the expression corresponding to eq. (3.57) for the additional deflection would be

$$v_1 = \frac{\delta_0}{[(8\pi/l)^2(BI/P) - 1]},$$

so that the critical load at which the deflection becomes very large would be

$$P_{\text{crit}} = BI\left(\frac{8\pi}{l}\right)^2,$$

which is 8^2 or 64 times as large as the critical load for the member initially deflected into a half sine-wave. Thus the critical load increases as the square of the order of the initial waves.

Crooked members.—We see that members containing initial deflections of shorter wavelengths are amplified much more slowly with an increase of axial load than are members containing initial deflections of longer wavelengths. But a member that is initially deflected into waves of eighth order, for example, can buckle in at least two ways: either the entire member can buckle or the eighth-order waves can buckle. The critical load for the eighth-order waves is 64 times that for the first-order waves, so that if the member is ever so slightly deflected in the first-order waves, we would expect the first-order waves to dominate the fold pattern.

For example, Fig. 3.21A shows a member initially deflected in both first- and eighth-order waves. The initial amplitudes of the two orders are the same, δ_0. As a load is applied to such a member, amplitudes of waves of both orders are increased, but the growth of the amplitude of the first order far outstrips that of the eighth-order waves as the axial load approaches the critical load for the first-order waves (Figs. 3.21B and 3.21C).

Figure 3.22A shows a member initially deflected into first- through fourth-order waves; the amplitudes of each of the four orders are the same, so that the expression for the initial form of the member is:

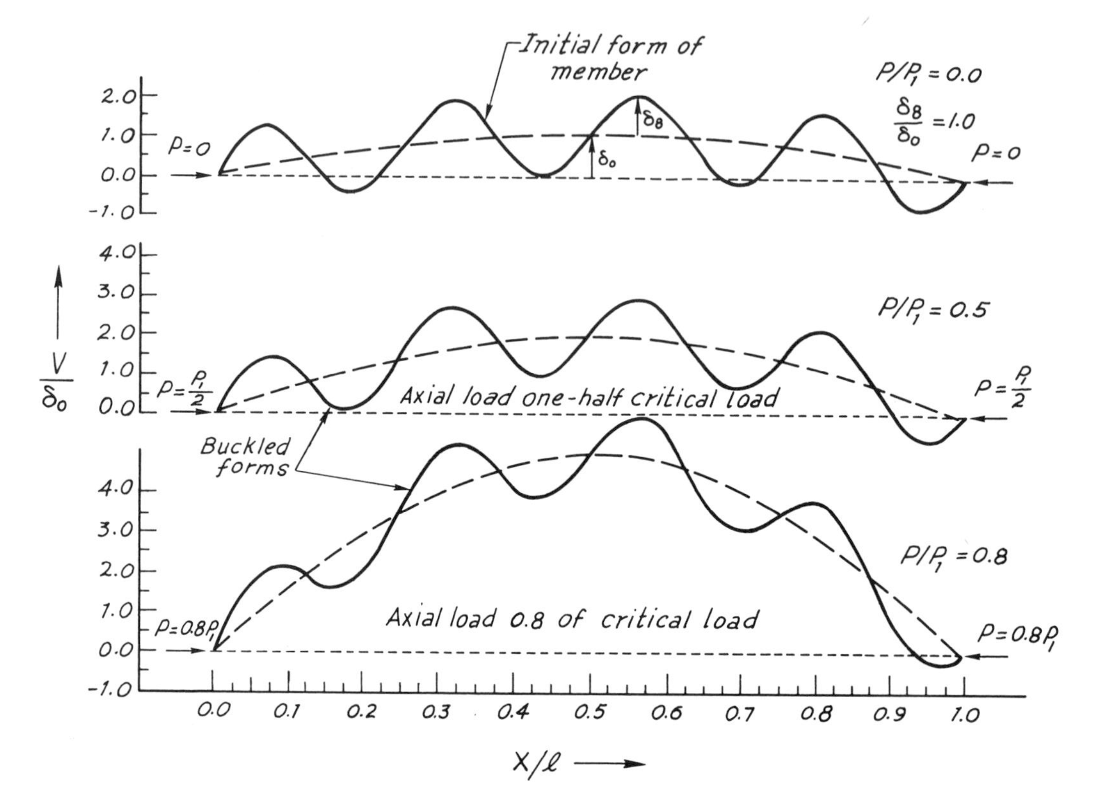

Figure 3.21. Further bending of member initially deflected in first- and eighth-order waves.

A. Initial form of members. Axial load is zero.

B. Small waves of eighth order and long wave of first order grow slightly when axial load is half the critical buckling load, P_1.

C. Growth of first-order waves has greatly outstripped growth of eighth-order waves when axial load is eight-tenths of critical axial load.

$$v_0 = \delta_0 \sin\left(\frac{\pi x}{l}\right) + \delta_0 \sin\left(\frac{2\pi x}{l}\right) + \delta_0 \sin\left(\frac{3\pi x}{l}\right) + \delta_0 \sin\left(\frac{4\pi x}{l}\right),$$

where δ_0 is the amplitude of the initial waves and l is the total length of the member. If we substitute this expression for initial deflection into eq. (3.52) and proceed toward a solution as before, we find that the total deflection is (7)

$$v = \frac{\delta_0}{[1 - (P/P_1)]} \sin\left(\frac{\pi x}{l}\right) + \frac{\delta_0 2^2}{[2^2 - (P/P_1)]} \sin\left(\frac{2\pi x}{l}\right) + \frac{3^2\delta_0}{[3^2 - (P/P_1)]} \sin\left(\frac{3\pi x}{l}\right) + \frac{4^2\delta_0}{[4^2 - (P/P_1)]} \sin\left(\frac{4\pi x}{l}\right). \quad (3.60)$$

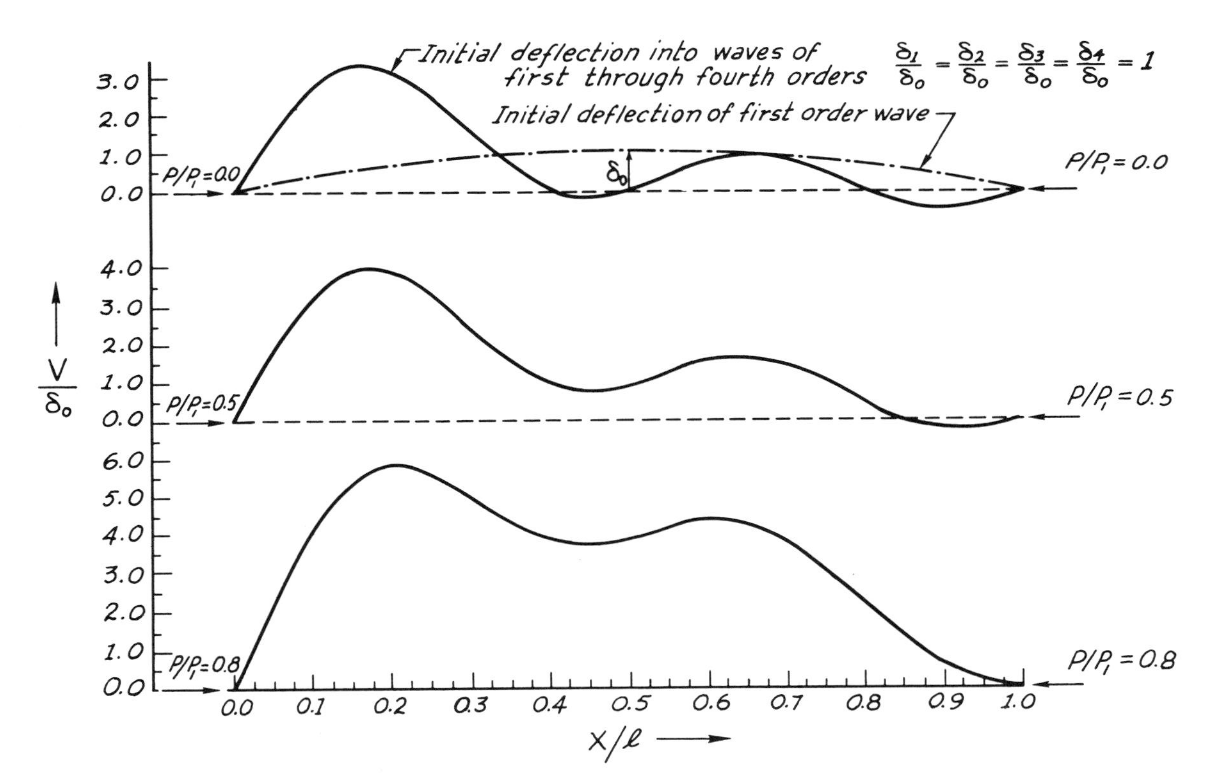

Figure 3.22. Further deflection of member initially deflected in first- through fourth-order waves. Growth of first-order wave is much greater than growth of waves of higher order.

Equation (3.60) is plotted in Fig. 3.22 for axial loads ranging from zero to eight-tenths of the critical load for first-order waves.

We can see in Fig. 3.22C that the deflection of the entire member in the first-order wave is much more rapid than the deflection in the other three orders. As a result, even though the first-order wave is not clearly visible in the initial deflection curve shown in Fig. 3.22A, representing a combination of the four orders, the first-order wave becomes visible when an axial load is applied to the member.

The reason that the first-order wave usually is predominant can be seen in eq. (3.60). The denominator,

$$\left(1 - \frac{P}{P_1}\right),$$

of the first term in eq. (3.60) approaches zero much more rapidly, as the axial load approaches P_1, than do the denominators of the following three terms. Thus the first term usually is predominant, unless, for example, the initial deflection of the second- or highest-order waves is much greater than the initial deflection of the first-order wave. It follows that, in general, the deflection of a crooked member can be understood in tèrms of a member with a simple sinusoidal initial deflection.

Even if the member were highly crooked, we could still apply these conclusions. The shape of a crooked member can be described exactly by the sum of an infinite number of sine terms, called Fourier Series (*6*). Fourier Series will be discussed more fully in later pages, but for the present application we can simply state that such a series is:

$$v_0 = (\delta_0)_1 \sin\left(\frac{\pi x}{l}\right) + (\delta_0)_2 \sin\left(\frac{2\pi x}{l}\right) + \cdots + (\delta_0)_n \sin\left(\frac{n\pi x}{l}\right), \qquad (3.61)$$

on to infinity. The initial deflection curve of a crooked member could be described by such a series, where the magnitudes of the initial deflections, $(\delta_0)_1, \ldots, (\delta_0)_n, \ldots,$ depend upon the shape of the member. If we substituted the series into eq. (3.52) and proceeded as we did before to derive an expression for the total deflection, we would find that the total deflection is:

$$v = \frac{(\delta_0)_1}{[1 - (P/P_1)]} \sin\left(\frac{\pi x}{l}\right) + \frac{(\delta_0)_2\, 2^2}{[2^2 - (P/P_1)]} \sin\left(\frac{2\pi x}{l}\right) + \cdots + \frac{(\delta_0)_n\, n^2}{[n^2 - (P/P_1)]} \sin\left(\frac{n\pi x}{l}\right). \qquad (3.62)$$

As we concluded in our analysis of the member described by eq. (3.60), the first term of eq. (3.62) usually will be larger than succeeding terms. As a result, even a very crooked member usually will buckle in the first-order wave and the initial deflection of smaller wavelengths will have little effect on the gross bending of the member.

Exercise

Derive eq. (3.62), explaining each step in your derivation.

Interpretation of theory.—The theory of the buckling of unconfined, elastic members allows us to make several useful generalizations. We have shown that if a member has even the slightest initial deflection it will begin to flex under the slightest axial load. On the other hand, if it were perfectly straight, it would not flex until the axial load reached the critical buckling load. As the axial load acting on the initially deflected member approaches the critical load for the straight member, the deflection becomes very large; our theory states that the deflection is boundless, but this simply indicates that the theory is not valid for large axial loads, as we will show in the next chapter when we analyze large deflections and ptygmatic folds. Thus our theory helps us to interpret the critical axial load of a member.

Our theory tells us that if an unconfined, elastic member is irregular in shape, or even crooked, certain wavelengths will become more prominently developed than others during buckling. Usually the first mode, the one with a wavelength equal to twice the length of the member, will grow in amplitude most rapidly. If waves of other lengths have much larger initial deflections than the wave of the first mode, the other waves may amplify more rapidly for low axial loads, but as the axial load increases the first mode will begin to dominate. Thus the initial deflections do not have to be large for them to dominate buckling. Even the slightest, perhaps even invisible, initial deflections can control the location and, sometimes, the size of wavelengths we see in experiments and in the natural situation.

Figure 3.23 shows another of Willis' experimental folds. A prominent initial dip was built into the layers at a distance of 24 inches from the piston, on the right. A small initial syncline is visible about 12 inches from the piston. The larger initial dip determined the position of an anticline which formed in the thinly bedded materials. The smaller initial dip, probably in combination with a shear force transmitted by the piston to the layers, caused an anticline to form later, near the piston. These results are grossly explained by our theory of initial deflections: The larger deflection began to buckle appreciably under a smaller load than did the smaller deflection. However, there are several features of this experiment that are not accounted for in our theory. The reason for the difference between theory and experiment is that the conditions assumed by the theory were not realized in the experiment.

The most important difference between the conditions assumed by our theory and the conditions actually imposed on the layers is that the experimental layers were confined above and below. They were confined above by 1100 pounds of shot

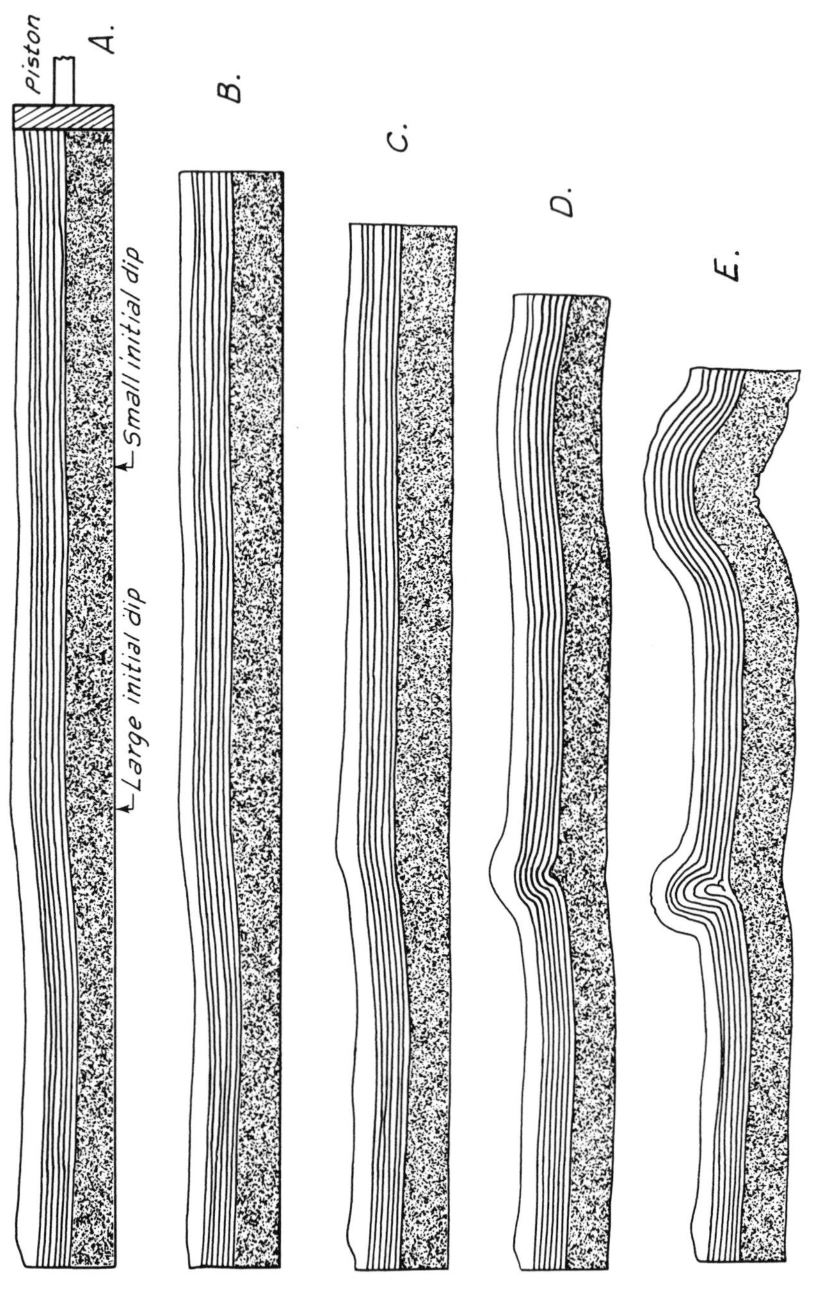

Figure 3.23. Experimental folds showing effect of amplitude of initial dip on growth of fold (after Willis, ref. *8*).

and they were confined below by a layer of soft wax that was $2\frac{1}{2}$ inches thick, which, in turn, was confined below by the rigid base of the box (see Fig. 3.11).

The effects of confinement on buckling are twofold. First, the layered materials drag on the confinements above and below, causing the axial load within the stack of strata to decrease in magnitude with increasing distance from the place where the load is applied. Second, the favored wavelengths of confined and unconfined members are different. Whereas for an unconfined member the wavelength requiring the smallest axial load for its growth is equal to twice the length of the member, for a confined member it is usually less than twice the length of the member. We will discuss this idea in much more detail in the next chapter. However, here we can note that the expected wavelength, L_u, for the unconfined member is twice the length, l, of the member:

$$L_u = 2l.$$

The wavelength, L_c, for the confined member is a function of the elastic modulus of the member, B, the modulus of the medium, B_0, and the thickness, T, of the member (*1*):

$$L_c = 2\pi T\left(\frac{B}{B_0 6}\right)^{1/3}.$$

The relation would be identical in form if the member and medium were viscous (*1*):

$$L_c = 2\pi T\left(\frac{\eta}{6\eta_0}\right)^{1/3},$$

except that the viscosities, η and η_0, replace the moduli, B and B_0.

We would not expect the wavelengths to equal the entire length of Willis' confined strata, therefore, because the properties of the wax strata and the slightly softer wax substrata probably do not differ by even an order of magnitude. For example, if the layers are 48 times more competent than the medium, that is, if

$$\frac{B}{B_0} = 48 = \frac{\eta}{\eta_0},$$

the expected wavelength is about twelve times the thickness of the member. If the properties of the member and medium are more nearly similar, say, if

$$\frac{B}{B_0} \text{ or } \frac{\eta}{\eta_0} = 6,$$

the expected wavelength is about six times the thickness of the member. The length of Willis' squeeze box was about one meter and the thickness of the strata was about nine cm, so that the wavelength of the first buckling mode would be about two meters and the ratio of the wavelength to thickness for the first mode would be about 22. Buckling into a wavelength with this ratio would require the layers to be 100 times more competent than the softer substratum. The properties of wax could not have this large range, so that the experimental observation by Willis that the fold wavelengths are much shorter than the length of the experimental strata is not surprising.

Another understandable feature of the experimental folds is that their wavelengths are related to the thicknesses of strata being buckled. The flexure in the center of Fig. 3.23D involves only the upper layers; its wavelength is shorter. The flexure on the right involves the soft material below as well as the upper layers; its wavelength is longer, as we would predict from the equations above. We will discuss this relationship in more detail in the next chapter.

References cited in Chapter 3

1. Biot, M. A., 1961, "Theory of Folding of Stratified Viscoelastic Media and its Implications in Tectonics and Orogenesis": *Geol. Soc. America Bull.*, V. 72, p. 1595–1632.
2. Goodier, S. N., 1967, "Lectures on Elastic-Plastic Instabilities": Stanford University (unpub. lecture notes).
3. Hall, Sir James, 1815, *On the Vertical Position and Convolution of Certain Strata and Their Relation with Granite*: *Trans. Roy. Soc. Edinburgh*, V. 7.
4. Hetenyi, M., 1946, *Beams on Elastic Foundation*: Univ. Mich. Press, Ann Arbor.
5. Ipsen, D. C., 1960, *Units, Dimensions, and Dimensionless Numbers*: McGraw-Hill Book Co., Inc.
6. Kreyszig, E., 1962, *Advanced Engineering Mathematics*: John Wiley and Sons, Inc., N.Y.
7. Timoshenko, S. P., and Gere, J. M., 1961, *Theory of Elastic Stability*: McGraw-Hill Book Co., Inc., N.Y.
8. Willis, Bailey, 1894, "Mechanics of Appalachian Structure": Thirteenth Annual Report of U.S. Geological Survey, 1891–92, p. 213–281.
9. Willis, Bailey, and Willis, Robin, 1934, *Geologic Structures*: McGraw-Hill Book, Co., Inc., N.Y.

Other references on folding

Anderson, T. B., 1964, "Kink Bands and Related Geological Structures": *Nature*, V. 202, p. 272–274.

Bain, G. W., 1931, "Flowage Folding": *Am. Jour. Sci.*, V. 22, p. 503–530.

Bayly, M. B., 1964, "A Theory of Similar Folding in Viscous Materials": *Amer. Journ. Sci.*, V. 262, p. 753–766.

Biot, M. A., 1937, "Bending of an Infinite Beam on an Elastic Foundation": *Trans. Amer. Soc. Mechanical Engr.*, V. 59, p. A1–A7.

———, 1959, "The Influence of Gravity on the Folding of a Layered Viscoelastic Medium under Compression": *Jour. Franklin Inst.*, V. 269, p. 211–228.

———, 1960, "Instability of a Continuously Inhomogeneous Viscoelastic Half-Space under Initial Stress": *Jour. Franklin Inst.*, V. 270, p. 190–201.

———, 1961, "Theory of Folding of Stratified Viscoelastic Media and Its Implications in Tectonics and Orogenesis": *Geol. Soc. Amer. Bull.*, V. 72, p. 1595–1632.

———, 1964, "Theory of Viscous Buckling of Multilayered Fluids Undergoing Finite Strain": *The Physics of Fluids*, V. 7, p. 855–859.

———, 1964, "Theory of Buckling of a Porous Slab and Its Thermoelastic Analogy": *Jour. Appl. Mechanics*, Ser. E, V. 31, p. 194–198.

———, 1965, "Theory of Similar Folding of the First and Second Kind": *Geol. Soc. America Bull.*, V. 76, p. 251–258.

———, 1965, *Mechanics of Incremental Deformations*: John Wiley & Sons, Inc., N.Y.

———, 1965, "Theory of Viscous Buckling and Gravity Instability of Multilayers with Large Deformation": *Geol. Soc. Amer. Bull.*, V. 76, p. 371–378.

———, 1965, "Further Development of the Theory of Internal Buckling of Multilayers": *Geol. Soc. Amer. Bull.*, V. 76, p. 833–840.

———, and Odé, H., 1962, "On the Folding of a Viscoelastic Medium with Adhering Layer Under Compressive Initial Stress": *Quarterly Appl. Mathematics*, V. 11, p. 351–355.

Burmister, D., 1945, "The General Theory of Stresses and Displacements in Layered Systems": *Jour. Applied Physics*, V. 16, p. 89–94, 126–127, 296–302.

Busk, H. G., 1929, *Earth Flexures*: Cambridge Univ. Press, London.

Campbell, J. D., 1958, "En-Echelon Folding": *Econ. Geol.*, V. 53, p. 448–472.

Campbell, J. W., 1951, "Some Aspects of Rock Folding by Shear Deformation": *Am. Jour. Sci.*, V. 249, p. 625–639.

Currie, Patrode, and Trump, 1962, "Development of Folds in Sedimentary Strata": *Geol. Soc. Amer. Bull.*, V. 73.

DeSitter, L. U., 1956, *Structural Geology*: McGraw-Hill Book Co., Inc., N.Y.

Dewey, J. F., 1966, "Kink Bands in Lower Carboniferous Slates of Rush Co., Dublin": *Geol. Mag.*, V. 103, p. 138–142.

Donath, F. A., and Parker, R. D., 1964, "Folds and Folding": *Geol. Soc. Amer. Bull.*, V. 75, p. 45–62.

Essenburg, F., 1962, "Shear Deformations in Beams on Elastic Foundations": *Jour. Appl Mech.*, June, V. 29, Series E. No. 2.

Flinn, D., 1962, "On Folding During Three Dimensional Progressive Deformation" *Quart. Jour. Geol. Soc.*, V. 118, p. 385–433.

Freedman, J., Wise, D. U., and Beutley, R. D., 1964, "Pattern of Folded Folds in the Appalachian Piedmont Along Susquehanna River": *Geol. Soc. Amer. Bull.*, V. 75, p. 621–638.

Freudnethal, A. M., and Lorsch, H. G., 1957, "The Infinite Beam on a Linear Visco-elastic Foundation": *Proc. ASCE, Eng. Mech. Div.*, V. 83, No. EM1, Jan.

Goguel, J., 1962, *Tectonics*: W. H. Freeman and Company, San Francisco, 304 p. (Trans. by H. Thalmann.)

Heard, H. C., Turner, F. J., and Weiss, L. E., 1965, "Studies of Heterogenous Strain in Experimentally Deformed Calcite, Marble, and Phyllite": *Univ. of Calif. Pubs. in Geol. Sci.*, V. 46, p. 81–152.

Hetenyi, M., 1946, *Beams on Elastic Foundations*: Univ. Michigan Press, Ann Arbor.

———, 1950, "A General Solution for the Bending of Beams on Elastic Foundations for Arbitrary Continuity": *Jour. Appl. Physics*, V. 21, p. 55–58.

Hobbs, B. E., and Talbot, J. L., 1966, "The Analysis of Strain in Deformed Rocks": *Jour. Geol.*, V. 74, p. 500–513.

Kerr, A. D., 1961, "Viscoelastic Winkler Foundation with Shear Interactions": *Proc. ASCE, Eng. Mech. Div.*, No. 3, p. 13–31.

Love, A. E. H., 1944, *A Treatise on the Mathematical Theory of Elasticity*: Dover Publications, Inc., N.Y.

Marin, J., and Sauer, J. A., 1954, *Strength of Materials*: Macmillan Co., N.Y.

McBirney, A. R., and Best, M. G., 1961, "Experimental Deformation of Viscous Layers in Oblique Stress Fields": *Geol. Soc. Amer. Bull.*, V. 72, p. 495–498.

Patterson, M. S., and Weiss, L. E., 1962, "Experimental Folding in Rocks": *Nature*, V. 195, p. 1046–1048.

———, 1966, "Experimental Deformation and Folding in Phyllite": *Geol. Soc. Amer. Bull.*, V. 77, p. 343–374.

Pickett, G., et al., 1951, "Deflections, Moments, and Reactive Pressures for Concrete Pavements": Kansas State College, Manhattan, Kansas, *Bull.* 65.

Prager, W., and Hodge, P. G., 1951, *Perfectly Plastic Solids*: John Wiley and Sons, Inc., N.Y.

Ramberg, Hans, 1959, "Evolution of Ptygmatic Folding": *Norsk Geologisk Tidsskrift*, V. 39, p. 2–3, 99–151.

———, 1960, "Relationships Between Length of Arc and Thickness of Ptygmatically Folded Veins": *Amer. Jour. Sci.*, V. 258, p. 36–46.

———, 1961, "Relationship Between Concentric Longitudinal Strain and Concentric Shearing Strain During Folding of Homogeneous Sheets of Rock": *Amer. Jour. Sci.*, V. 259, p. 382–390.

———, 1963, "Strain Distribution and Geometry of Folds": Geological Institutions of the Univ. Uppsala, *Bull.*, V. 42, p. 1–20.

———, 1963, "Fluid Dynamics of Viscous Buckling Applicable to Folding of Layered Rocks": *Bull. American Assoc. Petrol. Geologists*, V. 47, p. 484–505.

———, 1963, "Evolution of Drag Folds": *Geological Magazine*, V. 100, p. 97–106.

———, 1964, "Selective Buckling of Composite Layers with Contrasted Rheological Properties, A Theory for Simultaneous Formation of Several Orders of Folds": *Tectonophysics*, V. 1, p. 307–341.

———, and Stephansson, Ove., 1964, "Compression of Floating Elastic and Viscous Plates Affected by Gravity, A Basis for Discussing Crustal Buckling": *Tectonophysics*, V. 1, p. 101–120.

Ramsay, J. G., 1960, "The Geometry and Mechanics of Formation of 'Similar' Type Folds": *Jour. Geol.*, V. 70, p. 309–329.

———, 1967, *Folding and Fracturing of Rocks*: McGraw-Hill Book Co., Inc., N.Y.

Reisser, E., 1944, "On the Theory of Elastic Plates": *Jour. Math. and Physics*, V. 23, p. 184–191.

———, 1945, "Effect of Transverse Shear Deformations on the Bending of Elastic Plates": *Jour. Appl. Mech., ASME*, V. 67, p. 69–77.

———, 1947, "On the Bending of Elastic Plates": *Quarterly of Appl. Mech.*, V. 5, p. 55–60.

———, 1955, "Stresses in Elastic Plates over Flexible Subgrades": *Proc. ASCE*, V. 81, No. 690.

———, 1958, "A Note on the Deflection of Plates on Viscoelastic Foundations": *Jour. Appl. Mech., ASME*, V. 80, p. 144–145.

Shanley, F. R., 1947, "Inelastic Column Theory": *Jour. Aeronautical Sciences*, V. 14, p. 261–268.

Sokolnikoff, I. S., and Redheffer, R. M., 1958, *Mathematics of Physics and Modern Engineering*: McGraw-Hill Book Co., Inc., N.Y.

Turner, F. J., 1962, "Rotation of the Crystal Lattice in Kink Bands, Deformation Bands, and Twin Lamellae of Strained Crystals": *Proc. National Academy of Sciences*, V. 48, p. 955–963.

———, and Weiss, L. E., 1965, "Deformational Kinks in Brucite and Gypsum": *Proc. National Academy of Sciences*, V. 54, p. 359–364.

Turner, F. J., and Weiss, L. E., 1963, *Structural Analysis of Metamorphic Tectonites*: McGraw-Hill Book Co., Inc., N.Y., 545 p.

Van Hise, C. R., 1896, "Principles of North American Pre-Cambrian Geology": *16th Ann. Rept. U.S. Geol. Survey*, pt. 1, p. 581–843.

Vesic, A. B., 1962, "Bending of Beams Resting on Elastic Isotropic Solid": *Proc. ASCE, Eng. Mech. Div.*, V. 87, EM2, Part I.

Whitten, E. H. T., 1966, *Structural Geology of Folded Rocks*: Rand McNally and Co., Chicago.

Williams, Emyr, 1966, "An Analysis of the Deformation in a Fold in North-East Tasmania": *Geol. Mag.*, V. 103, p. 115–119.

Williams, G., 1961, "The Deformation of Confined Incompetent Layers in Folding": *Geol. Mag.*, V. 98, 317–323.

Williams, G., 1965, "The Deformation of Competent Granular Layers in Folding": *Am. Jour. Sci.*, V. 263, p. 229–237.

Wynne-Edwards, H. R., 1963, "Flow Folding": *Am. Jour. Sci.*, V. 261, p. 793–814.

Recapitulation

We have seen several illustrations of the method introduced almost 100 years ago by G. K. Gilbert, the method of physical processes. Gilbert used the mechanical theory that was available at his time to explain some peculiarities of laccolithic intrusive bodies he discovered in the Henry Mountains in Utah. His mechanical analysis seems to have influenced his concept of the ideal laccolith, a mushroom-shaped body with steep sides, smoothly rounded upper surface, and circular plan shape. Also, with a simple mathematical model, he was able to explain remarkably well the lack of small laccoliths and the tendency for the laccoliths to increase in size with increasing depths of emplacement.

Perhaps Gilbert's most remarkable quality was his ability to interpret into mechanical theory what he observed in the field. For example, he recognized elements of an ideal laccolithic form after he had examined details of several laccoliths. He inferred that laccoliths will tend to be circular in plan if they are intruded into undisturbed, flat-lying strata. Nowhere did he actually see the ideal, mushroom-shaped laccolith that he sketched; rather, he *deduced* it while examining field occurrences, using his intuitive feel for things of mechanical origin. Thus, he was able to see a form that actually does not exist in nature, because in nature there are always complicating factors not accounted for in either our concepts or our mathematical theories.

About 100 years later, in 1968, David Pollard applied Gilbert's method to understand features of laccoliths that Gilbert probably did not see during his brief field study of the Henry Mountains. For example, Pollard was able to deduce a plausible concept of the evolution of laccolithic bodies through a program of detailed field observations, ingenious experimentation, and theoretical analyses. He recognized that, during the initial stage of growth, a thin body of magma being insinuated between sedimentary strata propagates much as a pressurized crack. When the crack-like sheet of magma becomes large enough to feel the finite thickness of the overburden, it begins to lift and bend the overburden. The sheet of magma, according to Pollard, expands into a laccolith when the overburden fails, by one of two modes: either a dike or a nearly vertical fault forms near the periphery of the intrusive, allowing the overburden to be lifted upwards, producing a thick body of intrusive rock that we would call a laccolith.

Pollard's investigation is a fascinating interplay of theory, experiment, and field observations. All three proceeded simultaneously so that one time the field observations guided the theoretical and experimental studies, another time the theoretical conclusions gave direction to his field study and to his selection of experimental materials, and another time the experiments helped him to select boundary conditions for his mathematical analyses and to study situations that are too complicated to analyze mathematically. For example, the exact theory of the transition from sill-like bodies to laccolithic-type bodies would be very difficult to study theoretically, but the essence of the transition is relatively simple to elucidate experimentally.

G. K. Gilbert was instrumental in persuading Bailey Willis to apply the method of mechanical processes to a study of folds in the Appalachian Mountains. In the course of Willis' study, he introduced half a dozen fundamental concepts of folding, such as the effect of initial dips of sedimentary strata and of vertical forces on shapes and distributions of folds, as well as the concept of the strut member, which we will investigate in the next chapter.

Willis used very little mathematical theory during his study, but he was an unusually capable experimentalist; rather than use mathematics to check an idea, he performed a series of experiments. For example, in order to be sure that he correctly deduced the concept of initial dips, which states that a change in dip of strata can localize a fold, he performed tens of experiments with his squeezing machine. He built initial dips into the experimental strata at different distances from the driving piston. Each time a fold appeared in a predictable position, so that his concept was verified.

Willis was moderately concerned about producing a scale model of the folds in the Appalachian Mountains, and some critics claim that his study is not especially valuable because he did not produce a scale model of the natural folds. His use of lead shot to load the experimental materials, for example, has been stated to be unrealistic because similar loading must have been absent in the Appalachian Mountains at the time the folds formed. The conclusion is that all of his results are suspect. These critics, however, do not understand what Willis was trying to do, nor do they understand experimentation.

Willis used the experiments to test and to illustrate his theories about the occurrence of certain anomalously large folds, such as Nittany arch, distant from the presumed source of the load applied in the southeast. Also, he wanted to test his theory about the effect of competence of strata on the transferral of loads across a wide area such as the Appalachians. It is unimportant whether or not he used a scale model of the Appalachians to check these theories experimentally. He might just as well have used materials such as corn syrup and a light oil, or hard and soft rubber strips, instead of the hard and soft wax layers he did use in his experiments. His theories relate to boundary conditions and mechanics, not to rheology, so that the rheological behavior of the experimental materials has no real bearing on the validity of experiments designed to test and develop his theories. An initial dip, for example, is a geometric feature, not a rheological property.

When we perform experiments and begin to worry about scaling factors and scale model theory, we would do well to ask ourselves this question: Does scale really have anything to do with the problem? Usually it does not. Also, contrary to much that has been written in the geological literature, the usual procedure of trying to guess what variables are important in a situation and of performing certain dimensional operations with the variables seems to be so much bunk for many if not most problems in geology. First, if we do not know enough about a problem to write the differential equations and the boundary conditions that describe it, we probably cannot *guess* which variables are important. Second, if we can write the differential equations and the boundary conditions, we need not bother ourselves with guesswork. All the variables are contained in the equations and they can be manipulated fairly easily to produce scale factors. For example, see the excellent book on dimensional analysis by Ipsen (*5*). Finally, if we can solve the differential equations for the conditions of both the model and the prototype, dimensional analysis will tell us nothing more than is already contained in the solution.

In my opinion, the value of experimentation in the mechanics of earth materials is not to produce scale models but, rather, to investigate certain *concepts*, such as Willis' concept of the effect of initial dips on folding, and to stimulate the imagination of the experimenter. Thus, David Pollard developed his theory of the transition from sills to laccoliths by carefully studying experiments in which he used oil as the intruding material and layers of common gelatin as the medium being intruded. After observing the growth and evolution of his experimental intrusions, he was able to recognize results of the same general behaviors in the field occurrences, thereby deducing much about the transition from sills to laccoliths.

4

Chapter Sections

Introduction
Folding of Single Strut Member in Elastic Medium
- Formulation of Mathematical Problem
- Reexamination of Folded Quartz Veins
- Effect of Finite Thickness of Soft, Confining Medium

Effect of Bedding on Flexibility of Strata
Effect of Gravity on Fold Wavelength of a Strut Member
The Strut Member Concept
Ptygmatic Folds — Large Deflections
- Ways Ptygmatic Folds Might Form
- The *Elastica*
- Relation between Theoretical and Actual Forms of Ptygmatic Folds

References Cited in Chapter 4

4

Folding of Single Members

Introduction

We have seen that Bailey Willis made outstanding contributions to our understanding of folding processes. He provided insights into the effects that shear forces and initial deflections have on determining places where folds initiate. He was able to explain why large folds might form at great distances from the driving force and, by using his concepts, we were able to understand drag flexures such as the one adjacent to a fault in Utah. But Willis contributed much more to our understanding of folding. He introduced the *strut member concept* to geology and recognized the fact that wavelengths of folds in the Appalachian Mountains and in his experimental apparatus depend upon thicknesses and rigidities of certain strut members, or sequences of strut members, in the strata.

Exactly what he meant by a *strut member* is stated in his report on mechanics of folding in the Appalachian Mountains, published in 1894, and is further explained in his textbook on structural geology, published in 1934:

> The stratigraphic column in this district includes all the Paleozoic formations from Cambrian to Carboniferous, and its total varies from 18,000 to 27,000 feet. At the base is the Cambro-Silurian limestone, and above are the sandstone and shales of Upper Silurian, Devonian, and Carboniferous, in all their variety and development [Fig. 3.2]. Stratigraphically the beds present in composition and color variations of great interest; but in relation to structural problems they fall into only two principal divisions—the great limestone and the greater shale-sandstone series. The former is massive, little divided by vague bedding planes—*a rigid unit in folding.* The latter is thin bedded, and, though some sandstone and calcareous strata are by themselves thick and hard, the entire second division resisted folding as a mass of weak beds (ref. *13*, p. 225). [*Italics mine.*]

> In the course of time...they are subjected to compressive stresses in the direction of bedding.... Under these conditions a stratum is a *strut,* which is subjected to the shearing and bending stresses incident to its mechanical situation. *If it be short and stiff, it may shear before it can bend.* If it be rigidly supported below and confined by a sufficiently heavy load above, it also may shear before it can bend. *But if it be of considerable extent in comparison with its thickness* and be not too rigidly restrained from below, *it will bend rather than shear....* (ref. *14*, p. 78 and 79). [*Italics mine.*]
>
> Where limestones, sandstones, and shales occur in thick deposits, they will commonly differ greatly in strength, and it is in such sequences that the typical structures of folded mountain ranges are developed. Often there is one formation, usually a thick limestone, which is *the* competent stratum of the series.... (ref. *14*, p. 82).

Two features of folds in stratified rocks that contain strut members invite special attention. First, flexures in a given strut member are often periodic; spacings between anticlinal crests and synclinal troughs in the member tend to be equal in some series of folds such as we see in the Appalachian Mountains or on the much smaller scale of minor folds in quartz veins. Second, the wavelengths of many folds seem to be controlled by thicknesses of strut members.

The wavelengths of folds shown in the geologic cross-section of Pennsylvania (Fig. 3.3) are quite variable, but certain wavelengths seem to recur. For example, the anticlines that form Tuscarora Mountain and Nittany Mountain are both about five miles in width. Also, smaller folds with wavelengths of about two miles and involving a few stratigraphic units are abundant. Thus the folds in the Appalachians tend to be periodic.

The tiny folds in the quartz veins shown in Fig. 4.1 are in a calcareous conglomerate, exposed in the upper end of Wildrose Canyon, Panamint Mountains, California. The thicker quartz vein, shown near the top of Fig. 4.1, is buckled into larger wavelengths than the thinner vein. This is but one example of many which support Willis' contention that wavelengths of folds increase with increasing thicknesses of the strut members in the folds. Figure 3.23, which shows the development of two folds during one of Bailey Willis' experiments, illustrates the same relationship between widths of folds and thicknesses of members being folded. The fold involving only the upper layers is not as wide as the fold involving both the upper layers and the soft substratum.

Thus, two questions we want to answer with beam and plate theory are: Why do spacings between axes of folds in strut members tend to be nearly constant for each strut? How are wavelengths of folds in strut members related to thicknesses and to physical properties of the rocks forming the struts?

In order to investigate the influence of strut members on folding, we must not

Figure 4.1. Folded quartz veins in carbonate exposed in Wildrose Canyon, Panamint Mountains, California. Dime indicates scale.

only assign certain physical properties and boundary conditions to the strut but also explicitly define the confining medium. We will develop idealized pictures of the properties of confining media and relate the resulting folds to these properties. For example, we will first treat the confining medium as an infinitely thick, elastic body in which the strut is embedded. Heretofore we have simply stated that there is a medium with a certain resistance, k, to deformation. Now we will define k. Also, we will determine the effects of layering within the strut and of gravity on folding of strut members.

Thus, three types of confining media and two types of strut members will be discussed:

Struts:

1. A single member forming the strut.
2. Many members comprising the strut. The contacts between the members are frictionless.

Media:

1. An elastic medium of infinite thickness above and below the strut. The infinite medium probably is a good first approximation to folding situations where the strut member is embedded in softer material that is much thicker than the strut, such as a quartz vein in shale.
2. An elastic medium of finite thickness above and below the strut. Folding within a medium of finite thickness probably has a natural analog where a thin sandstone is embedded in moderately thick shale beds, which, in turn, are bounded by very thick, relatively rigid sandstone beds.
3. A dense, elastic medium. If folds form near the surface of the earth, the buoyancy of the substratum can be appreciable and should be considered.

Folding of single strut member in elastic medium

Formulation of Mathematical Problem.—The two quartz veins shown in Fig. 4.1 are widely separated, so that as a first approximation we probably can assume that the quartz veins buckled independently, as though each of them buckled in a medium that was very thick. Thus we will represent folds shown in Fig. 4.1 by an isolated, buckled member in an infinitely thick, elastic medium (Fig. 4.2A), which is supporting an axial load, P. We will further idealize the folding situation by assuming that the strut member is slightly deflected prior to buckling (Fig. 4.2B).

The differential equation relating the deflection, v, to the axial load, P, acting on the member, and the rigidity, B, of the member, and the resistance, k, of the medium is [see eq. (3.17)]:

$$BI\frac{d^4v}{dx^4} + P\frac{d^2v}{dx^2} + kv = 0. \tag{4.1}$$

The pertinent value for the resistance, k, of the confining medium of infinite thickness was derived by Biot (*1*) by methods that are beyond the scope of the strength-of-materials approach that we are following now. We will derive the relation after we have introduced the theory of elasticity. According to Biot (*2*) and to an equation that we will derive in Chapter 7,

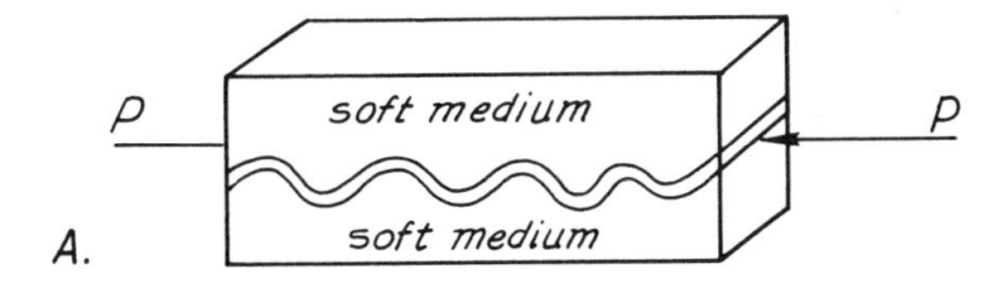

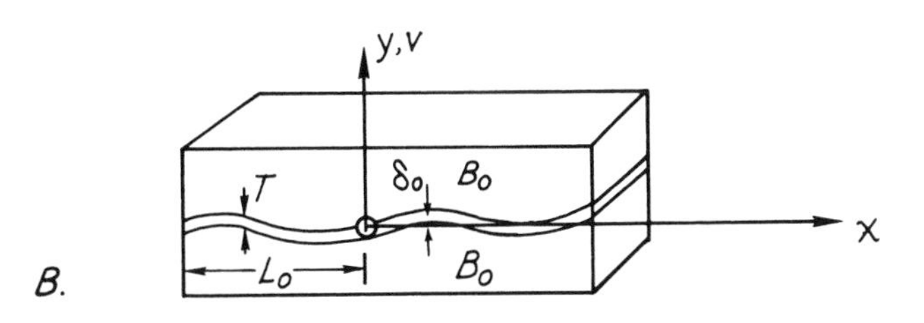

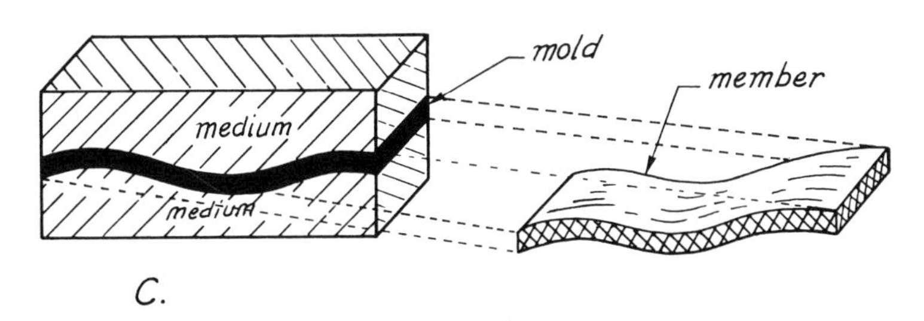

Figure 4.2. Buckling of initially deflected member in soft medium.
A. Final shape of folded member.
B. Initial deflection of member.
C. Member and medium are unstressed in their initially deflected states. If the member were removed from the medium, the member would maintain its initial deflections and the medium would be a mold of the member.

$$k = \frac{2B_0\pi b}{L_0}, \tag{4.2}$$

where B_0 is the elastic modulus of the confining medium. The modulus is

$$B_0 = E_0$$

for a beam, and

$$B_0 = \frac{E_0}{1 - \nu_0^2}$$

for a plate. Here L_0 is the wavelength of the slight deflection of the member prior to buckling. Thus eq. (4.1) becomes

$$BI\frac{d^4v}{dx^4} + P\frac{d^2v}{dx^2} + \frac{2B_0\pi b}{L_0}v = 0. \tag{4.3}$$

As we showed earlier, when we discussed the buckling of an unconfined, initially deflected member, the total deflection can be split into two parts: One part, v_0, is the deflection of the member in the unloaded state,

$$v_0 = \delta_0 \sin\left(\frac{2\pi x}{L_0}\right). \tag{4.4}$$

We will assume it is sinusoidal in form. The other part of the deflection, v_1, is the deflection caused by the axial load. Thus the total deflection, v, is

$$v = v_0 + v_1. \tag{4.5}$$

Let us further assume that the member and the medium were unstressed in their initially deflected states. Thus, if the member were removed from the medium, the member would retain its initial deflection and the gap formed in the medium by removal of the member would be a mold of the member (see Fig. 4.2D). In this case, eq. (4.3) would be written

$$BI\frac{d^4v_1}{dx^4} + P\frac{d^2(v_0 + v_1)}{dx^2} + \frac{2B_0\pi b}{L_0}v_1 = 0, \tag{4.6}$$

so that the bending moment and the resistance of the medium are functions of v_1, the deflection caused by the axial load. The bending moment induced by the axial load,

$$P(v_0 + v_1),$$

however, is a function of the total deflection.

Now we can solve eq. (4.6) for the deflection. Substituting eq. (4.4) into eq. (4.6),

$$BI\frac{d^4v_1}{dx^4} + P\frac{d^2v_1}{dx^2} + \frac{2B_0\pi b}{L_0}v_1 = P\delta_0\left(\frac{2\pi}{L_0}\right)^2 \sin\left(\frac{2\pi x}{L_0}\right). \tag{4.7}$$

Equation (4.7) is a fourth-order, nonhomogeneous differential equation. It is nonhomogeneous because there is one term, the one to the right of the equal sign, that does not include the dependent variable, v_1. This type of equation can be solved by finding a solution to the homogeneous part, the part to the left of the equal sign, and adding to the solution so derived a *particular solution* (*11*). The particular solution is any expression for v_1 that satisfies the entire differential equation, eq. (4.7). The method will clarify as we proceed.

First consider the homogeneous part of eq. (4.7):

$$BI\frac{d^4v_1}{dx^4} + P\frac{d^2v_1}{dx^2} + \frac{2B_0\pi b}{L_0}v_1 = 0.$$

We have already derived the general solution to the homogeneous part and it is [see eq. (3.29)]:

$$(v_1)_{\text{homo}} = (C_1e^{\beta x} + C_2e^{-\beta x}) \cos \alpha x + (C_3e^{\beta x} + C_4e^{-\beta x}) \sin \alpha x,$$

where

$$\left.\begin{aligned} \alpha &= \left(\left[\frac{k}{4BI}\right]^{1/2} + \frac{P}{4BI}\right)^{1/2}, \\ \beta &= \left(\left[\frac{k}{4BI}\right]^{1/2} - \frac{P}{4BI}\right)^{1/2}, \end{aligned}\right\} \tag{4.8}$$

and where

$$k = \frac{2B_0\pi b}{L_0},$$

if v_1 is sinusoidal. If v_1 were not sinusoidal, we might not have an exact expression for k.

A *particular solution* of the entire differential equation, eq. (4.7), can be derived by trying a solution of the form:

$$(v_1)_{\text{part}} = A \sin\left(\frac{2\pi x}{L_0}\right), \tag{4.9}$$

where A is a constant. By substituting eq. (4.9) into the differential equation, eq. (4.7), we can determine A. Thus

$$BI\frac{d^4v_1}{dx^4} = ABI\left(\frac{2\pi}{L_0}\right)^4 \sin\left(\frac{2\pi x}{L_0}\right),$$

and

$$P\frac{d^2v_1}{dx^2} = -AP\left(\frac{2\pi}{L_0}\right)^2 \sin\left(\frac{2\pi x}{L_0}\right),$$

and eq. (4.7) becomes, after dividing by $\sin(2\pi x/L_0)$,

$$ABI\left(\frac{2\pi}{L_0}\right)^4 - AP\left(\frac{2\pi}{L_0}\right)^2 + A\frac{2B_0\pi b}{L_0} = P\delta_0\left(\frac{2\pi}{L_0}\right)^2.$$

Solving for A,

$$A = \frac{P\delta_0}{BI(2\pi/L_0)^2 - P + (B_0bL_0/2\pi)}. \tag{4.10}$$

Therefore, the particular solution is, according to eqs. (4.9) and (4.10),

$$(v_1)_{\text{part}} = \frac{P\delta_0 \sin(2\pi x/L_0)}{BI(2\pi/L_0)^2 + (B_0bL_0/2\pi) - P}. \tag{4.11}$$

The general solution is the sum of the particular solution and the homogeneous solution (*10*), so that the general solution of eq. (4.7) is, according to eqs. (4.8) and (4.11),

$$(v_1)_{\text{general}} = (C_1 e^{\beta x} + C_2 e^{-\beta x}) \cos \alpha x + (C_3 e^{\beta x} + C_4 e^{-\beta x}) \sin \alpha x + \frac{P\delta_0 \sin (2\pi x/L_0)}{BI(2\pi/L_0)^2 + (B_0 b L_0/2\pi) - P}. \quad (4.12)$$

The arbitrary constants in eq. (4.12) can be evaluated with the aid of boundary conditions. All four constants, $C_1, \ldots, C_4$, must be zero in order to satisfy the condition that the total deflection is equal to the intial deflection of the member when the axial load is zero:

For all possible values of x,

$$v = v_1 + v_0 = v_0, \qquad \text{when} \quad P = 0.$$

Therefore,

$$v_1 = 0, \qquad \text{when} \quad P = 0, \quad \text{for any value of } x.$$

However, according to eqs. (4.8),

$$\alpha \neq 0$$

and

$$\beta \neq 0, \qquad \text{when} \quad P = 0,$$

for all values of x, so that $C_1 = C_2 = C_3 = C_4 = 0$.

We have shown, therefore, that the solution which applies to an initially deflected member under an axial load is:

$$v = v_0 + v_1 = \delta_0 \sin\left(\frac{2\pi x}{L_0}\right) + \frac{P\delta_0 \sin (2\pi x/L_0)}{BI(2\pi/L_0)^2 + (B_0 b L_0/2\pi) - P},$$

or

$$v = \left[\frac{BI(2\pi/L_0)^2 + (B_0 b L_0/2\pi)}{BI(2\pi/L_0)^2 + (B_0 b L_0/2\pi) - P}\right] \delta_0 \sin\left(\frac{2\pi x}{L_0}\right).$$

The term $BI(2\pi/L_0)^2$ is identical to the critical load for buckling of an unconfined member of length $L_0/2$. If we designate the critical load for the unconfined member as P_1, where

$$P_1 = BI\left(\frac{2\pi}{L_0}\right)^2,$$

the equation for the deflection can be written as

$$v = \frac{[(P_1/P) + (B_0 b L_0/2\pi P)]}{[(P_1/P) + (B_0 b L_0/2\pi P) - 1]} \delta_0 \sin\left(\frac{2\pi x}{L_0}\right). \quad (4.13)$$

Reexamination of Folded Quartz Veins.—Comparing our solution for the deflection, eq. (4.13), and an earlier solution that we derived for the unconfined member [eq. (3.58)]:

$$v = \frac{P_1/P}{[(P_1/P) - 1]} \delta_0 \sin \left(\frac{2\pi x}{L_0}\right), \tag{4.14}$$

we see that the axial load at which the deflection becomes boundless for the confined member is different from that of the unconfined member by the term $(B_0 b L_0)/2\pi P$. This is the contribution of the medium to the stability of the member. Also, in light of the observation that the load, P_1, is the critical load for buckling of the perfectly straight, unconfined member of length $L_0/2$, the value

$$P_{\text{crit}} = BI\left(\frac{2\pi}{L_0}\right)^2 + \frac{B_0 b L_0}{2\pi} \tag{4.15}$$

is apparently the critical load for buckling of a very long, perfectly straight, confined member.

Thus far we have assumed that the member contains a single initial deflection with a wavelength of L_0. We can determine the wavelength for which the buckling load, P_{crit}, is a minimum by differentiating P_{crit} with respect to L_0 and setting the result equal to zero. Thus, from eq. (4.15),

$$\frac{dP_{\text{crit}}}{dL_0} = -2\frac{BI(2\pi)^2}{(L_0)^3} + \frac{B_0 b}{2\pi} = 0.$$

Solving for the wavelength,

$$(L_0)^3 = \frac{2BI(2\pi)^3}{B_0 b} = \frac{B(2\pi T)^3}{6B_0},$$

because $I = bT^3/12$. It follows that

$$(L_0)_m = 2\pi T\left(\frac{B}{6B_0}\right)^{1/3} \tag{4.16}$$

is the wavelength for which the critical load is a minimum.

Earlier we established that an unconfined elastic member will usually deflect into a half-wave, that is, into the first mode, even if the member is initially very crooked, because the first mode required the lowest axial load for buckling. The first mode, however, usually is not the one that requires the least axial load for buckling of a confined member; the wavelength that requires the least load is given by eq. (4.16), which depends upon the elastic constants of the medium and member and upon the thickness of the member. If an initially deflected member contains initial waves of the length given by eq. (4.16), that wavelength is the one we would expect to be developed into a fold or train of folds. Even though the member is deflected in a complex manner initially, folds with the wavelength expressed by eq. (4.16) will tend to dominate the resulting fold pattern because they will grow in amplitude most rapidly with increasing loads.

Exercises

1. Show that $(L_0)_m$ is, indeed, the wavelength for which the axial load, P, is a minimum.
2. Show that if a member embedded in an infinite elastic medium has an initial deflection described by the sine series,

$$v_0 = \delta_1 \sin\left[\frac{\pi x}{2(L_0)_m}\right] + \delta_2 \sin\left[\frac{\pi x}{(L_0)_m}\right] + \delta_3 \sin\left[\frac{2\pi x}{(L_0)_m}\right] + \delta_4 \sin\left[\frac{3\pi x}{(L_0)_m}\right] + \cdots + \delta_n \sin\left[\frac{(n-1)\pi x}{(L_0)_m}\right],$$

the initial deflection with a wavelength equal to $(L_0)_m$ will increase in amplitude most rapidly with an increase in load. Here $(L_0)_m$ is the wavelength for which the axial load is a minimum in members initially deflected into sine waves [see derivations under "crooked member" in earlier pages].

The value of the critical load necessary for buckling of the member can be derived by solving eq. (4.16) for B_0 and substituting the result into eq. (4.15):

$$P_{\text{crit}} = 12BI\left[\frac{\pi}{(L_0)_m}\right]^2. \tag{4.17}$$

According to our theory of the buckling of an elastic strut in an elastic medium of infinite thickness, the wavelength that we would expect to see in nature, because it would grow most rapidly, is the one given by eq. (4.16), which states that the wavelength is directly proportional to the thickness of the member and to the cube root of the ratio of the moduli of the member and the medium.

Figure 4.1 shows small folds in quartz veins in a calcareous conglomerate in the Panamint Mountains, California. A thin quartz vein has been buckled into minute folds and a thicker vein has been buckled into somewhat larger folds. Our theory, therefore, checks out qualitatively: The thicker the member, the longer the wavelength of folds in the member.

We can check the applicability of our theory to this natural example by making certain measurements. We need to know the wavelength of the folds and the thicknesses of the beds. The wavelength of interest, however, is the wavelength at the time the folds begin to form, because our theory is valid only for small deflections. We presume that it is applicable to folds of large amplitudes if arc lengths remain constant during the buckling process. Thus our theory should be valid for folds of large amplitudes if the wavelengths of the folds were determined at an early stage and if arc lengths remained equal to this wavelength throughout folding.

The pertinent measurement, therefore, is the arc length of the folds. Fortunately Currie, Patnode, and Trump (*4*) have greatly simplified the job of measuring arc lengths, because they have prepared a graph relating arc length, s, and final wavelength, L (see Fig. 4.3).

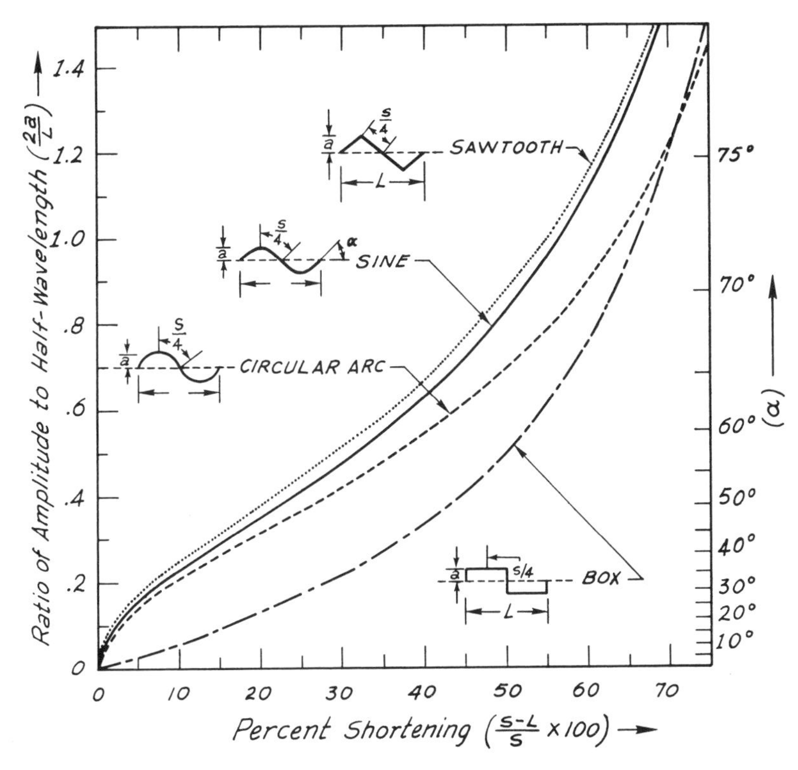

Figure 4.3. Relations among shortening, amplitude, and final wavelength of box, circular, sine, and sawtooth waveforms (after Currie et al., ref. *4*).

If we define L_0 as the wavelength of the folds at inception of buckling, then according to Fig. 4.3,

$$s = L_0 = \frac{L}{1 - sh}, \tag{4.18}$$

where sh is the shortening, given on the bottom of the graph. The folds in the larger quartz veins are asymmetric so that their geometry is best described by the sawtooth shape shown in Fig. 4.3.

The following data were collected by measuring the amplitude, a, and final wavelength, L, of several of the folds shown in Fig. 4.1.

We see that ratios of initial wavelengths to thicknesses for the thicker vein range from 5 to 11 and for the thinner vein range from 3.3 to 10. As far as we can determine, therefore, the two layers folded with similar wavelength-to-thickness ratios, as the theory predicts [see eq. (4.16)].

If the quartz veins and the medium behaved as simple elastic substances during initial stages of buckling, we could use our theory to estimate the ratios of the elastic moduli of members and medium. They should be related through eq. (4.16). Using that equation, the computed ratios of moduli for the thicker layer and the confining medium range from about 3 to 33 and for the thinner member they range from about 1 to 25. These ratios are in accord with laboratory determinations of elastic constants of rock. Young's moduli of very few rocks range more widely than an order of magnitude; ratios of moduli for most rock that has been tested in the laboratory is less than 10 [see Table 5.1, Chapter 5].

TABLE 4.1

Wavelength L (mm)	Amplitude a (mm)	Shortening sh (Fig. 4.3)	Vein thickness T (mm)	Initial wavelength L_0 (mm)	L_0/T	B/B_0 or η/η_0
Single Folds						
25	13	0.55	6.0	55.5	9.3	20
25	7	0.32	6.0	36.8	6.1	6
21	3	0.12	5.0	24.0	5.0	3
40	8	0.21	4.0	45.0	11.0	33
Average of Several Small Folds						
3.2	1	0.36	1.0	5.0	5.0	3
2.3	0.8	0.42	0.7	4.0	5.5	4
8.5	2	0.16	1.0	10.0	10.0	25
2.5	1	0.47	0.7	4.7	7.0	9
1.5	0.7	0.55	1.0	3.3	3.3	1

A viscosity ratio is also indicated in Table 4.1. We will show later, in Chapter 7, that the wavelength of a stiff, viscous member embedded in a soft, viscous medium is given by an equation that is identical in form to eq. (4.16):

$$L_d = 2\pi T\left(\frac{\eta}{6\eta_0}\right)^{1/3}, \tag{4.19}$$

where η is the coefficient of viscosity of the member and η_0 is the coefficient of viscosity of the medium. In general it probably is impossible to decide if a rock deformed elastically or viscously during initial stages of geologic deformation, so that the ratios presented in Table 4.1 could represent ratios of either elastic moduli or viscosity coefficients.

Sherwin and Chapple (*9*) measured 800 folds in single-layer struts similar to those shown in Fig. 4.1. Most of the struts were quartz veins in slate and phyllite. The ratios of wavelength to thickness of the folds in sandy and phyllitic matrices had a mean of about 5; those in slate were slightly larger, about 6; and those in shale were about 4. Thus our data are in accord with observations of many other folds in quartz veins. Sherwin and Chapple (*9*) stated that the viscosity ratios computed with eq. (4.19) are unreasonably low, so that they slightly modified the theory to account for thickening and shortening of the quartz veins during folding. I refer you to their paper for details of the simple corrections.

Currie, Patnode, and Trump (*4*), whose clearly written paper on elastic buckling of strata first stimulated my interest in the theory of folding, produced experimental folds to study further the behaviors of strut members. They buckled gum rubber strips, ranging from 0.4 to 1.6 millimeters in thickness, embedded in household gelatin (Fig. 4.4). The rubber strips had an elastic modulus of about 100 psi and the modulus of the gelatin ranged from about 1 to 10 psi. The gelatin and embedded rubber strips were compressed axially and small folds were produced. According to Currie et al. (*4*), the wavelengths were predicted remarkably well with the theory, eq. (4.16).

Figures 4.4A, B, and C show rubber strips ranging in thickness from 0.4 to 1.6 mm embedded in gelatin and compressed in the vertical direction. As is predicted by theory, the wavelength increases with the thickness of the strut member.

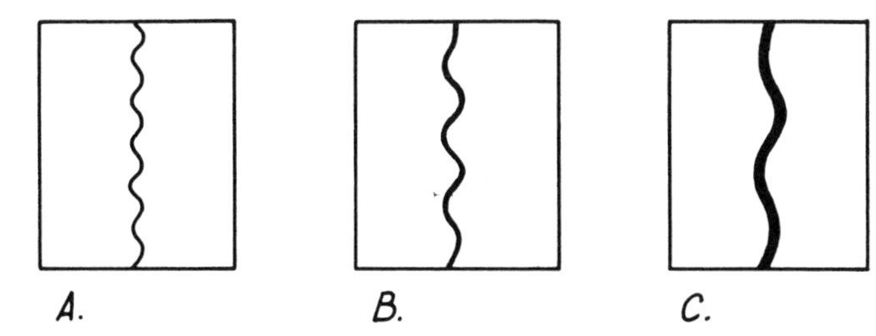

Figure 4.4. Experiment of folds of rubber strips in gelatin. Wavelengths increase with thicknesses of strips.
A. Rubber strip is 0.4 mm thick.
B. Rubber strip is 0.8 mm thick.
C. Rubber strip is 1.6 mm thick.
(Drawn from photograph in ref. *4*.)

Exercise

By measuring the folds shown in Figs. 4.4A, B, and C, check the conclusion by Currie et al. (*4*) that eq. (4.16) predicts the wavelengths of the experimental folds. Use Fig. 4.3 to determine wavelengths of the experimental folds at small amplitudes.

Effect of Finite Thickness of Soft, Confining Medium.—An obvious difference between the conditions assumed in our theory and the conditions of the experiments of Currie et al. is that we assumed an elastic medium of infinite thickness. In the experiments the medium had a finite thickness, of course. The question is how thin the medium can be for the theory of buckling of members in infinite media to be approximately valid. Consider the same competent member of thickness T, embedded between two relatively incompetent elastic sheets of thickness J, against which are placed two rigid boundaries (Fig. 4.5). This idealization might

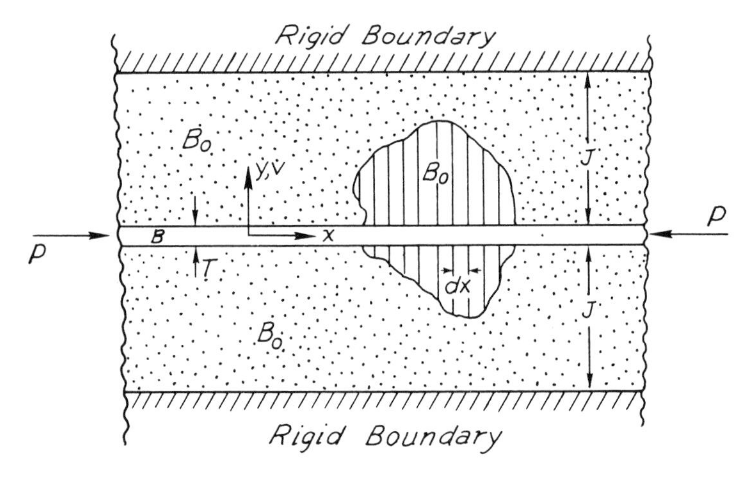

Figure 4.5. Member confined by elastic medium of thickness, J, bounded rigidly (after ref. *4*).

represent the boundary conditions for the buckling of a thin sandstone bed sandwiched between two shale beds, which, in turn, are bounded by two thick sandstone beds.

The finite thickness of the incompetent medium requires that the reaction, k, be modified. For a first approximation, the incompetent elastic medium will be represented by a series of individual columns, of width dx, that are not connected to each other (Fig. 4.5). Thus, the constraint behaves as a large number of elastic springs, each with an elastic modulus of B_0. The force per unit of length required to deform each pair of columns, the pairs consisting of one column below and one above the competent stratum, is (*4*)

$$k = \frac{2B_0b}{J}, \qquad q = kv = 2B_0b\left(\frac{v}{J}\right), \tag{4.20}$$

where B_0 is the modulus of the medium, and v/J is the strain in each column.

The differential equation and the solution for this problem are the same as those for the problem of buckling of a strut in an infinite medium. Therefore, to determine effects of the finite thickness on deflection, we can simply replace

$$\frac{B_0bL_0}{2\pi}$$

in eq. (4.12) by

$$\frac{B_0b}{2J}\left(\frac{L_0}{\pi}\right)^2.$$

The same substitution is made in the expression for the critical load, eq. (4.15):

$$P_{\text{crit}} = BI\left(\frac{2\pi}{L_0}\right)^2 + \frac{B_0b}{2J}\left(\frac{L_0}{\pi}\right)^2.$$

The wavelength for which the critical load is a minimum can be calculated by differentiating P_{crit} with respect to L_0 and setting the result equal to zero:

$$\frac{dP_{\text{crit}}}{dL_0} = 0 = -\frac{2BI(2\pi)^2}{L_0^3} + \frac{2B_0bL_0}{2J\pi^2}.$$

Solving for L_0, after substituting $I = bT^3/12$,

$$(L_0)_m^4 = \left(\frac{2\pi^4T^3JB}{3B_0}\right). \tag{4.21}$$

Rewriting eq. (4.16) for the wavelength of a member in an infinite medium,

$$(L_0)_m^3 = \frac{4\pi^3T^3B}{3B_0}, \tag{4.22}$$

we see that the wavelengths in infinite and finite media would be equivalent if $J = 2L_0/\pi$. As a first approximation, therefore, if the thickness of the incompetent confining medium is equal to or greater than the wavelength, the equation for the most likely wavelength of a member in a medium of infinite thickness, eq. (4.16) and other equations derived from it, should predict wavelengths of folds.

The problem of calculating the effect of incompetent surrounding materials with finite thicknesses will be solved exactly in Chapter 7, after we have discussed the mathematical theory of elasticity. But it is of interest to compare the exact solution with the one derived by means of the spring analogy. According to the exact solution for the displacement, v_i, of an incompetent medium of infinite thickness, derived earlier,

$$v_i = \frac{qL_0}{2B_0\pi b}$$

However, according to the exact solution for the displacement, v_f, of a medium of finite thickness, J, to be derived subsequently,

$$v_f = \frac{q}{CE_0\lambda b}\{[D(1 - \nu_0^2) + \nu_0(1 + \nu_0)A][\sinh(\lambda J)] - H(1 + \nu_0)[\lambda J\cosh(\lambda J) - \sinh(\lambda J)]\}, \quad (4.23)$$

where

$$A = \lambda J\cosh(\lambda J) - \sinh(\lambda J),$$

$$H = \sinh(\lambda J),$$

$$C = 2\lambda J + \sinh(2\lambda J),$$

$$D = \lambda J\cosh(\lambda J) + \sinh(\lambda J),$$

$$\lambda = \frac{2\pi}{L_0},$$

$$\sinh(\lambda J) = \frac{e^{\lambda J} - e^{-\lambda J}}{2},$$

and

$$\cosh(\lambda J) = \frac{e^{\lambda J} + e^{-\lambda J}}{2}.$$

Figure 4.6 shows the relation between the thickness of the medium and the deflection, predicted by eq. (4.23), the solution for the confining medium of finite thickness. The deflection is plotted as a ratio of the deflection predicted with the solution for the infinite medium. We see that if the ratio of the thickness of the medium to the wavelength of the folds is greater than about 0.6, the deflections predicted by the different solutions become nearly identical. In most folding situations where the relatively incompetent medium is at least one-half the wavelength of the folds in the strut, therefore, the medium and the member interact as though the medium were infinitely thick. Of course this conclusion applies only to small deflections. The greater the deflection, the greater the thickness of the medium would have to be for the finite thickness to have no importance. We cannot exactly solve this problem, because it would involve consideration of large deflections and large strains.

Exercises

A sequence of strata consists of two thick sandstone beds, of thickness T, embedded in very thick, incompetent shale. The two sandstone beds are separated

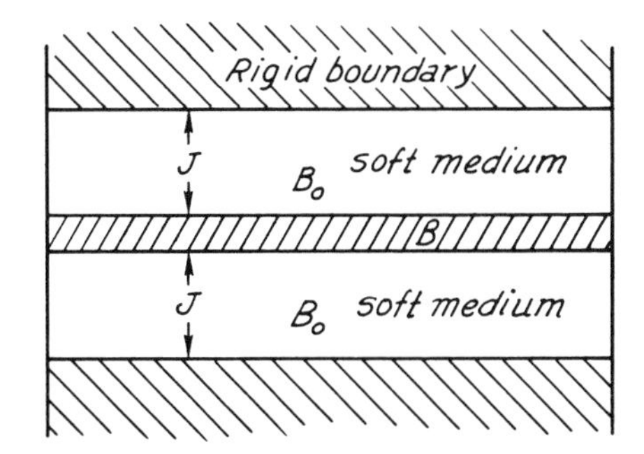

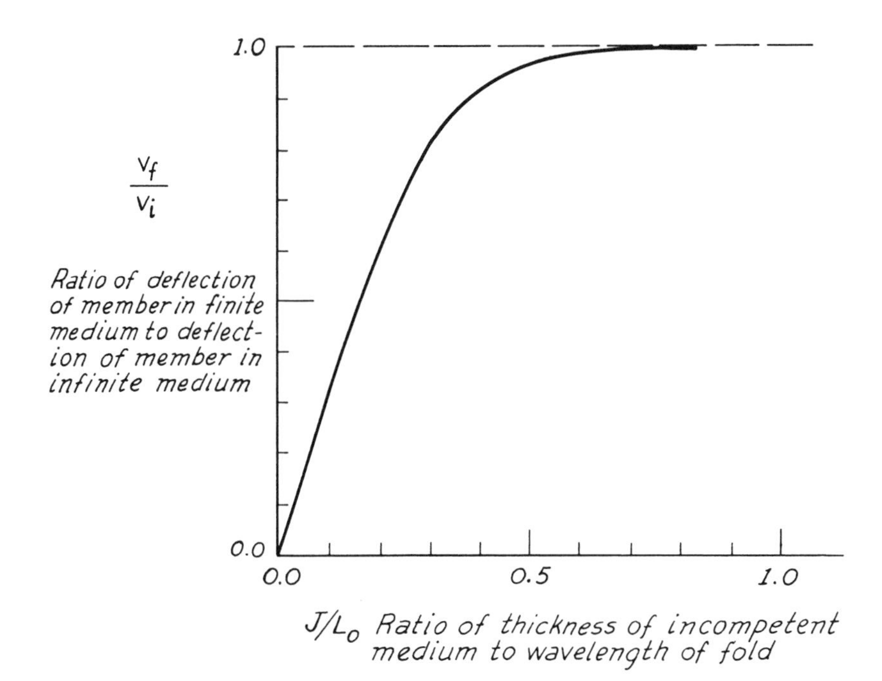

Figure 4.6. Relation between thickness of soft medium and deflection of competent member.

by an incompetent shale of thickness T. Within the incompetent shale, between the thick sandstone beds, is a thin sandstone bed with a thickness of $T/50$. The ratio of the moduli of the sandstone and shale beds is estimated to be

$$\frac{B}{B_0} = 500.$$

1. Calculate the wavelength-to-thickness ratios of the sandstone beds if they are buckled by axial loads.

2. Sketch the buckled sequence, showing the contrast in wavelengths of thick and thin sandstone beds.
3. Modify the sketch you made of the buckled members by qualitatively accounting for the shear caused by buckling of the thick sandstone beds and explain how and where drag folds might form on the flanks of major folds (see ref. *7*).

Two or more strut members buckling simultaneously in the same medium will buckle independently if they are separated widely enough. Perhaps we can conclude, on the basis of our study of a medium of finite thickness, that if the two members are separated by a distance of at least a wavelength, the two struts will buckle essentially independently. Again, this conclusion applies to the inception of folding, which, presumably, is the time when wavelengths are determined. Experiments by Curie, Patnode, and Trump (*4*) indicate that our conclusion is approximately correct for elastic materials. Figure 4.7A shows three struts of rubber embedded in a soft gelatin medium. The struts have buckled under an axial load. They are separated by a distance of about $1\frac{1}{2}$ wavelengths of the thicker strut, in the center. The waves seem to have formed independently because the buckles in the smaller struts are smooth waves, whose amplitudes are nearly constant whether an anticline or a syncline in the thick layer is nearby.

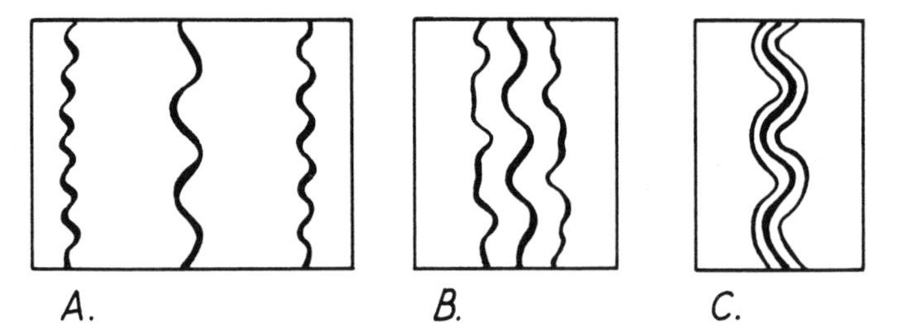

Figure 4.7. Interaction of members folded experimentally.

A. Rubber strip of 0.8 mm and two strips of 0.4 mm thickness widely separated. Negligible interference.

B. Same strips brought closer together. Folds in thick strip strongly interfere with fold patterns in thin strips, complex fold pattern.

C. Simple fold pattern emerges once more when layers are brought quite close together.

(Drawn from photographs in ref. *4*.)

Effect of bedding on flexibility of strata

If the rubber struts, or strips, are separated by about $\frac{1}{2}$ wavelength of the thicker member, as is shown in Fig. 4.7B, the waveforms in the thinner strips are definitely influenced by the proximity of the waves in the thicker strip. The wave pattern in the thinner strips is complicated if the rubber strips are rather close together.

However, if the rubber strips are brought very close together they buckle into a simple waveform, the wavelength of which is slightly larger than that of the thicker strip alone (Fig. 4.7C). Thus, if struts of different thicknesses are widely separated or very close together they buckle into simple wave trains. If they are separated only slightly, say between $\frac{1}{3}$ and one wavelength, they interfere with each other and buckle into a complicated waveform. We have considered the theory of buckling of isolated members; now we will analyze members that are close together.

One of the many contributions that Bailey Willis (*13*) made to our understanding of folding was recognition of the effect of stratification on flexibility of a sequence of strata such as we see in the Appalachian Mountains. According to Willis, stratified rock will buckle more easily than will massive rock. He explains (ref. *13*, p. 239):

> To deform stratified rocks demands that beds shall slip past one another and bend; the friction among beds and the interstitial resistances of different beds to folding are much less than the cohesive forces of a solid mass. It follows that strata are more easily deformed than masses...
>
> This idea of flexibility of rocks in the Appalachians is strengthened by two considerations: such a mass of strata is not divided into a hundred but into thousands of layers, and great subdivision weakens it; and furthermore, in folding, the strata do not yield as parts of a simple mass, but resist individually and irregularly. Reference to the columnar sections of strata in the Appalachian province will show how heterogeneous is the pile in a vertical direction, and how varied are the deposits in adjacent parts of the same district. There is every class of sedimentary deposit: Conglomerate, sandstone, shale, and limestone, with graduations from one into another, giving an indefinitely varied series. Each bed of such deposits is, in relation to others, more or less flexible, ... and the relative flexibility...of the principal members of a series has an important influence in determining the result of deformation. Flexibility is a direct function of lamination and toughness of the layers; its opposite, frangibility, is directly proportional to the thickness and incoherence of the stratum.

We can appreciate the tremendous effect stratification has on buckling by considering a special case: A sequence of beds between which the friction is zero (*4*) is embedded in a soft, elastic medium that is very thick (Fig. 4.8). Suppose that the sequence consists of an unspecified number of members of different thicknesses,

$$t_1, t_2, t_3, \ldots, t_n,$$

and different elastic moduli,

$$B_1, B_2, B_3, \ldots, B_n.$$

If we define, for shorthand notation,

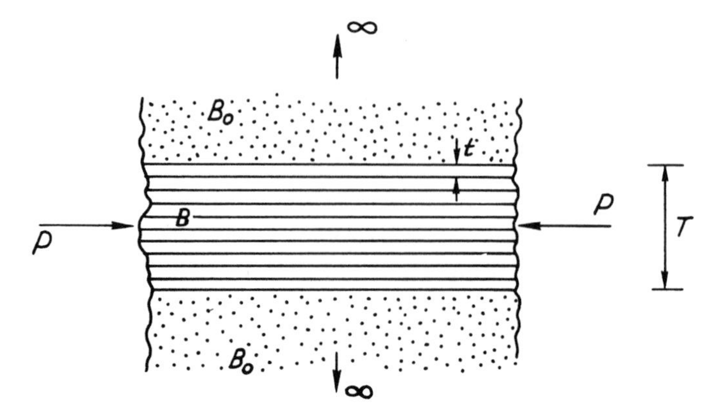

Figure 4.8. Sequence of members with frictionless contacts embedded in a medium of infinite thickness. T = total thickness, t = thickness of each bed.

$$\sum BI = B_1 I_1 + B_2 I_2 + \cdots + B_n I_n,$$

then

$$M = \sum BI \frac{d^2v}{dx^2}.$$

The equation for the bending moment of the sequence of beds is

$$M = \sum BI \frac{d^2v}{dx^2} = \left[\frac{B_1 bt_1^3}{12} + \frac{B_2 bt_2^3}{12} + \cdots + \frac{B_n bt_n^3}{12}\right] \frac{d^2v}{dx^2}, \tag{4.24}$$

because the total bending moment of the stack of layers is the sum of the bending moments of the individual layers.

It follows that all the preceding solutions apply to the buckling of a sequence of layers if we replace BI in the equations by $\sum BI$, defined above. Accordingly, the buckling wavelength of a stack of layers in an elastic medium is

$$[(L_0)_m]^3 = 2\,\frac{[\sum BI](2\pi)^3}{B_0 b}. \tag{4.25}$$

If each strut has the same elastic modulus and thickness, then $t_1 = t_2 = \cdots = t_n = t$, and $B_1 = B_2 + \cdots + B_n = B$, so that

$$[(L_0)_m]^3 = \frac{n(2\pi t)^3 B}{6B_0}$$

or

$$(L_0)_m = 2\pi t\left(\frac{nB}{6B_0}\right)^{1/3}. \tag{4.26}$$

According to eq. (4.26), the more layers there are, the larger is n, and the larger is the expected wavelength of folding. For a given total thickness, T, of strata, however, the wavelength is *reduced* as the number of strata comprising the thickness is increased. Thus,

$$nt = T,$$

so that

$$(L_0)_m = 2\pi T\left(\frac{B}{6B_0 n^2}\right)^{1/3}, \tag{4.27}$$

which states that the wavelength decreases nearly linearly with an increase in the number of beds aggregating the total thickness of T.

Our conclusion is verified qualitatively by the experiments shown in Figs. 4.9A, B, and C. In Fig. 4.9A, two widely separated struts of 0.4 mm thickness have buckled into folds with a wavelength of about 4.1 mm, or a wavelength-to-thickness ratio of about 10. The layers were brought closer together for the experiment shown in Fig. 4.9B, where the wavelength is about 5.0 mm. According to eq. (4.25) the wavelength of the two layers close together should be about 4.9 mm, so that the equation quite accurately predicts the wavelength of the two layers buckling together.

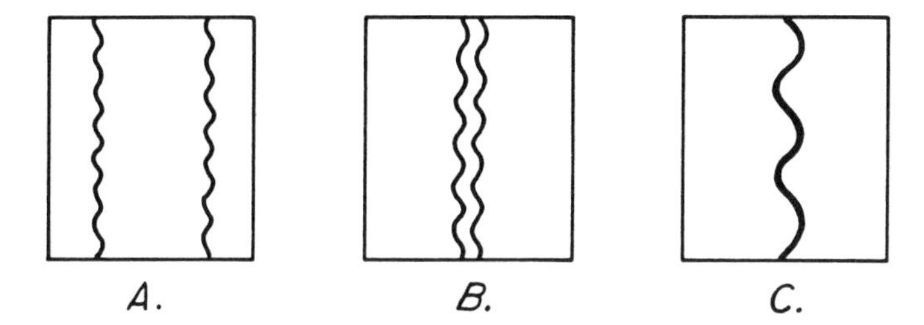

Figure 4.9. Comparison of wavelengths of experimental folds in two strips and in a single strip with a thickness equal to the combined thickness of the two strips.
A. Two widely separated strips of thickness 0.4 mm. Wavelength controlled by thickness of each strip.
B. Two interfering strips. Wavelength slightly increased.
C. Single strip of thickness 0.8 mm. Wavelength markedly increased.
(Drawn from photographs in ref. *4*.)

Exercises

A. Calculate the ratio of the moduli of the folded struts and medium shown in Fig. 4.7A. Using the calculated ratio, determine if the wavelengths of folds shown in Fig. 4.7C are predicted by eq. (4.25).

B. The theoretical and experimental investigations reported in preceding pages indicate that wavelengths of folds produced by axial loading should be controlled by competent members or groups of competent members. Figure 4.10 shows a plot of fold wavelengths and thicknesses of competent members for a wide range

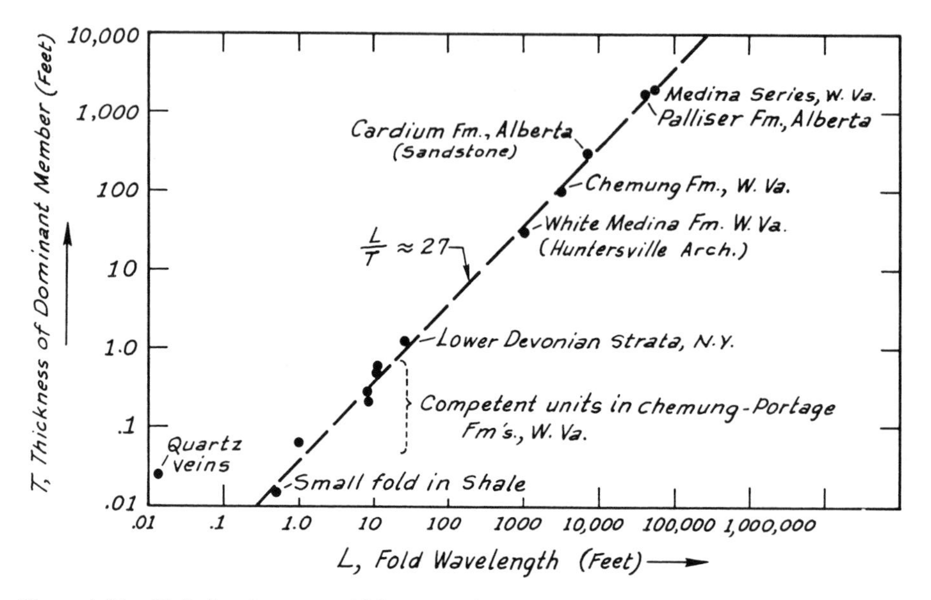

Figure 4.10. Relation between thickness and wavelength of natural folds (after Currie et al., ref. *4*).

of fold sizes. The dashed line, which approximately generalizes the data, can be represented by an expression of the form

$$\log L = \log C + \log T,$$

where C is a constant. The equation is identical to

$$\log L = \log (TC).$$

Thus $L = TC$, or $L/T = C$. Now according to eq. (4.16),

$$C = 2\pi\left(\frac{B}{6B_0}\right)^{1/3},$$

and the dashed line in Fig. 4.10 is equivalent to a L/T ratio of about 27, so that $C = 27$, and

$$\frac{B}{B_0} = 6\left(\frac{27}{2\pi}\right)^3 \approx 474.$$

Thus, the modulus of the competent members must have been about 470 times the modulus of the incompetent members if the rocks behaved elastically and as single members during folding.

1. Is this a reasonable ratio of moduli for rocks tested in laboratories? (See Table 5.1, Chapter 5.)
2. Why should the L/T ratios be approximately constant for folds measured by Currie, Patnode, and Trump (*4*) (Fig. 4.10)?
3. Suppose there were several competent members of roughly the same thickness buckling as a unit. What effect would this have on the above estimate of B/B_0?

Effect of gravity on fold wavelength of a strut member

We ignored effects of gravity in preceding discussions of folding because we assumed that the strut members were embedded in thick confining media. As an opposite extreme, suppose a strut member rests on an incompetent medium of infinite thickness (Fig. 4.11).

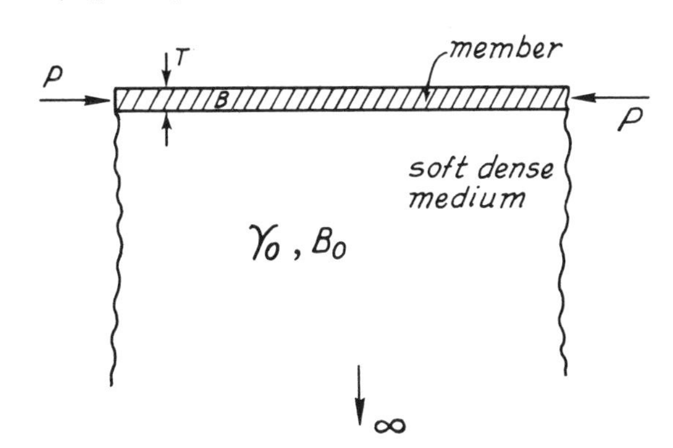

Figure 4.11. Member resting on dense soft medium of infinite thickness. Unit weight of medium is γ_0 (density times acceleration of gravity).

The resistance to bending of a strut resting on an infinite confining medium is

$$q = \frac{B_0 \pi b v}{L} = kv, \tag{4.28}$$

which is one-half that for a strut within an elastic medium [eq. (4.2)]. We are assuming that the strut is supported on only one side.

The constraint due to the weight and elasticity of a dense, elastic medium can be derived as follows. We will ignore initially the elastic resistance of the dense medium and will assume that the medium behaves as a fluid. The pressure exerted by such a medium on the member is proportional to the distance of the neutral plane of the member after deflection below or above the position of the neutral plane of the straight member. The pressure is hydrostatic, so that it can be written as

$$p_0 = \gamma_0 v, \tag{4.29}$$

where γ_0 is the unit weight, or density of the medium times acceleration of gravity, and where v is the total displacement of the neutral plane of the member. Thus in synclines the pressure is increased; the medium tends to push the member back into its original position. In anticlines the pressure is decreased and may become tensile. There need not be tension in the medium in the anticlines, however, because the absolute value of the pressure depends upon the state of pressure, prior to buckling, at the interface of the member and the medium. The initial pressure will be compressive because the medium supports the weight of the member.

The total resistance of the medium to deflection of the member, therefore, contains a pressure term,

$$q = \gamma_0 b v,$$

and an elastic term, eq. (4.28). Thus,

$$q = \left(\gamma_0 + \frac{B_0 \pi}{L}\right) b v. \tag{4.30}$$

The differential equation for the member on an infinitely thick, dense medium becomes

$$BI\frac{d^4 v}{dx^4} + P\frac{d^2 v}{dx^2} + \left(\gamma_0 + \frac{B_0}{L}\right) b v = 0. \tag{4.31}$$

The total deflection, v, is the sum of the initial deflection, v_0, and the deflection caused by the axial load, v_1, so that

$$v = v_0 + v_1.$$

If we assume a simple initial deflection of sinusoidal form,

$$v_0 = \delta_0 \sin\left(\frac{2\pi x}{L_0}\right),$$

where δ_0 is its amplitude and L_0 is its wavelength, the problem is basically the same as the one we solved earlier when we considered buckling of an initially deflected strut in an elastic medium. The differential equation becomes

$$BI\frac{d^4 v_1}{dx^4} + \left(\gamma_0 + \frac{B_0 \pi}{L_0}\right) b v_1 + P\frac{d^2 v_1}{dx^2} = -P\frac{d^2 v_0}{dx^2} - \gamma_0 b v_0$$

$$= \left[P\left(\frac{2\pi}{L_0}\right)^2 - \gamma_0 b\right] \delta_0 \sin\left(\frac{2\pi x}{L_0}\right). \tag{4.32}$$

We already have derived the general solution to the homogeneous part of eq. (4.32), so we will turn immediately to a search for a particular solution. A particular solution can be derived by substituting the solution

$$v_1 = A \sin\left(\frac{2\pi x}{L_0}\right) \tag{4.33}$$

into eq. (4.32) and then solving for A. If we do this, we find that

$$A = \frac{P - (L_0/2\pi)^2(b\gamma_0)}{BI(2\pi/L_0)^2 - P + (L_0/2\pi)^2[\gamma_0 + (B_0\pi/L_0)]b}. \tag{4.34}$$

We see that if the denominator of eq. (4.34) approaches zero, the deflection of the member, eq. (4.33), approaches infinity. The load at which the deflection would become boundless is, therefore,

$$P_{\text{crit}} = BI\left(\frac{2\pi}{L_0}\right)^2 + \gamma_0 b\left(\frac{L_0}{2\pi}\right)^2 + \frac{B_0L_0b}{4\pi}. \tag{4.35}$$

We will be unable to derive a simple expression for the wavelength for which the buckling load is minimum. We can, however, understand the effect of gravity on wavelength and axial load by studying graphically the solution for the critical axial load, eq. (4.35).

Let us plot axial load versus wavelength for four conditions of constraint: unconfined; elastic, weightless medium; dense, fluid medium; and dense, elastic medium. Comparison of these graphs will allow us to understand clearly the effects of different kinds of media on fold wavelengths.

A. Unconfined, elastic member.—The first folding situation we considered was buckling of an unconfined member. According to eq. (3.9), the relation between critical axial load, P_1, and wavelength, L_0, is

$$P_1 = BI\left(\frac{2\pi}{L_0}\right)^2,$$

which states that the critical load is inversely proportional to the square of the wavelength. The longer the wavelength, the smaller the load required for buckling (Fig. 4.12A). No finite wavelength gives a minimum buckling load.

B. Member resting on elastic, weightless medium.—If the member rests on an infinitely thick, elastic medium, the critical load for buckling is [eq. (4.15), with resistance divided by two]

$$P_{\text{crit}} = BI\left(\frac{2\pi}{L_0}\right)^2 + \frac{bB_0}{4}\left(\frac{L_0}{\pi}\right).$$

As the wavelength becomes large, the first term, that due to the rigidity of the member, becomes insignificant and the second term, that due to the rigidity of the medium, dominates. For large wavelengths, as is shown in Fig. 4.12B, the relation between load and wavelength is nearly linear, because the linear term dominates. There is a wavelength, L_m, for which the critical axial load for buckling is a minimum.

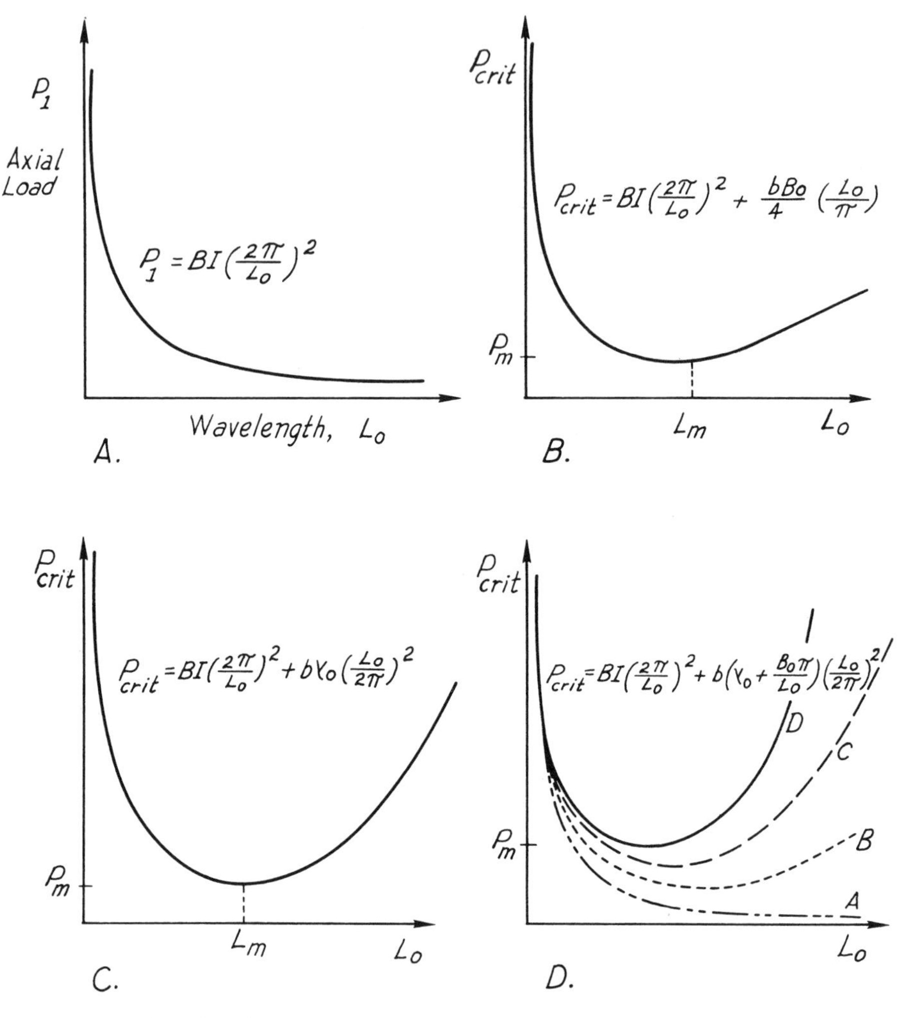

Figure 4.12. Relations between axial load and deflection of member resting on various kinds of media.

A. Unconfined member.

B. Elastic medium.

C. Dense, fluid medium.

D. Dense, elastic medium. Curves shown in A, B, and C also shown in D.

C. Member resting on dense, fluid medium. The critical load for buckling is essentially proportional to the square of the wavelength [eq. (4.35) with $B_0 = 0$], for large wavelengths,

$$P_{\text{crit}} = BI\left(\frac{2\pi}{L_0}\right)^2 + b\gamma_0\left(\frac{L_0}{2\pi}\right)^2,$$

if the medium is a dense fluid. The constraint of the dense fluid is proportional to the square of the wavelength and increases much more rapidly with increasing wavelength than does the constraint of an elastic medium, which is directly proportional to the wavelength. For large wavelengths, the relation between load and wavelength is nonlinear, as is shown in Fig. 4.12C. There is a wavelength for which the critical axial load is a minimum; it depends on the density of the medium and on the rigidity of the member.

D. Member confined by dense, elastic medium. The solid line in Fig. 4.12D is the relation between load and wavelength for a dense, elastic medium [eq. (4.35)]. Relations between buckling load and wavelength for other types of media are shown by dotted lines in Fig. 4.12D for comparison. Note that the minimum in the curve relating axial load and wavelength shifts to the left, that is, to a smaller wavelength, as first the elasticity, then the weight of the medium, are added as constraints.

Exercises

1. Calculate the wavelength, L_m, and the minimum axial load, P_m, for the buckling of an elastic member on a dense, elastic substratum.

   Ans: $$\left(\frac{L_m}{T}\right)^3 (2\gamma_0 L_m + B_0 \pi) = \frac{B}{6} (2\pi)^4,$$

   $$P_m = 3P_1 - b\gamma_0 \left(\frac{L_m}{2\pi}\right)^2,$$

   $$\text{where } P_1 = BI\left(\frac{2\pi}{L_m}\right)^2.$$

2. Compute the magnitude of the critical axial load and the wavelength, L_m, for the buckling of an elastic member on a dense, elastic substratum, where

   $$B = 3 \times 10^{11} \frac{\text{dn}}{\text{cm}^2},$$

   $$B_0 = \frac{B}{500} \frac{\text{dn}}{\text{cm}^2},$$

   $$T = 100 \text{ m},$$

   $$b = \text{unity},$$

   S.G. (specific gravity) of the incompetent medium = 2.4.

3. Compare the wavelength of a member on a dense, elastic medium with the wavelength of a member on a weightless, elastic medium using the data given in problem No. 2.

The strut member concept

The tendency for the small folds in each of the quartz veins shown in Fig. 4.1 to have approximately the same wavelengths and for the folds in the two quartz veins of different thicknesses to have different wavelengths is accounted for by our extension of Willis' strut member concept. According to our derivations, the wavelengths of folds in isolated layers in a soft medium are determined by elasticity or viscosity properties and by thicknesses of the layers. The thicker and more competent the layer and the softer the medium, the longer the fold wavelength; the thinner the layer and the more competent the medium, the shorter the fold wavelength. Under some conditions we should be able to measure fold wavelengths and thicknesses of isolated members in the field and calculate ratios of elastic moduli or of viscosity coefficients of the members and the confining media.

Sometimes we can recognize particularly thick layers or struts that seem to have controlled forms of folds. Usually, however, a group of beds served as the strut and it is difficult to decide which layers comprise the strut. For example, the two experimental folds shown in Fig. 3.23 involve several layers of relatively firm wax. Probably each of the layers influenced the size of the resulting folds, but we have no way of knowing how much friction there was among adjacent layers, so that we can estimate only maximum or minimum ratios of elastic moduli or viscosity coefficients of the layers and substratum, depending upon whether we assume zero friction or perfect bonding among the layers. We can see that the thickness of the buckled strata influences the wavelengths of the folds, because the fold involving the firm layers and the soft substratum is larger than the fold involving only the firm layers. Thus, even though we cannot necessarily identify a single strut member in a folded sequence of strata, we can still understand the relative sizes of folds.

The elastic resistance and the density of the confining medium, if increased, both tend to reduce the size of folds we would expect to find in nature. Both reduce the size of the wavelength of the initial deflections that requires the least axial load for significant amplification. As noted by Biot (*2*), theory predicts that there will be a practical limit to the sizes of folds that can occur in nature. Very large folds, with wavelengths of many miles, would require axial loads of such magnitudes that the strata would crush before they could be deflected significantly. The upper limit of fold sizes is a function of the thickness and the crushing strength of the struts.

In all our derivations we have assumed that deflections of buckled members are small. However, the folds in the quartz veins have large amplitudes, relative to wavelengths, and the slopes of the buckled veins far exceed the value of 20 degrees, which we adopted as the upper limit for our theory to be valid. For larger slopes, we showed, the approximate expression we have used for radius of curvature of the members is seriously in error. When we discuss ptygmatic folds we will be considering contortions that outrageously violate the assumption of small deflections that we have accepted heretofore. We can correct this error, however, by using the exact expression for the curvature of a member.

Ptygmatic folds—large deflections

Ptygmatic folds are highly contorted quartzo-feldspathic veins or dikes, usually in granitic or metamorphic rocks (*8*). Some examples are shown in Figure 4.13.

The meandering form of ptygmatic folds has at least two different origins: Some of the folds apparently grew while the magma forming them was being injected; others appear to be related to various structures indicating compression of the rock. These probably were caused by gross shortening (*15*). Thus, some ptygmatic folds seem to have been formed by buckling of a dike when the dike and its host rock were compressed, whereas others seem to have been formed by buckling of a dike while the dike was being injected into a soft medium.

Sederholm (*8*) studied many ptygmatic folds in granitic and metamorphic rocks on islands in southwestern Finland. Figure 4.13 shows some of the folded dikes Sederholm found on the island of Stickellandet. The folded aplite dikes are in metamorphic rocks. According to Sederholm, the aplite dikes he studied were

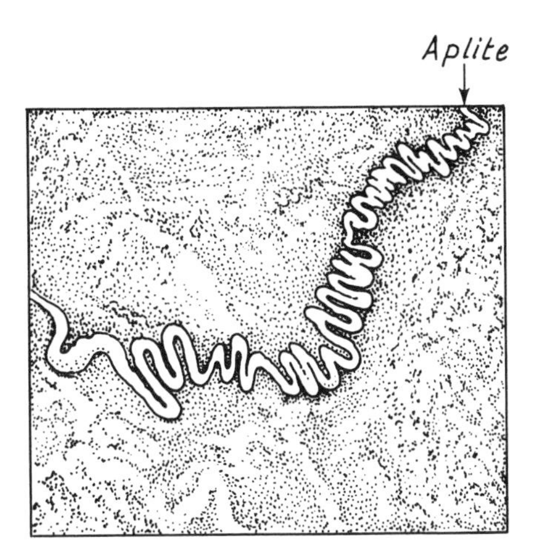

Figure 4.13. Ptygmatic folds in aplite dike (from Sederholm, ref. *8*).

folded before the granitic magma was completely solid, because folded dikes commonly are transected by other dikes that are straight. Thus, he reasoned, the folded dikes were crumpled while others were being injected. He also observed that folded aplite dikes may be distinctly preserved on one face of a specimen, but a few centimeters away, on the opposite face, they may have entirely lost their outlines and merged into a cloudlike aplite (*8*).

Ways Ptygmatic Folds Might Form.—Folds that bear a marked resemblance to ptygmatic folds have been produced experimentally under conditions that seem to be compatible with field relations of ptygmatic folds. For example, Ramberg (*7*) produced ptygmatic-like buckles by compressing layers composed of alternating thin, stiff rubber strips and thick, soft rubber strips. He has produced the same form with stiff rubber strips in a soft clay medium.

Wilson (*15*) produced ptygmatic-like buckles by injecting putty into soft gelatin. The conditions of his experiments presumably correspond roughly with the conditions under which a magma is folded while being injected. He found that the forms of the putty intrusions depend upon the relative consistency of the putty and the gelatin. When the gelatin was slightly more firm than the putty, the putty simply formed an irregular blob at the end of the tube Wilson used to inject the putty. When the gelatin was slightly softer than the putty, a ribbon of putty cut its way through the gelatin, only deflecting slightly. When the putty reached the edge of the tank of gelatin, it bent near its lower end and evolved into a series of tight buckles (Fig. 4.14). Wilson's experiments demonstrate a way ptygmatic folds could form in soft rocks that are not themselves being squeezed.

Whether ptygmatic folds are formed by general compression of the dike and its host or by buckling of the dike only, the peculiar feature of ptygmatic folds is that the member forming them was intensely buckled. In all our previous analyses we have assumed that deflections are small, so that none of them can apply to ptygmatic folds.

The Elastica.—One of the earliest problems of the theory of elasticity, solved by Euler in 1744, was the bending of a thin beam or rod (*6*). Euler laid the foundations of modern beam theory and solved many of the problems in beam theory, including the critical load for buckling of an unconfined member, which we derived in preceding pages. In addition, he derived the exact expression for the form of a thin plate or rod bent with large deflections. The form is called the *elastica*. We will follow the first few steps in the derivation of the *elastica*.

Consider a very thin member of length l, to the ends of which are applied axial loads, P, in the x-direction (Fig. 4.15A). The loads are greater than the critical load, P_1, so that a large deflection is expected. Let s be the distance along the arc and θ be the angle of inclination of the arc at any point along the bent member. Let α be the angle between the x-axis and the arc at the origin or coordinates. Thus, α is the maximum slope angle of the arc (Fig. 4.15A).

The radius of curvature, ρ, at any point along the arc is

$$\rho = \frac{[1 + (dv/dx)^2]^{3/2}}{|d^2v/dx^2|}, \qquad (4.36)$$

in terms of the displacement, v, in the y-direction. All the differential equations we

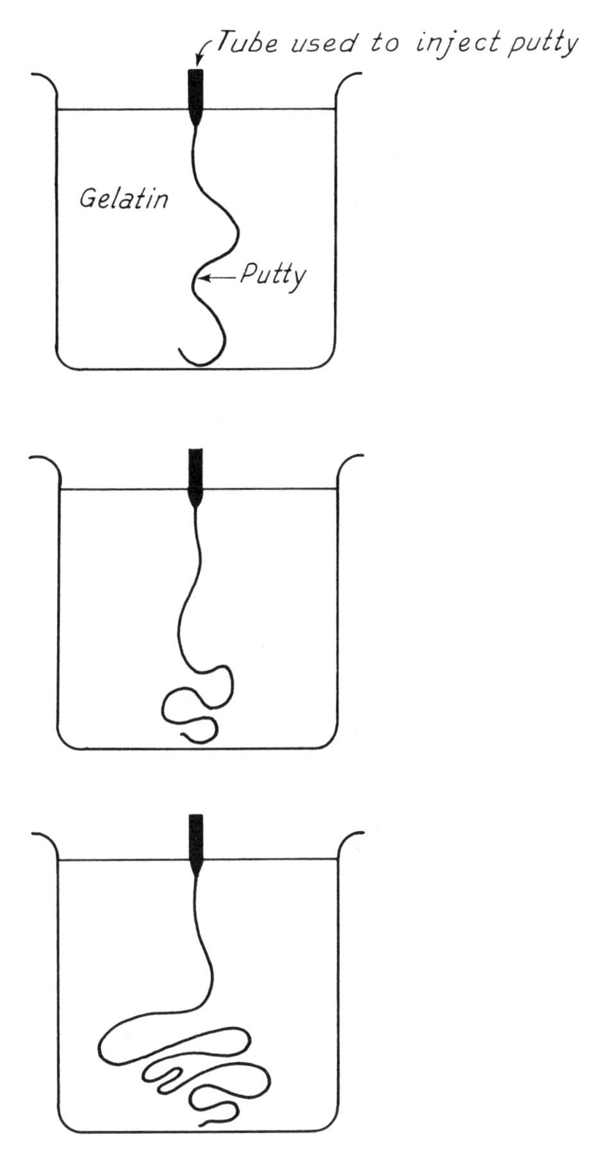

Figure 4.14. Meandering pattern formed by putty ribbon injected into soft gelatin (after Wilson, ref. *15*).

have derived until now incorporate the approximation that the curvature of members being bent is

$$\frac{d^2v}{dx^2}.$$

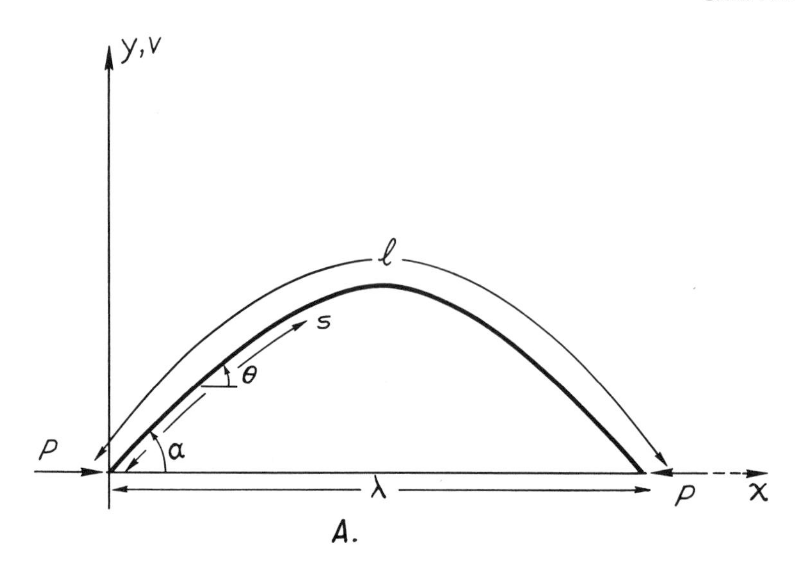

Figure 4.15. Coordinate system and radius of curvature of highly deflected member.
A. Coordinate system. s is the distance along the member, θ is the slope angle of the member.
B. Radius of curvature of an increment of arc length.

Thus, until now, we have assumed that the term

$$\left(\frac{dv}{dx}\right)^2$$

in eq. (4.36) is negligible. If the exact expression for curvature is used, the deflection will be determined precisely as long as strains within the member are small. This latter condition can be met if the thickness of the buckled member is very small.

The radius of curvature also can be expressed in terms of θ- and s-coordinates. If δs is the length of an element of arc and $\delta\theta$ is the angle between two radii of curvature of the arc (Fig. 4.15B), then $\delta s = \rho\, \delta\theta$, if the arc is circular. Thus, the radius of curvature for a circular arc is

$$\rho = \frac{\delta s}{\delta \theta}.$$

But this equation is valid for the radius of curvature of a curve of any form if the limit is taken of the right-hand side. Thus, in general,

$$\rho = \frac{ds}{d\theta}. \tag{4.37}$$

We showed earlier that for infinitesimal strain the radius of curvature of a bent, elastic member is

$$\rho = \frac{BI}{M}, \tag{4.38}$$

where M is the bending moment, B is the elastic modulus, and I is $bT^3/12$. The radius of curvature is proportional to the product of the modulus and moment of inertia of the cross-section of the member, and inversely proportional to the bending moment.

Combining eqs. (4.37) and (4.38), we have

$$BI\frac{d\theta}{ds} = M \tag{4.39}$$

as the fundamental differential equation for the deflection of an elastic member.

But for any section of the member,

$$M = -Pv, \tag{4.40}$$

where P is the axial load and v is the deflection. Combining eqs. (4.39) and (4.40),

$$BI\frac{d\theta}{ds} + Pv = 0, \tag{4.41}$$

we derive the differential equation for the deflection of an elastic member by axial loads.

But $dv/ds = \sin\theta$, so that if eq. (4.41) is differentiated with respect to s, we can write the differential equation in terms of θ and s:

$$\frac{d}{ds}\left[BI\frac{d\theta}{ds} + Pv\right] = 0,$$

$$BI\frac{d^2\theta}{ds^2} + P\frac{dv}{ds} = 0,$$

or

$$BI \frac{d^2\theta}{ds^2} + P \sin \theta = 0. \tag{4.42}$$

Equation (4.42) is a *nonlinear differential equation*, whose solution is an elliptic integral (*10,12*). It can be integrated as follows:

$$\frac{d}{ds}(-2 \cos \theta) = 2 \frac{d\theta}{ds} \sin (\theta), \tag{4.43}$$

and

$$\frac{d}{ds}\left(\frac{d\theta}{ds}\right)^2 = 2\left(\frac{d\theta}{ds}\right)\left(\frac{d^2\theta}{ds^2}\right). \tag{4.44}$$

Therefore, multiply each term in eq. (4.42) by 2 $d\theta/ds$. We find that

$$2\left(\frac{d\theta}{ds}\right)\left(\frac{d^2\theta}{ds^2}\right) + 2 \frac{P}{BI}\frac{d\theta}{ds} \sin \theta = 0,$$

so that, according to identities (4.43) and (4.44), we can rewrite eq. (4.42) as

$$\frac{d}{ds}\left[\left(\frac{d\theta}{ds}\right)^2 - 2 \frac{P}{BI} \cos \theta\right] = 0.$$

It follows that a solution is:

$$\left(\frac{d\theta}{ds}\right)^2 - 2 \frac{P}{BI} \cos \theta = C_1,$$

where C_1 is an arbitrary constant.

Now, according to Fig. 4.15A, the curvature vanishes at the ends of the member, where

$$\theta = +\alpha \quad \text{and} \quad -\alpha,$$

so that

$$C_1 = -2 \frac{P}{BI} \cos \alpha,$$

and

$$\left(\frac{d\theta}{ds}\right)^2 = 2 \frac{P}{BI} (\cos \theta - \cos \alpha). \tag{4.45}$$

Integrating once more,

$$\left(2 \frac{P}{BI}\right)^{1/2} \int_0^{l/2} ds = \int_\alpha^\theta \frac{d\theta}{(\cos \theta - \cos \alpha)^{1/2}}. \tag{4.46}$$

Equation (4.46) is an expression for the length, l, of the member being bent.

With it we can compute the slope angle, α, of the member at its origin for a given axial load, P. We will derive an expression for the deflection of the member subsequently.

According to Sokolnikoff and Redheffer (*10*), any integral of the form

$$\int [a \sin \beta + b \cos \beta + c]^{\pm 1/2} \, d\beta$$

can be transformed into a combination of an elliptic integral and other integrals that can be evaluated with elementary methods.

There are three kinds of elliptic integrals (*10*):

$$\left.\begin{array}{ll} \text{First kind:} & \displaystyle\int_0^{\phi} \frac{d\beta}{(1 - k^2 \sin^2 \beta)^{1/2}} = K(k) \\ \text{Second kind:} & \displaystyle\int_0^{\phi} (1 - k^2 \sin^2 \beta)^{1/2} \, d\beta = E(k) \\ \text{Third kind:} & \displaystyle\int_0^{\phi} \frac{d\beta}{(1 + n \sin^2 \beta)(1 - k^2 \sin^2 \beta)^{1/2}} \end{array}\right\} \quad (4.47)$$

Each of these can be expanded in an infinite series and can be integrated, term by term, to the accuracy required for a problem. However, elliptic integrals occur so frequently that this work has already been done for us and the values of the integrals are contained in many handbooks.

The term we want to integrate, eq. (4.46), is not obviously any of the above three kinds. However, it can be transformed into the form of an elliptic integral of the first kind by substituting new variables into eq. (4.46) and by using trigonometric identities. We will not follow the details of the lengthy transformation, which is clearly described by Timoshenko and Gere (*12*).

Timoshenko and Gere show that the solutions for axial load, P, and distance between the free ends of the member, λ, (Fig. 4.15) are

$$P = \frac{4BI}{l^2} [K(k)]^2$$

and (4.48)

$$\lambda = 2\left[2\left(\frac{BI}{P}\right)^{1/2} E(k) - 1\right],$$

where $K(k)$ is the value of the elliptic integral of the first kind and $E(k)$ is the value of the elliptic integral of the second kind.

Values of $K(k)$ and $E(k)$ are tabulated in many handbooks (*11*). Part of the table of elliptic integrals given in the Handbook of Chemistry and Physics is given in Table 4.2.

TABLE 4.2
Elliptic Integrals

Angle (α)	$K(k)$, First Kind	$E(k)$, Second Kind
0	1.5708	1.5708
10	1.5738	1.5678
20	1.5828	1.5589
30	1.5981	1.5442
40	1.6200	1.5238
50	1.6490	1.4981
60	1.6858	1.4675
70	1.7312	1.4323
80	1.7868	1.3931
90	1.8541	1.3506
120	2.1565	1.2111

The relation between deflection, δ, and axial load, P, is shown graphically in Fig. 4.16A. The curve is tangent to the horizontal line, $P = P_1$, at point A, where the deflection is zero. Thus a small increase of P above P_1 produces a large deflection of the member. This explains why the load was found to be independent of the magnitude of the deflection when we used the approximate relation for the curvature of the member.

The deflection reaches its maximum value, point B in Fig. 4.16A, when segments of the member are nearly in contact with one another. Three stages in the buckling process are shown in Fig. 4.17.

TABLE 4.3

$\alpha =$	0°	20°	40°	60°	80°	100°	120°	140°	160°	176°
$\frac{P}{P_1} =$	1	1.015	1.063	1.152	1.293	1.518	1.884	2.541	4.029	9.116
$\frac{\lambda}{l} =$	1	0.970	0.881	0.741	0.560	0.349	0.123	−0.107	−0.340	−0.577
$\frac{\delta_0}{l} =$	1	0.220	0.422	0.593	0.719	0.792	0.805	0.750	0.625	0.421

Table 4.3 relates angle of deflection, α, at the origin of coordinates; applied load, P, and distance, λ, between two ends of the buckled member (*12*) (p. 79). Here l is the length of the member and P_1 is given by

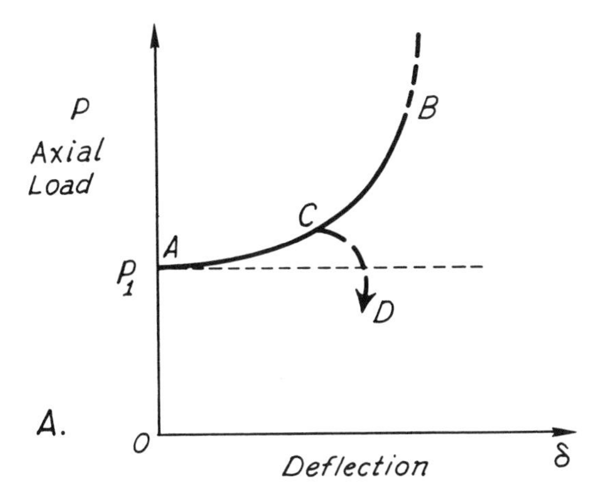

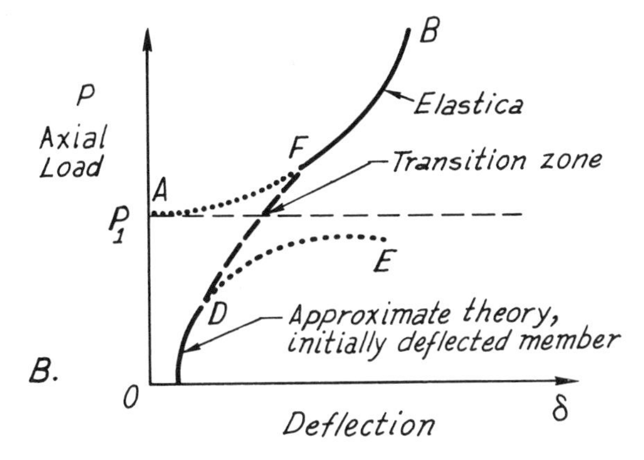

Figure 4.16. Relation between axial load and deflection of highly deflected member.
A. Initially straight member.
B. Approximate relation for initially deflected member.

$$P_1 = BI\left(\frac{\pi}{l}\right)^2.$$

Figure 4.16B shows a composite diagram of the plot of deflection versus load for initially deflected members. Curve *CDE* relates deflection to load for a member with an initial deflection of amplitude δ_0. Curve *AFB* is the relation between deflection and axial load for the elastica. At large bending, initial deflections in members do not markedly affect the relation between load and deflection, so that

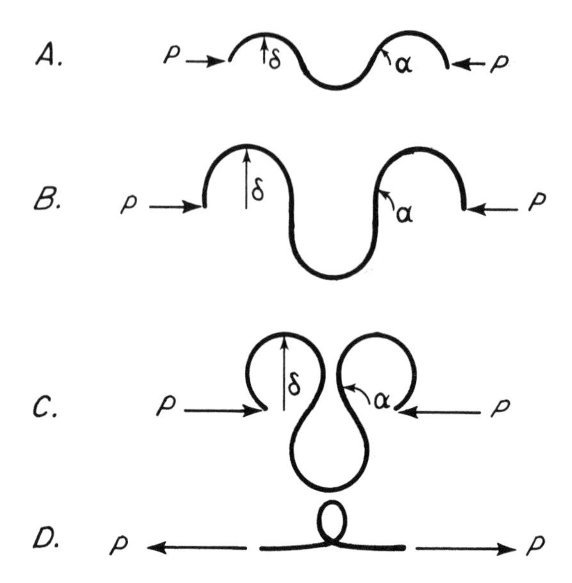

Figure 4.17. The elastica. Sketch of form of highly bent member. In C the loops nearly are in contact with each other. In D the ends of the elastic strip have actually passed and the strip forms a closed loop (after Love, ref. *6*, p. 404).

curve *CDFB* probably describes the behavior of crooked members, or members with initial deflections. All real members have initial deflections.

Relation between Theoretical and Actual Forms of Ptygmatic Folds.— The resemblance between the form of the highly bent, thin, elastic member and the forms of ptygmatic folds, such as those shown in Figs. 4.13 and 4.14, is truly remarkable. Both have a meandering form. We have ignored the effect of a confining medium, but by analogy with our analysis of small deflections of unconfined and confined members we probably can speculate safely that the form of a highly bent member in a soft medium is very similar to the form of the same member buckled in air.

One obvious discrepancy between our model and the actual members buckled into ptygmatic folds is the mechanical behavior of the members. Our mathematical model consists of a very thin, elastic member. The actual members consist of magma or aplite dikes, which probably do not bend elastically. We could show, however, that the form of an isolated viscous member that is highly bent is identical to that of an isolated elastic member (*3*). We will show in Chapter 7 that we simply replace the modulus, B, in the differential equation for deflection,

$$BI\frac{d\theta}{ds}+Pv = 0,$$

with A, where

$$A = 4\eta \frac{\partial}{\partial t}$$

and η is the viscosity of the member. Thus, the differential equation becomes

$$4\eta I \frac{\partial^2 \theta}{\partial t\, \partial s} + Pv = 0.$$

The solution to the latter equation is nearly identical to that of the former. The primary difference is that the deflection of the viscous member is a function of time as well as of the axial load. Also, if the axial load is small, the amount of thickening of the viscous member might overshadow the amount of buckling. Otherwise the problems are essentially identical and there probably is no obvious way of distinguishing forms of buckles in elastic and viscous rock members.

If the member possesses a definite yield point, however, its deflection is not described by our theory. For example, the relation between axial load and amplitude shown by the solid lines in Fig. 4.16A is valid only to the point where the strength of the member is reached. When the deflection is described by point C in that figure, the member is beginning to buckle inelastically. At that point the member begins to collapse so that it cannot support a larger axial load. In fact, the greater the deflection beyond that point, the less axial load it can support, and the load-deflection relation is approximately as is indicated by the dashed line, C–D, in Fig. 4.16A. The existence of a yield strength probably could cause the form of the buckled member to be more peaked than is indicated in Figs. 4.17. Instead of the crest of anticlines and the troughs of synclines being smooth curves, they would be bent sharply as is shown in Fig. 1.2. This behavior and viscous buckling will be discussed in detail in Chapter 8.

The problems of intense buckling of an elastic or a viscous member in a medium have not been solved exactly. Neither has the problem of the intense bending of a member with initial deflections been solved. However, Chapple (*3*) has determined the forms of intensely buckled viscous members in viscous media by means of numerical approximations and a computer. He showed that the forms of intensely buckled members depend upon the relative viscosities of the medium and the member, but only slightly. The reader is referred to the original paper (*3*) for details.

References cited in Chapter 4

1. Biot, M. A., 1937, "Bending of an Infinite Beam on an Elastic Foundation": *Jour. Applied Mechanics*, V. 4, p. A1–A7.
2. ———, 1961, "Theory of Folding of Stratified Viscoelastic Media and its Implication in Tectonics and Orogenesis": *Geol. Soc. Amer. Bull.*, V. 72, p. 1595–1632.
3. Chapple, W. M., 1968, "A Mathematical Theory of Finite-Amplitude Rock-Folding": *Geol. Soc. Amer. Bull.*, V. 79, p. 47–68.

4. Currie, J. B., Patnode, H. W., and Trump, R. P., 1962, "Development of Folds in Sedimentary Strata": *Geol. Soc. Amer. Bull.*, V. 73, p. 655–674.
5. Goodier, S. N., 1967, "Lectures on Elastic-Plastic Instability": Stanford University (unpub.).
6. Love, A. E. H., 1944, *The Mathematical Theory of Elasticity*: Dover Publications, N.Y.
7. Ramberg, Hans, 1963, "Evolution of Drag Folds": *Geological Magazine*, V. 100, p. 97–110.
8. Sederholm, J. J., 1926, "On Migmatites and Associated Pre-Cambrian Rocks of Southwestern Finland": *Bull. De La Comm. Geologique De Finlande*, no. 77.
9. Sherwin, Jo-Ann, and Chapple, W. M., 1968, "Wavelengths of Single Layer Folds: A Comparison Between Theory and Observation": *Amer. Jour. Sci.*, V. 266, p. 167–179.
10. Sokolnikoff, I. S., and Redheffer, R. M., 1958, *Mathematics of Physics and Modern Engineering*: McGraw-Hill Book Co., Inc., N.Y.
11. Synge, J. L., and Griffith, B. A., 1959, *Principles of Mechanics*: McGraw-Hill Book Co., Inc., N.Y., p. 327–332.
12. Timoshenko, S. P., and Gere, J. M., 1961, *Theory of Elastic Stability*: McGraw-Hill Book Co., Inc., N.Y.
13. Willis, Bailey, 1894, "Mechanics of Appalachian Structure": *Thirteenth Annual Report, U.S. Geol. Survey*, 1891–92, p. 213–281.
14. ———, 1934, *Geologic Structures*: McGraw-Hill Book Co., Inc., N.Y.
15. Wilson, G., 1952, "Ptygmatic Structures and Their Formation": *Geological Magazine*, V. LXXXIX, p. 1–21.

5

Chapter Sections

Stress
- Internal Forces
- Concept of Stress
- Stress as a Vector Quantity
- Stresses on Planes with Various Orientations
- Reason for Nine Stress Components
- Notation for Stresses
- Signs of Stresses

Strain
- Displacement and Strain
- Concept of Strain
- Components of Displacement and Infinitesimal Strain
  - Translation
  - Rotation
  - Elongation and Contraction
  - Shear
- Finite Strain
- Superposition of Strains

Elastic Constants and Stress-Strain Relations
- Young's Modulus and Poisson's Ratio
- Relations among Normal Stresses and Normal Strains
- Young's Moduli and Poisson's Ratios of Some Earth Materials
- Modulus of Rigidity

Plane Stress and Plane Strain — Important Special Cases
- Plane Stress
- Plane Strain
- Differences Between Bending of Beams and Plates

References Cited in Chapter 5

5

Theoretical Interlude: Stress, Strain, and Elastic Constants

Stress

Internal Forces.—We have shown that internal shear and normal forces and internal bending moments are required to balance external forces on members such as beams, plates, and rock layers. Through analysis of combinations of these internal forces and moments acting on internal cross-sections of layers we have been able to study the deformation of country rock around laccoliths and the formation of certain kinds of folds. We did mention stresses when we assumed a distribution of fiber stresses and related the stresses to bending moment, but we did not step back to examine precisely what we meant by "stress."

We can ignore stresses no longer, however, because in following problems we need to relate deformation of an object to its material properties. We need to determine intensities of forces, that is, stresses, rather than forces within materials, because resistance to deformation and capacity to resist forces depend directly upon these intensities. For example, we know from experience that a wooden 4 × 4 post can resist far greater forces than can a 2 × 4 board. The difference is that, for the same loading conditions, the intensity of forces within the 2 × 4 is far greater than the intensity of forces within the 4 × 4. The intensity of forces, or stress, is determined by such things as the cross-sectional area of the wooden member and the magnitude of the applied force. Another factor is the way in which the force is applied to the member, viz., the different rigidities of a 2 × 4 bent by pushing on its 4-inch side and on its 2-inch side. Yet, experience shows that the maximum intensities of internal forces the 2 × 4 and the 4 × 4 can withstand are the same if they are made of identical materials.

Concept of Stress.—The intensity of force on a small area is the essence of our concept of *stress*. Stresses are measured in units of force divided by units of area.

Accordingly, stresses multiplied by the respective areas on which they act are increments of force. Sums of these force increments maintain the equilibrium of a body, and these sums appear in the equations of equilibrium with which we have been dealing in preceding pages. There are shear forces and normal forces and there are shear stresses and normal stresses. The meanings are similar. A shear force and a shear stress act tangentially to a surface. A normal force and a normal stress act normally to a surface.

In order to define stresses more completely, let us consider a beam loaded simply by axial forces (Fig. 5.1A). On any internal plane parallel to the ends of the beam, there acts only a force, F, normal to the plane. On an internal plane with a different

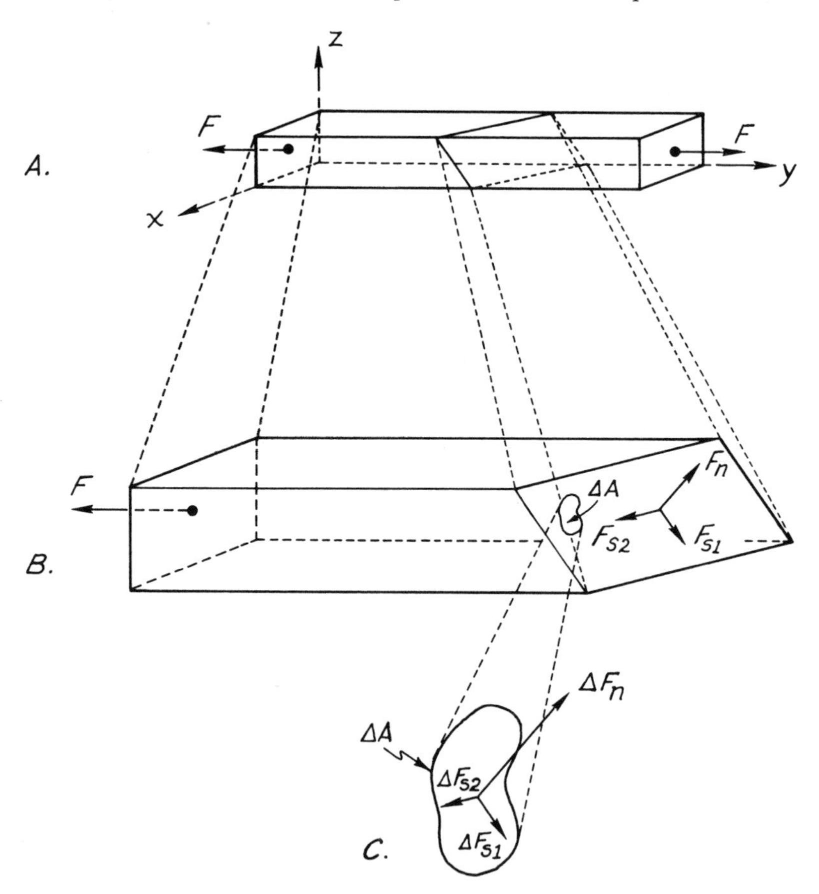

Figure 5.1. Intensities of forces on an arbitrary internal plane of a beam.
A. Beam loaded by forces distributed over ends.
B. Free-body diagram of part of beam.
C. Force intensities on small area, ΔA, of plane.

orientation, however, we can resolve the force, F, into three components, a normal force, F_n, acting normally to the plane and two shear forces, F_{s1} and F_{s2}, acting tangentially to the plane.

The average components of shear and normal stresses are simply the respective shear and normal forces divided by the area of the cross-section on which the forces act. In general, however, the force intensities vary from point to point on a cross-section. For example, if the beam is supporting a bending moment, we know that the material at the outer surfaces of the beam supports greater force intensities than does material at the neutral plane of the beam. Indeed, the intensity at the neutral plane is zero, as the term "neutral plane" implies. Accordingly, stress, that is, force intensity near a point in a body, is defined as *the limit of the force intensity as the area approaches zero*. We will use the symbol sigma (σ) for stresses. Thus, we select a small area, ΔA, near some point on a section of the beam, Fig. 5.1B, and consider the parts, ΔF_n, ΔF_{s1}, and ΔF_{s2}, of the forces distributed over that small area. Then the normal stress component is defined as the force intensity, $\Delta F_n/\Delta A$, as the area approaches zero, $\Delta A \rightarrow 0$:

$$\sigma_n = \lim_{\Delta A \rightarrow 0} \frac{\Delta F_n}{\Delta A}. \qquad \text{Normal stress near a point.} \tag{5.1}$$

Similarly, the shear stresses are:

$$\left.\begin{aligned} \sigma_{s1} &= \lim_{\Delta A \rightarrow 0} \frac{\Delta F_{s1}}{\Delta A}, \\ \sigma_{s2} &= \lim_{\Delta A \rightarrow 0} \frac{\Delta F_{s2}}{\Delta A}. \end{aligned}\right\} \qquad \text{Shear stresses near a point.} \tag{5.2}$$

The limit, $\Delta A \rightarrow 0$, cannot be realized in actual materials because the surfaces of actual materials are not actually continuous. For example, when we consider stresses on or within crystals, the limital area, ΔA, must remain large with respect to atomic distances. For granular solids, such as sands, the limital area must be on the order of several square millimeters. For coarsely crystalline materials, such as granite, the limited area must be on the order of several square centimeters; in this case, the stress acting on a large granite body would be defined in terms of force intensities averaged over an area of perhaps one hundred square centimeters. In fractured rock, such as a jointed basalt flow on the Columbia River Plateau, the limital area may be on the order of several square meters. The area of a "point" in jointed basalt, therefore, may be ten square meters. The area of a "point" in any material depends upon the scale of the discontinuities within the material and upon the scale of the deformation being considered. Whenever we use the symbol for the limiting process, $\lim_{\Delta A \rightarrow 0}$, therefore, we refer to the limit as the incremental area approaches the smallest continuous area of the system we are studying.

Stress as a Vector Quantity.—In order to illustrate the vectorial quality of stress, let us examine internal forces and stresses of a beam externally loaded so that an internal cross-section of the beam is subjected to normal, N, and shear, Q, forces and a bending moment, M (Fig. 5.2A). The components of internal forces are of varying magnitudes at different cross-sections of the beam so that their resultants, R (Fig. 5.2B), are of varying directions and magnitudes. The internal forces and moments are vectorial.

The internal forces and bending moments for a particular cross-section can be represented in terms of equivalent stress distributions, as is shown in Fig. 5.2C. For example, the shear stress, σ_s, might be uniform over the cross-section so that it can be replaced by small arrows of uniform length and direction. Similarly, the normal force may be represented by the sum of uniform normal stresses, σ_N, times small areas. The bending moment might be represented by a stress distribution, σ_M, such as that

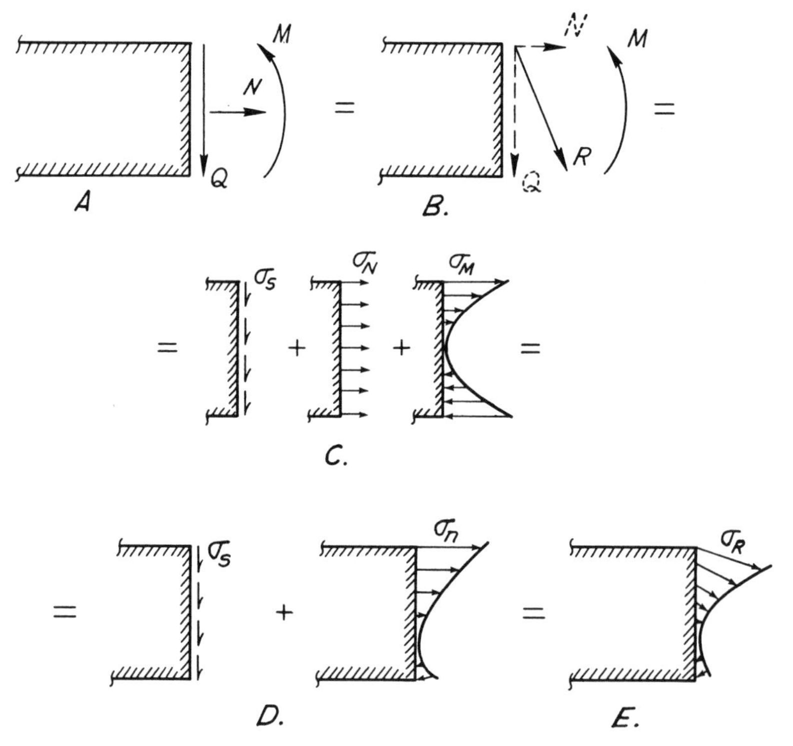

Figure 5.2. Internal forces and bending moment represented by equivalent stress distributions.

A. Force-components and bending-moment vector.
B. Force and bending-moment vectors.
C. Normal- and shear-stress components.
D. Normal- and shear-stress components.
E. Resultant stress vectors.

shown in Fig. 5.2C, where the stresses causing the moment are maximum at each edge of the beam and zero at the center.

The two distributions of normal stress can be added so that we have simply a distribution of normal stresses and a distribution of shear stresses (Fig. 5.2D). Finally, the components of shear and normal stresses on a particular plane can be added vectorially to yield a distribution of stress vectors acting on the plane (Fig. 5.2E). The vector diagram for stresses, Fig. 5.2E, is equivalent to the vector diagram for force, R, and moment, M, in Fig. 5.2B.

Stresses on Planes with Various Orientations.—A peculiarity of stresses acting on internal cross-sections, or *planes*, of a body, such as a beam, is that their magnitudes are functions of the inclinations of the planes. Consider a beam simply loaded by axial tensile forces, F_y (Fig. 5.3A). The force intensity on each end of the

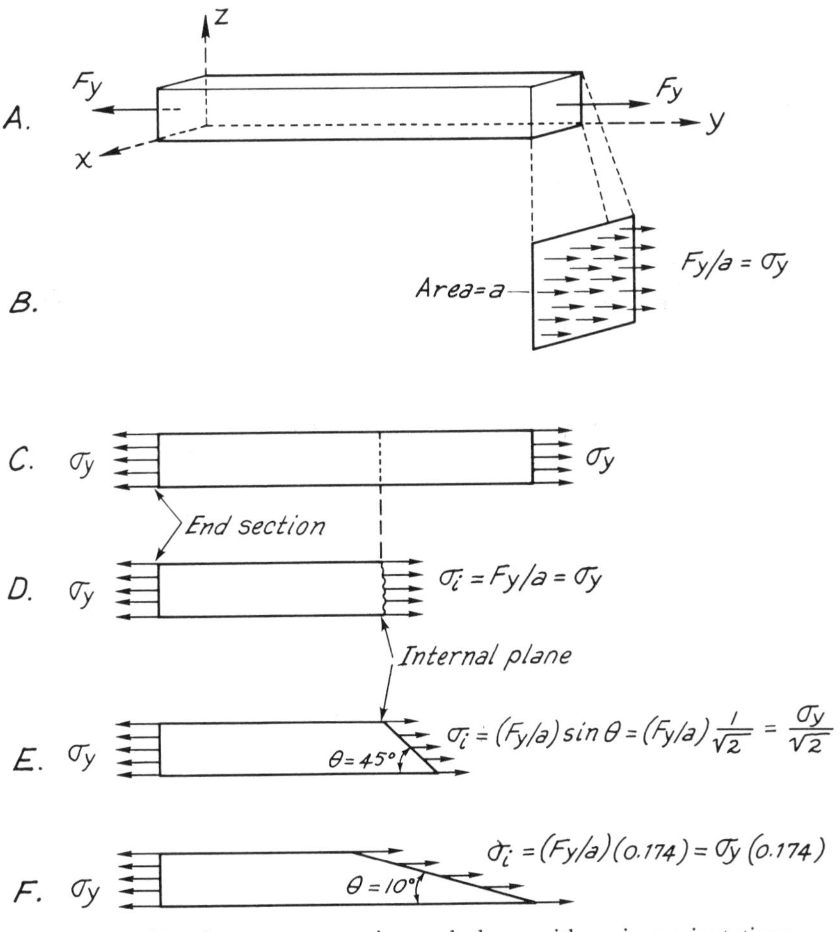

Figure 5.3. Stress vectors on internal planes with various orientations.

beam is the force, F_y, divided by the cross-sectional area of the beam, a, if the force is uniformly distributed over the end of the beam. Therefore, the force intensity, or average stress, σ_y, is (F_y/a). The stress on an internal plane, *parallel to the ends*, Fig. 5.3C, is the same as the stress on the ends. If the internal section is inclined at an angle of $\theta = 45$ degrees to the end section, however, the area of the internal section is $1/\sin\theta$, or $(2)^{1/2}$ times the area of the end section. Now, the stress vector, σ_i, on

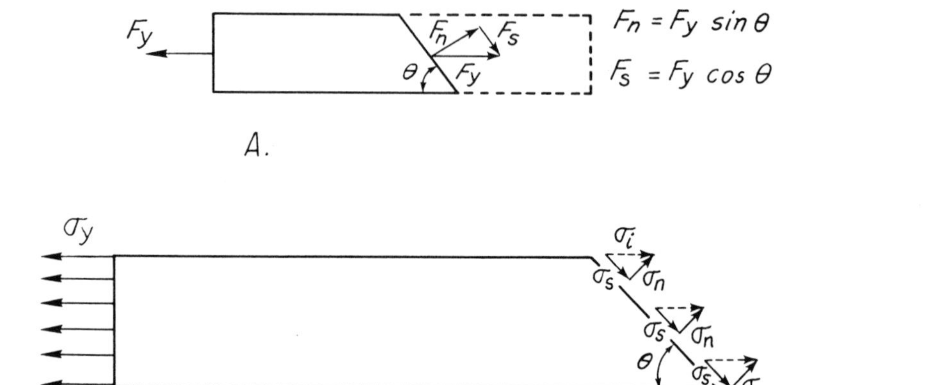

Figure 5.4. Components of forces and stresses on inclined internal plane of beam.

the internal plane is inversely proportional to the area of the plane, so that its magnitude is *reduced* to $\sigma_i = \sigma_y \sin\theta = \sigma_y/(2)^{1/2}$. If the internal plane is inclined even farther, say to an angle of 10 degrees, the stress vector is reduced to about one-sixth of that on the end section (Fig. 5.3F). Thus, different values of stress are obtained for planes with different orientations; the force remains the same but the force intensities, or stresses, vary. *This is a fundamental, if not the fundamental, difference between stresses and forces.*

The stress vectors, just like the force vectors themselves, can be resolved into components parallel to and normal to the planes on which they act. For example, consider the force on the internal plane shown in Fig. 5.3E. Each of the force vectors can be resolved into normal, F_n, and shear, F_s, components, as is shown in Fig. 5.4A. The magnitudes of the force components depend upon the angle of inclination of the plane, so that

$$F_n = F_y \sin\theta \quad \text{and} \quad F_s = F_y \cos\theta.$$

The magnitudes of the components of stress depend also upon the angle of inclination of the plane, as is shown in Fig. 5.4B. A significant difference between components of stresses and components of forces, however, is that components of stress vectors depend upon the product of two trigonometric functions whereas the components of force vectors depend upon a single trigonometric function. Therefore, we cannot determine components of a stress vector by the method used to determine components of a force vector. The reason is that magnitudes of vectors of stress depend upon the areas of the planes on which they act as well as their directions. We will see in Chapter 9 that the dependence of stress components upon the product of two trigonometric functions causes Mohr's Circle, a useful graph of stresses, to be constructed in terms of 2θ instead of θ. Here the important observation is simply that components of forces and components of stresses must be manipulated in significantly different ways (*4*).

Exercises

A cube of unit dimensions is compressed uniformly in one direction by an axial stress, σ_a. Derive expressions for shear stress, σ_s, and normal stress, σ_n, acting on a plane inclined to the y-axis at an angle, θ.

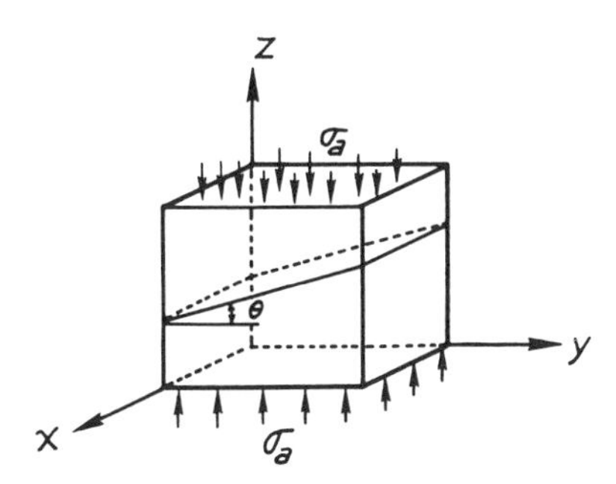

Construct one graph showing the relation between shear stress and θ and another graph showing the relation between normal stress and θ, where θ ranges from zero to 180 degrees. On which planes are σ_n and σ_s maximum? Minimum? Is the maximum normal stress larger than the maximum shear stress?

Reason for Nine Stress Components.—We showed that, even in the simplified case of a uniformly loaded beam, Fig. 5.4, the magnitudes of stresses acting on a cross-section, or plane, of a beam, indeed a plane through any body, depend upon the orientation of the plane. For each orientation of plane there are different stress components. There are unlimited planes of different orientation passing through a point in a body, and the stresses acting on each plane passing through the point will have different stress components. Accordingly, knowledge of the stress components on only one of the planes is incomplete knowledge of the state of stress at the point. A complete description of the state of stress at the point requires that we specify the stress components or their resultant vectors on all possible planes through the point.

It turns out, fortunately, that we need not write expressions for stresses on an unlimited number of planes. Rather, we can completely specify the state of stress at the point by indicating the stresses on any three planes passing through the point. The stress components on any other plane can be derived, with trigonometry, from the stress components on the three planes. We will prove this when we derive Mohr's Circle in Chapter 9. Mohr's Circle is an especially useful graphical method for two-dimensional states of stress; we can use it for deriving stresses on any plane if we are given maximum and minimum principal stresses, or normal and shear stresses on two mutually perpendicular planes, or even stresses on any two planes.

There are nine stress components of three-dimensional systems because each stress vector acting on each of three planes can be resolved into three components of stress, one normal and two shear components. Three times three makes nine, so that nine stress components are required to define completely the state of stress near a point (see Fung, ref. *3*, p. 41, for a discussion of the stress tensor).

Notation for Stresses.—Consider a small square element, Δx and Δy, with sides parallel to the coordinate axes (Fig. 5.5). A normal stress acting in the y-

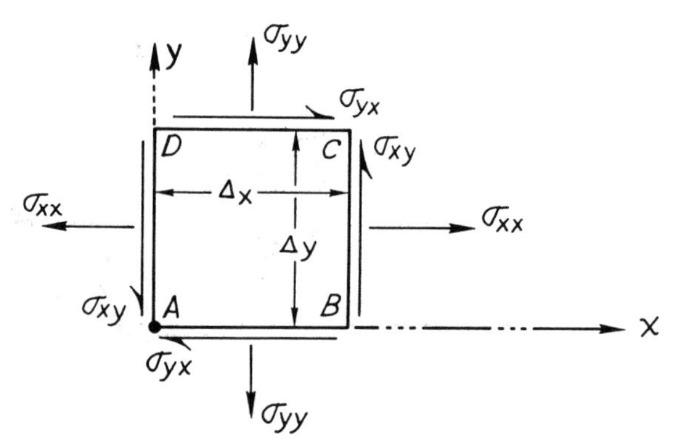

Figure 5.5. Notation for stresses in two dimensions. The convention used here is that a normal stress is positive if it is tensile. Two subscripts are used to designate a stress. The first subscript letter indicates the direction of the normal to the plane on which the stress acts. The second indicates the direction of the component of stress, that is, the direction of the arrow.

direction is applied to sides *DC* and *AB*, Fig. 5.5. We will denote this stress by σ_{yy}, using two subscript letters. The first subscript indicates the direction of the normal to the plane on which the stress acts and the second indicates the direction of the component of stress, that is, the direction of the arrow. Thus the normal stress acting on sides *AD* and *BC* of the square is denoted by σ_{xx}. Also, the component of shear stress acting in the x-direction, on sides perpendicular to the y-axis, sides DC and AB, is denoted by σ_{yx}. By convention, the shear stresses that act *counterclockwise* with respect to a point in the center of the unit square are *positive*.

The convention used in this book is that a *normal stress is positive if it is tensile.* Although this convention is opposite to that generally used in geological literature, it is the convention adopted in engineering mechanics and materials science (*3,4,5*). We will continue for many years to draw heavily on engineering literature for the fundamentals of mechanics; we should, therefore, be familiar with the conventions used by engineers.

The designation of two shear stresses, σ_{xy} and σ_{yx}, usually is an unnecessary complication. This may be demonstrated as follows. Assume that the element in Fig. 5.5. is in static equilibrium. Then, the summation of moments about any point is zero. If moments are taken around one corner of the unit square, say point A, the forces times their lever arms must sum to zero. Thus,

$$\left[\sum M_A = 0\right] b\sigma_{xx}\,\Delta y\,\frac{\Delta y}{2} - \sigma_{xx}b\,\Delta y\,\frac{\Delta y}{2} + \sigma_{yy}b\,\Delta x\,\frac{\Delta x}{2} - \sigma_{yy}b\,\Delta x\,\frac{\Delta x}{2} - \sigma_{yx}b\,\Delta x\,\Delta y + \sigma_{xy}b\,\Delta y\,\Delta x = 0,$$

from which it follows that

$$\sigma_{xy} - \sigma_{yx} = 0.$$

Here Δy or Δx is the lever arm and b is the thickness of the element. Therefore, the magnitude of σ_{xy} is equal to the magnitude of σ_{yx}.

The notation I have adopted for this book is merely one of several variously used in engineering, physics, geophysics, geology, and mathematics. Different authors use different systems of notation for stress components. For example, Sokolnikoff (*8*) uses the notation:

$$\begin{vmatrix} \tau_{11} & \tau_{12} & \tau_{13} \\ \tau_{21} & \tau_{22} & \tau_{23} \\ \tau_{31} & \tau_{32} & \tau_{33} \end{vmatrix} \equiv \tau_{ij}; \quad \begin{matrix} i = 1, 2, 3 \\ j = 1, 2, 3 \end{matrix}, \tag{5.3}$$

instead of the notation we are using:

$$\begin{vmatrix} \sigma_{xx} & \sigma_{xy} & \sigma_{xz} \\ \sigma_{yx} & \sigma_{yy} & \sigma_{yz} \\ \sigma_{zx} & \sigma_{zy} & \sigma_{zz} \end{vmatrix} \equiv \sigma_{ij}; \quad \begin{matrix} i = x, y, z \\ j = x, y, z \end{matrix}. \tag{5.4}$$

Some other authors (*3,4,5*) use,

$$X_x \equiv \sigma_{xx}; \qquad X_y \equiv \sigma_{xy}, \qquad \text{etc.,} \tag{5.5}$$

$$\widehat{xx} \equiv \sigma_{xx}; \qquad \widehat{xy} \equiv \sigma_{xy}, \qquad \text{etc.,} \tag{5.6}$$

or

$$\sigma_x \equiv \sigma_{xx}; \qquad \tau_{xy} \equiv \sigma_{xy}, \qquad \text{etc.} \tag{5.7}$$

Perhaps the most commonly used notation in geological literature is the last example, where τ (tau) implies shear stress and σ implies normal stress. I have adopted the notation σ_{xx}, σ_{xy}, etc. because it is intermediate between tensor notation and the notation commonly used in geological literature. Once you have learned one notation, however, it is a simple matter to decipher other notations by studying them for a few minutes.

Signs of Stresses.—The type of sign convention we adopted for force vectors in beam and plate theory is also useful for sign conventions of stresses. The sign convention we will use for stresses depends upon definition of positive and negative sides of the stress element shown in Fig. 5.6. We define a side facing a positive co-

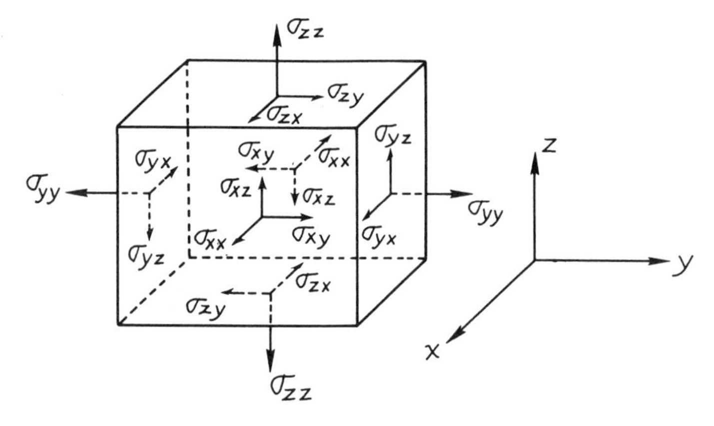

Figure 5.6. Notation for stresses in three dimensions. The notation and directions of positive and negative stresses are the same as those for two dimensions.

ordinate direction as a positive side, and a side facing a negative coordinate direction as a negative side (Fig. 5.7A). Then the convention for signs of stresses becomes remarkably simple: Positive stress components act on positive sides in positive directions or they act in negative directions on negative sides. The reader should examine Figs. 5.5 and 5.6 in detail, verifying that each stress component shown is positive. Figs. 5.7B and 5.7C show examples of positive and negative stress components.

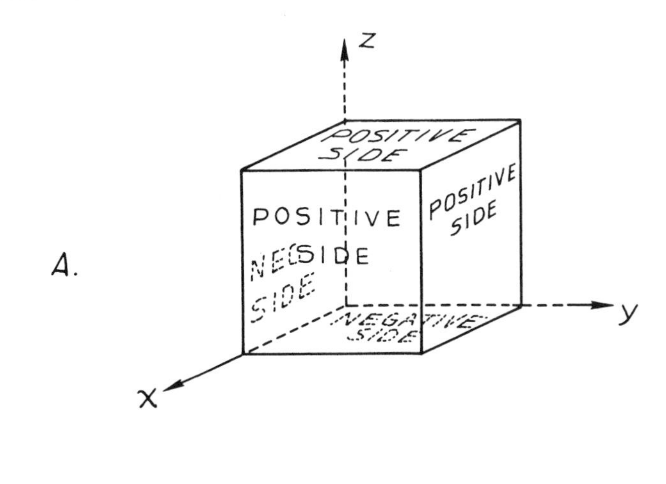

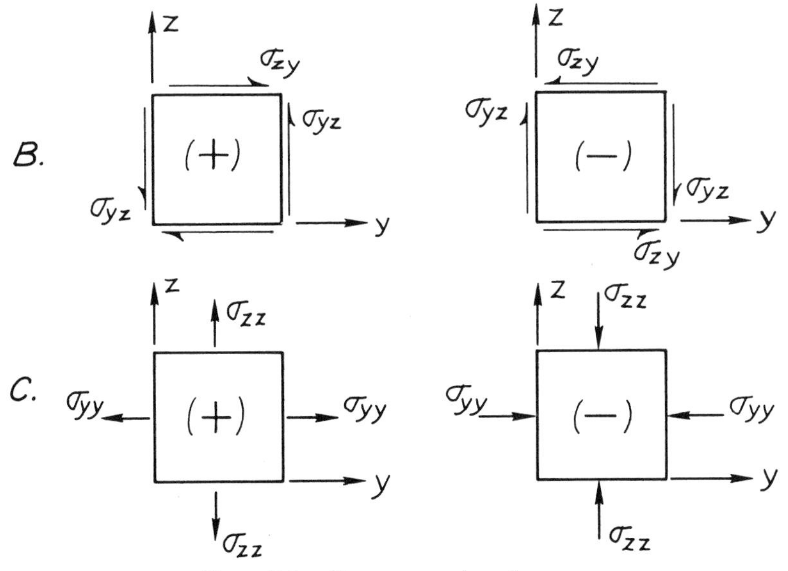

Figure 5.7. Sign conventions for stresses.

Strain

Displacement and Strain.—We have shown that forces can be divided into little increments and expressed in terms of force intensities, or stresses, which can be manipulated in order to increase our detailed understanding of the way in which small particles of a body, such as a layer overlying a laccolith, are squeezed, pulled, or sheared. Sums of the force intensities times units of area give the familiar forces and bending moments which are described by Newton's laws. In a similar way we can increase our understanding of displacements and distortions of a body by considering the deformations of small elements of the body. Sums of these small deformations within a body give displacements of points within the body. Whereas stresses are related to forces, *strains* are related to displacements, and whereas normal stress is force divided by area, normal strain is change in length divided by the original length of the element.

Strains become important when we are concerned with internal deformations of bodies, when we wish to compare deformation intensities of bodies of different sizes, and, above all, when we wish to relate deformation of bodies to stresses acting on and within the bodies. Relations between internal stresses and internal strains or strain rates are the all-important rheological properties of materials.

There are two types of displacements which we must separate when we contemplate strain of a body. One type of displacement involves *translation* as a rigid body, without internal deformation and, therefore, without strain of the body. For example, when we drive an automobile from one town to another, the auto is moved laterally and vertically and even rotated about various axes, but the driving causes little permanent distortion of the auto. If we manage to avoid colliding with other vehicles or with additional obstacles along our route, the auto will reach the other town with the same length, width, height, and shape as when it left the first town. Thus, a body can be translated without deformation by moving laterally and by rotation. The other type of displacement involves relative change of position of particles within the body. This deformation we call strain. It can be divided into two general categories: changes in length of lines within the body, which is *normal strain*, and changes in angles between interesecting lines, which is *shear strain*. Normal strains are related to normal stresses and shear strains are related to shear stresses.

Most displacements of geological interest are a combination of translation and deformation and both types of displacement must be recognized in order to understand the final displacements. A landslide, for example, may result in several feet or even tens of feet of lateral translation of a mass of soil or rock. Slippage may be along a single plane, however, so that the internal deformation is entirely restricted to a thin film of clay along the slip plane. In this case, translation is responsible for displacement of most of the slide mass. The lateral displacement of a glacier, on the other hand, is predominantly by internal deformation, which accumulates from the base to the surface of the ice mass, causing the surface to move at an appreciable rate. Lastly, it is becoming increasingly clear that orientations of megascopic fractures of rock are as likely to be reflections of possible directions of translation of the rock parts separated by the fractures as to have been caused by the stresses and strains within the rocks. This is the primary reason why gross failure of rock, such as faulting, normally cannot be predicted simply with failure criteria, which are in terms of stresses or strains alone. More about this later.

Concept of Strain.—A physical test used in the design of underground structures in rock, such as mines and tunnels, is the uniaxial compression test (*6*). In this test, a sample, usually a rock core, is placed between two platens of a loading machine (Fig. 5.8A) and an increasing force is applied through the platens to the sample until the sample breaks. The force at which the sample breaks divided by the area of the end of the sample is an average stress, called the compressive strength of the sample. The compressive strength is different for different kinds of rocks and it gives the design engineer an indication of rock behavior during excavation. In addition to compressive strength, the breaking strain of the sample usually is observed. As the sample is subjected to larger and larger loads, the change in length of the sample is measured, for example by placing a dial gauge between the two loading platens (Fig. 5.8B), and the difference between the length of the sample before loading and at failure is recorded. The changes in length are different for test specimens of different lengths. If, however, we divide the change in sample length at

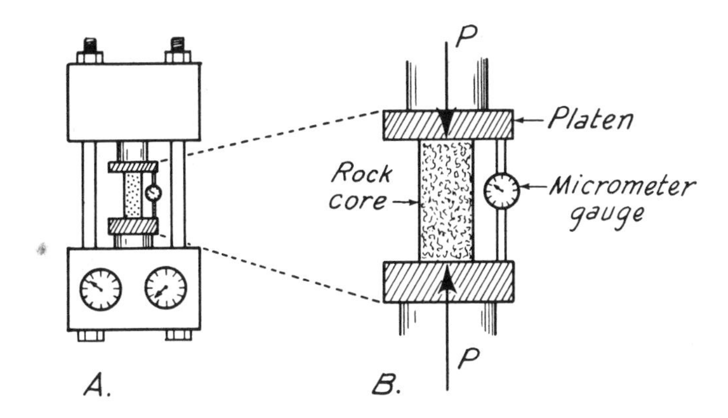

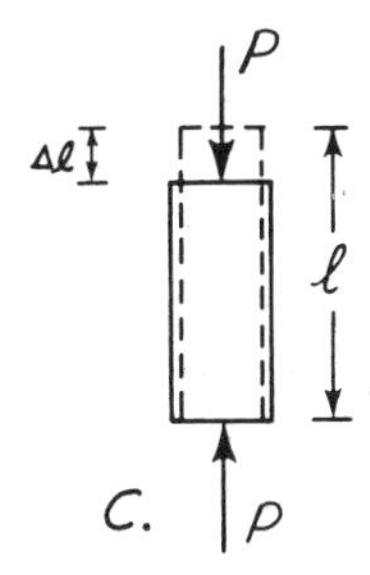

Figure 5.8. Compression of rock core.
A. Loading machine.
B. Micrometer gauge used to measure displacements.
C. Deformation of rock core.

failure by the original sample length, we find that the values of breaking strain are essentially constant. If Δl is the change in length of a sample and l is the original length of the sample, the breaking strain, ϵ_b, is

$$\epsilon_b = \frac{\Delta l}{l}.$$

The breaking strains of rocks are indicators of rock ductility, which is important in tunnel design. The breaking strain is an average strain of the entire sample.

Strain is a dimensionless quantity, because its definition is in terms of length divided by length, but we usually refer to strain as inches per inch, or centimeters per centimeter, or meters per meter. Sometimes strain is recorded as a percent. Strains for most solids, with a few exceptions, such as rubber, are extremely small. For example, the strain at which granite breaks is on the order of 10^{-4}, so that a ten-centimeter core of granite will shorten imperceptibly to our eyes, about $\frac{1}{100}$ of a millimeter, by the time it fails in a compression test.

If we were to measure strains at various places on the surface of the sample core we would find that the strains are quite variable and significantly different from the average strain measured between the platens. Strain, just like stress, is defined near a point on or within a body. If l is the original distance between two points within a body, and Δl is the change in distance after deformation, then we would define strain near a point in terms of a limiting process,

$$\epsilon = \lim_{l \to 0} \left(\frac{\Delta l}{l} \right).$$

The limiting process used to define strain is an idealization which is only approximately realized in actual materials. When we assign a value of strain to a point in a real body, we assume that the body is continuous. That is, we assume that the body is a *continuum* (*4*). A small volume of a material forming a continuum has all the characteristics of the entire volume of material, and other small volumes of the material are not appreciably different. These small volumes are large relative to discontinuous properties of materials, such as atoms of crystals, crystals of granite, and joint blocks of sandstone. The gauge length, l, therefore, is always considerably greater than zero in real materials.

Components of Displacement and Infinitesimal Strain.—In discussing the strain of the rock sample we used the symbol l to indicate the original length of the sample and Δl to indicate the change in length of the sample. Thus, the length of the sample at failure was $l - \Delta l$. We defined the strain as the change in length divided by the original length, so that the strain is $\Delta l/l$. Because this type of notation would become impossible to work with when we deal with differential quantities of strain, we will introduce a new notation. We will substitute δx for the original length in the x-direction and δu for the change in length in the x-direction. Thus, the strain is

$$\epsilon = \lim_{l \to 0} \frac{\Delta l}{l} = \lim_{\delta x \to 0} \frac{\delta u}{\delta x} = \frac{\partial u}{\partial x}.$$

The partial derivative simply indicates that the displacement, u, may vary in the y- and z-directions as well as the x-direction.

In all our further discussion of strain, therefore, we will use u for displacement in the x-direction, v for the y-direction, and w for the z-direction.

Displacement of points within a body may result from several types of motion, some of which are strain. To illustrate these types of motion, consider a square with sides parallel to x- and y-axes (Fig. 5.9A).

Translation.—One type of motion consists of lateral *translation*, in which the square displaces as a rigid body in the x-direction or y-direction (Fig. 5.9B), or in a direction with both x- and y-components (Fig. 5.9C). Translation involves displacement of points on the square, but it does not involve changes in relative positions of points on the square, so that it involves no strain. Translation can be expressed in terms of displacement components, u and v, and is shown in Figs. 5.9B and 5.9C.

Rotation.—Another possible type of translation of the square as a whole is *rotation*, which we shall designate by omega (ω). Figure 5.9D shows rotation of a square about the origin of coordinates, O. Points on the square are displaced by rotation so that, for example, in rotating point A to A', there is a displacement in the y-direction of

$$v = (\delta x) \tan \omega,$$

where ω is the angle of rotation and δx is the width of the square. If the rotation is small, the angle itself, in radians, is equal to the tangent of the angle, so that, to a first approximation,

$$v = (\delta x)\omega.$$

Exercises

1. Write the infinite series for the tangent and explain why, for small values of an angle, the tangent of the angle is essentially equal to the angle itself, in radians.
2. The component of shear strain in the $x - y$ plane usually is expressed in terms of displacements as

$$\epsilon_s = \left(\frac{\partial v}{\partial x} + \frac{\partial u}{\partial y}\right),$$

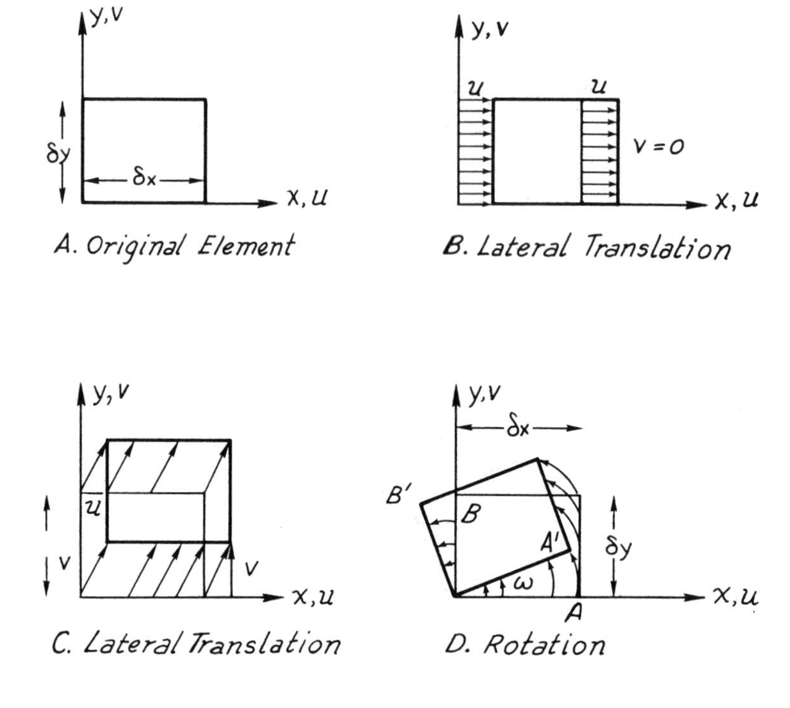

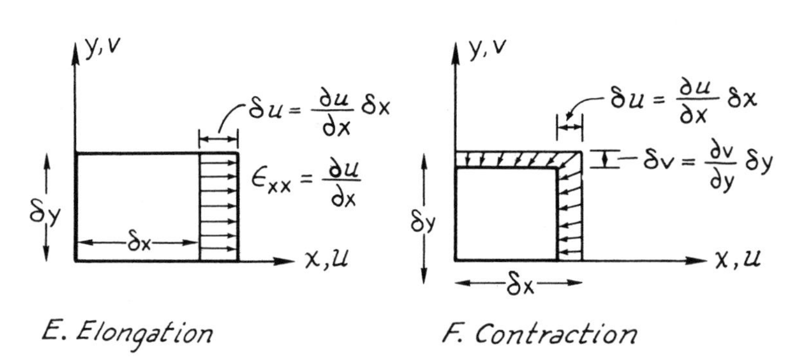

Figure 5.9. Types of displacement and strain.

and the rotation is expressed as

$$\omega = \frac{1}{2}\left(\frac{\partial v}{\partial x} - \frac{\partial u}{\partial y}\right).$$

Assuming very small displacements, derive the equation for rotation by examining Fig. 5.9D.

The distances between points within the square are unchanged by rotation, so that rotation causes no strain of the square.

Elongation and contraction.—Two types of motion of the square that cause strains are elongation and contraction. Figure 5.9E shows elongation of original square in the x-direction and Fig. 5.9F shows contraction in two directions. Elongation of the original square in Fig. 5.9E causes a net relative displacement, δu, of points on opposite sides of the square. The relative displacement related to normal strain in the x-direction is defined by the limiting process,

$$\epsilon_{xx} = \lim_{\delta x \to 0} \frac{\delta u}{\delta x} = \frac{\partial u}{\partial x}. \tag{5.8}$$

Normal strain in the y-direction would be similarly defined,

$$\epsilon_{yy} = \lim_{\delta y \to 0} \frac{\delta v}{\delta y} = \frac{\partial v}{\partial y}. \tag{5.9}$$

An elongation or extension normally is defined as positive normal strain and a contraction is defined as negative normal strain.

The change in area of the unit square is defined as *dilatation*, e. Thus,

$$-\delta x\, \delta y + (\delta x + \epsilon_{xx}\, \delta x)(\delta y + \epsilon_{yy}\, \delta y) = e\, \delta x\, \delta y,$$

or the dilatation is, to the first approximation, in two dimensions,

$$e = \epsilon_{xx} + \epsilon_{yy}. \tag{5.10}$$

In three dimensions the dilatation is

$$e = \epsilon_{xx} + \epsilon_{yy} + \epsilon_{zz}. \tag{5.11}$$

Shear.—The other type of motion that produces strain within a body is *distortion.* One type of distortion is known as pure shear and is illustrated in Fig. 5.10A. The square is strained when subjected to pure shear because distances between some points within the sheared square are different from distances between corresponding points within the original square; one diagonal has increased in length and the other has decreased in length. The total shear strain, ϵ_{xy}, can be defined as an angular deformation, ε_{xy}, of the side δx plus the equal angular deformation, ε_{yx}, of the side δy. The components ε_{xy} and ε_{yx} are equal because there is no rotation of the element involved in pure shear. The total shear strain, ϵ_{xy}, is defined as the angular change from the original right angle of the corners of the element. Thus,

$$\epsilon_{xy} = \varepsilon_{xy} + \varepsilon_{yx}.$$

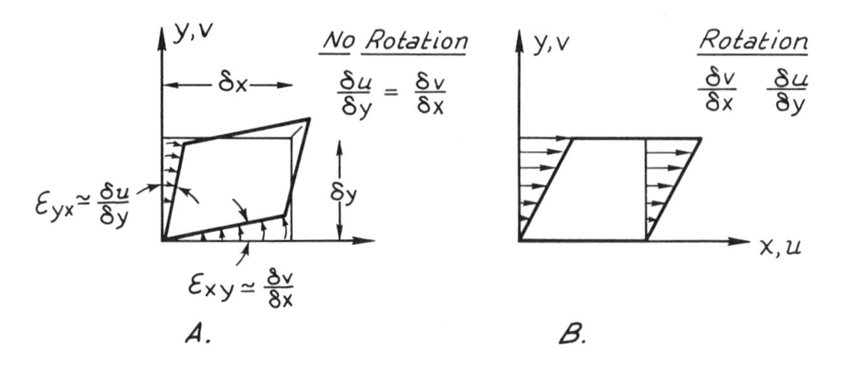

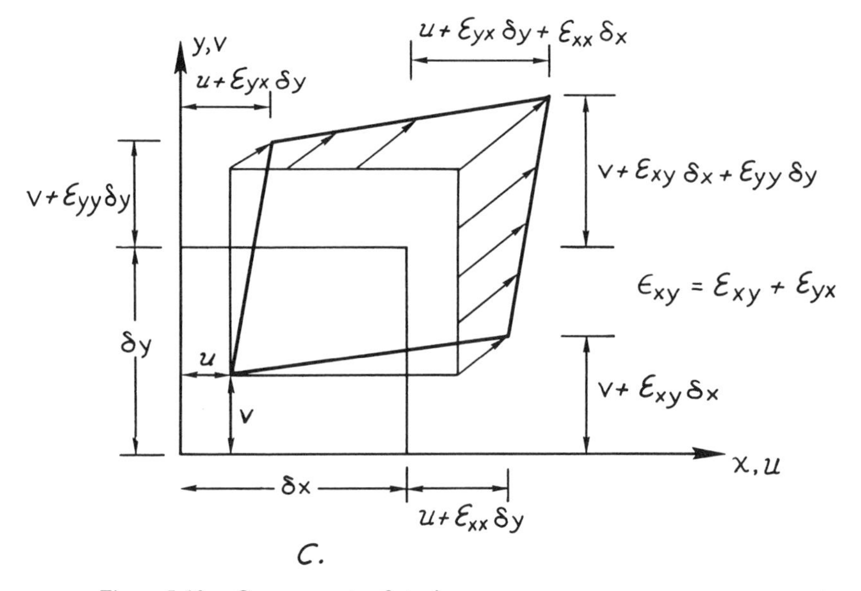

Figure 5.10. Components of strain.
A. Pure shear.
B. Simple shear.
C. Combination of translation, elongation, and pure shear.

If the angular strain is very small, we can express it in terms of displacements,

$$\frac{\partial u}{\partial y} = \tan\,(\varepsilon_{yx}) \approx \varepsilon_{yx},$$

and

$$\frac{\partial v}{\partial x} \approx \varepsilon_{xy},$$

so that for infinitesimal strain,

$$\epsilon_{xy} = \varepsilon_{xy} + \varepsilon_{yx} = \frac{\partial v}{\partial x} + \frac{\partial u}{\partial y}. \tag{5.12}$$

The tensor components of shear strain are different from the total, or engineering, components by a factor of two. Thus, the tensor components are

$$\varepsilon_{xy} = \frac{1}{2}\epsilon_{xy} = \frac{1}{2}\left(\frac{\partial v}{\partial x} + \frac{\partial u}{\partial y}\right). \tag{5.13}$$

The tensor components, ε_{xy} and ε_{yx}, are equal, by definition, because shearing strain, by definition, does not involve rotation and because rotation does not involve shearing strain.

Each of the types of infinitesimal strain can be expressed in terms of displacement components. A unique and powerful aspect of infinitesimal strain is that displacements caused by different strains and translations can be added or separated. For example, Fig. 5.10C shows a square that has been translated and extended in two directions and subjected to pure shear. Expressions for displacements of each of the four corners are indicated on the figure.

Exercises

1. Write expressions for the displacements of the center of the deformed square (Fig. 5.10C).
2. Write expressions for shear components of the strain tensor for a square subjected to *simple shear* (Fig. 5.10B).

The components of the strain tensor are

$$\begin{vmatrix} \varepsilon_{xx} & \varepsilon_{xy} & \varepsilon_{xz} \\ \varepsilon_{yx} & \varepsilon_{yy} & \varepsilon_{yz} \\ \varepsilon_{zx} & \varepsilon_{zy} & \varepsilon_{zz} \end{vmatrix} \equiv \varepsilon_{ij}; \qquad \begin{matrix} i = x, y, z \\ j = x, y, z \end{matrix} \tag{5.14}$$

where

$$\varepsilon_{xx} \equiv \epsilon_{xx}, \qquad \varepsilon_{yy} \equiv \epsilon_{yy}, \qquad \text{etc.}$$

and

$$\varepsilon_{xy} = \tfrac{1}{2}\epsilon_{xy}, \qquad \varepsilon_{zx} = \tfrac{1}{2}\epsilon_{zx}, \qquad \text{etc.}$$

We will consistently use the engineering definition of shear strain, the three components of which are

$$\left.\begin{aligned} \epsilon_{xy} &= \left(\frac{\partial v}{\partial x} + \frac{\partial u}{\partial y}\right) = \epsilon_{yx}, \\ \epsilon_{yz} &= \left(\frac{\partial w}{\partial y} + \frac{\partial v}{\partial z}\right) = \epsilon_{zy}, \\ \epsilon_{xz} &= \left(\frac{\partial w}{\partial x} + \frac{\partial u}{\partial z}\right) = \epsilon_{zx}. \end{aligned}\right\} \tag{5.15}$$

The three normal strain components,

$$\left.\begin{aligned} \epsilon_{xx} &= \frac{\partial u}{\partial x}, \\ \epsilon_{yy} &= \frac{\partial v}{\partial y}, \\ \epsilon_{zz} &= \frac{\partial w}{\partial z}, \end{aligned}\right\} \tag{5.16}$$

and the six shear strain components in eqs. (5.15) can be written compactly using tensor notation (*3*):

$$\varepsilon_{ij} = \frac{1}{2}\left[\frac{\partial u_j}{\partial x_i} + \frac{\partial u_i}{\partial x_j}\right]; \qquad \begin{aligned} i &= 1, 2, 3 \\ j &= 1, 2, 3 \end{aligned} \tag{5.17}$$

where

Tensor Notation		Our Notation
u_1	=	u
u_2	=	v
u_3	=	w
x_1	=	x
x_2	=	y
x_3	=	z
ε_{11}	=	ϵ_{xx}
ε_{22}	=	ϵ_{yy}
	etc.	

and

ε_{12}	=	$\frac{1}{2}\epsilon_{xy}$
ε_{13}	=	$\frac{1}{2}\epsilon_{xz}$
	etc.	

Exercise

Expand eq. (5.17) step by step to yield eqs. (5.15) and (5.16).

Finite Strain.—It is no simple matter to deduce whether the formation of a certain structure involved large or small strains, that is, finite or infinitesimal strains. The strains are finite if they exceed an amount of about 10^{-2}, or about one percent (*7*). We cannot use the amplitude of a fold, for example, to indicate the amount of strain. Material forming highly contorted, ptygmatic folds may have been subjected to infinitesimal strains if the folded layers were quite thin. In contrast, material in a fold with a wavelength of a few meters and a bed with a thickness of one meter was subjected to finite strain if the fold is perceptible at all. The amount of strain can be deduced by examining objects such as fossils which deformed passively with the rock.

The study of finite strains of rock is quite involved and requires a great deal of study. Perhaps it will prove to be a fruitful line of research. I refer you to the book by Ramsay (*7*) for a detailed and exceptionally lucid treatment of the subject.

Many products of geologic processes probably reflect finite strain of rocks but analyses of the processes need not be restricted to finite strain. We can understand initial stages of the processes in terms of infinitesimal strains and many times we probably can imagine, even though we cannot precisely describe, successive stages of deformation that led from where infinitesimal strain theory becomes invalid to the final structure. Further, the mathematical analysis of finite deformation of even very simple materials is nearly intractable.

The analyses in this book are restricted to consideration of infinitesimal strain and strain rates but it would be advisable for us to have some understanding of the upper limits of applicability of the analyses.

Essentially, finite strains are large and infinitesimal strains are small. The actual division between these two types of strains is difficult to define other than in terms of the expressions for finite strain which we will derive. The following derivation of the exact expressions for the change in length of a line, in two-dimensional space, is a geometrical interpretation of finite strain. Let us calculate the two-dimensional strain of a line element on the x-axis. One end of the line remains fixed, and the other end is displaced a distance, du in the x-direction and dv in the y-direction (Figs. 5.11A and 5.11B). The original length of the line element is dx and the finite length is ds. The finite strain of the line element, e_{xx}, is

$$e_{xx} = \frac{ds - dx}{dx}. \tag{5.18}$$

We need to derive an expression for ds in terms of displacements du and dv.

According to Fig. 5.11B,

$$(ds)^2 = (dx + du)^2 + (dv)^2. \tag{5.19}$$

However, according to the chain rule of calculus,

$$du = \frac{\partial u}{\partial x} dx + \frac{\partial u}{\partial y} dy$$

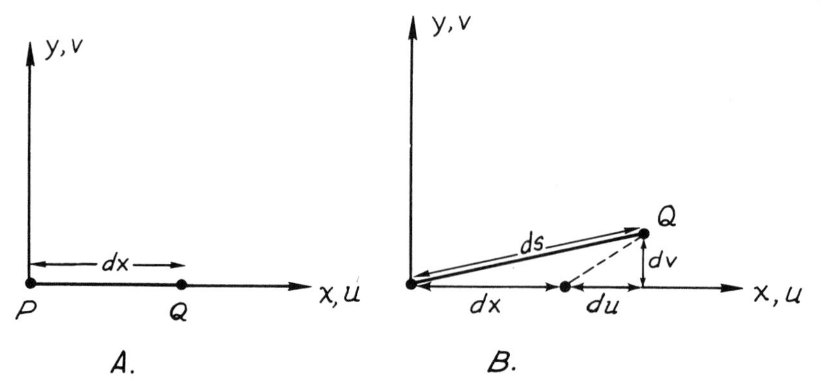

Figure 5.11. Finite strain of linear element.

$$= \frac{\partial u}{\partial x} dx, \tag{5.20}$$

because dy is zero. Similarly,

$$dv = \frac{\partial v}{\partial x} dx. \tag{5.21}$$

Substituting eqs. (5.20) and (5.21) into eq. (5.19),

$$(ds)^2 = \left(dx + \frac{\partial u}{\partial x} dx\right)^2 + \left(\frac{\partial v}{\partial x} dx\right)^2$$

$$= (dx)^2\left[1 + 2\frac{\partial u}{\partial x} + \left(\frac{\partial u}{\partial x}\right)^2 + \left(\frac{\partial v}{\partial x}\right)^2\right],$$

so that

$$ds = dx\left[1 + 2\frac{\partial u}{\partial x} + \left(\frac{\partial u}{\partial x}\right)^2 + \left(\frac{\partial v}{\partial x}\right)^2\right]^{1/2}. \tag{5.22}$$

Substituting eq. (5.22) into eq. (5.18), we have an expression for the strain of the line element,

$$e_{xx} = \left[1 + 2\frac{\partial y}{\partial x} + \left(\frac{\partial y}{\partial x}\right)^2 + \left(\frac{\partial v}{\partial x}\right)^2\right]^{1/2} - 1. \tag{5.23}$$

Expanding this last equation as a binomial series, in which

$$(a + b)^n = a^n + na^{n-1}b + \frac{n(n-1)}{2!} a^{n-2}b^2 + \cdots + [a^2 > b^2],$$

$$e_{xx} = \frac{\partial u}{\partial x} + \frac{1}{2}\left[\left(\frac{\partial u}{\partial x}\right)^2 + \left(\frac{\partial v}{\partial x}\right)^2\right] - \frac{1}{8}\left[2\frac{\partial u}{\partial x} + \left(\frac{\partial u}{\partial x}\right)^2 + \left(\frac{\partial v}{\partial x}\right)^2\right]^2 + \cdots \tag{5.24}$$

Thus we see that the first-order approximation of infinitesimal strain involves only the first term, $\partial u/\partial x$, in the exact expression, eq. (5.24). A second-order approximation would retain the first and second terms. Higher-order approximations would retain the first, second, third, and, perhaps, other terms in the infinite series. If we assume infinitesimal strains, we assume that the squared terms, and the products, are negligible. Thus, *whether measured strains are finite or infinitesimal depends upon the magnitudes of the displacement gradients and upon the accuracy of the measurements of the displacement gradients.*

The components of finite strain, to the second approximation, are (*5*)

$$\left.\begin{aligned}
e_{xx} &= \frac{\partial u}{\partial x} + \frac{1}{2}\left[\left(\frac{\partial u}{\partial x}\right)^2 + \left(\frac{\partial v}{\partial x}\right)^2 + \left(\frac{\partial w}{\partial x}\right)^2\right],\\
e_{yy} &= \frac{\partial v}{\partial y} + \frac{1}{2}\left[\left(\frac{\partial u}{\partial y}\right)^2 + \left(\frac{\partial v}{\partial y}\right)^2 + \left(\frac{\partial w}{\partial y}\right)^2\right],\\
e_{zz} &= \frac{\partial w}{\partial z} + \frac{1}{2}\left[\left(\frac{\partial u}{\partial z}\right)^2 + \left(\frac{\partial v}{\partial z}\right)^2 + \left(\frac{\partial w}{\partial z}\right)^2\right],\\
e_{xy} &= \frac{\partial u}{\partial y} + \frac{\partial v}{\partial x} + \left[\frac{\partial u}{\partial x}\frac{\partial u}{\partial y} + \frac{\partial v}{\partial x}\frac{\partial v}{\partial y} + \frac{\partial w}{\partial x}\frac{\partial w}{\partial y}\right],\\
e_{yz} &= \frac{\partial w}{\partial y} + \frac{\partial v}{\partial z} + \left[\frac{\partial u}{\partial y}\frac{\partial u}{\partial z} + \frac{\partial v}{\partial y}\frac{\partial v}{\partial z} + \frac{\partial w}{\partial y}\frac{\partial w}{\partial z}\right],\\
&\text{and}\\
e_{xz} &= \frac{\partial w}{\partial x} + \frac{\partial u}{\partial z} + \left[\frac{\partial u}{\partial x}\frac{\partial u}{\partial z} + \frac{\partial v}{\partial x}\frac{\partial v}{\partial z} + \frac{\partial w}{\partial x}\frac{\partial w}{\partial z}\right].
\end{aligned}\right\} \tag{5.25}$$

In terms of tensor notation, eqs. (5.25) can be reduced to one equation (*3*),

$$e_{ij} = \frac{1}{2}\left[\frac{\partial u_j}{\partial x_i} + \frac{\partial u_i}{\partial x_j} + \frac{\partial u_k}{\partial x_i}\frac{\partial u_k}{\partial x_j}\right]; \quad \begin{aligned} i &= 1, 2, 3\\ j &= 1, 2, 3\\ k &= 1, 2, 3 \end{aligned} \tag{5.26}$$

where, for example,

$$\frac{\partial u_k}{\partial x_1}\frac{\partial u_k}{\partial x_2} = \left(\frac{\partial u_1}{\partial x_1}\frac{\partial u_1}{\partial x_2}\right) + \left(\frac{\partial u_2}{\partial x_1}\frac{\partial u_2}{\partial x_2}\right) + \left(\frac{\partial u_3}{\partial x_1}\frac{\partial u_3}{\partial x_2}\right).$$

Thus, if we assume infinitesimal strain, we assume that the squared terms and the products are negligible. (See Ramsay (*7*) and Love (*5*) for discussions of the equations of finite strain.)

Superposition of Strains.—A powerful tool for solving complicated mechanical problems is the principle of superposition. According to this principle, we can add the effects of several forces or stresses and the final deformation is independent of the order in which we add the effects. We will use the principle several times. For example, when we explain the radial dike pattern at Spanish Peaks, Colorado, we determine the strains and stresses caused by one source of pressure in an infinite body and then add the strains caused by a second source of pressure, a short distance from the first, within the same body.

When strains in the body of interest are small, we can apply the principle of superposition, but when they are finite, we cannot. For example, consider the superposition of two finite shear strains on a body that is initially square. One strain is a result of simple shear in which the sides of the square that are parallel to the y-direction are moved positively or negatively in the y-direction (Fig. 5.12B). The other strain involves simple shear in the direction of the x-axis (Fig. 5.12C). As is shown in Figs. 5.12D and 5.12E, the final form of, as well as the displacement of points on, the object which was initially a square depends upon the order in which these simple shear strains are superimposed (7).

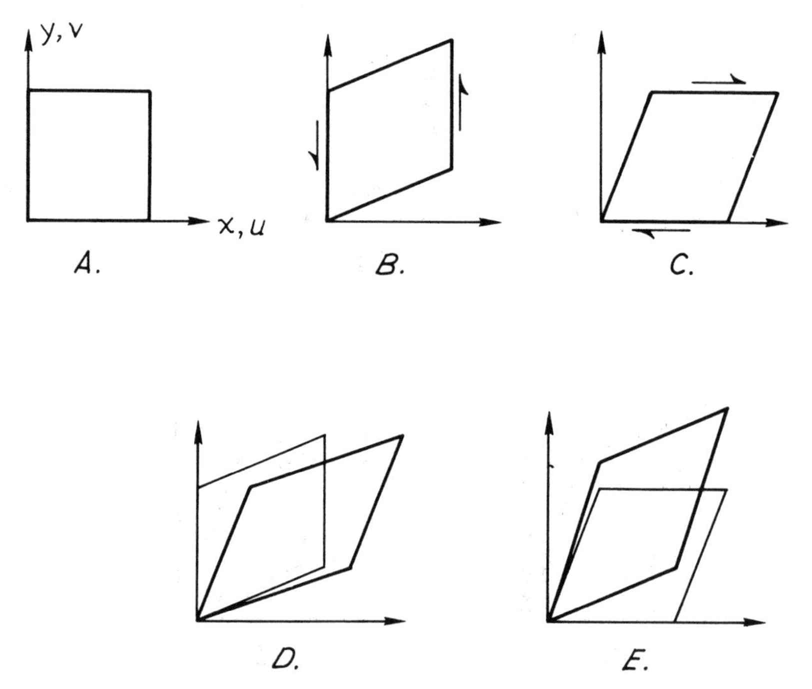

Figure 5.12. Superposition of two simple-shear strains. Final shape of deformed square depends upon order of superposition.

Exercise

By means of sketches, show that the deformed shape of a rectangle subjected to finite simple shear and then to rotation is different from the shape of the same rectangle subjected to rotation and then to finite simple shear.

Elastic constants and stress-strain relations

The concepts of stress and strain which we have developed in preceding pages are valid for all materials that are continua. Relations between stress and strain, however, are different for different materials. Several rheological models have been invented to describe these relations and some of the models yield remarkably good descriptions of behaviors of actual materials. One of the simplest rheological models, the Hookean body, was invented in a crude form by Hooke in the late 1600's when he stated that extensions of elastic bodies are proportional to forces applied to them. The Hookean model of elasticity was generalized to three-dimensional states of stress in the early 1800's by Cauchy and Poisson (*5*).

Young's Modulus and Poisson's Ratio.—In order to specify relations between all possible states of stress and states of strain for the simple, Hookean elastic substance, we need to specify only two elastic constants. All other elastic constants can be defined in terms of the two. We will select Young's modulus and Poisson's ratio. Young's modulus is a constant of proportionality between normal stress and the resulting normal strain, and Poisson's ratio is the ratio of the normal strain at right angles to the direction of applied stress to the normal strain parallel to the direction of applied stress. When an elastic body, such as a rubber band, is stretched in one direction, it contracts in the opposite direction. The ratio of the contraction to the stretching is Poisson's ratio. It normally has a value ranging between zero and 0.5. Steel has a Poisson's ratio of about 0.3. An incompressible substance has a ratio of 0.5.

Exercise

Determine experimentally Poisson's ratio for rubber in a wide rubber band.

We will show in following pages that the stress-strain relations for the Hookean body, which is an isotropic, homogeneous, linearly elastic substance, are:

Normal Components *Shear Components*

$$\begin{aligned}
\epsilon_{xx} &= \frac{1}{E}[\sigma_{xx} - \nu(\sigma_{yy} + \sigma_{zz})] & \epsilon_{xy} &= \frac{1}{G}\sigma_{xy} \\
\epsilon_{yy} &= \frac{1}{E}[\sigma_{yy} - \nu(\sigma_{xx} + \sigma_{zz})] & \epsilon_{yz} &= \frac{1}{G}\sigma_{yz} \qquad (5.27) \\
\epsilon_{zz} &= \frac{1}{E}[\sigma_{zz} - \nu(\sigma_{xx} + \sigma_{yy})] & \epsilon_{xz} &= \frac{1}{G}\sigma_{xz}
\end{aligned}$$

where E is *Young's modulus*, ν is *Poisson's ratio*, and G is *modulus of rigidity*.

The meaning of Young's modulus and Poisson's ratio can be made clear by studying the deformation of a cube of ideal elastic material. If the elastic cube, shown in Fig. 5.13, is subjected to an axial tensile stress, σ_{yy}, uniformly distributed

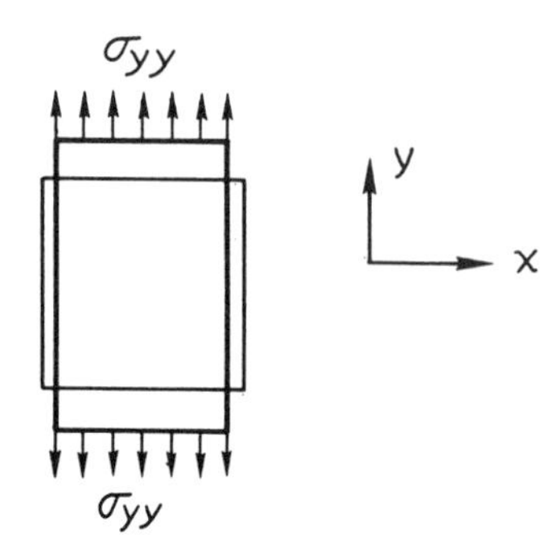

Figure 5.13. Deformation of elastic cube by normal stresses in y-direction.

over opposite ends, the cube will extend in the y-direction, parallel to the applied stress, in an amount that is proportional to the stress:

$$\epsilon_{yy} = \frac{1}{E} \sigma_{yy}, \tag{5.28}$$

where E, the constant of proportionality, is Young's modulus. Young's modulus has dimensions of stress, that is, dimensions of force per unit of area. Its value for several rocks is given in Table 5.1. Steel has a modulus of about 21×10^{10} N/m^2 (30×10^6 psi).

Accompanying the axial elongation of an elastic body, such as the cube in Fig. 5.13, is contraction in the x-direction, which is related to the axial strain and axial stress by the equations:

$$\epsilon_{xx} = -\nu\epsilon_{yy} = -\frac{\nu}{E} \sigma_{yy}, \tag{5.29}$$

where ν is the proportionality constant, called Poisson's ratio. There is the same contraction in the z-direction:

$$\epsilon_{zz} = -\nu\epsilon_{yy}. \tag{5.30}$$

Relations among Normal Stresses and Normal Strains.—Now suppose that the cube shown in Fig. 5.14 is subjected to three tensile stresses, σ_{xx}, σ_{yy}, and σ_{zz}, uniformly distributed over its sides. The resultant components of strain can be derived by using equations of forms (5.28) and (5.29).

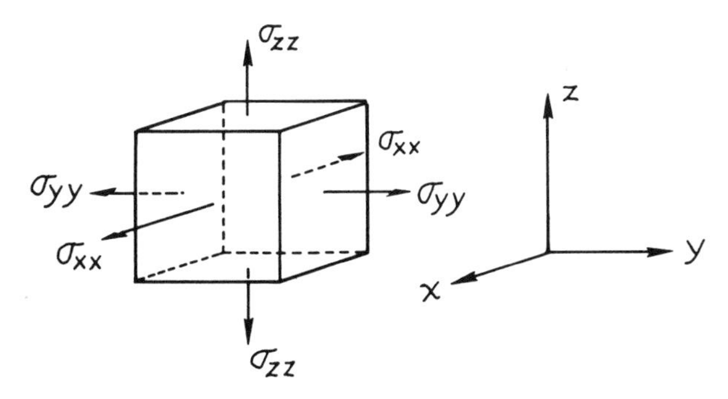

Figure 5.14. Elastic cube supporting normal stresses.

Strains due to Normal Stress in x-Direction	Strains due to Normal Stress in y-Direction	Strains due to Normal Stress in z-Direction
$\epsilon_{xx} = \frac{1}{E}\sigma_{xx}$	$\epsilon_{xx} = \frac{\nu}{E}\sigma_{yy}$	$\epsilon_{xx} = -\frac{\nu}{E}\sigma_{zz}$
$\epsilon_{yy} = -\frac{\nu}{E}\sigma_{xx}$	$\epsilon_{yy} = \frac{1}{E}\sigma_{yy}$	$\epsilon_{yy} = -\frac{\nu}{E}\sigma_{zz}$
$\epsilon_{zz} = -\frac{\nu}{E}\sigma_{xx}$	$\epsilon_{zz} = -\frac{\nu}{E}\sigma_{yy}$	$\epsilon_{zz} = \frac{1}{E}\sigma_{zz}$

Summing the increments of strain caused by each of the normal stresses, the net normal strain in each of the three directions is

$$\left.\begin{aligned} \epsilon_{xx} &= \frac{1}{E}[\sigma_{xx} - \nu(\sigma_{yy} + \sigma_{zz})], \\ \epsilon_{yy} &= \frac{1}{E}[\sigma_{yy} - \nu(\sigma_{xx} + \sigma_{zz})], \\ \text{and} \quad \epsilon_{zz} &= \frac{1}{E}[\sigma_{zz} - \nu(\sigma_{xx} + \sigma_{yy})]. \end{aligned}\right\} \tag{5.31}$$

Young's Moduli and Poisson's Ratios of Some Earth Materials.—The ideal Hookean elastic substance is elastically isotropic so that all of its elastic properties can be defined in terms of two fundamental constants. In engineering practice, which we are following, the fundamental constants are Poisson's ratio and Young's modulus. In geophysics, the fundamental constants are called Lamé's constants, which are related to Young's modulus and Poisson's ratio by the equations (*4*)

TABLE 5.1

Mechanical Properties of Some Earth Materials

Rock Type	Compressive Strength* (mult. by 10^8 N/m^2)			Tensile Strength† (mult. by 10^8 N/m^2)		Young's Modulus‡ (mult. by 10^{10} N/m^2)		Poisson's Ratio‡		
	Max.	Min.	50% of data within	Max.	Min.	Max.	Min.	Max.	Min.	Ave.
Basalt	3.6	0.81	...	...	...	11.2	4.9	0.42	−0.04	0.17
Diabase	3.6	1.60	2.8–3.5	...	...	10.7	7.2	0.28	0.10	0.26
Diorite	3.3	1.60	1.7–2.4	...	...	8.7	5.5	0.32	0.17	0.26
Dolomite	3.6	0.62	...	...	...	9.3	4.9	0.18	−0.09	0.04
Gneiss	2.5	1.50	1.7–2.4	...	...	7.0	0.3	...	0.02	0.26
Granite	2.9	1.60	1.7–2.4	0.56	0.28	7.6	1.7	0.55	−0.30	0.04
Limestone	2.6	0.37	1.4–2.1	0.40	0.14	8.2	2.9	0.50	−0.37	0.17
London Clay‖	...	...	...	...	...	0.015	0.001	...	...	...
Marble	2.4	0.46	2.1–2.4	...	...	8.7	2.8	0.31	0.10	0.25
Marlestone	1.9	0.72	0.69–1.4	...	...	3.3	0.41	0.94	−0.30	0.26
Quartzite	6.3	1.40	...	...	...	10.0	4.2	0.29	0.07	0.12
Sandstone	2.3	0.33	...	0.19	0.07	5.0	0.97	0.61	−0.51	0.02
Shale	2.3	0.76	0.69–1.4	...	...	5.2	1.2	0.92	−0.51	0.10
Siltstone	3.2	0.34	...	...	...	4.4	1.3	...	...	0.12

* After Obert and Duvall (*6*).

† After U.S. Bur. Mines Repts. Invests. 4459, 4727, 5130, and 5244.

‡ After Obert and Duvall (*6*) and Clark (*2*).

‖ After Bishop et al. (*1*).

$$\left.\begin{aligned} \lambda &= \frac{E\nu}{(1+\nu)(1-2\nu)} \\ &\text{and} \\ G &= \frac{E}{2(1+\nu)}. \end{aligned}\right\} \tag{5.32}$$

Lamé's constants can be derived by solving eqs. (5.31) simultaneously for the normal stress, so that, for example,

$$\sigma_{xx} = 2G\epsilon_{xx} + \lambda(\epsilon_{xx} + \epsilon_{yy} + \epsilon_{zz}). \tag{5.33}$$

Regardless of which constants we assume to be fundamental, however, a rock must be isotropic or nearly isotropic for its behavior to be described in terms of two constants. If an elastic rock is definitely anisotropic, 21 different elastic moduli may be required to describe completely its elastic properties (*2*). This is the case for triclinic crystals. Monoclinic crystals require 13 independent constants to describe their elastic properties (*2*). The only truly isotropic earth materials would be non-crystalline rocks, such as glass. However, we can treat polycrystalline rocks as isotropic materials if their minerals are randomly oriented and if the problem in which we are interested involves bodies of rock that are large relative to the crystals within the rock. These are the kinds of assumptions we are making when we use elasticity theory to solve geological problems. Such assumptions are always open to question until the directional properties and homogeneity of actual materials are examined.

The elastic constants or rocks usually are measured with electrical strain gauges mounted on rock cores about two inches in diameter and four inches in length (*6*). An electrical strain gauge is mounted near mid-height on the core so that the gauge measures the change in length of the core over the gauge length, usually 0.5 to 7 cm. The electrical resistance strain gauge is essentially a wire resistor wound so that it is highly sensitive to changes in length in one direction and insensitive to changes in length in the other direction. The change in resistance of the gauge is proportional to the strain of the gauge. Thus, in order to measure strains in two directions, one gauge is mounted so that it senses strains in the direction parallel to the long axis of the core, parallel to the direction of squeezing by the loading machine. This is the longitudinal gauge. Another gauge is mounted at right angles, that is, transverse, to the axis of the core. This is the transverse gauge.

Figure 5.15 shows graphs relating longitudinal and transverse strain to axial compression of a sample of Chelmsford granite from Massachusetts. The average axial stress acting on the sample is determined by dividing the axial load by the area of one end of the core. Then the slope of the graph relating axial stress and longitudinal strain is Young's modulus and the ratio of the transverse strain to the longitudinal strain, at the same axial stress, is Poisson's ratio of the rock core.

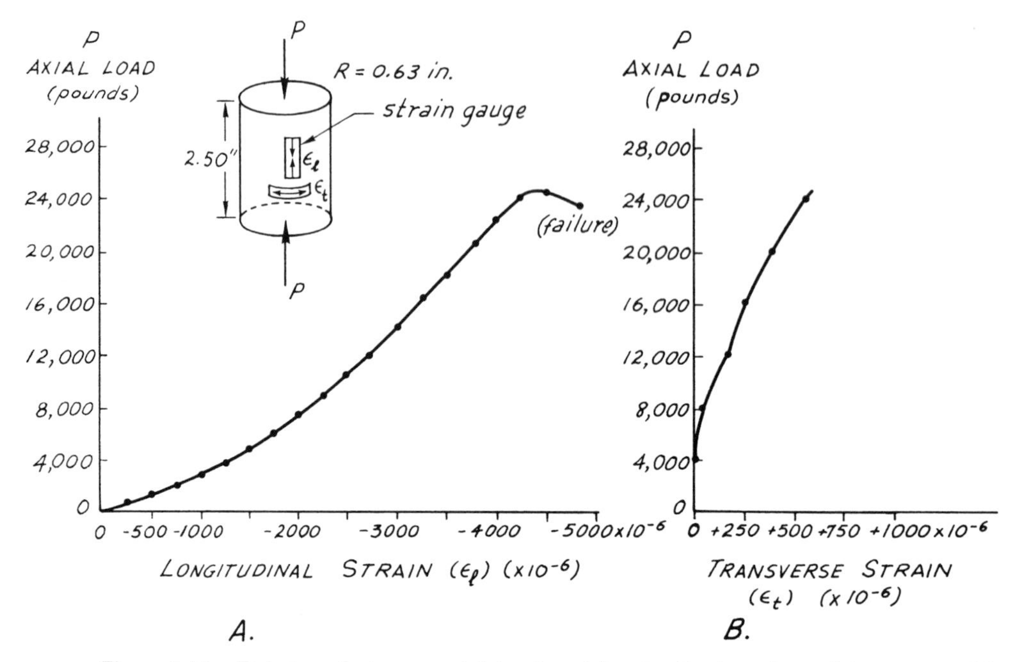

Figure 5.15. Relations between axial load and longitudinal strain and between axial load and transverse strain for a cylindrical sample of Chelmsford granite. (Data collected by S. D. Peng, Stanford University, 1969.)

Exercises

Determine Young's modulus and Poisson's ratio from the graphs shown in Fig. 5.15. Are Young's modulus and Poisson's ratio actually constant? Construct graphs showing relations between average axial stress and Young's modulus and Poisson's ratio.

Values of Young's modulus of several rock types are indicated in Table 5.1. The range of moduli is nearly an order of magnitude (order of magnitude = 10 or $\frac{1}{10}$) for several of the rock types and the moduli for each rock type have values varying by a factor of two. Also, the difference between average Young's moduli of shale and granite is only about a factor of two. Therefore, we can probably learn to estimate Young's modulus quite accurately for rocks in the field. Rocks which ring under the blow of a hammer probably have a modulus of 9 to 12 $\times$ 10^{10} N/m^2 and rocks that respond with a dull thud to the blow of a hammer probably have a modulus of 0.5 to 2 $\times$ 10^{10} N/m^2. None of the rocks has a modulus as high as that of steel, which is about 21 $\times$ 10^{10} N/m^2.

Poisson's ratios of rocks are highly variable and rather surprising. The values of Table 5.1 range from -0.5 to $+0.9$. Both of these extremes are peculiar because an incompressible substance has a Poisson's ratio of 0.5 and a perfectly compressible substance has a ratio of 0.0. A ratio of 1.0 would indicate that, for example, the material expands twice as much as it is compressed. Cork has a Poisson's ratio of about zero, so that when it is compressed in one direction it does not expand significantly in other directions; it simply decreases in volume. A ratio of -0.5 would indicate that a substance expands in all directions when it is pulled in one direction, or that it contracts in all directions when it is compressed in one direction. Although we would normally expect a rock core to increase slightly in diameter when it is compressed axially, rock with a negative Poisson's ratio would decrease in diameter when it is compressed axially.

Exercises

1. Why do incompressible substances have Poisson's ratios of 0.5?
2. How might negative Poisson's ratios be explained in crystalline, microfractured rock? In sand?

Modulus of Rigidity.—The relations among normal strains and normal stresses in three directions are completely defined by means of two constants, ν and E. Combinations of these constants can be shown to define the relations between shear strain and shear stress. Consider a block that is loaded in the x- and y-directions so that (Fig. 5.16*B*)

$$\sigma_{xx} = -\sigma_{yy}; \qquad \sigma_{zz} = 0.$$

Cut out an element, *abcd*, by planes parallel to the z-axis and inclined at 45 degrees to the x- and y-directions. We will show that the state of stress on the element is one of *pure shear*. That is, the normal stresses and strains acting on sides *ab*, *bc*, *ad*, and *dc* are zero. Also, we will show that the shear stress, σ_s, acting tangentially to those sides is equal in magnitude to the externally applied normal stresses, σ_{xx} and σ_{yy}.

The easiest method of demonstrating the relations is with Mohr's Circle of stress. We will derive Mohr's Circle and discuss it in detail in a later section of the book. Most readers will have been previously introduced to Mohr's Circle; therefore, we will make only a brief digression here to refresh memories.

Relations between stresses on a plane with any orientation through a point and stresses on other planes through the same point can be represented by a graph known as Mohr's Circle. Let σ_1 represent the maximum principal stress and σ_3 represent the minimum principal stress acting on a small cubic element (Fig. 5.17A). The stresses acting on any other plane containing the z-axis can be determined from Mohr's Circle. In Fig. 5.17B the plane is shown to be inclined at an angle of θ from the direction of σ_3.

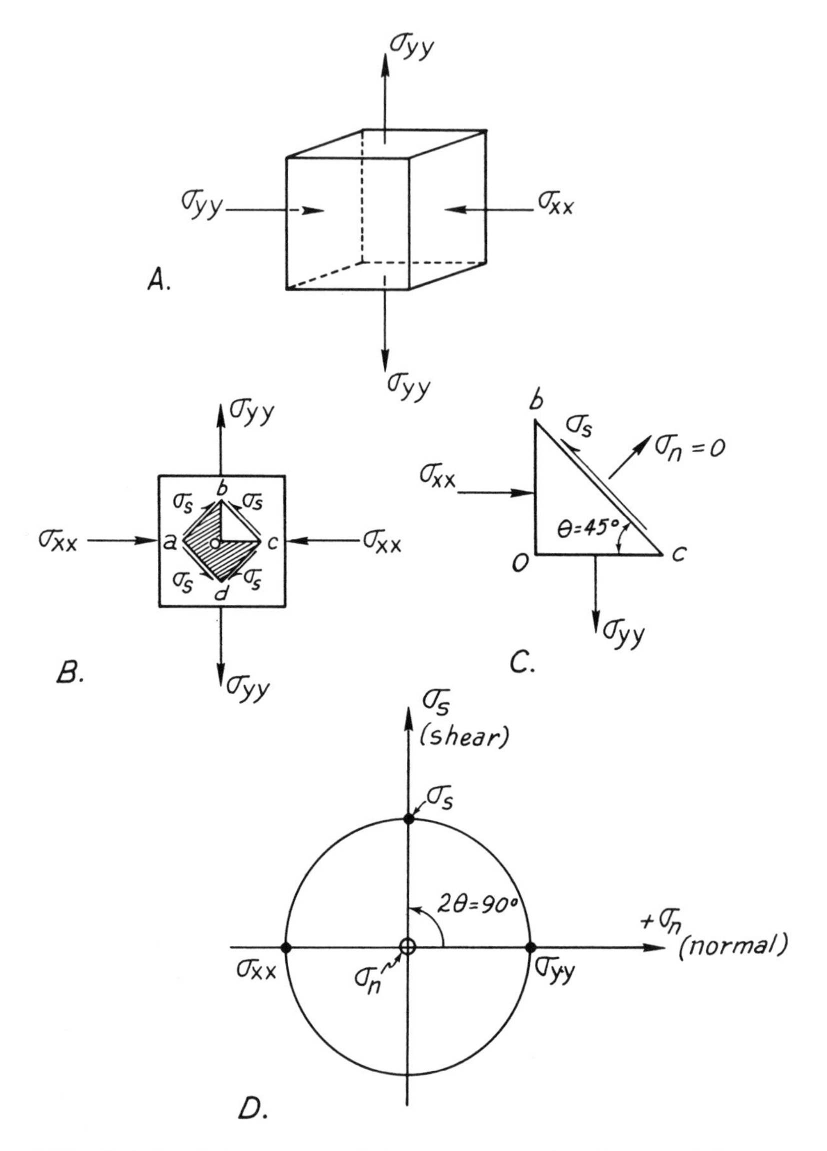

Figure 5.16. Relation between normal stresses on *x*- and *y*-planes and shear stress on plane inclined at angle of 45 degrees to *x*-direction.

Mohr's Circle is constructed by selecting an arbitrary origin, *O*, Fig. 5.17C, and scaling distances in the horizontal direction, proportional to the magnitudes of the maximum principal stress, σ_1, and the minimum principal stress, σ_3. The directions in which we measure σ_1 and σ_3 depend upon their signs: compressive stresses are

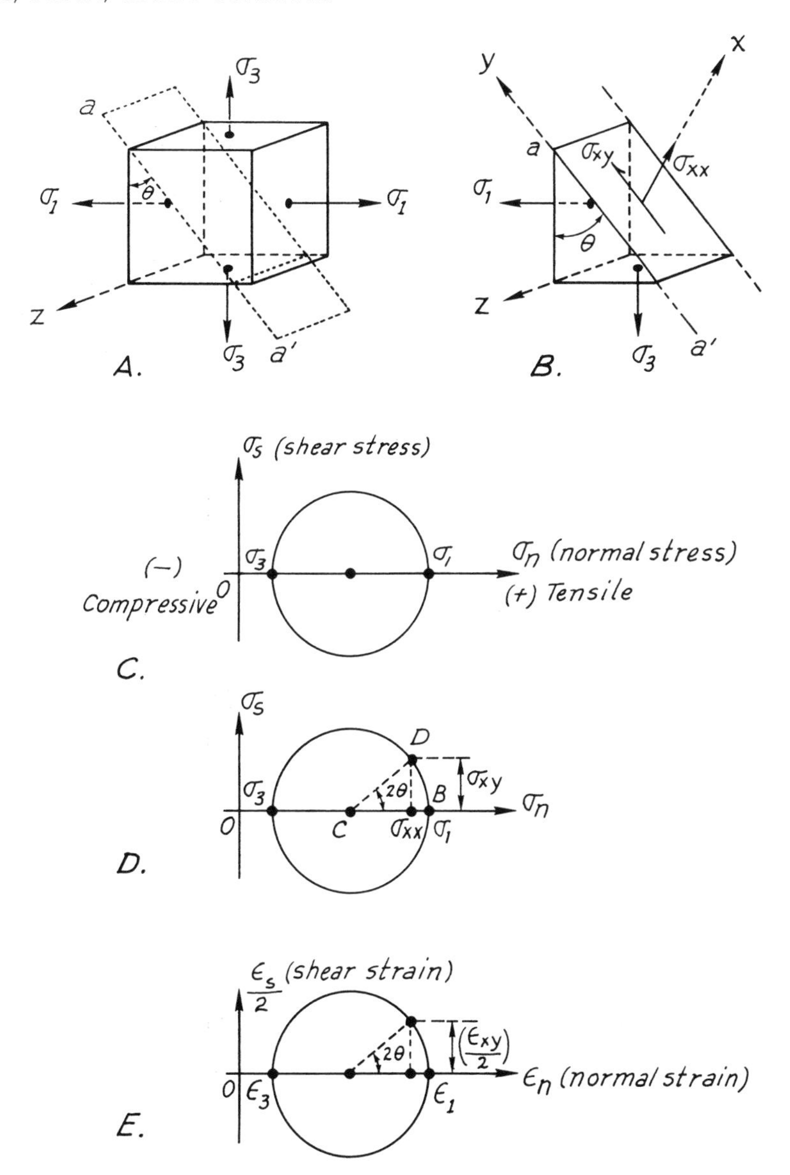

Figure 5.17. Mohr's Circles of stress and strain.

scaled off to the left (negative side) of point O and tensile stresses are scaled off to the right. Distance OA corresponds with the magnitude of σ_3 and distance OB corresponds with the magnitude of σ_1. We find the midpoint, C, between point A and B and construct a circle with a radius of CB.

We shall prove later on that the coordinates of each point on the circle represent the state of stress on a certain plane through the element in Fig. 5.17A. For example, in order to find the stresses on the plane *a–a'*, Fig. 5.17A, the normal to which is inclined at an angle, θ, to the direction of σ_1, construct the angle *BCD*, Fig. 5.17D, *equal to* 2θ and locate point *D*, the coordinates of which represent the stresses on plane *a–a'*. The horizontal coordinate of point *D* is σ_{xx}, the normal stress acting on the plane, and the vertical coordinate is σ_{xy}, the shear stress acting on the plane. This is all we need to know, for the present, about Mohr's Circle of stress.

Mohr's Circle of strain is nearly identical to the circle of stress (Fig. 5.17E). The two circles are read in the same way except that the magnitudes of shear strains plotted are half their actual values (Fig. 5.17E).

The Mohr Circle for the state of stress shown in Fig. 5.16A is shown in Fig. 5.16D. The normal stress in the *y*-direction is equal in magnitude but of opposite sign to that in the *x*-direction, so that the corresponding Mohr Circle is centered on the origin of stress coordinates. We can see, therefore, that the shear stresses acting on planes inclined 45 degrees to the *x*-direction ($2\theta = 90$ degrees in Fig. 5.16D) are maximum and that the normal stresses on these planes are zero. Further, we can see that the magnitudes of the shear stresses are equal to the magnitudes of the applied normal stresses, which are principal stresses; they are all proportional to the radius of the Mohr Circle (Fig. 5.16D). Thus,

$$\sigma_s = \sigma_{yy} = -\sigma_{xx}. \tag{5.34}$$

As a result of the compressive stress, σ_{xx}, and the tensile stress, σ_{yy}, of equal magnitude, the small element of elastic material shown in Fig. 5.16A is compressed by an amount ϵ_{xx} and extended by an equal amount, ϵ_{yy}, in the opposite direction. These strains are shown plotted on the Mohr strain diagram in Fig. 5.18. We see that

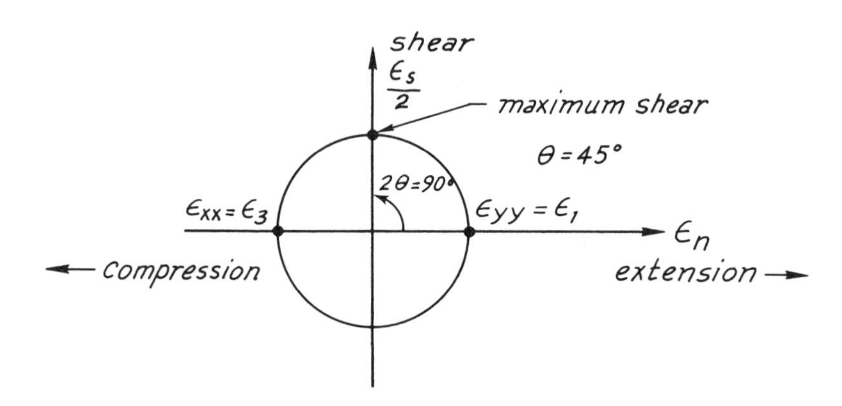

Figure 5.18. Mohr Circle of strain, showing condition of pure shear.

the strain on planes inclined 45 degrees from the direction of compression ($2\theta = 90$ degrees) is pure shear. The magnitude of the pure shear strain is proportional to twice the radius of the Mohr Circle, so that

$$\epsilon_s = 2\epsilon_{yy} = -2\epsilon_{xx}, \tag{5.35}$$

the shear strain is twice the extension of the element.

Now we can derive the expression for the elastic modulus of shear, that is, the modulus of rigidity; we know the relations between normal stresses and shear stresses, between normal strains and shear strains, and between normal stresses and normal strains, and thus we can write an equation relating shear stress and shear strain, in terms of Young's modulus and Poisson's ratio. According to eq. (5.35),

$$\epsilon_s = 2\epsilon_{yy}.$$

However, according to the second of the stress-strain relations in eqs. (5.31), if $\sigma_{zz} = 0$,

$$\epsilon_{yy} = \frac{1}{E}[\sigma_{yy} - \nu\sigma_{xx}],$$

so that

$$\epsilon_s = \frac{2}{E}[\sigma_{yy} - \nu\sigma_{xx}].$$

However, according to eq. (5.34),

$$\sigma_{yy} = -\sigma_{xx} = \sigma_s,$$

so that the stress-strain relation for shear is

$$\epsilon_s = \frac{2}{E}\sigma_s[1 + \nu]. \tag{5.36}$$

Therefore, the relation between shear stress and shear strain is

$$\sigma_s = \frac{E}{2(1 + \nu)}\epsilon_s.$$

The combination of elastic constants usually is denoted by G and called the *modulus of rigidity* of the Hookean material,

$$G = \frac{E}{2(1 + \nu)}. \qquad \text{Modulus of Rigidity} \tag{5.37}$$

If only shear stresses act on the sides of an element, the distortions of the right angles of the element depend only upon the shear stresses, so that the stress-strain relations for shear are:

$$\left.\begin{aligned} \epsilon_{xy} &= \frac{1}{G}\sigma_{xy} \\ \epsilon_{xz} &= \frac{1}{G}\sigma_{xz} \\ \epsilon_{yz} &= \frac{1}{G}\sigma_{yz}. \end{aligned}\right\} \tag{5.38}$$

Other elastic constants can be derived by manipulating the stress-strain relations. For example, the Bulk modulus relates volumetric strain to the mean stress, or pressure, acting on an elastic material, and Lamé's elasticity constants can be derived by solving the stress-strain relations, eqs. (5.31), for the stresses.

Plane stress and plane strain—important special cases

Three-dimensional relations between stress and strain are relatively complicated; hence they rarely are used to solve problems. Complex problems involving changes in stresses in three directions, such as the problem of calculating the stresses within high-pressure vessels used in experimental petrology, usually are reduced to ideal problems involving changes in stresses in only two dimensions. Thereby the problems are greatly simplified, results are relatively easy to obtain, yet the correlation between theoretical and actual stress distributions can be quite high. For example, the stresses in a pressure vessel consisting of a long, thick-walled tube, permanently sealed at one end and sealed by a cap at the other end, can be calculated quite accurately for the region in the vicinity of the mid-length of the vessel by assuming that the vessel is infinitely long. There the stresses vary only with respect to directions normal to the long axis of the vessel. Such a problem is called two-dimensional.

Many two-dimensional problems in stress analysis, whether we are concerned with elastic, plastic, or viscous behavior, can be classified as one of two types: Problems of plane *stress* and problems of plane *strain*.

Plane Stress.—An example of a body that is in a state of plane stress is a thin, plate-like body that is loaded only along its edges (Fig. 5.19A). In this case, the stresses on two opposite faces of the body in plane stress are zero, and if we let the z-axis be normal to the stress-free surfaces,

$$\sigma_{zz} = \sigma_{xz} = \sigma_{yz} = 0,$$

and the state of stress of the body is specified entirely by three stress components, such as σ_{xx}, σ_{yy}, and σ_{xy}.

If the x- and y-coordinates are in the plane of the plate, $\sigma_{zz} = 0$, so that the normal stress-strain relations, eqs. (5.31), reduce to:

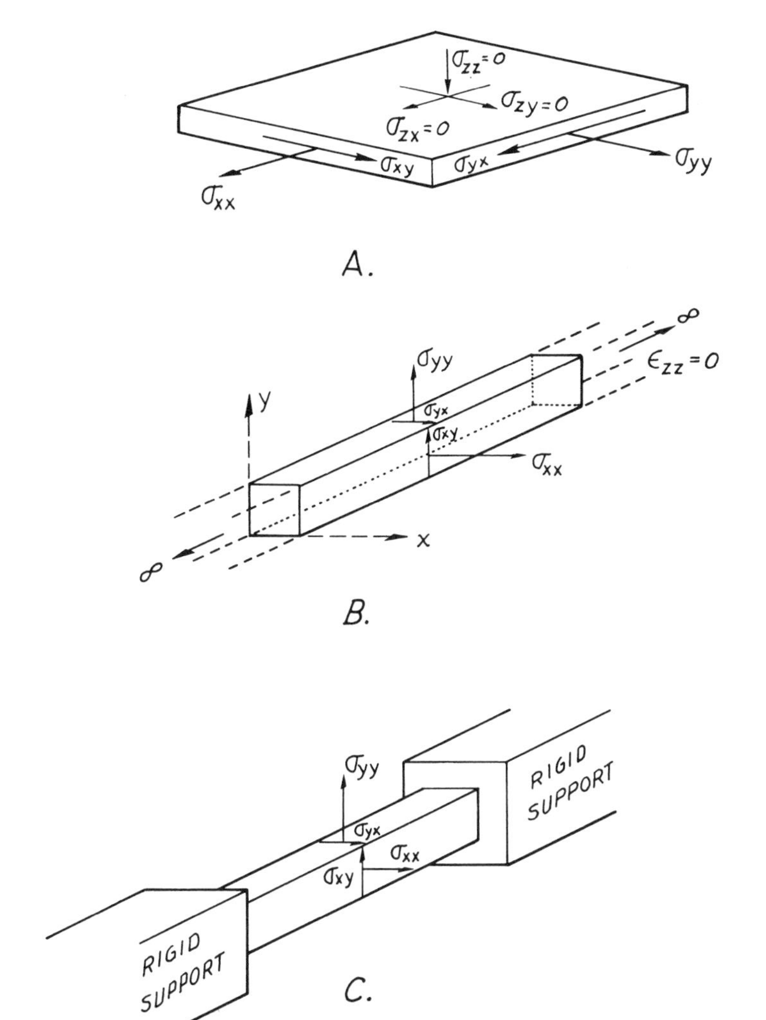

Figure 5.19. Examples of plane stress and plane strain.
A. Plane stress. Stresses normal and tangential to surface of plate are zero.
B. Plane strain. Beam is infinitely long in z-direction.
C. Plane strain. Beam is rigidly confined in z-direction.

$$\left.\begin{aligned}\epsilon_{xx} &= \frac{1}{E}[\sigma_{xx} - \nu\sigma_{yy}] \\ \epsilon_{yy} &= \frac{1}{E}[\sigma_{yy} - \nu\sigma_{xx}] \\ \epsilon_{zz} &= -\frac{\nu}{E}[\sigma_{xx} + \sigma_{yy}]\end{aligned}\right\} \quad \textit{Plane Stress} \tag{5.39}$$

Exercises

Given the following displacements:

$$u = 0.01\,x \quad \text{and} \quad v = -0.002\,y$$

1. Calculate normal and shear strains in the x–y plane.
2. If the displacements describe the deformation of a weightless rectangular plate with a thickness of one centimeter and with lateral dimensions of 5 by 10 meters, where the long dimension is in the x-direction, compute the displacements of the edge of the plate.
3. If the plate was deformed by stresses of

$$\sigma_{xx} = 10^8\ \text{N/m}^2$$

and

$$\sigma_{yy} = \sigma_{zz} = \sigma_{xy} = 0,$$

what are Young's modulus and Poisson's ratio for the material of which the plate is composed?

Ans: $E = 10^{10}\ \text{N/m}^2$,

$\nu = 0.20.$

4. What might be the material of which the plate is composed? (Note strains as well as modulus and Poisson's ratio.)
5. What was the thickness of the plate before the plate was stressed?

Plane Strain.—Whereas a body is in a state of plane stress if the components of *stress* acting on one plane are zero, a body is in a state of plane strain if the components of *strain* on one plane are zero. For example, if a body is very long in one direction and if each cross-section at right angles to its long dimension is subjected to the same state of stress, there are no strains in the long direction (Fig. 5.19B). Or, if two ends of a body are confined by rigid, lubricated surfaces, and if each cross-section is stressed the same way, there are no strains normal to the cross-section (Fig. 5.19C). These are physical examples of the theoretical concept of plane strain.

If the direction of zero strain is in the z-direction, the components of displacement, u and v, are functions of the x- and y-directions but they are independent of the z-direction. Also, the displacement in the z-direction is zero. The result is that three components of strain are zero:

$$\epsilon_{yz} = \frac{\partial v}{\partial z} + \frac{\partial w}{\partial y} = 0,$$

$$\epsilon_{xz} = \frac{\partial u}{\partial z} + \frac{\partial w}{\partial x} = 0,$$

and

$$\epsilon_{zz} = \frac{\partial w}{\partial z} = 0.$$

For plane strain, therefore, the stress-strain relations reduce to:

$$\left.\begin{aligned} \epsilon_{xx} &= \frac{1}{E}[\sigma_{xx}(1 - \nu^2) - \nu(1 + \nu)\sigma_{yy}], \\ \epsilon_{yy} &= \frac{1}{E}[\sigma_{yy}(1 - \nu^2) - \nu(1 + \nu)\sigma_{xx}], \\ \epsilon_{zz} &= 0, \\ \sigma_{zz} &= \nu(\sigma_{xx} + \sigma_{yy}). \end{aligned}\right\} \quad \textit{Plane Strain} \qquad (5.40)$$

Note that plane stress involves the assumption that the stresses do not vary through the thickness of the body; a body approaches a state of plane stress as it becomes very thin. Plane strain involves no such assumption. Therefore, plane strain is exact whereas plane stress is approximate.

Exercises

1. Assume that we can model the earth by a body of infinite extent horizontally and of infinite extent below a flat surface, that is, by an infinite half-space.
   a. Ignoring tectonics and assuming that the vertical stress in our model of the earth is due solely to the weight of the overburden, write an expression for the vertical stress as a function of depth.

   Ans. $\sigma_{yy} \approx 2$ to 3×10^4 N/m^2 per meter of depth (or 1.2 psi per foot of depth).

   b. Assuming a reasonable value for Poisson's ratio of rock, say 0.25, what would be the magnitudes of horizontal stresses, σ_{xx} and σ_{zz}, in our model of the earth (*7*)?

   Ans: $\sigma_{xx} = \sigma_{zz} = \left(\frac{\nu}{1 - \nu}\right) \sigma_{yy} = \frac{\sigma_{yy}}{3}.$

2. Granite is quarried in the Fletcher quarry, near Westford, Massachusetts, by a combination of burning and wire-saw cutting. Vertical slots or channels are cut ten to fifteen meters deep and about ten meters laterally into the sides of the quarry. The channels are bounded on each side by a burn cut, a sinuous slot about six inches wide, formed by moving a large torch up and down a side of the quarry, slowly forming the slot. The sides of the burn cuts have a rippled surface such as is shown in the faces in the background, oriented parallel to the plane of the photograph in Fig. 5.20. The blocks of granite within the channels, bounded by the burn cuts, are removed to form the channel, which is about eight feet wide. Remnants of two channels are visible in Fig. 5.20, one in the foreground and one in the middle distance.

Figure 5.20. Vertical slabs formed by wire-saw cutting. Bottom of H. E. Fletcher Co. quarry, near Westford, Mass.

After the channels have been cut, a gang of wire saws cuts vertically downward, at right angles to the channels, through the block of granite between the channels, to form ten to fifteen slabs of granite about one-half to one meter thick, ten to fifteen meters tall, and about thirty meters long. Remnants of slabs are visible in the floor of the quarry. One of the two steel towers which support the wires used to cut the slabs is standing in the channel shown in the middle distance in Fig. 5.20.

Figure 5.21A shows a plan view sketch of the quarry at the time the remnant of the saw block shown in the middle foreground in Fig. 5.20 was isolated on two sides by means of channels but before the block was wire-sawed. The faces on the left-hand side of the photograph in Fig. 5.20 form the northeastern edge of the deepest part of the quarry. The view in the photograph is toward the southeast.

Suppose that three pins were placed in the sawblock of granite, before it was sawed, in the positions shown in Fig. 5.21B. Two pins were cemented in holes drilled one meter apart at the northwestern edge of the sawblock and one pin was centered in a small hole drilled near the eastern corner of the sawblock (Fig. 5.21C).

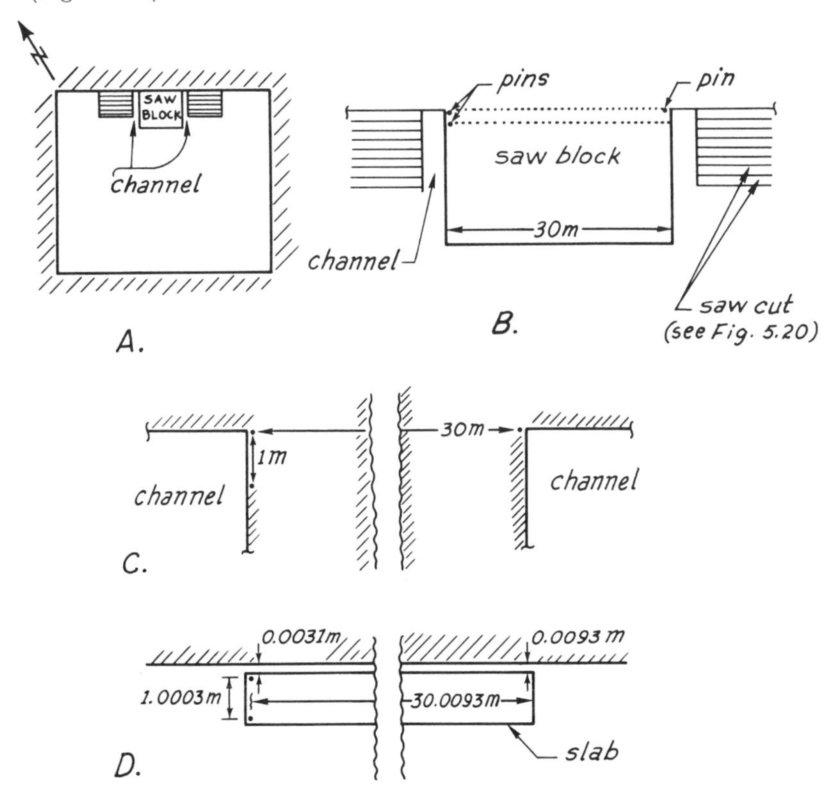

Figure 5.21. Deformation of granite slab during quarrying.
A. Sketch of quarry.
B. Locations of reference pins in saw block.
C. Distances between pins prior to quarrying.
D. Distances between pins after quarrying.

a. When the sawblock was wire-sawed, the slab of granite containing the pins lengthened, thickened, and distorted slightly as is shown in Fig. 5.21D. Compute the strains in the x–z plane.

Ans. $\varepsilon_{xz} \approx 2 \times 10^{-4}$

$$\epsilon_{xz} = 2\varepsilon_{xz} \approx 4 \times 10^{-4}$$

$$\epsilon_{xx} \approx 3.1 \times 10^{-4}$$

$$\epsilon_{zz} \approx 3.0 \times 10^{-4}$$

b. Are the strains finite?

c. With the values of Young's modulus and Poisson's ratio that you calculated from the graphs in Fig. 5.15, which are test results for a sample of the granite in the Fletcher quarry, compute the normal and shear stresses that acted on the granite slab at the time the pins were placed.

d. What were the magnitudes and directions of the principal stresses acting in the granite slab prior to quarrying? (Use Mohr's Circle.)

3. Most of the granite slabs in the Fletcher quarry strain about 700 micrometers per meter when they are sawed, so that a slab of granite 30 meters long increases in length by about 2.1 centimeters when it is quarried.

a. Assume that the shear strain, ϵ_{xy}, in the slabs is zero, and that $\sigma_{xx} = \sigma_{zz}$, and that the vertical stress, σ_{yy}, is zero. What is the magnitude of the compressive stress acting on a slab?

b. At what depth was the granite emplaced? (This problem may require considerable thought.) Assume hydrostatic pressure in the granite immediately after it solidified, and prior to the time that erosion brought the surface of the earth down to the level of the present quarry. What is questionable about your estimates of regional stresses and of original depth of the granite?

Ans: $\text{depth} = \dfrac{\sigma_{xx}}{\rho g}\left[\dfrac{1-\nu}{1-2\nu}\right].$

Differences Between Bending of Beams and Plates.—The bending moment tending to resist deflection of an elastic beam was shown to be, approximately,

$$M = EI\left(\frac{d^2v}{dx^2}\right),$$

where E is Young's modulus and v is the deflection of the beam in the y-direction. What would be the corresponding moment for a plate?

According to eqs. (5.27), if $\sigma_{yy} = \sigma_{zz} = 0$, that is, for a beam,

$$\epsilon_{xx} = \frac{\sigma_{xx}}{E}. \tag{5.41}$$

But for a plate, σ_{zz} in the plane of the plate cannot be zero, because the problem is one of plane strain; the plate is infinitely wide. The stress-strain relations for plane strain are

$$\left.\begin{aligned} \sigma_{zz} &= \nu(\sigma_{xx} + \sigma_{yy}) \\ &\text{and} \\ \epsilon_{xx} &= \frac{1}{E}\left[\sigma_{xx}(1 - \nu^2) - \sigma_{yy}(\nu + \nu^2)\right]. \end{aligned}\right\} \tag{5.42}$$

But for a plate, the stress normal to the surface is zero, that is, $\sigma_{yy} = 0$, so that eqs. (5.42) become

$$\left.\begin{aligned} \sigma_{zz} &= \nu\sigma_{xx} \\ &\text{and} \\ \epsilon_{xx} &= \frac{\sigma_{xx}}{E}(1 - \nu^2). \end{aligned}\right\} \tag{5.43}$$

Comparing eqs. (5.43) and (5.41) we see why the moment equation for a beam can be converted to the moment equation for a plate if $E/(1 - \nu^2)$ is substituted for E. Thus, the bending moment for a plate or a beam, referred to generally as a *member*, is

$$M = BI\left(\frac{d^2v}{dx^2}\right),$$

where $B = E$ for a beam and $B = E/(1 - \nu^2)$ for a plate.

References cited in Chapter 5

1. Bishop, A. W., Webb, D. L., and Lewin, P. I., 1965, "Undisturbed Samples of London Clay from the Ashford Common Shaft: Strength: Effective Stress Relationships": *Geotechnique*, V. 15, p. 1–31.
2. Clark, S. P., Jr., ed., 1966, *Handbook of Physical Constants*: The Geological Society of America, Memoir 97.
3. Fung, Y. C., 1965, *Foundations of Solid Mechanics*: Prentice-Hall, Inc., Englewood Cliffs, N.J.
4. Housner, G. W., and Vreeland, Thad, Jr., 1966, *The Analysis of Stress and Deformation*: The Macmillan Co., N.Y.
5. Love, A. E. H., 1944, *A Treatise on the Mathematical Theory of Elasticity*: Dover Publications, N.Y.
6. Obert, L., and Duvall, W. I., 1967, *Rock Mechanics and the Design of Structures in Rock*: John Wiley and Sons, Inc., N.Y.
7. Ramsay, J. G., 1967, *Folding and Fracturing of Rocks*: McGraw-Hill Book Co., Inc., N.Y.
8. Sokolnikoff, I. S., 1956, *Mathematical Theory of Elasticity*: McGraw-Hill Book Co., Inc., N.Y.

6

Chapter Sections

Folds Near Moab, Utah
Conditions at Time of Folding of Carmel
Biot Theory of Buckling of Interlayered Soft and Stiff Materials
  Idealization of Rock Sequence
  Explanation of the Theory
  Differential Equation of Deflection of Multilayer
  Elastic Constants of Multilayer
  Review of Assumptions
  Buckling of Multilayer Confined by Rigid Boundaries
Comparison of Theoretical Folds with Those in Carmel Formation
References Cited in Chapter 6

6

Development of Folds in Carmel Formation, Near Moab, Utah

Folds near Moab, Utah*

Not far northeast of the Henry Mountains, near Moab, Utah, are some fascinating folds. The folds are visible around the peripheries of the numerous buttes and mesas in Arches National Monument, a few miles north of Moab. For example, Fig. 6.1A shows a series of folds at the base of a cliff in the Windows section of the Monument. They are in the Carmel Formation, which comprises the crenulated beds between the massive, light-toned rocks forming the apron at the base of the cliff and the massive rocks forming the upper part of the cliff. The rocks above the Carmel are part of the Entrada Sandstone and those below are part of the Navajo Sandstone (see stratigraphic section, Fig. 6.2).

There are two types of distortions of bedding planes within the Carmel Formation, shown in the right-hand half of the photograph in Fig. 6.1A: upward and downward deflection of beds without appreciable thickening or thinning; and thickening and thinning beds, resulting in diminishing amplitudes of upward and downward deflections of bedding planes with increasing distances normal to bedding planes. Both types of distortions of bedding planes are repeated at nearly uniform intervals along the cliff face shown in Fig. 6.1A so that the distortions are periodic. I have interpreted them to be folds (*6*).

Conditions at time of folding of Carmel

The conditions of the Carmel Formation at the time of folding are suggested by several lines of evidence. Some of the folds must have formed after the Entrada

* This chapter is largely based on ref. *6*.

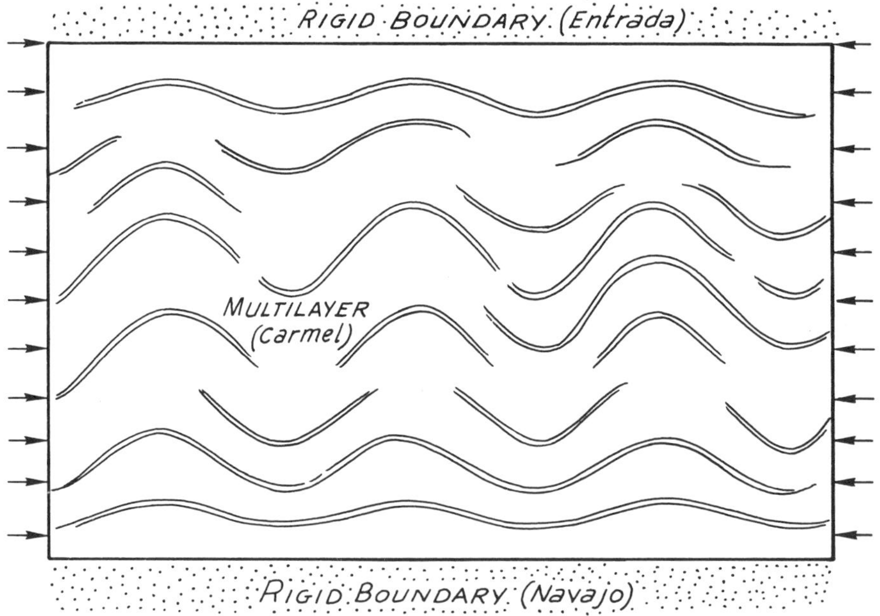

Figure 6.1. Folds in multilayer confined by rigid boundaries.

A. Folds in Carmel Formation, in Windows section of Arches National Monument, near Moab, Utah. Upper surface of Navajo Sandstone and lower surface of

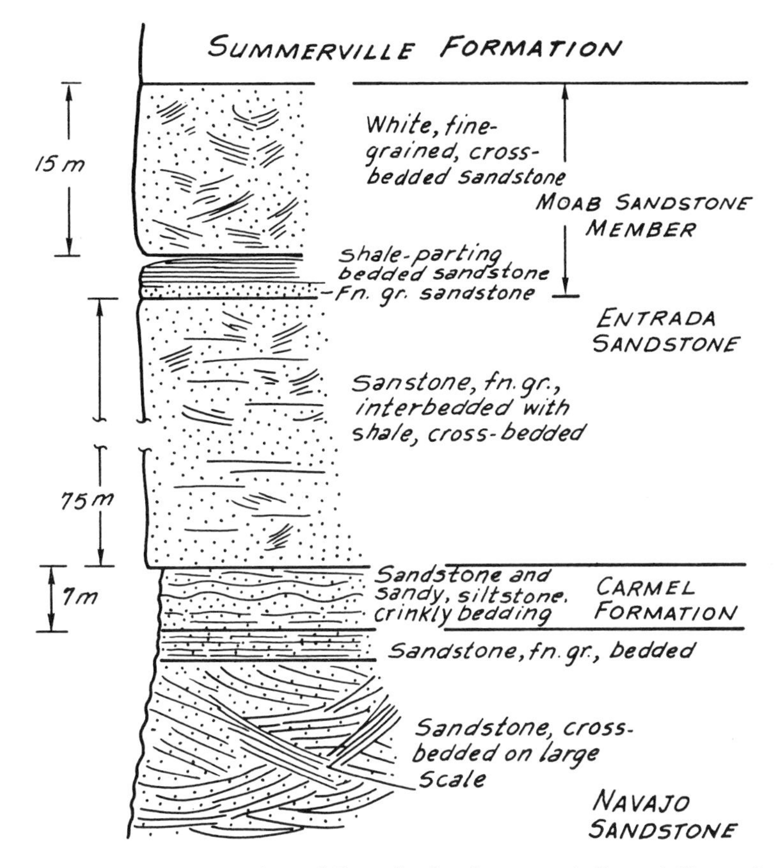

Figure 6.2. Stratigraphic section of Entrada Sandstone and Carmel Formation, near Dewey, Utah (after Dane, ref. *5*, p. 95).

Sandstone was deposited upon the Carmel, because in many places they extend through the Carmel into the overlying Entrada. They diminish in amplitude upwards into the Entrada but, according to Dane (*5*), they extend as far as 20 meters into the Entrada in some places.

Both Dane (*5*) and McKnight (*7*), who studied the Carmel Formation in the vicinity of Moab, mention that upturned bedding planes within some of the folded

---

Entrada Sandstone are relatively undeformed, so that folds within Carmel Formation diminish in amplitude upward and downward. Relief on cliff face of Carmel is caused by differential weathering of sandy siltstone, which forms indentations, and muddy sandstone.

B. Form of theoretical folds in a multilayer consisting of interbedded soft and stiff elastic layers. Upper and lower boundaries are rigid and correspond to Entrada and Navajo sandstones in Fig. 6.1A.

rocks are truncated and that horizontally bedded rocks of the Entrada cut across tops of the folds. Apparently, therefore, part of the deformation we see in the Carmel could have occurred prior to deposition of the Entrada, so that the beds of the Carmel were initially deflected, prior to folding.

Marked changes of thickness of sandstone beds associated with the folds in the Carmel probably are indications that the muddy sand of the Carmel was soft at the time the folds formed. It is difficult to imagine that the thick sandstone bed near the base of the Carmel, shown in Fig. 6.1A, could have been firm at the time units overlying it were deflected into the series of waves we see today. The thick sandstone ranges in thickness back and forth between two and four meters over horizontal distances of about eight meters. Dane also believed that the Carmel was soft at the time the folds formed, because he referred to the "water-soaked sandy mud of the Carmel" in discussing the origin of the crenulations (*5*) (p. 100).

Biot Theory of buckling of interlayered soft and stiff materials

Idealization of Rock Sequence.—The folds shown in the right-hand side of Fig. 6.1A are gentle waves with a wavelength of about 12 meters and a maximum amplitude of one meter. The amplitudes die out upwards and downwards, so that within a vertical distance of about 12 meters the folds originate, grow to a maximum amplitude, and die out (Fig. 6.3A).

The deflection of beds within the Navajo and the Entrada appears to have been minor, so that we probably can represent mechanical behaviors of these two formations by rigid units, bounding relatively flexible beds of the Carmel (Fig. 6.3C). This condition is not precisely realized in nature because the lower boundary of the Entrada commonly is deflected somewhat and the folds die out upwards within the Entrada. However, the Entrada was more nearly rigid to deflection than the Carmel and to a first approximation it was rigid.

The Carmel Formation is an interlayered sequence of thin, sandy siltstone beds and thicker sandstone beds, as is shown in Fig. 6.3A. The siltstone beds were probably relatively soft and the sandstone beds relatively stiff at the time the folds formed. The siltstone beds apparently provided places for sliding adjustments between adjacent sandstone beds. Thus, we might idealize the sequence of beds within the Carmel as thin, interbedded soft and stiff layers, called a *multilayer* by Biot (*1*), confined above and below by thick, rigid layers (Figs. 6.3B and 6.3C). This assumption of many soft and stiff layers can be only a first approximation to the actual situation, because the siltstone beds generally are discontinuous layers that thicken and thin laterally.

Explanation of the Theory.—Among the many important contributions Maurice Biot has made to our understanding of the theory of folding is his analysis of the bending of sequences of soft and stiff layers. He has shown (*3*) that the bend-

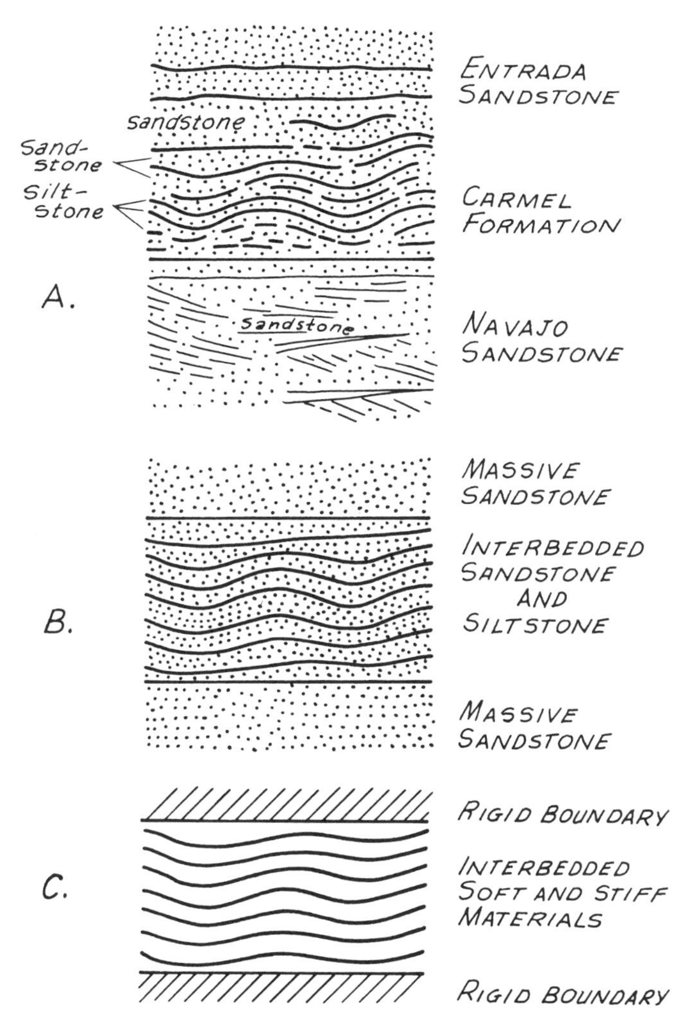

Figure 6.3. Stages of idealization of folds in Windows section of Arches Monument (*6*).

A. Map of sandstone and siltstone beds exposed in cliff face.

B. Entrada and Navajo sandstones idealized by massive sandstones, and Carmel Formation idealized as uniformly interbedded sandstone and siltstone.

C. Behavior of massive sandstones idealized by rigid boundaries and behavior of Carmel idealized by interbedded soft and stiff elastic materials.

ing of such sequences can be approximately represented by an equation that is a special case of equations he has derived for deformation of initially stressed anisotropic bodies. We can derive an equation that is identical to his special case, however, by slightly modifying the theory of bending that we have already developed. The theory is intermediate in rigor between the exact theory of elasticity which we will develop in the next chapter and the approximate theory of beams and plates which we have developed in preceding chapters. Biot's intermediate theory allows us to solve problems that cannot be simplified to the standard beam theory but that are intractable when expressed in terms of elasticity theory.

We will idealize systems of interbedded rocks by representing the beds with alternating layers of soft and stiff materials. The soft materials have thickness t_2 and modulus B_2, where

$$B_2 = \frac{E_2}{1 - \nu_2^2}.$$

The stiff layers have a modulus of B_1 and a thickness of t_1. The total thickness of a system of beds, that is, of a *multilayer*, is T (Fig. 6.4A).

Differential Equation of Deflection of Multilayer.—Now suppose that a multilayer is bent under the action of axial loads, shear forces, and bending moments (Fig. 6.4B). We can relate the bending of the multilayer to its elastic properties and the applied load as follows. Remove from the multilayer a small element (Fig. 6.4C) which consists of one stiff layer and one soft layer. The element has a thickness of t, where

$$t = t_1 + t_2,$$

and a length of δx.

Acting on the element are shear forces, Q_y and $Q_y + \delta Q_y$; bending moments, M and $M + \delta M$; distributed loads, q and $q + \delta q$; and axial loads, P and $P + \delta P$; all of which we have already considered in our previous theory. The only other forces we must consider for the multilayer are the shear forces of interaction between adjacent layers, Q_x and $Q_x + \delta Q_x$.

We use the equations of equilibrium of moments and forces in order to derive relations between the various forces and moments. Thus, if we sum moments about the center of the right-hand end of the element, Fig. 6.4C, we find that: $[\sum M = 0]$,

$$-M + (M + \delta M) + Q_y\,\delta x - Q_x\frac{t}{2} - (Q_x + \delta Q_x)\frac{t}{2} + P(\delta v) - q\frac{(\delta x)^2}{2} + (q + \delta q)\frac{(\delta x)^2}{2} = 0. \tag{6.1}$$

We can derive an equivalent relation in terms of shear stresses rather than shear forces and in terms of normal stress rather than distributed load. The disadvantage of the form of the equilibrium equation given above is that shear forces are not readily related to elastic constants, which are constants of proportionality between stress and strain. Figure 6.4D shows the same small element of the multilayer shown in Fig. 6.4C. Acting on the element are shear stresses, σ_{xy} and σ_{yx}, and a normal stress, σ_{yy}, in addition to the axial load, P, and the bending moment, M. If we com-

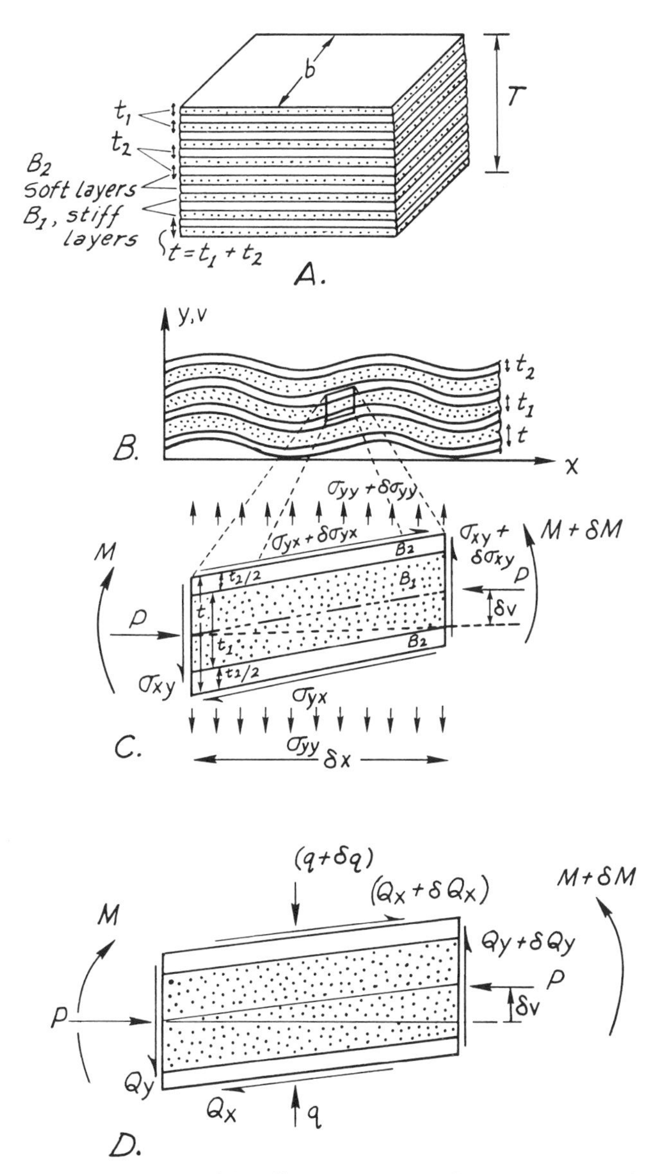

Figure 6.4. Forces, stresses, and bending moments acting on element of multilayer (*6*).
A. Multilayer consisting of interbedded soft and stiff materials.
B. Position of element within bent multilayer.
C. Forces on element.
D. Stresses on element.

pare Figs. 6.4C and 6.4D we see that the shear forces are related to the shear stresses by the equations

$$\left.\begin{aligned} Q_x &= \sigma_{yx}(\delta x)b \\ \text{and} \\ Q_y &= bt\sigma_{xy}. \\ \text{The distributed load, } q\text{, is given by} \\ q &= -\sigma_{yy}b, \end{aligned}\right\} \quad (6.2)$$

where the negative sign is required because positive values of q are compressional, whereas positive values of σ_{yy} are tensile.

Both the bending moment, M, and the axial load, P, contribute to the normal stress, σ_{xx}, in the x-direction, as is indicated pictorially in Fig. 6.5. However, the

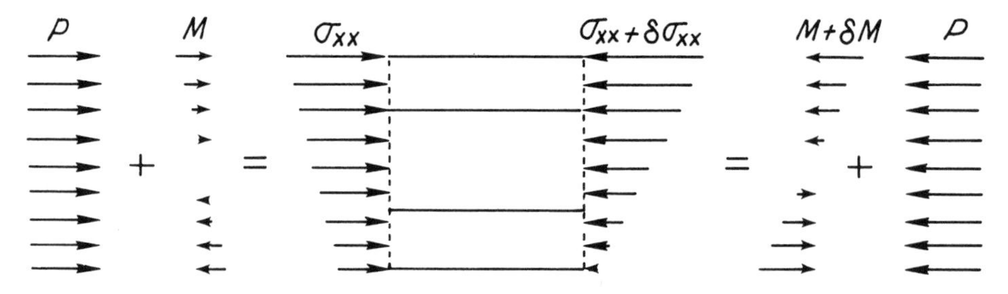

Figure 6.5. Alternative representations of axial load and bending moment in terms of stresses.

effects of bending moment and axial load are more readily understood than are the effects of axial stresses; for that reason we will maintain our old notation for the bending moment and axial load.

Thus, according to eqs. (6.2), the equation of equilibrium of moments, eq. (6.1), is equivalent to

$$\delta M + \sigma_{xy}\,\delta xtb - \sigma_{yx}t\,\delta xb - \frac{\delta\sigma_{yx}tb\delta x}{2} + P\,\delta v - \delta\sigma_{yy}\frac{(\delta x)^2}{2}b = 0. \quad (6.3)$$

Some of the terms in this equation are much smaller than others; therefore, to a first approximation we can ignore them. That is,

$$\sigma_{yx}\,\delta x \gg \frac{\delta\sigma_{yx}\,\delta x}{2} \to 0,$$

and

$$\delta\sigma_{yy}(\delta x)^2 \to 0,$$

because each of these terms involves the product of two small quantities. To a first approximation, therefore, the equation of equilibrium of moments becomes

$$\delta M + Q_y\,\delta x + Q_x t + P\,\delta v = 0,$$

or in terms of stresses,

$$\delta M + \sigma_{xy}\,\delta x t b - \sigma_{yx} t\,\delta x b + P\,\delta v = 0.$$

Dividing through by δx,

$$\frac{\delta M}{\delta x} + Q_y + \frac{Q_x t}{\delta x} + P\frac{\delta v}{\delta x} = 0, \tag{6.4}$$

or

$$\frac{\delta M}{\delta x} + (\sigma_{xy} - \sigma_{yx})tb + P\frac{\delta v}{\delta x} = 0. \tag{6.5}$$

At this point there is no advantage in proceeding with the derivation of the equation in two forms. We will continue the derivation with equation (6.5). If we take the limit of $\delta M/\delta x$ and $\delta v/\delta x$ as δx approaches zero, we have

$$\frac{\partial M}{\partial x} + (\sigma_{xy} - \sigma_{yx})tb + P\frac{\partial v}{\partial x} = 0. \tag{6.6}$$

The differential equation we derived in Chapter 3 for the relation between the bending moment and the deflection of an axially loaded member is slightly different [eq. (3.16)],

$$\frac{dM}{dx} - Q + P\frac{dv}{dx} = 0.$$

Q in this equation is equivalent to the negative of $\sigma_{xy}tb$ in eq. (6.6), so that the only additional term in eq. (6.6) is the shear force, $\sigma_{yx}tb$. The sign of the shear force, Q, is negative because we have changed sign conventions. The remaining difference between the two equations is that we now use partial derivatives instead of total derivatives. The reason for this change is that the vertical displacement of a multilayer is a function of both the x- and y-coordinates, whereas the vertical displacement of the members we considered earlier was a function of only the distance along the members, the x-coordinate.

We can relate the shear stress, σ_{xy}, to the normal stress, σ_{yy}, by summing forces in the y-direction acting on the small element of the multilayer (Fig. 6.4D): $[\sum F_y = 0]$

$$-\sigma_{yy}\,\delta x b + (\sigma_{yy} + \delta\sigma_{yy})\,\delta x b - \sigma_{xy} t b + (\sigma_{xy} + \delta\sigma_{xy}) t b = 0,$$

or (6.7)

$$\delta\sigma_{yy}\,\delta x + \delta\sigma_{xy} t = 0.$$

We are assuming that the bending of the multilayer is so slight that σ_{yx} is parallel to the x-direction. If we assume that the vertical normal stress, σ_{yy}, varies continuously in the vertical direction,

$$\delta\sigma_{yy} = \frac{t\,\partial\sigma_{yy}}{\partial y}, \tag{6.8}$$

so that, from eqs. (6.7), and (6.8),

$$\frac{\partial\sigma_{yy}}{\partial y}\,t\,\delta x + \delta\sigma_{xy}t = 0,$$

or

$$\frac{\delta\sigma_{yy}}{\delta y} = -\frac{\delta\sigma_{xy}}{\delta x}.$$

In the limit, as $\delta x \to 0$ and $\delta y \to 0$,

$$\frac{\partial\sigma_{yy}}{\partial y} = -\frac{\partial\sigma_{xy}}{\partial x}. \tag{6.9}$$

In terms of shear force, Q_y, and distributed load, q, eq. (6.9) becomes:

$$t\frac{\partial q}{\partial y} = \frac{\partial Q_y}{\partial x}. \tag{6.10}$$

Equations (6.9) and (6.6) can be combined by differentiating eq. (6.6) with respect to x and by substituting eq. (6.9) into eq. (6.6):

$$\frac{\partial^2 M}{\partial x^2} - \left(\frac{\partial\sigma_{yy}}{\partial y} + \frac{\partial\sigma_{yx}}{\partial x}\right)tb + P\frac{\partial^2 v}{\partial x^2} = 0. \tag{6.11}$$

Elastic Constants of Multilayer.—In order to put eq. (6.11) into a form that we can solve, we must substitute functions of v for the variables, M, σ_{yy}, and σ_{yx}. To do this we will relate stresses and displacements to the elastic constants of the multilayer. For a homogeneous elastic member, we know that, to a first approximation,

$$M = BI\frac{\partial^2 v}{\partial x^2}.$$

That is, the bending moment within a single elastic member is proportional to the modulus, B, the moment of inertia, I, and the second derivative of the displacement, v, in the x-direction. The problem is to find expressions for B and I for a multilayer.

If we assume that the resistance to bending of a multilayer arises from both the soft and stiff layers, we can determine the modulus and the moment of inertia of the multilayer as follows. Figure 6.6A shows a small element of the multilayer. It

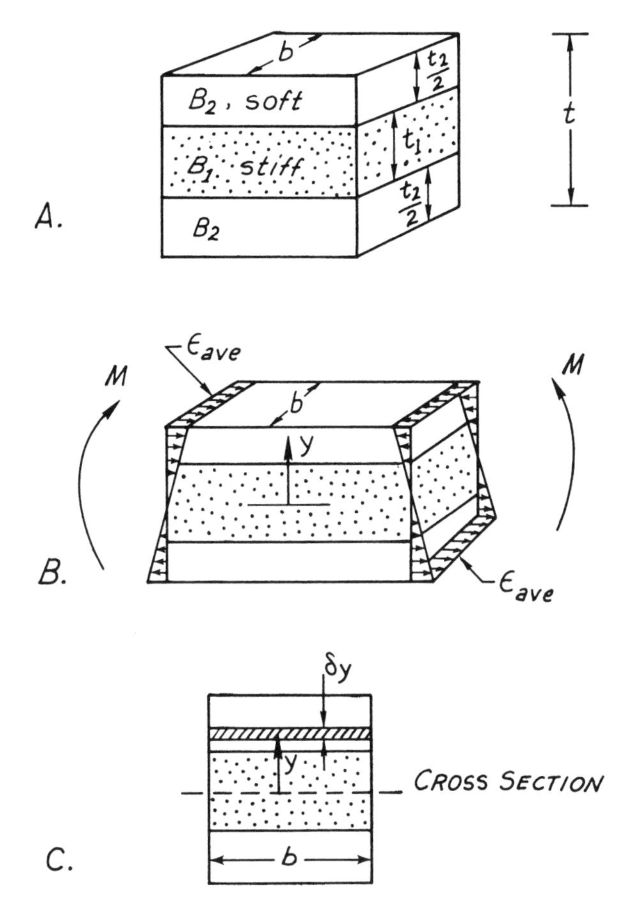

Figure 6.6. Distortion of element of multilayer (*6*).
A. Element of multilayer.
B. Linear distribution of fiber strains is assumed.
C. Cross-section of element.

consists of one soft and one stiff layer, half of the soft layer being on each side of the stiff layer. If the element is bent, the soft and stiff layers are strained in the horizontal direction. We will assume that the strain is linearly distributed relative to the center of the stiff layer so that the strain increases linearly from zero at the center to a maximum at the top and bottom edges. As we have already shown, the strain is related to the radius of curvature, ρ, of the neutral plane and the distance from the neutral plane of a member by (Fig. 6.6B)

$$\epsilon_f = \frac{y}{\rho}. \tag{6.12}$$

Also, we have already shown that the bending moment, M, is related to the fiber stress by the equation (Fig. 6.6C)

$$M = b\int \sigma_f y \, dy. \tag{6.13}$$

However, the fiber stress is related to the fiber strain by the equations

$$\sigma_f = \epsilon_f B_1, \qquad \text{within stiff layers,}$$

and (6.14)

$$\sigma_f = \epsilon_f B_2, \qquad \text{within soft layers.}$$

Therefore, combining eqs. (6.12), (6.13), and (6.14), the bending moment is

$$\begin{aligned} M &= b\int \epsilon_f B_1 y \, dy + b\int \epsilon_f B_2 y \, dy \\ &= \frac{B_1 b}{\rho}\int y^2 \, dy + \frac{B_2 b}{\rho}\int y^2 \, dy, \end{aligned} \tag{6.15}$$

in which the first integral applies to the stiff layer and the second to the soft layer. Integrating through the stiff and soft layers, we find that

$$M = \frac{2B_1 b}{\rho}\int_0^{t_1/2} y^2 \, dy + \frac{2B_2 b}{\rho}\int_{t_1/2}^{t/2} y^2 \, dy,$$

where t is the combined thickness of one stiff and one soft layer. Thus,

$$M = \frac{B_1 b}{12\rho} t_1^3 + \frac{B_2 b}{12\rho}[t^3 - t_1^3],$$

or

$$M = \frac{t_1^3 b}{12\rho}\left\{B_1 + B_2\left[\left(\frac{t}{t_1}\right)^3 - 1\right]\right\}. \tag{6.16}$$

Therefore, if we define two new constants,

$$I_h = \frac{t_1^3 b}{12}, \qquad \text{Moment of Inertia,} \tag{6.17}$$

and

$$B_h = B_1 + B_2\left[\left(\frac{t}{t_1}\right)^3 - 1\right], \qquad \text{Horizontal Elastic Modulus,} \tag{6.18}$$

we can write the equation relating the bending moment to the radius of curvature in our standard form,

$$M = \frac{B_h I_h}{\rho},$$

or

$$M = B_h I_h \frac{\partial^2 v}{\partial x^2}. \tag{6.19}$$

For a multilayer, then, the differential equation of equilibrium, eq. (6.11), becomes

$$B_h I_h \frac{\partial^4 v}{\partial x^4} - \left(\frac{\partial \sigma_{yy}}{\partial y} + \frac{\partial \sigma_{yx}}{\partial x}\right) tb + P \frac{\partial^2 v}{\partial x^2} = 0. \tag{6.20}$$

Now we will relate the shear stress, σ_{yx}, to the shear modulus and the deflection of the multilayer. We will assume that the soft and stiff layers are subjected to the same shear stress when the multilayer is bent and that the layers attempt to slip relative to each other. The resulting shear strain in the stiff layer is $(\epsilon_{yx})_1$, so that the stress-strain relation for shear in the stiff layer is

$$(\epsilon_{yx})_1 = \frac{1}{G_1} \sigma_{yx}, \tag{6.21}$$

where G_1 is the shear modulus of elasticity of the stiff layer. Similarly, the shear strain in the soft layer is

$$(\epsilon_{yx})_2 = \frac{1}{G_2} \sigma_{yx}, \tag{6.22}$$

where G_2 is the shear modulus of the soft layer.

The total horizontal displacement due to shear of the surfaces of the unit element of the multilayer, shown in Fig. 6.6B, is the "average" shear strain, ϵ_{ave}, times the thickness, t, of the unit element. It is equal to the sum of the displacements in the soft and stiff layers, so that

$$t\epsilon_{ave} = t_1(\epsilon_{yx})_1 + t_2(\epsilon_{yx})_2,$$

or (6.23)

$$\epsilon_{ave} = \frac{t_1}{t}(\epsilon_{yx})_1 + \frac{t_2}{t}(\epsilon_{yx})_2.$$

Substituting eqs. (6.21) and (6.22) into eq. (6.23),

$$\epsilon_{ave} = \frac{\sigma_{yx}}{t}\left[\frac{t_1}{G_1} + \frac{t_2}{G_2}\right],$$

so that we can define an "average" shear strain of the multilayer as

$$\epsilon_{ave} = \frac{\sigma_{yx}}{G_a}, \tag{6.24}$$

where the "average" shear modulus is

$$G_a = \frac{1}{[(t_1/tG_1) + (t_2/tG_2)]} = \frac{tG_1G_2}{G_2t_1 + G_1t_2}. \tag{6.25}$$

Maurice Biot (*3*) calls G_a the "sliding modulus".

We showed that the shear strain, ϵ_{yx}, can be defined in terms of displacements, u and v, in the x- and y-directions, respectively:

$$\epsilon_{yx} = \left(\frac{\partial u}{\partial y} + \frac{\partial v}{\partial x}\right).$$

Let us define the average shear strain in the same way so that $\epsilon_{ave} = \epsilon_{yx}$. If the multilayer is deflected mostly in the y-direction, so that the layers move upwards but not horizontally, the displacement in the x-direction, u, is zero, and

$$\frac{\partial u}{\partial y} = 0.*$$

The average shear strain reduces to

$$\epsilon_{ave} = \frac{\partial v}{\partial x}, \tag{6.26}$$

the change in vertical displacement with respect to the horizontal direction.

Substituting eq. (6.26) into eq. (6.24), the shear stress is

$$\sigma_{yx} = G_a \frac{\partial v}{\partial x}.$$

Thus, the differential equation, eq. (6.20), becomes

$$B_h I_h \frac{\partial^4 v}{\partial x^4} - \left(\frac{\partial \sigma_{yy}}{\partial y} + G_a \frac{\partial^2 v}{\partial x^2}\right) tb + P \frac{\partial^2 v}{\partial x^2} = 0. \tag{6.27}$$

Now we have a differential equation in which all but one of the terms are expressed as functions of the displacement, v.

The differential equation can be completely expressed in terms of displacement in the y-direction if we can relate the vertical normal stress, σ_{yy}, to the deflection, v. In order to derive an expression for σ_{yy} in terms of v, we will have to assume a general boundary condition for our problem of folding. If folding can occur only by

* In some applications the change of displacement, u, with respect to the y-direction might be appreciable, but constant. If $\partial u/\partial y$ is constant, the differential equation, eq. (6.27), is unchanged so that the solution we will derive in following pages will be applicable. Such a situation might arise where a multilayer is confined between two thick, competent layers that are being folded on a larger scale than are the multilayers. As the two thick layers bend they slip relative to each other, transferring a shear strain through the multilayer. The result would be a differential horizontal displacement of the crests of the minor folds in the multilayer, presumably producing drag flexures.

the compression of the multilayers, the vertical normal stress imposed by neighboring members on any layer is

$$\sigma_{yy} = B_v \epsilon_{yy},$$

where B_v is a modulus, which we will derive, and where the vertical normal strain is

$$\epsilon_{yy} = \frac{\partial v}{\partial y}.$$

Thus,

$$\frac{\partial \sigma_{yy}}{\partial y} = B_v \frac{\partial^2 v}{\partial y^2}. \tag{6.28}$$

If we assume that strains in the vertical direction are the same in adjacent soft and stiff layers,

$$\epsilon_{yy} = \frac{(\sigma_{yy})_1}{B_1} = \frac{(\sigma_{yy})_2}{B_2}$$

at the boundaries. Then the weighted average of the vertical normal stress, σ_{yy}, for the soft and stiff layers is

$$\sigma_{yy} = \frac{t_1}{t}(\sigma_{yy})_1 + \frac{t_2}{t}(\sigma_{yy})_2.$$

Thus,

$$\sigma_{yy} = \epsilon_{yy}\left[\frac{t_1}{t} B_1 + \frac{t_2}{t} B_2\right],$$

and we can define B_v, the elastic modulus in the vertical direction, normal to bedding, as (*3*)

$$B_v = \left[\frac{t_1}{t} B_1 + \frac{t_2}{t} B_2\right]. \tag{6.29}$$

When we substitute the expression for the constraint of the adjacent layers, eq. (6.28), into the differential equation, eq. (6.27), we have an equation that is entirely in terms of the vertical displacement, v,

$$B_h I_h \frac{\partial^4 v}{\partial x^4} + (P - tbG_a)\frac{\partial^2 v}{\partial x^2} = tbB_v \frac{\partial^2 v}{\partial y^2}. \tag{6.30}$$

The coefficients are all constants, independent of the three variables, x, y, and v.

Review of Assumptions.—Before we proceed to solve differential equation (6.30) and to derive the buckling form of multilayers confined by rigid boundaries, we will state explicitly the assumptions we used in order to derive the differential equation. Some of the assumptions probably will have to be changed to solve problems other than the one of interest to us, so they should be clearly understood.

1. The soft and stiff layers of the multilayer are ideal elastic substances. Biot has shown that the differential equations and their solutions are essentially the same for elastic and viscous substances.
2. Strains are infinitesimal.
3. Deflections are small enough so that $(dv/dx)^2$ in the expression for radius of curvature is much smaller than unity.
4. The thicknesses of the individual beds, t_1, t_2, and of the unit element of the multilayer, t, are small relative to the total thickness of the multilayer.
5. The axial load is constant from one end to the other of each buckled member. If the axial load were applied to one end of the multilayers, we would proceed toward a solution just as we did when we analyzed the buckling of a single member near a fault, in Chapter 3.
6. Buckling is caused solely by the axial load. There are no external shear forces applied to the boundaries of the multilayer.
7. The vertical normal stress, σ_{yy}, varies continuously in the y-direction so that the change in constraint over the entire thickness of a soft and stiff layer $(\delta\sigma_{yy}/t)$ is the same as the change of constraint at any point within the multilayer, $\delta q/\delta y b$, where b is the width of the members [see eq. (6.8)].
8. The fiber strains in stiff and soft layers are linearly related to the distance from the center of the stiff layer [see Fig. 6.7A]. This assumption led to the following definitions of the modulus and the moment of inertia of the multilayer, eqs. (6.17) and (6.18),

$$I_h = \frac{t_1^3 b}{12},$$

$$B_h = B_1 + B_2\left[\left(\frac{t}{t_1}\right)^2 - 1\right].$$

If we had assumed, on the contrary, that the stresses are linearly related to the distances from the centers of the stiff layers, the horizontal strains would have to be discontinuous. This is necessary because the constants of proportionality between stress and strain are different in the two materials. Thus, if the strains are continuous, the stresses are discontinuous; if the stresses are continuous, the strains are discontinuous.

The degree of bonding between adjacent soft and stiff layers is another important consideration. For example, if the stresses were a linear function of y and the stiff and soft layers were firmly bonded together, the end of the element of the multilayer would assume a configuration similar to that shown in Fig. 6.7B. On the other hand, if the contacts between the layers were friction-

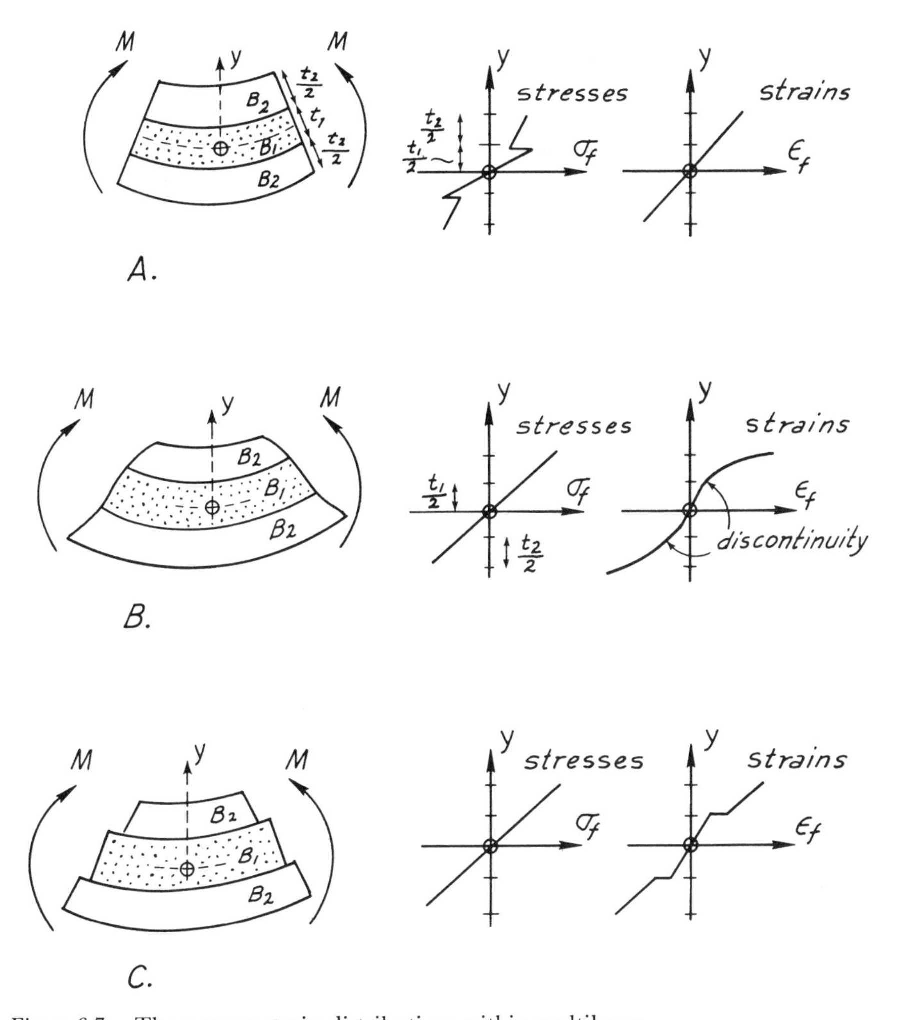

Figure 6.7. Three stress-strain distributions within multilayer.

A. Strains are linear functions of distance from center of unit element of multilayer. Stresses are discontinuous. Displacements are continuous.

B. Stresses are linear functions of y. Beds are glued together. Displacements are continuous. Strains are discontinuous.

C. Stresses are linear functions of y. Beds are free to slide relative to each other. Displacements are discontinuous. Strains are discontinuous. Radii of curvature are discontinuous.

less so that the layers could slide easily relative to each other, the ends of the element of the multilayer would have shapes similar to the ones shown in Fig. 6.7C. In either case, the expressions for constants B_h and I_h are different from those we have derived [eqs. (6.17) and (6.18)].

9. There is high friction between layers so that the shear stress is transferred from one layer to the other. Further, the shear stress is the same in the soft and stiff layers. With these assumptions we derived the shear modulus, which is Biot's "slide modulus," given by eq. (6.25). If we had assumed, instead, that the shear strain was the same in adjacent soft and stiff layers, we would replace eq. (6.23) with

$$(\sigma_{xy})_{\text{ave}} = \frac{t_1}{t}(\sigma_{yx})_1 + \frac{t_2}{t}(\sigma_{yx})_2,$$

where

$$(\sigma_{yx})_1 = \epsilon_{yx}G_1,$$

and

$$(\sigma_{yx})_2 = \epsilon_{yx}G_2,$$

so that the elastic modulus would be

$$G_a = \frac{t_1}{t}G_1 + \frac{t_2}{t}G_2 \tag{6.31}$$

instead of the value given by eq. (6.25).

10. We assumed that the vertical normal strains are the same in adjacent soft and stiff layers. If we had assumed, instead, that the vertical normal stress was the same in the soft and stiff layers,

$$\sigma_{yy} = (\epsilon_{yy})_1 B_1 = (\epsilon_{yy})_2 B_2,$$

and

$$(\epsilon_{yy})_{\text{ave}} = \sigma_{yy}\left[\frac{t_1}{tB_1} + \frac{t_2}{tB_2}\right],$$

and

$$(\epsilon_{yy})_{\text{ave}} = \sigma_{yy}\left[\frac{t_1}{tB_1} + \frac{t_2}{tB_2}\right],$$

so that B_v would have been

$$B_v = \frac{tB_1B_2}{t_2B_2 + t_1B_1}, \tag{6.32}$$

rather than the value given by eq. (6.29).

Buckling of Multilayer Confined by Rigid Boundaries.—We now have a differential equation that we can solve in order to describe folds of the type that developed in the Carmel Formation, near Moab, Utah (Fig. 6.1A). Field relations of the folds seem to indicate that, at the time of folding, the Carmel was overlain by at least 200 feet of relatively massive sandstone of the Entrada Sandstone and underlain by a thick section of massive Navajo Sandstone. Thus the Carmel apparently was crenulated while it was trapped between rocks forming nearly rigid boundaries (*5*). The actual cause of the loads that produced the folds is unknown (*6*); Dane (*5*) states that water-soaked sandy mud of the Carmel was deformed under "differential loading." We will assume that the folds formed under axial loading.

The boundary conditions that we have deduced from field observations and that we have assumed in order to solve the differential equation are illustrated in Fig. 6.8. A uniform horizontal load, P, acts on each member in the multilayer. Thus the boundary conditions are:

$$\begin{aligned} &\text{At } x = 0, && v = 0 \quad \text{for all } y. \\ &\text{At } y = T, \text{ and } y = 0, && v = 0 \quad \text{for all } x. \end{aligned} \tag{6.33}$$

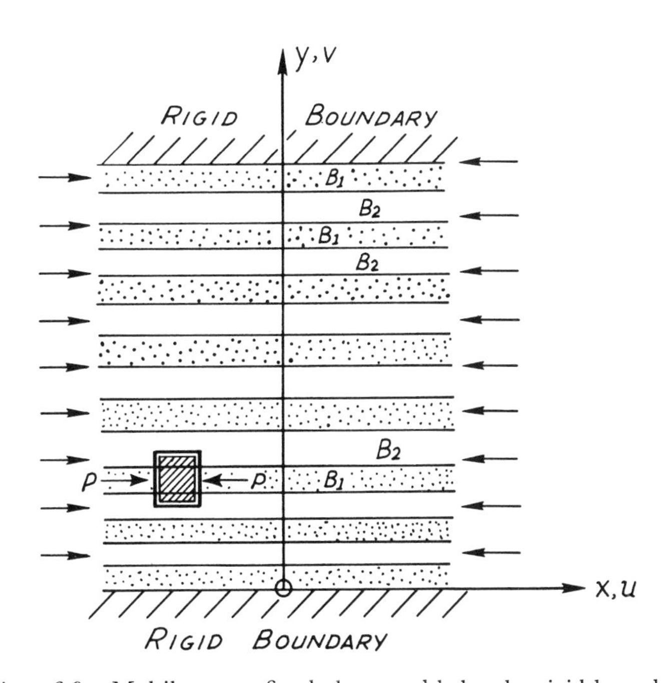

Figure 6.8. Multilayer confined above and below by rigid boundaries.

We will use the method of *separation of variables* to solve the differential equation, eq. (6.30):

$$B_h I_h \frac{\partial^4 v}{\partial x^4} + (P - tG_a b) \frac{\partial^2 v}{\partial x^2} = tbB_v \frac{\partial^2 v}{\partial y^2}. \tag{6.34}$$

By this method, we select an expression for the displacement,

$$v = f(x)g(y), \tag{6.35}$$

which is a function of x, $f(x)$ and a function of y, $g(y)$. The product of the two functions is the displacement. The problem is to find the functions.

Substituting eq. (6.35) into (6.34),

$$B_h I_h g(y) \frac{\partial^4 f(x)}{\partial x^4} + (P - tbG_a) g(y) \frac{\partial^2 f(x)}{\partial x^2} = bB_v t f(x) \frac{\partial^2 g(y)}{\partial y^2}.$$

Dividing each side of the equation by the product $f(x)g(y)$,

$$B_h I_h \frac{1}{f(x)} \frac{\partial^4 f(x)}{\partial x^4} + (P - tbG_a) \frac{1}{f(x)} \frac{\partial^2 f(x)}{\partial x^2} = btB_v \frac{1}{g(y)} \frac{\partial^2 g(y)}{\partial y^2}. \tag{6.36}$$

Thus we have "separated the variables." The terms on the left-hand side of the equation depend only upon x and the one on the right-hand side depends only upon y. The right-hand side must equal the left-hand side of eq. (6.36) for any value of x and y, so that each side of eq. (6.36) must be constant. Let the constant be C_0. Then we can rewrite eq. (6.36) as two equations:

$$B_h I_h \frac{1}{f(x)} \frac{d^4 f(x)}{dx^4} + (P - tbG_a) \frac{1}{f(x)} \frac{d^2 f(x)}{dx^2} = -C_0 \tag{6.37}$$

and

$$tbB_v \frac{1}{g(y)} \frac{d^2 g(y)}{dy^2} = -C_0. \tag{6.38}$$

We can write the differentials as total derivatives because $f(x)$ is independent of y and $g(y)$ is independent of x, according to our definition of the functions.

We will solve eq. (6.38) first. Try

$$g(y) = e^{my}$$

as a solution. Then eq. (6.38) becomes

$$btB_v m^2 e^{my} + C_0 e^{my} = 0,$$

so that if m and y are finite,

$$m = \pm \left(\frac{-C_0}{btB_v}\right)^{1/2}.$$

It follows that

$$g(y) = C_3 \exp\left[\left(\frac{-C_0}{btB_v}\right)^{1/2} y\right] + C_4 \exp\left[-\left(\frac{-C_0}{btB_v}\right)^{1/2} y\right],$$

where, for example, exp $[y]$ means e^y. In terms of trigonometric functions,

$$g(y) = C_1 \sin\left[\left(\frac{C_0}{btB_v}\right)^{1/2} y\right] + C_2 \cos\left[\left(\frac{C_0}{btB_v}\right)^{1/2} y\right] \tag{6.39}$$

is the general solution of eq. (6.38). If the term under the brackets is positive, therefore, the displacement,

$$v = f(x)g(y),$$

is a sinusoidal function of y.

We can evaluate the constant, C_0, before proceeding with the solution. In order for the equation to satisfy the boundary condition [eq. (6.33)] that $v = f(x)g(y) = 0$ at $y = 0$, for all x, the constant preceding the cosine term in eq. (6.38) must be zero. Thus,

$$g(y) = C_1 \sin\left(\left[\frac{C_0}{btB_v}\right]^{1/2} y\right). \tag{6.40}$$

In order for eq. (6.40) to satisfy the boundary condition that $v = f(x)g(y) = 0$, at $y = T$, for all x,

$$g(y) = 0 = C_1 \sin\left(\left[\frac{C_0}{btB_v}\right]^{1/2} T\right),$$

either $C_1 = 0$ or the term

$$\left(\frac{C_0}{btB_v}\right)^{1/2} T$$

must be a multiple of π. The possibility that $C_1 = 0$ gives a trivial solution, that is, $v = 0$ for all y and there are no folds. Obviously this is of no interest. Therefore,

$$n\pi = \left(\frac{C_0}{btB_v}\right)^{1/2} T,$$

or

$$C_0 = \left(\frac{n\pi}{T}\right)^2 B_v bt, \tag{6.41}$$

where n is 1, 2, 3, 4, ... We see that the constant, C_0, is always a positive quantity because b, t, and B_v are always positive.

Now we will solve eq. (6.37). Try

$$f(x) = e^{mx}$$

for a solution. Then eq. (6.37) becomes

$$B_h I_h m^4 e^{mx} + (P - tbG_a)m^2 e^{mx} + C_0 e^{mx} = 0,$$

or

$$m^4 + \left(\frac{P - tbG_a}{B_h I_h}\right)m^2 + \frac{C_0}{B_h I_h} = 0.$$

Solving for m^2,

$$m^2 = -r \pm (r^2 - s)^{1/2},$$

where

$$r = \frac{P - bG_a t}{2B_h I_h},$$

and (6.42)

$$s = \frac{C_0}{B_h I_h}.$$

Solving for m,

$$m_1 = -m_2 = [-r + i(s - r^2)^{1/2}]^{1/2},$$
$$m_3 = -m_4 = [-r - i(s - r^2)^{1/2}]^{1/2}. \tag{6.43}$$

We must know certain things about the quantities under the radical signs in eqs. (6.43) in order to evaluate the solution further. The type of solution will depend upon whether the expression under the smaller radical sign is positive or negative. First we will assume that

$$s - r^2 > 0.$$

Then eqs. (6.43) can be rewritten in the form

$$m = \pm(\alpha \pm i\beta), \tag{6.44}$$

where

$$\alpha = \left[\left(\frac{s}{4}\right)^{1/2} - \frac{r}{2}\right]^{1/2},$$

and (6.45)

$$\beta = \left[\left(\frac{s}{4}\right)^{1/2} + \frac{r}{2}\right]^{1/2}.$$

We could verify that eq. (6.44) is equivalent to eqs. (6.43) by squaring and then taking the square root of eq. (6.44).

The general solution to the differential equation, eq. (6.37), is, therefore,

$$f(x) = C_5 e^{m_1 x} + C_6 e^{-m_1 x} + C_7 e^{m_3 x} + C_8 e^{-m_3 x},$$

or

$$f(x) = (C_9 e^{\alpha x} + C_{10} e^{-\alpha x}) \cos \beta x + (C_{11} e^{\alpha x} + C_{12} e^{-\alpha x}) \sin \beta x, \tag{6.46}$$

as we have shown in earlier pages.

The arbitrary constants in eq. (6.46), $C_9, \ldots, C_{12}$, can be evaluated by means of the boundary conditions, eqs. (6.33):

$$\text{At } x = 0, \quad v = 0 = f(x)g(y) = 0, \quad \text{for all } y, \tag{6.47}$$

$$\text{at } y = T \text{ and } y = 0, \quad v = 0, \quad \text{for all } x.$$

If the folds maintain constant amplitudes throughout the length of the buckled layers, we can further state that, as

$$x \to \pm\infty, \quad v \text{ is finite,}$$

that is, the deflection remains finite even though x becomes very large. This condition requires that α in eq. (6.46) equals zero. The alternative, that all four arbitrary constants are zero, is uninteresting.

Thus,

$$\alpha = 0,$$

and according to the first of eqs. (6.45) and the definitions of r and s in eqs. (6.42),

$$\frac{P - tbG_a}{4B_hI_h} = \left(\frac{C_0}{4B_hI_h}\right)^{1/2}, \tag{6.48}$$

or

$$P = 2(C_0B_hI_h)^{1/2} + G_abt. \tag{6.49}$$

Equation (6.49) presumably expresses the critical axial load for buckling of the multilayer.

Because alpha is zero, the solution, eq. (6.46), reduces to:

$$f(x) = C_{13}\cos\beta x + C_{14}\sin\beta x. \tag{6.50}$$

According to the first of boundary conditions (6.47), the deflection is zero at the origin of coordinates. Thus,

$$\text{at } x = 0, \quad v = 0 = f(x)g(y),$$

so that, according to eq. (6.50),

$$f(x) = C_{13} + 0 = 0,$$

or $C_{13} = 0$. Thus, the solution for v, according to eqs. (6.40), (6.50), and (6.35), becomes

$$v = f(x)g(y),$$

or

$$v = C_{15}\sin(\beta x)\sin\left(\left\{\frac{C_0}{btB_v}\right\}^{1/2} y\right). \tag{6.51}$$

Substituting eq. (6.41) into eq. (6.51),

$$v = C_{15} \sin(\beta x) \sin\left(\frac{n\pi y}{T}\right), \tag{6.52}$$

we see that our solution predicts that the multilayer will buckle into a series of waves that change in amplitude, sinusoidally, with distance from the rigid boundaries. The maximum deflection is along the center line of the multilayer, equidistant from the rigid boundaries. The form of the theoretical folds is shown in Figs. 6.1B and 6.9. The theoretical form is remarkably similar to the forms of the actual folds in the Carmel Formation, shown in Fig. 6.1A.

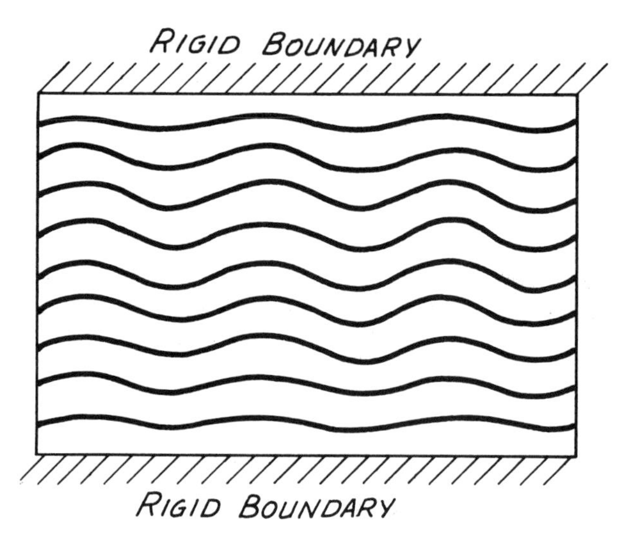

Figure 6.9. Theoretical form of multilayer buckled internally (*6*). Vertical deflections vary sinusoidally in vertical and horizontal directions.

We can calculate the most likely wavelength of the folds by examining β in eq. (6.52). In order for v to be zero at the origin and at any number of nodal points,

$$\beta x = \frac{2\pi x}{L},$$

because $\sin(\pi) = 0$. Here L is the wavelength of the folds. Thus, using eq. (6.45) and the definitions of r and s given by eqs. (6.42),

$$\left(\frac{2\pi}{L}\right)^2 = \left(\frac{C_0}{4B_h I_h}\right)^{1/2} + \frac{P - btG_a}{4B_h I_h}. \tag{6.53}$$

According to eq. (6.48), the two terms on the right-hand side of eq. (6.53) are equal, so that

$$\left(\frac{2\pi}{L}\right)^2 = 2\left(\frac{C_0}{4B_hI_h}\right)^{1/2} \tag{6.54}$$

Substituting the value of C_0 from eq. (6.41) into eq. (6.54),

$$\left(\frac{2\pi}{L}\right)^2 = 2\left(\frac{n\pi}{T}\right)\left(\frac{btB_v}{4B_hI_h}\right)^{1/2}$$

Inverting, taking the square root of each side, and solving for L gives

$$L_B = 2\pi\left(\frac{T}{n\pi}\right)^{1/2}\left(\frac{B_hI_h}{btB_v}\right)^{1/4}, \tag{6.55}$$

the *Biot wavelength*, which is the wavelength of folds in the multilayer. Waves of this length presumably form when the axial load applied to the ends of each stiff layer is given by eq. (6.49),

$$P_m = 2\left(\frac{n\pi}{T}\right)(btB_vB_hI_h)^{1/2} + G_atb. \tag{6.56}$$

We see that the wavelength, L_B, is independent of the shear modulus, G_a, which is the constant of proportionality between the shear stress and the shear strain of soft and stiff layers.

Exercises

1. Derive an expression for the total deflection of a multilayer that contains an initial deflection of the form:

$$v_0 = \delta_0 \sin\left(\frac{n\pi y}{T}\right) \sin\left(\frac{2\pi x}{L_0}\right)$$

Ans. $v = v_0 + v_1$

$$v_1 = \frac{\delta_0 P \sin\left(\frac{n\pi}{T}\right) \sin\left(\frac{2\pi x}{L_0}\right)}{B_hI_h\left(\frac{2\pi}{L_0}\right)^2 - P + tb\left[G_a + B_v\left(\frac{nL_0}{2T}\right)^2\right]}.$$

2. Show that L_m in eq. (6.55) is the wavelength of initial deflection requiring the minimum axial load for significant amplification.

If the elastic modulus of the soft layers is much less than the modulus of the stiff layers, the elastic moduli, B_h and B_v, reduce to [see eqs. (6.18) and (6.29)]:

$$B_h = B_1 + B_2\left[\left(\frac{t}{t_1}\right)^3 - 1\right] \approx B_1, \qquad \text{if } B_1 \gg B_2.$$

Also,

$$B_v = \left[\frac{t_1}{t} B_1 + \frac{t_2}{t} B_2\right] \approx \frac{t_1}{t} B_1, \qquad \text{if } B_1 \gg B_2.$$

In this case the Biot wavelength reduces to

$$L_B \approx 2\pi\left(\frac{T}{n\pi}\right)^{1/2}\left(\frac{B_1 t_1^3 b}{12btB_1(t_1/t)}\right)^{1/4} = 2\left(\frac{T\pi t_1}{(12)^{1/2}n}\right)^{1/2}$$

or

$$L_B \approx 1.90\left(\frac{Tt_1}{n}\right)^{1/2}. \tag{6.57}$$

According to eq. (6.57), the expected wavelength is solely a function of the thickness, t_1, of the stiff layers, the total thickness, T, of the sequence of layers, and the number, n, of cycles of folds stacked one on top of the other. There is one cycle of folds shown in the field example, Fig. 6.1A, and the theoretical folds, Figs. 6.1B and 6.9, so that, for them, $n = 1$, and eq. (6.57) becomes (*1*)

$$L_B \approx 1.90(Tt_1)^{1/2}. \tag{6.58}$$

Comparison of theoretical fold with those in Carmel Formation

Our solution, eq. (6.58), expresses the wavelength that should be most commonly found in nature, where stiff layers are separated by soft layers, because this wavelength is the one that requires the least axial load. We assumed that the layers were initially perfectly straight, but, as has been indicated in previous pages, if the layers have slight initial deflections, the deflections with the wavelength indicated by eq. (6.55) would grow in amplitude most rapidly under an increasing axial load.

We can determine whether our solution is approximately applicable to the natural situation at Moab, Utah, by measuring the folds in the Carmel, shown in Fig. 6.1A. First we will rewrite eq. (6.58) in another form:

$$\frac{L_B}{T} \approx 1.90\left(\frac{t_1}{T}\right)^{1/2},$$

which is an expression for the ratio of wavelength to total thickness of the folded sequence, T. The ratios of wavelength of folds to total thickness of the folded sequence shown in Fig. 6.1A range from about 0.9 to 1.3. The sandy siltstone interbeds are thin, so that to a first approximation the total thickness, T, divided by the thickness, t_1, of the stiff layers is equal to the number of stiff layers. Therefore,

$$\frac{L_B}{T} \approx 1.90\left(\frac{1}{\text{no. of stiff beds}}\right)^{1/2} \approx 0.9 \text{ to } 1.3.$$

In order for L_B/T to range from 0.9 to 1.3, the number of stiff layers should range between 2 and 3 if the stiff layers are of equal thickness.

The sandstone beds shown in Fig. 6.1A are of varying thickness at different places in the cliff face, because the sandy siltstone beds are discontinuous, but there appears to be a thick bed near the top and bottom and a thinner bed near the middle of the Carmel Formation. The folded beds comprise two thick beds and one thin bed, so that the estimate of 2 to 3 stiff beds predicted by the Biot theory seems to describe remarkably well the natural occurrence. The correlation between the theoretical and actual folds is particularly amazing, because the theory we developed assumed that there were many stiff and soft layers being folded, whereas in the part of the Carmel Formation shown in Fig. 6.1A there are only a few beds.

Biot (*1*) has shown that eq. (6.58) is valid for the buckling of viscous multilayers as well as elastic multilayers. He also has determined the effect of flow of material within the soft layers. If the ratio of the viscosities of the stiff and soft layers is less than about 3.5 times the number of layers, the correction for interstitial flow is unnecessary (*1*).

In another paper, Biot (*2*) demonstrates that the pattern of folds shown in Fig. 6.9 can develop even in a thick sequence of beds, where the rigid boundaries are widely separated. In this case, which Biot calls "self confinement," the amplitudes of the folds vary sinusoidally in the vertical direction as well as in the horizontal direction and there are several cycles of folds stacked one on top of the other, with anticlines being developed directly on top of synclines as is shown in Fig. 6.10A. The equation for the most likely wavelength, L_B, given by eq. (6.57), should predict the wavelengths of such fold sequences. The denominator, n, in eq. (6.57) is the number of stacks of folds.

Another interesting problem that Biot (*4*) has solved is that of buckling of strata that are severed by a vertical discontinuity, such as a fault. The amplitudes of the folds diminish in the horizontal direction, away from the fault, exactly as does the single member in a soft, elastic medium, but the amplitudes also diminish in the vertical direction (Fig. 6.10B). For details of these derivations the reader is referred to Biot's excellent papers (*1,2,3,4*).

Exercises

1. What is the magnitude of the vertical normal stress, σ_{yy}, at the mid-height, $T/2$, of the multilayer confined between rigid boundaries? Ignore the weight of the overlying beds and assume small deflections of the beds.
2. In several places in Arches Monument the crenulations in the Carmel Formation are similar in form to those shown in Fig. 6.1A, where the amplitudes of anticlines and synclines diminish upwards and downwards and where the maximum amplitudes are near the center of the Carmel. In many places,

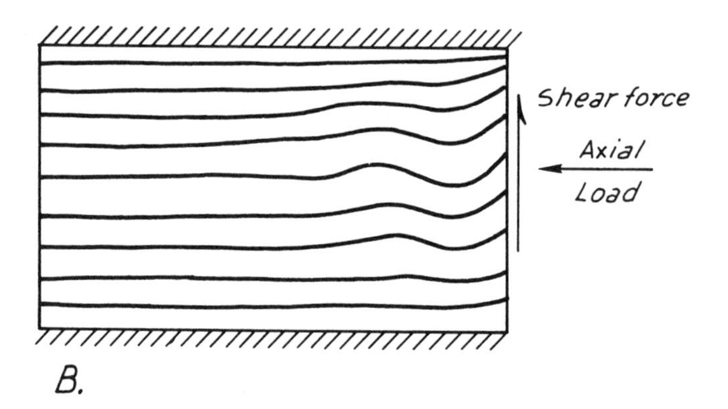

Figure 6.10. Two other modes of buckling of multilayers.
A. Self confinement.
B. Edge buckling.

however, the amplitudes are maximum near the base of the Carmel and they diminish much more rapidly downwards than upwards. The Navajo Sandstone, below the Carmel, apparently is undeflected, so that it presumably behaved as a rigid unit during folding.

What does this fold pattern tell us about the mechanical behavior of the lower beds of the Carmel relative to the mechanical behavior of the middle and upper beds? Explain your answer in detail. [Your answer to problem 1 should help you solve this problem.]

References cited in Chapter 6

1. Biot, M. A., 1964, "Theory of Internal Buckling of a Confined Multilayered Structure": *Geol. Soc. America, Bull.*, V. 75, p. 563–568.
2. ———, 1965, "Further Development of the Theory of Internal Buckling of Multilayers": *Geol. Soc. Amer. Bull.*, V. 76, 833–840.
3. ———, 1965, *Mechanics of Incremental Deformations*: John Wiley and Sons, N.Y.
4. ———, 1968, "Edge Buckling of a Laminated Medium": *Internat. Jour. Solids and Structures*, V. 4, p. 125–137.
5. Dane, C. H., 1935, "Geology of the Salt Valley Anticline and Adjacent Areas, Grand County, Utah": *U.S. Geological Survey, Bull.*, 863.
6. Johnson, A. M., 1969, "Development of Folds within Carmel Formation, Arches National Monument, Utah": *Tectonophysics*, V. 8, p. 31–77.
7. McKnight, E. T., 1940, "Geology of Area Between Green and Colorado Rivers, Grand and San Juan Counties, Utah": *U.S. Geological Survey, Bull.*, 908.

7

Chapter Sections

Introduction
Equations of Equilibrium in Terms of Stresses
Compatibility Equations for Two-Dimensional Problems
Compatibility Equation for Plane Stress of an Elastic Body
Compatibility Equation for Plane Strain of an Elastic Body
Airy's Stress Function
Reaction of an Elastic Medium of Infinite Thickness to Folding of a Member Resting on its Surface
The Problem
Selection of Stress Function
Evaluation of Arbitrary Constants
Derivation of Displacements
Relation between Distributed Load and Displacement
Reaction of a Soft Medium of Finite Thickness to Folding of Member Embedded in it
The Problem
Method of Images
Determination of Stresses and Displacements
Rheological Equations for Newtonian Viscous Substance
Strain Rate
Viscosity Coefficient
Relations between Normal Stresses and Normal Strain-Rates
Identity of Solutions to Some Problems in Elasticity and Viscosity
Folding of Firm, Viscous Strut Member Embedded in Soft, Viscous Medium
Bending Resistance of Viscous Member
Resistance of Viscous Medium
Wavelength of Folds
Similarities and Differences between Elastic and Viscous Folding
References Cited in Chapter 7

7

Theoretical Interlude: Introduction to Theories of Elasticity and Viscosity

Introduction

We left several loose ends in preceding chapters because we had not developed adequate theories to solve the problems. The reaction of a confining medium of finite thickness to the bending of a member embedded in it was determined approximately in Chapter 4 by representing the medium by a battery of elastic springs. In addition, the solution for the reaction of a confining medium of infinite thickness was quoted. Now we will derive these relations by means of the theory of elasticity of infinitesimal strain.

In several preceding chapters I asked the reader to take on faith the comment that our analyses, based on elasticity theory, were not restricted to this special theory, which may be an unrealistic representation of rock behaviors. Instead, it was stated, our analyses are actually valid for viscoelastic and viscous materials, so that the conclusions we derived from the analyses apply to a wide range of possible rock behaviors. Now we will ignore geological problems for a while so that we can demonstrate some of the important similarities between viscous and elastic materials.

We will examine each of the general conditions that must be satisfied for the solution of problems of elasticity. These conditions are equations of equilibrium, compatibility equations, and boundary conditions. Equations of equilibrium and boundary conditions have been discussed already, but in a form different from what we will develop here. The compatibility equation is new. Briefly, it is a relation between two components of displacement and it must be satisfied for displacements to be continuous functions of the coordinate system.

The concept of stress function will be introduced because it greatly simplifies the solution of problems in elasticity and viscosity. It reduces the number of equations to be solved from three to one.

Equations of equilibrium in terms of stresses

In order for a body to be in static equilibrium the sums of forces and sums of moments acting on the body must be zero. We can state these conditions of static equilibrium in six equations:

$$\sum F_x = 0 \qquad \sum M_x = 0$$
$$\sum F_y = 0 \qquad \sum M_y = 0$$
$$\sum F_z = 0 \qquad \sum M_z = 0$$

The equations for moments, those in the second column, are satisfied by stresses acting on a small element (Fig. 7.1), if the shear stresses, e.g., σ_{xy} and σ_{yx}, are equal.

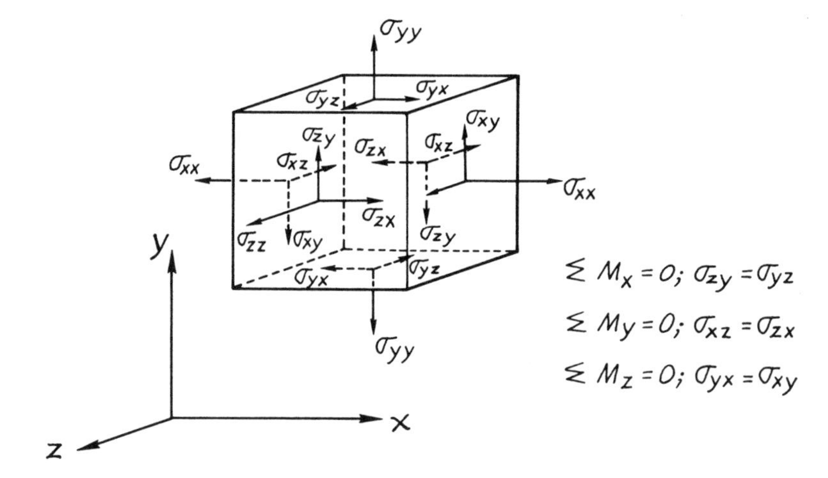

Figure 7.1. Stress components when stress variation and body forces are negligible.

Further, the equations of forces, those in the first column of equilibrium equations, were satisfied automatically when we assigned the same values to stresses acting on opposite faces of the element.

There are many conditions, however, where normal and shear stresses cannot be the same on opposite faces of the element. One condition is a result of body forces which act on all materials. One type of body force is weight; the pressure in a column of water changes with changing depth because of this weight body-force. For example, a result of the weight body-force in water is that the pressure on the top of a minute quantity or element of water is different from the pressure on the bottom of the element. If the pressure on the top is p, then the pressure on the bottom is $p + (\partial p/\partial y)\ \delta y$, or $p + \gamma\ \delta y$, where γ is the change in pressure with depth, which is the unit weight of water (Fig. 7.2).

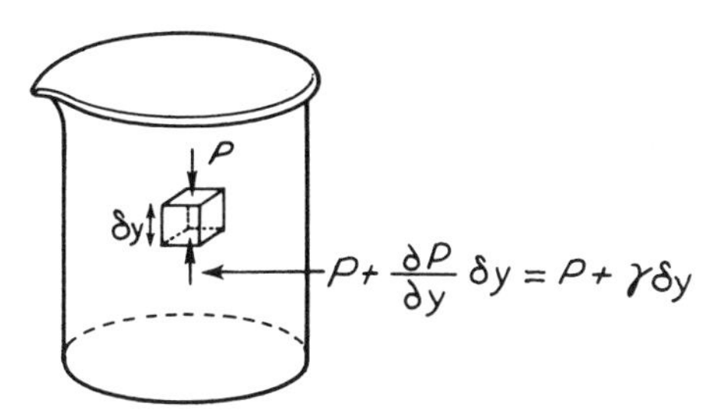

Figure 7.2. Small element within beaker of water. Weight body-force causes pressures to be different at top and bottom of element.

Another condition we can imagine, where normal and shear stresses are different on opposite sides of small elements within a body, is the condition of nonuniform distribution of surface forces. For example, a cantilever beam is supported firmly at one end and is free at the other end. If a vertical stress is applied near the free end, the stress must be translated through the beam to the supported end. The vertical stress on the part of the beam directly beneath the point of application of the load is equal to the applied vertical stress. However, the vertical stress decreases in magnitude through the depth of the beam until it is zero at the base, where no stress is applied. Thus, different stresses on opposite sides of elements can be caused by body forces and by nonuniform surface stresses.

The system of stresses acting on the small element, shown in Fig. 7.1, does not allow for effects of body forces or for effects of variation of stresses from point to point within a body. It is a simple matter to allow for such effects, however, and it is consideration of the variation of stresses that gives us a new formulation of the equations of equilibrium in terms of stresses.

Consider static equilibrium of a small element in two dimensions (Fig. 7.3). Only two-dimensional stress will be treated but the methods apply to three dimensions as well. The normal force, $\sigma_{xx}\ \delta y\ \delta z$, acts on the left-hand side of the element, and the different normal force, $[\sigma_{xx} + (\partial \sigma_{xx}/\partial x)\ \delta x]\ \delta y\ \delta z$, acts on the right-hand side of the element. The body force in the x-direction acts at the center of the element and it is $X\ \delta x\ \delta y\ \delta z$.

For the element to be in equilibrium, in two dimensions, the following equations must be satisfied:

$$\sum M_z = 0,$$

$$\sum F_x = 0,$$

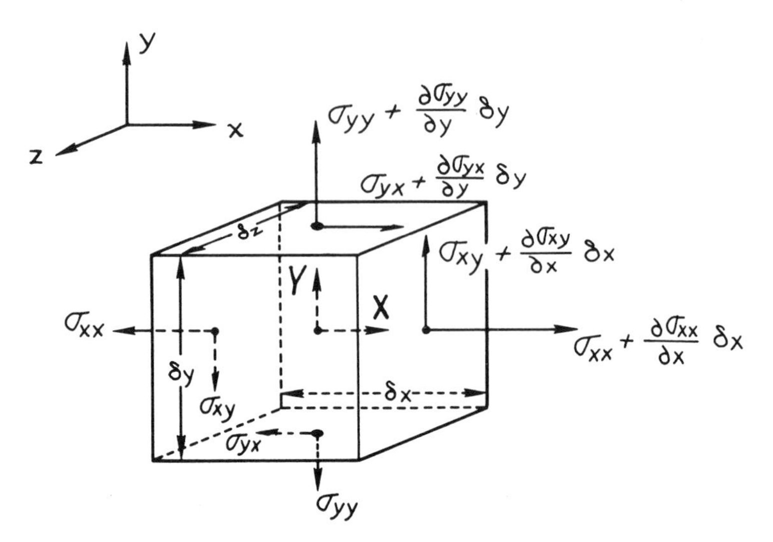

Figure 7.3. Cubic element supporting nonuniform surface stresses and being subjected to body forces, X and Y.

and

$$\sum F_y = 0.$$

The forces in the x-direction are

$$\left[\sum F_x = 0\right], \quad \left(\sigma_{xx} + \frac{\partial \sigma_{xx}}{\partial x}\frac{\delta x}{2}\right)\delta y\,\delta z + \left(\sigma_{yx} + \frac{\partial \sigma_{yx}}{\partial y}\frac{\delta y}{2}\right)\delta x\,\delta z$$
$$- \left(\sigma_{xx} - \frac{\partial \sigma_{xx}}{\partial x}\frac{\delta x}{2}\right)\delta y\,\delta z - \left(\sigma_{yx} - \frac{\partial \sigma_{yx}}{\partial y}\frac{\delta y}{2}\right)\delta x\,\delta z + X\,\delta x\,\delta y\,\delta z = 0. \tag{7.1}$$

And the forces in the y-direction are

$$\left[\sum F_y = 0\right], \quad \left(\sigma_{yy} + \frac{\partial \sigma_{yy}}{\partial y}\frac{\delta y}{2}\right)\delta x\,\delta z + \left(\sigma_{xy} + \frac{\partial \sigma_{xy}}{\partial x}\frac{\delta x}{2}\right)\delta y\,\delta z$$
$$- \left(\sigma_{yy} - \frac{\partial \sigma_{yy}}{\partial y}\frac{\delta y}{2}\right)\delta x\,\delta z - \left(\sigma_{xy} - \frac{\partial \sigma_{xy}}{\partial x}\frac{\delta x}{2}\right)\delta y\,\delta z + Y\,\delta x\,\delta y\,\delta z = 0. \tag{7.2}$$

Dividing eq. (7.1) by $\delta x\,\delta y\,\delta z$, the volume of the element,

$$\frac{\partial \sigma_{xx}}{\partial x} + \frac{\partial \sigma_{yx}}{\partial y} + X = 0. \tag{7.3}$$

Similarly, for the y-direction,

$$\frac{\partial \sigma_{yy}}{\partial y} + \frac{\partial \sigma_{xy}}{\partial x} + Y = 0. \tag{7.4}$$

With the moment equilibrium equation, $\sum M_z = 0$, we can show that

$$\sigma_{xy} = \sigma_{yx}.$$

Equations (7.3) and (7.4) comprise the equations of equilibrium in terms of stresses, for two-dimensional problems.

Exercises

1. Systematically derive the equations of equilibrium for three-dimensional states of stress. See Fig. 7.1 for the notation of the stresses.

Ans:
$$\frac{\partial \sigma_{xx}}{\partial x} + \frac{\partial \sigma_{xy}}{\partial y} + \frac{\partial \sigma_{xz}}{\partial z} + X = 0, \tag{7.5}$$

$$\frac{\partial \sigma_{xy}}{\partial x} + \frac{\partial \sigma_{yy}}{\partial y} + \frac{\partial \sigma_{yz}}{\partial z} + Y = 0, \tag{7.6}$$

and
$$\frac{\partial \sigma_{xz}}{\partial x} + \frac{\partial \sigma_{yz}}{\partial y} + \frac{\partial \sigma_{zz}}{\partial z} + Z = 0. \tag{7.7}$$

2. Given the condition that $\sigma_{zz} = \sigma_{xz} = \sigma_{yz} = Z = X = Y = 0$, show that the following stresses satisfy the equations of equilibrium:

$$\sigma_{xx} = x^4 + x^3y + 4xy^3,$$

$$\sigma_{yy} = 6x^2y^2 + xy^3,$$

and

$$\sigma_{xy} = -(4x^3y + 1.5x^2y^2 + y^4).$$

Compatibility equations for two-dimensional problems

The three components of infinitesimal strain of two-dimensional problems [eqs. (5.15) and (5.16)],

$$\epsilon_{xx} = \partial u/\partial x, \qquad \epsilon_{yy} = \partial v/\partial y, \qquad \epsilon_{xy} = \frac{\partial v}{\partial x} + \frac{\partial u}{\partial y},$$

are expressed in terms of two displacements, u and v, which are functions of x and y. There are three equations for two unknown functions, u and v. Therefore, we would expect the three strain components to be related. Indeed they are, as was shown by Saint Venant in 1860, about the time of the American Civil War. Saint Venant showed that a certain relationship among the strain components, called the *compatibility equation*, is necessary and sufficient to insure that the displacements are continuous and single-valued. Thus, satisfaction of the compatibility equation insures that the displacements of points within the body can be completely determined by integrating the equations of strain, except for displacements caused by rigid-body rotation or translation (see ref. *5*, p. 102, for proof).

The compatibility equation for two-dimensional states of strain can be derived as follows:

$$\frac{\partial^2}{\partial y^2}(\epsilon_{xx}) = \frac{\partial^3 u}{\partial y^2\,\partial x},$$

$$\frac{\partial^2}{\partial x^2}(\epsilon_{yy}) = \frac{\partial^3 v}{\partial x^2\,\partial y},$$

$$\frac{\partial}{\partial x}(\epsilon_{xy}) = \frac{\partial^2 u}{\partial x\,\partial y} + \frac{\partial^2 v}{\partial x^2},$$

and

$$\frac{\partial}{\partial y}\left(\frac{\partial \epsilon_{xy}}{\partial x}\right) = \frac{\partial^3 u}{\partial x\,\partial y^2} + \frac{\partial^3 v}{\partial x^2\,\partial y},$$

so that

$$\frac{\partial^2 \epsilon_{xx}}{\partial y^2} + \frac{\partial^2 \epsilon_{yy}}{\partial x^2} = \frac{\partial^2 \epsilon_{xy}}{\partial x\,\partial y}. \quad \text{Compatibility Equation} \tag{7.8}$$

This differential equation is the compatibility equation. It might not be satisfied under certain conditions of importance in geology—for example, under some conditions of thermal stress, initial stress, and discontinuous physical properties (see refs. *5* and *8*). We will study one problem where the compatibility equation is not satisfied when we consider initial stresses in granite, in Chapter 10.

There are six equations of compatibility for three-dimensional states of strain. The compatibility equations include three of the form (*5,8*)

$$\frac{\partial^2 \epsilon_{xx}}{\partial y^2} + \frac{\partial^2 \epsilon_{yy}}{\partial x^2} = \frac{\partial^2 \epsilon_{xy}}{\partial x\,\partial y} \tag{7.9}$$

and three of the form

$$2\frac{\partial^2 \epsilon_{xx}}{\partial y\,\partial z} = \frac{\partial}{\partial x}\left(-\frac{\partial \epsilon_{yz}}{\partial x} + \frac{\partial \epsilon_{zx}}{\partial y} + \frac{\partial \epsilon_{xy}}{\partial z}\right). \tag{7.10}$$

In tensor notation, the six equations of compatibility can be written as a single equation (*5*)

$$\frac{\partial^2 \varepsilon_{ij}}{\partial x_k\,\partial x_l} + \frac{\partial^2 \varepsilon_{kl}}{\partial x_i\,\partial x_j} - \frac{\partial^2 \varepsilon_{ik}}{\partial x_j\,\partial x_l} - \frac{\partial^2 \varepsilon_{jl}}{\partial x_i\,\partial x_k} = 0, \tag{7.11}$$

where $i = 1, 2, 3; j = 1, 2, 3; k = 1, 2, 3;$ and $l = 1, 2, 3$.

Compatibility Equation for Plane Stress of an Elastic Body.—The compatibility equation can be written in terms of stresses, also, by using the stress-strain relation of an elastic body. For plane stress, the relations are [eqs. (5.39)]

$$\epsilon_{xx} = \frac{1}{E}[\sigma_{xx} - \nu\sigma_{yy}],$$

$$\epsilon_{yy} = \frac{1}{E}[\sigma_{yy} - \nu\sigma_{xx}],$$

and

$$\epsilon_{xy} = \frac{\sigma_{xy}}{G} = \frac{2(1+\nu)}{E}\sigma_{xy}.$$

Substituting these equations into the compatibility equation for two-dimensions, eq. (7.8),

$$\frac{\partial^2}{\partial y^2}(\sigma_{xx} - \nu\sigma_{yy}) + \frac{\partial^2}{\partial x^2}(\sigma_{yy} - \nu\sigma_{xx}) = 2(1+\nu)\frac{\partial^2\sigma_{xy}}{\partial x\,\partial y}. \tag{7.12}$$

However, the stresses, σ_{xx}, σ_{yy}, and σ_{xy}, are interdependent according to the equations of equilibrium [eqs. (7.3) and (7.4)]. By differentiating eq. (7.3) with respect to x and eq. (7.4) with respect to y and adding the result,

$$2\frac{\partial^2\sigma_{xy}}{\partial x\,\partial y} = -\left(\frac{\partial X}{\partial x} + \frac{\partial Y}{\partial y}\right) - \frac{\partial^2\sigma_{xx}}{\partial x^2} - \frac{\partial^2\sigma_{yy}}{\partial y^2}. \tag{7.13}$$

By substituting eq. (7.13) into eq. (7.12), we have the compatibility equation in terms of stress components,

$$\left(\frac{\partial^2}{\partial x^2} + \frac{\partial^2}{\partial y^2}\right)(\sigma_{xx} + \sigma_{yy}) = -(1+\nu)\left(\frac{\partial X}{\partial x} + \frac{\partial Y}{\partial y}\right), \quad \text{for } \textit{Plane Stress}. \tag{7.14}$$

Compatibility Equation for Plane Strain of an Elastic Body.—If we substitute the stress-strain relations for plane strain, [eqs. (5.40)], into the compatibility equation, eq. (7.8), we find that the compatibility equation takes the following form:

$$\left(\frac{\partial^2}{\partial x^2} + \frac{\partial^2}{\partial y^2}\right)(\sigma_{xx} + \sigma_{yy}) = -\frac{1}{(1-\nu)}\left(\frac{\partial X}{\partial x} + \frac{\partial Y}{\partial y}\right), \quad \text{for } \textit{Plane Strain}. \tag{7.15}$$

The compatibility equations for plane strain and plane stress are similar in form. They reduce to the same equation if the body forces, X and Y, are independent of position, that is, if they are constant. Further, the compatibility equation in this case is *independent of the elastic constants* of materials. Thus the stress distribution is the same for all isotropic and homogeneous elastic materials that are subjected to the same boundary conditions. This conclusion is of great practical importance because it means that the stress distribution in a body can be determined experimentally, say by photoelastic methods (*8*), and the experimental results obtained are applicable to all elastic materials. Also, if body forces are constant, the compatibility equations for *plane stress and plane strain are identical.* This has important implications. It means that we can experimentally determine stress distributions for problems of plane strain, in which the body should theoretically be infinitely long in one dimension, by using models that have the shape of a plate (*8*).

Airy's stress function

The solution of two-dimensional problems reduces to integration of the differential equations of equilibrium with the compatibility equation for strains and the boundary conditions. If the unit weight, γ (gamma), is the only body force, the equations to be satisfied are:

$$\left.\begin{aligned} \frac{\partial \sigma_{xx}}{\partial x} + \frac{\partial \sigma_{xy}}{\partial y} &= 0, \\ \frac{\partial \sigma_{xy}}{\partial x} + \frac{\partial \sigma_{yy}}{\partial y} + \gamma &= 0, \end{aligned}\right\} \quad \text{Equations of Equilibrium} \tag{7.16}$$

$$\left(\frac{\partial^2}{\partial x^2} + \frac{\partial^2}{\partial y^2}\right)(\sigma_{xx} + \sigma_{yy}) = 0, \qquad \text{Compatibility Equation} \tag{7.15}$$

in addition to the boundary conditions, stated in terms of stress.

One method of solving these equations would be to guess values of stress components until we come up with components that satisfy the equations as well as the boundary conditions. However, it was discovered long ago that if we introduce a new variable, Φ (cap. phi), which is a function of x and y and is related to the stress components by the following equations,

$$\left.\begin{aligned} \sigma_{xx} &= \frac{\partial^2 \Phi}{\partial y^2} - \gamma y, \\ \sigma_{yy} &= \frac{\partial^2 \Phi}{\partial x^2} - \gamma y, \\ \text{and} \quad \sigma_{xy} &= -\frac{\partial^2 \Phi}{\partial x \, \partial y}, \end{aligned}\right\} \tag{7.16}$$

eqs. (7.14) and (7.15) reduce to one equation (*5,8*). Φ is called *Airy's stress function.* If eqs. (7.16) are substituted into eqs. (7.14), we see that they satisfy the equations of equilibrium. If eqs. (7.16) are substituted into the compatibility equation, e.g. (7.15), we find that the stress function must satisfy the equation

$$\frac{\partial^4 \Phi}{\partial x^4} + 2\frac{\partial^4 \Phi}{\partial x^2\, \partial y^2} + \frac{\partial^4 \Phi}{\partial y^4} = 0. \tag{7.17}$$

The solution of any two-dimensional problem, therefore, where the weight of the body is the only body force reduces to finding a solution to eq. (7.17) that satisfies the boundary conditions, in terms of stresses.

Equation (7.17) commonly is written in the following forms:

$$\left(\frac{\partial^2}{\partial x^2} + \frac{\partial^2}{\partial y^2}\right)\left(\frac{\partial^2 \Phi}{\partial x^2} + \frac{\partial^2 \Phi}{\partial y^2}\right) = 0, \tag{7.18}$$

$$\nabla^2 \nabla^2 \Phi = 0, \tag{7.19}$$

and

$$\nabla^4 \Phi = 0, \tag{7.20}$$

where (inverted delta) ∇ is called the *del operator* (*5,8*). Note that ∇^2 does *not* mean "del squared" but, rather, refers to second-order partial differentiation with respect to x and then with respect to y.

The solution of a problem in two dimensions, therefore, is a matter of finding a function, Φ, that satisfies the compatibility equation in terms of Φ, eq. (7.20). The results are independent of the elastic constants, E and ν, and plane stress and plane strain resolve themselves into the same problem. Normally we select a function, Φ; obtain values of σ_{xx}, σ_{yy}, and σ_{xy} with eqs. (7.16); determine the theoretical stress distribution at the boundaries; and compare the theoretical values of stress with the known values at the boundaries. If the calculated boundary stresses differ from the known boundary stresses, we modify Φ and try again (e.g., see ref. *8*, p. 29–130). The method should become clear as we proceed to solve problems.

Exercises

1. Show that the equation $\Phi = Bx^2y^2$ is a stress function. Derive the expressions for σ_{xx}, σ_{yy}, and σ_{xy} and interpret the results for a rectangle of depth $2c$ and length l (x-direction).
2. Show that the following stress functions satisfy the equations of compatibility. That is, show that they are stress functions. Verify the expressions for σ_{xx}, σ_{yy}, and σ_{xy} for each stress function. Interpret the stress distributions with respect to the rectangle described in problem 1.

$$\Phi = C_1 x^2; \qquad \sigma_{yy} = 2C_1, \qquad \sigma_{xx} = \sigma_{xy} = 0.$$

$$\Phi = C_2 xy; \qquad \sigma_{xx} = \sigma_{yy} = 0, \qquad \sigma_{xy} = -C_2.$$

$$\Phi = C_3 y^3; \qquad \sigma_{xx} = 6C_3 y, \qquad \sigma_{yy} = \sigma_{xy} = 0.$$

3. Show that the last stress function, $\Phi = C_3 y^3$, produces a resultant force of zero but a resultant moment of

$$M = 6C_3 \int_{-b}^{b} y^2 \, dy = 4C_3 b^3$$

on the ends of the rectangle, at $x = 0$ and $x = l$.

4. Show that

$$\Phi = -\frac{q}{8C^3}\left[x^3\left(\frac{2C^3}{3} + C^2 y - \frac{y^3}{3}\right) + x\left(\frac{C^4 y}{5} - \frac{2C^2 y^3}{5} + \frac{y^5}{5}\right)\right]$$

is a stress function. Evaluate the components of stress and interpret the stress for a rectangle bounded by $x = 0$, $x = l$, and $y = \pm C$.

Reaction of an elastic medium of infinite thickness to folding of a member resting on its surface

One of the models of folding that we considered in Chapter 4 was that of buckling of an elastic member in a medium of infinite thickness (Fig. 7.4). We used an equation derived by Biot (*1*),

$$q = \frac{2B_0 \pi b v}{L}, \tag{7.21}$$

for the reaction of the soft medium to displacement, v, of the member. We will now derive this equation.

The Problem.—The member of thickness T is embedded between two layers of soft, elastic material of infinite thickness (Fig. 7.4A). The problem is to calculate the deflection of the boundaries between the confining media and the member if the boundaries are loaded sinusoidally according to the expression

$$\sigma_{yy} = \sigma_0 \sin\left(\frac{2\pi x}{L}\right), \tag{7.22}$$

where L is the wavelength of the stress distribution (Fig. 7.4B). When we have derived an equation for the deflection, v, we can convert normal stress, σ_{yy}, to load per unit of length, q, and write an expression identical to eq. (7.21). We will first calculate the stresses within the medium. Then we will compute the deformation at the surface of the medium from the stress-strain relations of an elastic material.

Selection of Stress Function.—The first step in finding a solution is to derive a stress function that satisfies the conditions outlined above. According to eq. (7.16), if we ignore the body force, σ_{yy} and Airy's stress function, Φ, are related by

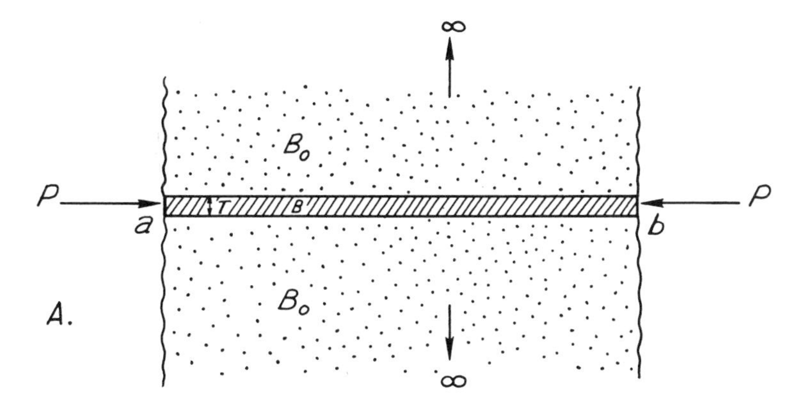

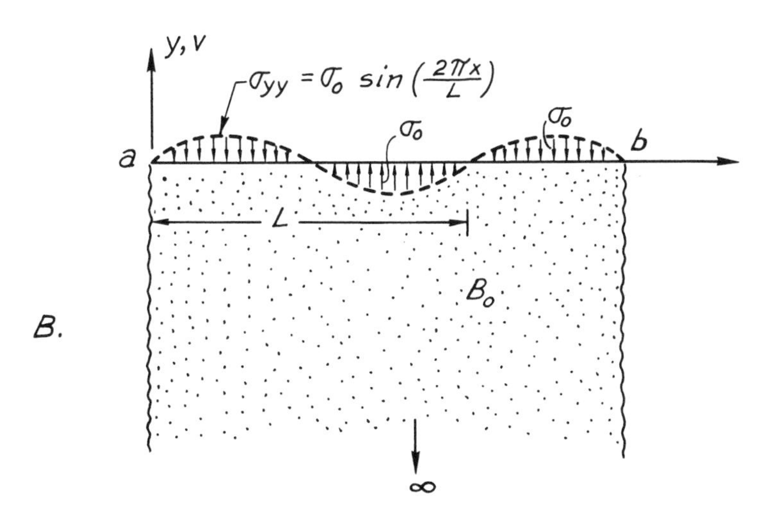

Figure 7.4. Reaction of elastic medium of infinite thickness to buckling of a member resting on its surface.

A. Member confined by elastic medium below and above it.

B. Isolation of medium below member. Assume that the medium is loaded by a sinusoidal stress distribution along its upper surface.

$$\sigma_{yy} = \frac{\partial^2 \Phi}{\partial x^2}.$$

Therefore, the vertical normal stresses, eq. (7.22), at the boundary are satisfied if the stress function has the form

$$\Phi = C_1 \sin(\lambda x),$$

where C_1 and λ (lambda) are constants. That is,

$$\frac{\partial^2\Phi}{\partial x^2} = \sigma_{yy} = -C_1\lambda^2 \sin(\lambda x)$$

is identical in form to eq. (7.22).

However, the stresses in the medium presumably are functions of y as well as x and therefore the stress function must contain a function of y. Let us try (*5,8*)

$$\Phi = f(y)\sin(\lambda x), \tag{7.23}$$

where

$$\lambda = \left(\frac{2\pi}{L}\right)$$

and $f(y)$ is some function of y.

We determine the value of the function of y by substituting eq. (7.23) into the compatibility equation, eq. (7.17),

$$\frac{\partial^4\Phi}{\partial x^4} + 2\frac{\partial^4\Phi}{\partial x^2\,\partial y^2} + \frac{\partial^4\Phi}{\partial y^4} = 0. \tag{7.24}$$

Accordingly,

$$\frac{\partial^4\Phi}{\partial x^4} = f(y)\lambda^4 \sin(\lambda x),$$

$$\frac{\partial^2\Phi}{\partial x^2} = -f(y)\lambda^2 \sin(\lambda x),$$

and

$$\frac{\partial^2}{\partial y^2}\left(\frac{\partial^2\Phi}{\partial x^2}\right) = -\frac{d^2f(y)}{dy^2}\lambda^2 \sin(\lambda x),$$

which yields a total differential because $f(y)$ is a function of y only, and

$$\frac{\partial^4\Phi}{\partial y^4} = \frac{d^4f(y)}{dy^4}\sin(\lambda x).$$

For this problem, then, the compatibility equation, eq. (7.24), reduces to a fourth-order, total differential equation,

$$\lambda^4 f(y) - 2\lambda^2\frac{d^2f(y)}{dy^2} + \frac{d^4f(y)}{dy^4} = 0. \tag{7.25}$$

We already have solved an equation similar in form to eq. (7.25). Following the same procedure, try

$$f(y) = e^{my}$$

as a solution, where m is some constant, the value of which is determined by substituting the trial solution into the differential equation:

$$\lambda^4 e^{my} - 2\lambda^2 m^2 e^{my} + m^4 e^{my} = 0$$

or

$$\lambda^4 - 2\lambda^2 m^2 + m^4 = 0 = (\lambda^2 - m^2)^2,$$

so that

$$m = \pm\lambda.$$

Thus the general solution would seem to be

$$f(y) = C_5 e^{\lambda y} + C_6 e^{-\lambda y}. \tag{7.26}$$

However, the differential equation is of fourth order so that there must be solutions not contained in eq. (7.26). Essentially, we have integrated four times; there must be four arbitrary constants. There are only two in eq. (7.26). In order to obtain the general solution, we will replace C_5 by $(C_1 + C_2 y)$ and C_6 by $(C_3 + C_4 y)$. Any book on ordinary differential equations explains why we make this substitution [e.g., see ref. *6*]. Equation (7.26), then, is replaced by

$$f(y) = (C_1 + C_2 y)e^{\lambda y} + (C_3 + C_4 y)e^{-\lambda y}, \tag{7.27}$$

where $C_1 \cdots C_4$ are constants of integration.

Now that we have determined the value of the arbitrary function of y, we combine it with the function of x in eq. (7.23):

$$\Phi = \sin(\lambda x)[(C_1 + C_2 y)e^{\lambda y} + (C_3 + C_4 y)e^{-\lambda y}]. \tag{7.28}$$

The stress components corresponding to eq. (7.28) are, if the density of the medium is ignored, from eqs. (7.16),

$$\sigma_{xx} = \frac{\partial^2\Phi}{\partial y^2} = \sin(\lambda x)\{[C_1\lambda^2 + C_2(2\lambda + \lambda^2 y)]e^{\lambda y} + [C_3\lambda^2 + C_4(\lambda^2 y - 2\lambda)]e^{-\lambda y}\}, \tag{7.29}$$

$$\sigma_{yy} = \frac{\partial^2\Phi}{\partial x^2} = -\lambda^2 \sin(\lambda x)[(C_1 + C_2 y)e^{\lambda y} + (C_3 + C_4 y)e^{-\lambda y}], \tag{7.30}$$

and

$$\sigma_{xy} = -\frac{\partial^2\Phi}{\partial x\,\partial y} = -\lambda\cos(\lambda x)\{[C_1\lambda + C_2(1 + \lambda y)]e^{\lambda y} - [C_3\lambda - C_4(1 - \lambda y)]e^{-\lambda y}\}, \tag{7.31}$$

where

$$\lambda = 2\pi/L.$$

Evaluation of Arbitrary Constants.—The next step is to evaluate the four arbitrary constants, $C_1 \cdots C_4$, in terms of boundary conditions. The boundary conditions are:

$$\left.\begin{aligned} &\text{at } y = 0, && \sigma_{yy} = \sigma_0 \sin(\lambda x), \\ &\text{and} && \sigma_{xy} = 0, \end{aligned}\right\} \tag{7.32}$$

and

$$\left.\begin{aligned} &\text{as } y \to -\infty, && \sigma_{xy} = 0, \\ & && \sigma_{yy} = 0, \\ &\text{and} && \sigma_{xx} = 0. \end{aligned}\right\} \tag{7.33}$$

The condition that the shear stress vanishes at the interface between the member and the medium, that is, at $y = 0$, assumes that the interface is frictionless.

The condition that the stresses vanish at an infinite distance from the interface between the member and the medium requires that two of the arbitrary constants, C_3 and C_4, be zero. This conclusion can be verified by inspecting eqs. (7.29) $\cdots$(7.31).

The other two arbitrary constants can be evaluated as follows: According to eqs. (7.32),

$$\text{at } y = 0,$$

$$\sigma_{yy} = \sigma_0 \sin(\lambda x) = -\lambda^2 C_1 \sin(\lambda x),$$

so that

$$C_1 = \frac{-\sigma_0}{\lambda^2}. \tag{7.34}$$

Also,

$$\sigma_{xy} = 0 = -\lambda[\cos(\lambda x)](C_1\lambda + C_2),$$

so that

$$C_2 = -C_1\lambda,$$

or

$$C_2 = \frac{\sigma_0}{\lambda}. \tag{7.35}$$

Thus the expressions for the stress components become

$$\left.\begin{aligned} \sigma_{xx} &= \sigma_0 \sin(\lambda x)[1 + \lambda y]e^{\lambda y}, \\ \sigma_{yy} &= \sigma_0 \sin(\lambda x)[1 - \lambda y]e^{\lambda y}, \\ \text{and} \\ \sigma_{xy} &= -\sigma_0 \cos(\lambda x)[\lambda y]e^{\lambda y}. \end{aligned}\right\} \tag{7.36}$$

Derivation of Displacements.—We have derived expressions for the stresses throughout the soft medium of infinite extent on one side of the member. The remainder of the problem of determining the deflection of the boundary between the medium and the member is to calculate the displacements, v and u in the y- and x-directions, respectively. We use the stress-strain relations for the calculation. The pertinent relations for plane stress are [eqs. (5.27) and (5.39)]

$$\left.\begin{aligned} \epsilon_{xx} &= \frac{1}{E}[\sigma_{xx} - \nu\sigma_{yy}] = \frac{\partial u}{\partial x}, \\ \epsilon_{yy} &= \frac{1}{E}[\sigma_{yy} - \nu\sigma_{xx}] = \frac{\partial v}{\partial y}, \\ \epsilon_{xy} &= \frac{1}{G}[\sigma_{xy}] = \frac{2(1+\nu)}{E}\sigma_{xy} = \frac{\partial u}{\partial y} + \frac{\partial v}{\partial x}. \end{aligned}\right\} \tag{7.37}$$

We will determine displacements for plane stress, even though the problem we posed is one of plane strain, because the notation for plane stress is somewhat simpler. A solution for plane stress is easily converted into a solution for plane strain.

The procedure for obtaining expressions for the displacements, u and v, consists of integrating eqs. (7.37) (*8*). First, we substitute eqs. (7.36) for the stresses into eqs. (7.37):

$$\frac{\partial u}{\partial x} = \frac{\sigma_0}{E}\sin(\lambda x)e^{\lambda y}[(1 + \lambda y) - \nu(1 - \lambda y)], \tag{7.38}$$

$$\frac{\partial v}{\partial y} = \frac{\sigma_0}{E}\sin(\lambda x)e^{\lambda y}[(1 - \lambda y) - \nu(1 + \lambda y)], \tag{7.39}$$

and

$$\frac{\partial u}{\partial y} + \frac{\partial v}{\partial x} = -2\frac{(1+\nu)}{E}\sigma_0 \cos(\lambda x)[\lambda y]e^{\lambda y}. \tag{7.40}$$

Integrating eq. (7.38), we derive an expression for u:

$$u = -\frac{\sigma_0 \cos(\lambda x)}{\lambda E}e^{\lambda y}[(1 + \lambda y) - \nu(1 - \lambda y)] + g(y), \tag{7.41}$$

where $g(y)$ is some function of y. When we integrate a partial differential equation with respect to x, the solution contains an unknown function of y because the dependent variable, u, is a function of both x and y. Thus, instead of an arbitrary constant of integration, there is an arbitrary function of integration.

Integrating eq. (7.39) we derive an expression for v:

$$v = \frac{\sigma_0 \sin(\lambda x)}{\lambda E} e^{\lambda y}[(1 - \nu) - (1 + \nu)(\lambda y - 1)] + f(x), \tag{7.42}$$

which contains an arbitrary function of x for the same reason that eq. (7.41) contains an arbitrary function of y.

We can solve for the arbitrary functions by using the expression for the shear strain, eq. (7.40). Substituting eqs. (7.41) and (7.42) into eq. (7.40),

$$\frac{\partial u}{\partial y} + \frac{\partial v}{\partial x} = -\frac{2(1 + \nu)}{E} \sigma_0 \cos(\lambda x)[\lambda y]e^{\lambda y}.$$

But, according to eq. (7.41), the derivative of u with respect to y is

$$\frac{\partial u}{\partial y} = \frac{-\sigma_0 \cos(\lambda x)}{E} e^{\lambda y}[(1 + \lambda y) - \nu(1 - \lambda y) + (1 + \nu)] + \frac{dg(y)}{dy}. \tag{7.43}$$

The derivative of g with respect to y is a total derivative, because by definition it is a function only of y.

Also, according to eq. (7.42),

$$\frac{\partial v}{\partial x} = \frac{\sigma_0 \cos(\lambda x)}{E} e^{\lambda y}[2 - \lambda y(1 + \nu)] + \frac{df(x)}{dx}. \tag{7.44}$$

Adding eqs. (7.43) and (7.44),

$$\frac{\partial v}{\partial x} + \frac{\partial u}{\partial y} = \frac{df(x)}{dx} + \frac{dg(y)}{dy} - \frac{\sigma_0 \cos(\lambda x)}{E} e^{\lambda y}[2\lambda y(1 + \nu)].$$

The last term in this equation is equal to eq. (7.40), which requires that

$$\frac{df(x)}{dx} + \frac{dg(y)}{dy} = 0, \tag{7.45}$$

which is a relation between the arbitrary functions, $f(x)$ and $g(y)$.

The first term in eq. (7.45) is a function of x and the second is a function of y. Therefore, they must each equal a constant, say C_0. Otherwise, for example, $f(x)$ would have to vary with y, which is not possible. Therefore,

$$\left.\begin{aligned} &\frac{df(x)}{dx} = C_0, \\ \text{and} \\ &\frac{dg(y)}{dy} = -C_0. \end{aligned}\right\} \tag{7.46}$$

Integrating each of eqs. (7.46),

$$\left.\begin{aligned} f(x) &= C_0x + C_1 \\ &\text{and} \\ g(y) &= -C_0y + C_2, \end{aligned}\right\} \tag{7.47}$$

where C_1 and C_2 are arbitrary constants of integration. We can evaluate the constants by substituting eqs. (7.47) back into the solutions for u and v, eqs. (7.41) and (7.42),

$$u = -\frac{\sigma_0 \cos(\lambda x)}{\lambda E} e^{\lambda y}[(1 + \lambda y) - \nu(1 - \lambda y)] - C_0y + C_2, \tag{7.48}$$

and

$$v = \frac{\sigma_0 \sin(\lambda x)}{\lambda E} e^{\lambda y}[(1 - \nu) - (1 + \nu)(\lambda y - 1)] + C_0x + C_1. \tag{7.49}$$

The boundary conditions are

$$v = 0 \quad \text{at} \quad x = 0, \qquad \text{for any } y, \tag{7.50}$$

$$u = v = 0 \quad \text{as} \quad y \to -\infty. \tag{7.51}$$

Therefore, for the displacement in the y-direction to satisfy condition (7.50),

$$C_0 = C_1 = 0.$$

Also, for the horizontal deflection, u, to satisfy condition (7.51),

$$C_2 = C_0 = 0.$$

The solutions for the deflections are, therefore,

$$\left.\begin{aligned} u &= -\frac{\sigma_0 \cos(\lambda x)}{\lambda E} e^{\lambda y}[(1 + \lambda y) - \nu(1 + \lambda y)], \\ v &= \frac{\sigma_0 \sin(\lambda x)}{\lambda E} e^{\lambda y}[(1 - \nu) - (1 + \nu)(\lambda y - 1)]. \end{aligned}\right\} \tag{7.52}$$

Relation between Distributed Load and Displacement.—At the surface of the elastic medium, that is, at $y = 0$, the deflection is, according to the second of eqs. (7.52),

$$v_{y=0} = \frac{2\sigma_0 \sin(\lambda x)}{\lambda E}.$$

But $\sigma_{yy} = \sigma_0 \sin(\lambda x)$ was the stress distribution applied at the boundary, so that

$$v_{y=0} = \frac{2\sigma_{yy}}{\lambda E}.$$

Also, $\lambda = 2\pi/L$, and $q = \sigma_{yy}b$, where b is the width of the member, so that

$$v_{y=0} = \frac{qL}{b\pi E},$$

or

$$q = \frac{b\pi E}{L} v. \tag{7.53}$$

Finally, the problem posed involves the constraint arising from the reaction of a medium of infinite extent below and above the member, whereas the preceding derivations are for a medium below the member only. The expression for q, therefore, should be doubled;

$$q = \frac{2bE\pi}{L} v. \tag{7.54}$$

Equation (7.54) was derived for plane stress conditions. However, it can be shown readily that for plane strain [see eq. (5.43)],

$$q = \frac{2bE\pi}{L(1 - \nu^2)} v, \tag{7.55}$$

so that, in general,

$$q = \frac{2bB_0\pi}{L} v, \tag{7.56}$$

where B_0 is Young's modulus for an elastic medium beneath a beam and $B_0 = E_0/(1 - \nu_0^2)$ for an elastic medium beneath a plate. Eq. (7.54) is the same as that derived by Biot (*1*).

Reaction of a soft medium of finite thickness to folding of member embedded in it

In earlier pages the effect of a confining medium of finite thickness on the buckling of a member was solved approximately by representing the medium by a series of springs. We showed that the effect of the medium of finite thickness on the most likely wavelength of folds is nearly identical to that of a medium of infinite thickness if the thickness of the medium is approximately equal to the wavelength of the folds. An exact solution quoted there supported this conclusion. Now we will derive the exact solution.

The Problem.—The member of thickness T is embedded between two layers of soft, elastic medium of thickness J. Each layer of soft medium is bounded rigidly on one side so that one side of it cannot be displaced (Fig. 7.5). The problem is to

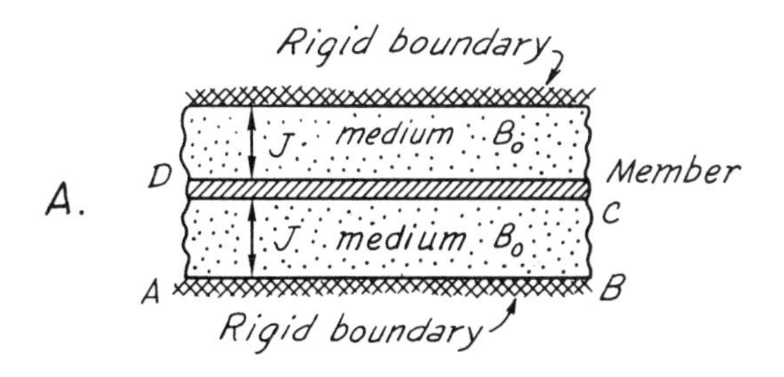

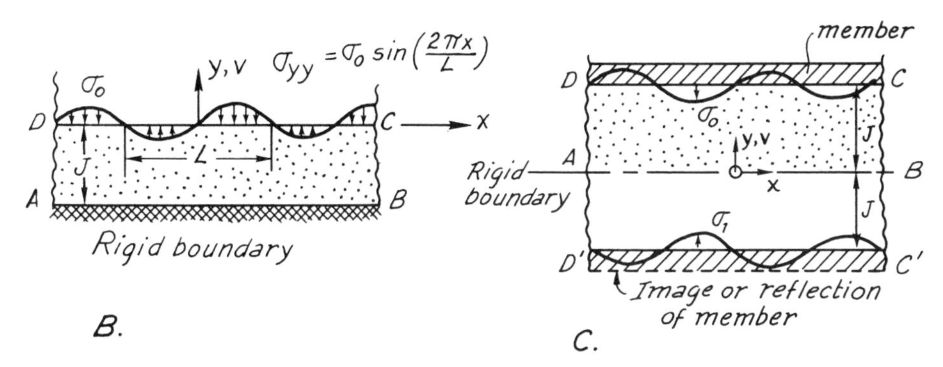

Figure 7.5. Simulation of reaction of medium of finite thickness by method of images.
A. Member confined between two layers of soft material of thickness, J. Medium is confined, in turn, by rigid boundaries.
B. Medium on one side of member is isolated for analysis.
C. Effect of rigid boundary is simulated by superimposing two stress distributions of equal magnitude but of opposite directions. The result is no displacement along AB, the rigid boundary.

calculate the deflection of the interface between the member and the soft medium when the interface is loaded sinusoidally according to the expression

$$\sigma_{yy} = \sigma_0 \sin\left(\frac{2\pi x}{L}\right), \tag{7.57}$$

where L is the wavelength of the stress distribution (Fig. 7.5B). One of the rigid boundaries is A-B in Fig. 7.5B.

Method of Images.—In order to derive the expression for the deflection of the boundary between the member and the soft medium, we will first calculate the stresses within the soft medium and then calculate the deformation from the elastic stress-strain relations.

The effect of the rigid boundary can be simulated by the *method of images*, or of *reflection* (*8*). By this method, a rigid boundary can be simulated by superimposing stress distributions of the same magnitude on opposite sides of the rigid boundary so that the displacements resulting from the two stress distributions cancel at the rigid boundary (Fig. 7.5C). Thus, the problem region is *reflected* across the rigid boundary so that there are now two loaded boundaries, *D-C* and *D'-C'*, shown in Fig. 7.5C. The stresses and displacements can be calculated for every point within the problem region, *ABCD*, because there are no boundaries with unknown states of stress.

Determination of Stresses and Displacements.—The problem is nearly the same as that of calculating the reaction of a medium of infinite thickness, discussed in preceding pages. The appropriate stress function already has been calculated [eqs. (7.23) and (7.27)]:

$$\left.\begin{aligned} &\Phi = f(y) \sin(\lambda x), \\ &\text{where} \\ &f(y) = (A + By)e^{\lambda y} + (C + Dy)e^{-\lambda y}, \end{aligned}\right\} \tag{7.58}$$

and where A, B, C, and D are constants.

Equations (7.58) can be rewritten in terms of hyperbolic sines and cosines, which are defined by the relations (*6*)

$$\left.\begin{aligned} &\cosh(\lambda y) = \frac{e^{\lambda y} + e^{-\lambda y}}{2} \\ &\text{and} \\ &\sinh(\lambda y) = \frac{e^{\lambda y} - e^{-\lambda y}}{2}. \end{aligned}\right\} \tag{7.59}$$

By manipulating these definitions of the hyperbolic functions we can show that

$$\left.\begin{aligned} &e^{\lambda y} = \cosh(\lambda y) + \sinh(\lambda y) \\ &\text{and} \\ &e^{-\lambda y} = \cosh(\lambda y) - \sinh(\lambda y). \end{aligned}\right\} \tag{7.60}$$

Therefore, eqs. (7.58) can be rewritten as

$$\Phi = \sin(\lambda x)[(C_1 + C_2 y)\cosh(\lambda y) + (C_3 + C_4 y)\sinh(\lambda y)], \tag{7.61}$$

where $C_1 \cdots C_4$ are new arbitrary constants.

Exercise

Derive eqs. (7.60) from eqs. (7.59).

The stress components corresponding with the stress function given by eq. (7.61) are:

$$\sigma_{xx} = \frac{\partial^2 \Phi}{\partial y^2} = \sin(\lambda x)\{C_1\lambda^2 \cosh(\lambda y) + C_2[\lambda^2 y \cosh(\lambda y) + 2\lambda \sinh(\lambda y)]$$

$$+ C_3\lambda^2 \sinh(\lambda y) + C_4[\lambda^2 y \sinh(\lambda y) + 2\lambda \cosh(\lambda y)]\} \qquad (7.62)$$

because

$$\frac{d[\cosh(\lambda y)]}{dy} = \lambda \sinh(\lambda y)$$

and

$$\frac{d[\sinh(\lambda y)]}{dy} = \lambda \cosh(\lambda y),$$

as can be verified by inspection of eqs. (7.59). Continuing,

$$\sigma_{yy} = \frac{\partial^2 \Phi}{\partial x^2} = -\lambda^2 \sin(\lambda x)[(C_1 + C_2 y)\cosh(\lambda y)$$

$$+ (C_3 + C_4 y)\sinh(\lambda y)] \qquad (7.63)$$

$$\sigma_{xy} = -\frac{\partial^2 \Phi}{\partial x\, \partial y} = \lambda \cos(\lambda x)\{C_1\lambda \sinh(\lambda y) + C_2[\lambda y \sinh(\lambda y)$$

$$+ \cosh(\lambda y)] + C_3\lambda \cosh(\lambda y) + C_4[\lambda y \cosh(\lambda y) + \sinh(\lambda y)]\}. \qquad (7.64)$$

The four arbitrary constants, $C_1 \cdots C_4$, must now be determined from the boundary conditions. Let us take $y = 0$ to be the rigid boundary, A-B, in Fig. 7.5C. The boundary conditions are

At $y = J$, $\quad \sigma_{xy} = 0$ and $\sigma_{yy} = -\sigma_0 \sin(\lambda x)$.

At $y = -J$, $\quad \sigma_{xy} = 0$ and $\sigma_{yy} = -\sigma_1 \sin(\lambda x)$.

The boundary between the member and the medium is assumed to be frictionless. In order for the boundary, A-B, Fig. 7.5C, to be rigid, $\sigma_1 = \sigma_0$.

Consider the equation for the normal stresses in the y-direction, eq. (7.63), and the boundary conditions. At $y = J$,

$$\sigma_{yy} = -\sigma_0 \sin(\lambda x) = -\lambda^2 \sin(\lambda x)[(C_1 + C_2 J)\cosh(\lambda J)$$

$$+ (C_3 + C_4 J)\sinh(\lambda J)]. \qquad (7.65)$$

Note that

$$\sinh(-\lambda J) = -\sinh(\lambda J)$$

and

$$\cosh(-\lambda J) = \cosh(\lambda J),$$

according to eqs. (7.59). Thus, at $y = -J$,

$$\sigma_{yy} = -\sigma_1 \sin(\lambda x) = -\lambda^2 \sin(\lambda x)[(C_1 - C_2 J)\cosh(\lambda J)$$

$$- (C_3 - C_4 J)\sinh(\lambda J)]. \qquad (7.66)$$

Now consider the equation for the shear stress, eq. (7.64). At $y = J$,

$$\sigma_{xy} = 0 = \lambda \cos(\lambda x)\{C_1\lambda \sinh(\lambda J) + C_2[\lambda J \sinh(\lambda J) + \cosh(\lambda J)] + C_3\lambda \cosh(\lambda J) + C_4[\lambda J \cosh(\lambda J) + \sinh(\lambda J)]\}. \quad (7.67)$$

Finally, at $y = -J$,

$$\sigma_{xy} = 0 = \lambda \cos(\lambda x)\{-C_1\lambda \sinh(\lambda J) + C_2[\lambda J \sinh(\lambda J) + \cosh(\lambda J)] + C_3\lambda \cosh(\lambda J) - C_4[\lambda J \cosh(\lambda J) + \sinh(\lambda J)]\}. \quad (7.68)$$

Adding eqs. (7.67) and (7.68),

$$2C_2[\lambda J \sinh(\lambda J) + \cosh(\lambda J)] + 2C_3\lambda \cosh(\lambda J) = 0,$$

from which

$$C_3 = -C_2\frac{[\lambda J \sinh(\lambda J) + \cosh(\lambda J)]}{\lambda \cosh(\lambda J)}. \quad (7.69)$$

Subtracting eq. (7.65) from eq. (7.66),

$$-\frac{(\sigma_0 - \sigma_1)}{\lambda^2} = 2C_2 J \cosh(\lambda J) + 2C_3 \sinh(\lambda J),$$

or

$$C_2 = \frac{[-(\sigma_0 - \sigma_1)/2\lambda^2] + C_3 \sinh(\lambda J)}{J \cosh(\lambda J)}. \quad (7.70)$$

Using the relations

$$\cosh(\lambda J) \sinh(\lambda J) = \tfrac{1}{2} \sinh(2\lambda J)$$

and

$$\cosh^2(\lambda J) - \sinh^2(\lambda J) = 1,$$

which are readily derived from eqs. (7.59), we can solve for C_3:

$$C_3 = \frac{[\sigma_0 - \sigma_1]}{\lambda^2}\frac{\lambda J \sinh(\lambda J) + \cosh(\lambda J)}{2\lambda J - \sinh(2\lambda J)}. \quad (7.71)$$

In the same tedious manner we could derive expressions for the other three constants (*8*):

$$C_1 = \frac{[\sigma_0 + \sigma_1]}{\lambda^2}\frac{\sinh(\lambda J) + \lambda J \cosh(\lambda J)}{\sin(2\lambda J + 2\lambda J}, \quad (7.72)$$

$$C_2 = \frac{[\sigma_0 - \sigma_1]}{\lambda^2}\frac{\lambda \cosh(\lambda J)}{\sinh(2\lambda J - 2\lambda J},$$

and

$$C_4 = -\frac{[\sigma_0 + \sigma_1]}{\lambda^2} \frac{\lambda \sinh(\lambda J)}{\sinh(2\lambda J) + 2\lambda J}. \tag{7.73}$$

For the problem we are considering, we are simulating a rigid boundary by reflecting a stress distribution across the boundary. In this case $\sigma_0 = \sigma_1$, $C_2 = C_3 = 0$, and the normal stress components become [eqs. (7.62) and (7.63)],

$$\sigma_{xx} = 2\sigma_0 \sin(\lambda x) \frac{A \cosh(\lambda y) - H\lambda y \sinh(\lambda y)}{C},$$

where

$$A = \lambda J \cosh(\lambda J) - \sinh(\lambda J),$$

$$H = \sinh(\lambda J),$$

and

$$C = 2\lambda J + \sinh(2\lambda J). \tag{7.74}$$

$$\sigma_{yy} = -2\sigma_0 \sin(\lambda x)\left(\frac{D \cosh(\lambda y) - H\lambda y \sinh(\lambda y)}{C}\right),$$

where

$$D = \lambda J \cosh(\lambda J) + \sinh(\lambda J$$

Finally, we have derived the expressions for the normal stress components throughout the problem region, $DCD'C'$, in Fig. 7.5C. The remaining problem in determining the shape of the boundary, DC, between the competent member and the incompetent medium is to calculate the displacements, v and u, in the y- and x-directions, respectively, of the boundary.

The stress-strain equations for plane stress are [eqs. (5.27) and (5.39)]

$$\epsilon_{xx} = \frac{\partial u}{\partial x} = \frac{1}{E}(\sigma_{xx} - \nu\sigma_{yy}),$$

$$\epsilon_{yy} = \frac{\partial v}{\partial y} = \frac{1}{E}(\sigma_{yy} - \nu\sigma_{xx}), \tag{7.75}$$

and

$$\epsilon_{xy} = \frac{2(1 + v)}{E}\sigma_{xy} = \frac{\partial v}{\partial x} + \frac{\partial u}{\partial y}.$$

By integrating eqs. (7.75) just as we did for the preceding problem of the deflection of an infinitely thick medium, we could show that the deflection at the surface of the finite medium, that is, at $y = J$, is

$$v_{y=J} = -\frac{2\sigma_0 \sin(\lambda x)}{CE\lambda}\{[D(1 - \nu^2) + \nu(1 + \nu)A][\sinh(\lambda J)] - H(1 + \nu)[\lambda J \cosh(\lambda J) - \sinh(\lambda J)]\}. \tag{7.76}$$

where A, C, D, and H are given by eqs. (7.74).

Equation (7.76) is similar to the one quoted in earlier pages when we were discussing the resistance of a medium of finite thickness to buckling of a member confined by it. However, there we replaced $b\sigma_0 \sin(\lambda x)$ by its identity, q, and the displacement in eq. (7.76) is different by a factor of two from the value quoted earlier. But thus far in the derivation of the exact solution we have considered only one half of the incompetent medium surrounding the competent member. The other half of the medium simply reduces by half the displacement for a given stress σ_{yy}. Or, stated inversely, with a medium on both sides of the member, it takes twice as much stress, σ_{yy}, to produce a given amount of displacement, v, as it would if one side of the member were not confined by the medium.

As was shown in earlier pages, if the thickness, J, of the incompetent medium is equal to, or greater than, the fold wavelength, the wavelength is essentially independent of the thickness of the incompetent medium.

Rheological equations for Newtonian viscous substance

The rheological equations of viscous and elastic materials are strikingly similar. For a viscous substance stress and strain rate are proportional, whereas for an elastic substance stress and strain are proportional. But not only the rheological equations are similar. The same compatibility equation, expressed in terms of Airy's stress function, is applicable to elastic and viscous substances. Therefore, if the boundary conditions are the same, the stresses within elastic and incompressible, viscous substances are identical.

But the similarities are more extensive and significant than we will show by our derivations. Not only are the viscous and elastic substances similar, but there is a complete analogy between elastic substances and viscoelastic substances, including the special case of the viscous substances. Biot (*2*) has shown that deformation of viscoelastic media is described by equations that can be derived from the theory of elasticity simply by replacing the elastic moduli by certain operators. The principle of relating the operators to viscosity coefficients and elasticity moduli he called the *correspondence principle*. We will not develop the theory of viscoelasticity here, but with the background we shall develop we can understand texts and papers on that subject (see refs. *2*, *4*, *5*, and *7*).

The Newtonian viscous substance is a rheological model in which the shear stress is linearly related to the rate of shear strain, that is, $\sigma_s = \eta\dot{\epsilon}_s$, where the dot signifies differentiation with respect to time. This simple equation is sufficient to describe shear deformation, but our concepts of the viscous substances must be generalized to study two- and three-dimensional deformations.

Strain Rate.—The equations of equilibrium, eqs. (7.3) and (7.4), are valid where a substance deforms elastically, viscously, or according to some other rheological model. Similarly, the equations for strain [eqs. (5.15) and (5.16)] also are valid and they can be converted to equations of strain rate simply by being differentiated with respect to time. If this differentiation is indicated by a dot, eqs. (5.16) become

$$\left.\begin{aligned}\dot{\epsilon}_{xx} &= \frac{\partial \dot{u}}{\partial x} = \frac{\partial^2 u}{\partial x\,\partial t},\\ \dot{\epsilon}_{yy} &= \frac{\partial \dot{v}}{\partial y} = \frac{\partial^2 v}{\partial y\,\partial t},\\ \dot{\epsilon}_{zz} &= \frac{\partial \dot{w}}{\partial z} = \frac{\partial^2 w}{\partial z\,\partial t},\end{aligned}\right\} \tag{7.77}$$

and eqs. (5.15) become

$$\left.\begin{aligned}\dot{\epsilon}_{xy} &= \dot{\epsilon}_{yx} = \frac{\partial \dot{v}}{\partial x} + \frac{\partial \dot{u}}{\partial y},\\ \dot{\epsilon}_{xz} &= \dot{\epsilon}_{zx} = \frac{\partial \dot{w}}{\partial x} + \frac{\partial \dot{u}}{\partial z},\\ \text{and}\quad \dot{\epsilon}_{yz} &= \dot{\epsilon}_{zy} = \frac{\partial \dot{w}}{\partial y} + \frac{\partial \dot{v}}{\partial z}.\end{aligned}\right\} \tag{7.78}$$

Viscosity Coefficient.—These equations can be verified by substituting the components of velocity, $\dot{u}$, $\dot{v}$, and $\dot{w}$, for the components of displacement, u, v, and w, in Fig. 5.10. The Newtonian equation for a viscous substance was defined as $\sigma_s = \eta\dot{\epsilon}_s$, so that, with the definitions of strain rate given by eqs. (7.88),

$$\sigma_{xy} = \eta\left(\frac{\partial \dot{v}}{\partial x} + \frac{\partial \dot{u}}{\partial y}\right), \tag{7.79}$$

$$\sigma_{yz} = \eta\left(\frac{\partial \dot{w}}{\partial y} + \frac{\partial \dot{v}}{\partial z}\right), \tag{7.80}$$

$$\sigma_{xz} = \eta\left(\frac{\partial \dot{w}}{\partial x} + \frac{\partial \dot{u}}{\partial z}\right), \tag{7.81}$$

where η (eta) is the coefficient of viscosity. Thus we have a rheological equation relating shear stresses to rates of shear strain.

Relations between Normal Stresses and Normal Strain Rates.—We can derive the rheological equations for normal stresses and normal strain rates by consideration of Mohr Circles of stress and strain rate. Suppose that an element (Fig. 7.6A) of viscous material is subjected to principal stresses, σ_{xx} and σ_{yy}, in the x- and y-directions. Let $\sigma_{xx} > \sigma_{yy}$ and $\sigma_{xy} = 0$. As a result of the applied normal stresses, the element deforms with strain rates, $\dot{\epsilon}_{xx}$ and $\dot{\epsilon}_{yy}$. The stresses and strain rates are shown in the Mohr Circles in Figs. 7.6B and 7.6C.

According to the definition of the Newtonian viscous substance, the rate of

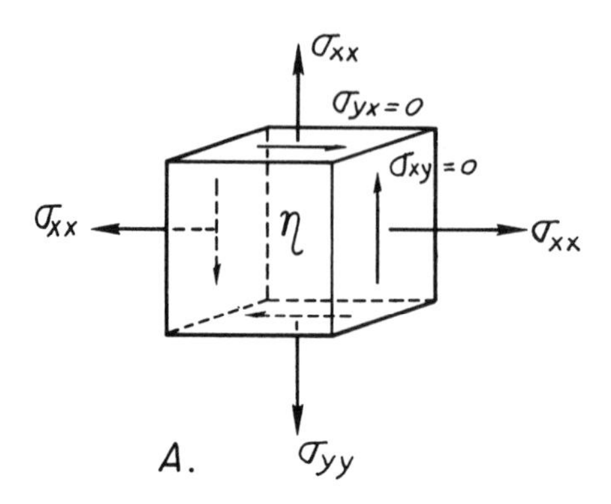

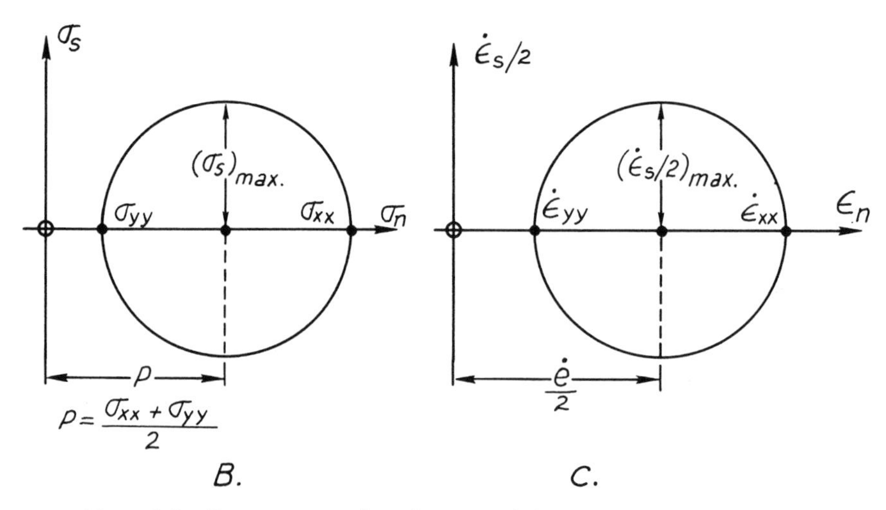

Figure 7.6. Stresses on, and strain rates of element of, viscous material.
A. Element of viscous material.
B. State of stress on element.
C. Mohr Circle of strain rates of element.

shear strain is proportional to the shear stress. Thus, the relation between maximum rate of shear and maximum shear stress is

$$\sigma_{s(\max)} = \eta \dot{\epsilon}_{s(\max)}. \tag{7.82}$$

According to Fig. 7.6B, the maximum shear stress is

$$\frac{\sigma_{xx} - \sigma_{yy}}{2} = \sigma_{s(\max)} \tag{7.83}$$

because σ_{xx} and σ_{yy} are principal stresses. Similarly, according to Fig. 7.6C, the maximum rate of shear strain is

$$\frac{\dot{\epsilon}_{s(\max)}}{2} = \frac{\dot{\epsilon}_{xx} - \dot{\epsilon}_{yy}}{2}, \tag{7.84}$$

or

$$\dot{\epsilon}_{s(\max)} = \dot{\epsilon}_{xx} - \dot{\epsilon}_{yy}. \tag{7.85}$$

The relation between normal stresses and rates of normal strain, therefore, is, according to eqs. (7.82), (7.83) and (7.85),

$$\frac{\sigma_{xx} - \sigma_{yy}}{2} = \sigma_{s(\max)} = \eta(\dot{\epsilon}_{xx} - \dot{\epsilon}_{yy}) = \eta\dot{\epsilon}_{s(\max)}. \tag{7.86}$$

Let us define the mean normal stress, or *pressure*, p, as

$$p = \frac{\sigma_{xx} + \sigma_{yy}}{2}, \tag{7.87}$$

and the rate of volumetric strain, $\dot{e}$, as

$$\dot{e} = \dot{\epsilon}_{xx} + \dot{\epsilon}_{yy} = \frac{\partial \dot{u}}{\partial x} + \frac{\partial \dot{v}}{\partial y}. \tag{7.88}$$

Then the rheological equations for normal stresses can be derived as follows, using eqs. (7.86), (7.87) and (7.88),

$$\sigma_{yy} = 2p - \sigma_{xx},$$

$$\sigma_{xx} = 2p - \sigma_{yy},$$

$$\dot{\epsilon}_{xx} = \dot{e} - \dot{\epsilon}_{yy},$$

$$\dot{\epsilon}_{yy} = \dot{e} - \dot{\epsilon}_{xx},$$

so that eq. (7.86) becomes

$$\frac{\sigma_{xx}}{2} - p + \frac{\sigma_{xx}}{2} = \eta(\dot{\epsilon}_{xx} - \dot{e} + \dot{\epsilon}_{xx})$$

or

$$\sigma_{xx} - p = 2\eta\left(\dot{\epsilon}_{xx} - \frac{\dot{e}}{2}\right) = 2\eta\frac{\partial \dot{u}}{\partial x} - \eta\left(\frac{\partial \dot{u}}{\partial x} + \frac{\partial \dot{v}}{\partial y}\right) \tag{7.89}$$

and

$$\sigma_{yy} - p = 2\eta\left(\dot{\epsilon}_{yy} - \frac{\dot{e}}{2}\right) = 2\eta\frac{\partial \dot{v}}{\partial y} - \eta\left(\frac{\partial \dot{u}}{\partial x} + \frac{\partial \dot{v}}{\partial y}\right). \tag{7.90}$$

These equations relate normal stress and strain rates of a viscous substance in plane strain, where $\dot{\epsilon}_{zz} = 0$.

For the general state of stress, the rheological equations for normal stresses are similar to those we derived above:

$$\sigma_{xx} - p = 2\eta\dot{\epsilon}_{xx} - \frac{2}{3}\eta(\dot{\epsilon}_{xx} + \dot{\epsilon}_{yy} + \dot{\epsilon}_{zz}), \tag{7.91}$$

$$\sigma_{yy} - p = 2\eta\dot{\epsilon}_{yy} - \frac{2}{3}\eta(\dot{\epsilon}_{xx} + \dot{\epsilon}_{yy} + \dot{\epsilon}_{zz}), \tag{7.92}$$

and

$$\sigma_{zz} - p = 2\eta\dot{\epsilon}_{zz} - \frac{2}{3}\eta(\dot{\epsilon}_{xx} + \dot{\epsilon}_{yy} + \dot{\epsilon}_{zz}). \tag{7.93}$$

In three dimensions, the volumetric strain rate is

$$\dot{e} = \dot{\epsilon}_{xx} + \dot{\epsilon}_{yy} + \dot{\epsilon}_{zz}$$

and the pressure is

$$p = \frac{\sigma_{xx} + \sigma_{yy} + \sigma_{zz}}{3}. \tag{7.94}$$

If the viscous substance is incompressible, the rate of volumetric strain, $\dot{e}$, is zero and the rheological equations for plane strain reduce to

$$\sigma_{xx} = p + 2\eta\frac{\partial\dot{u}}{\partial x}, \tag{7.95}$$

$$\sigma_{yy} = p + 2\eta\frac{\partial\dot{v}}{\partial y}, \tag{7.96}$$

and

$$\sigma_{xy} = \eta\left(\frac{\partial\dot{u}}{\partial y} + \frac{\partial\dot{v}}{\partial x}\right). \tag{7.97}$$

According to these rheological equations of the viscous substance, the rate of tangential or shear strain is proportional to the tangential stresses, and the rate of normal strain is proportional to the difference in pressure intensity or normal stresses in two directions.

Identity of solutions to some problems in elasticity and viscosity

The rheological equations for elastic and viscous substances are quite similar in form. Consequently, solutions for many problems in elasticity are solutions for analogous problems in viscosity.

The similarity between the rheological equations of a Hookean elastic body and of a Newtonian viscous body becomes apparent when we compare stress:strain and stress:strain-rate relations for the x-direction [from eqs. (5.33), (7.79), and (7.91)]:

$$\left.\begin{aligned}\sigma_{xx} &= 2G\epsilon_{xx} + \lambda(\epsilon_{xx} + \epsilon_{yy} + \epsilon_{zz}) \\ \sigma_{xy} &= G\epsilon_{xy}\end{aligned}\right\} \quad \text{Elastic}$$

$$\left.\begin{aligned}\sigma_{xx} - p &= 2\eta\dot{\epsilon}_{xx} - \tfrac{2}{3}\eta(\dot{\epsilon}_{xx} + \dot{\epsilon}_{yy} + \dot{\epsilon}_{zz}) \\ \sigma_{xy} &= \eta\dot{\epsilon}_{xy}\end{aligned}\right\} \quad \text{Viscous}$$

The forms of these equations are identical; one is written in terms of strain rate and the other in terms of strain.

More important similarities between elastic and viscous materials are made apparent by comparing compatibility equations, in terms of stresses, for the two materials. We showed that the displacements of an elastic body must satisfy the compatibility equation, eq. (7.8),

$$\frac{\partial^2 \epsilon_{xx}}{\partial y^2} + \frac{\partial^2 \epsilon_{yy}}{\partial x^2} = \frac{\partial^2 \epsilon_{xy}}{\partial x\, \partial y}, \tag{7.98}$$

to insure that there are continuous displacement functions, u and v. By analogy, the compatibility equation for flowing materials is

$$\frac{\partial^2 \dot{\epsilon}_{xx}}{\partial y^2} + \frac{\partial^2 \dot{\epsilon}_{yy}}{\partial x^2} = \frac{\partial^2 \dot{\epsilon}_{xy}}{\partial x\, \partial y}, \tag{7.99}$$

where the dots signify the operator, $\partial/\partial t$.

When we substituted the stress-strain relations for the elastic body into the compatibility equation, we found that the compatibility equation could be written in terms of stresses [eq. (7.15)],

$$\left(\frac{\partial^2}{\partial x^2} + \frac{\partial^2}{\partial y^2}\right)(\sigma_{xx} + \sigma_{yy}) = 0, \tag{7.100}$$

if the body forces are constant. Now we will show that this equation is valid also for an incompressible, viscous substance in plane strain.

The stress:strain-rate equations for a viscous substance in plane strain are [eqs. (7.95)–(7.97)]

$$\dot{\epsilon}_{xy} = \frac{\sigma_{xy}}{\eta},$$

$$\dot{\epsilon}_{xx} = \frac{\sigma_{xx} - p}{2\eta}, \tag{7.101}$$

and

$$\dot{\epsilon}_{yy} = \frac{\sigma_{yy} - p}{2\eta}.$$

Substituting eqs. (7.101) into eq. (7.100),

$$\frac{\partial^2}{\partial y^2}(\sigma_{xx} - p) + \frac{\partial^2}{\partial x^2}(\sigma_{yy} - p) = 2\frac{\partial^2 \sigma_{xy}}{\partial x\,\partial y}. \tag{7.102}$$

However, the stress components are not independent of one another. They are related through the differential equations of equilibrium, eqs. (7.3) and (7.4),

$$\frac{\partial \sigma_{xx}}{\partial_x} + \frac{\partial \sigma_{xy}}{\partial y} + X = 0 \tag{7.3}$$

and

$$\frac{\partial \sigma_{yy}}{\partial y} + \frac{\partial \sigma_{xy}}{\partial x} + Y = 0. \tag{7.4}$$

Differentiating eq. (7.3) with respect to x and eq. (7.4) with respect to y and adding,

$$2\frac{\partial^2 \sigma_{xy}}{\partial x\,\partial y} = -\left(\frac{\partial X}{\partial x} + \frac{\partial Y}{\partial y}\right) - \frac{\partial^2 \sigma_{xx}}{\partial x^2} - \frac{\partial^2 \sigma_{yy}}{\partial y^2}. \tag{7.103}$$

Assuming the body forces, X and Y, are constant, so that $\frac{\partial X}{\partial x} = \frac{\partial Y}{\partial y} = 0$, and substituting eq. (7.103) into eq. (7.102),

$$\left(\frac{\partial^2}{\partial x^2} + \frac{\partial^2}{\partial y^2}\right)(\sigma_{xx} + \sigma_{yy} - p) = 0. \tag{7.104}$$

But for plane strain,

$$\sigma_{zz} = p,$$

so that

$$p = \frac{\sigma_{xx} + \sigma_{yy} + p}{3} = \frac{\sigma_{xx} + \sigma_{yy}}{2}.$$

It follows that for plane strain eq. (7.104) is equivalent to

$$\left(\frac{\partial^2}{\partial x^2} + \frac{\partial^2}{\partial y^2}\right)\left(\frac{\sigma_{xx}}{2} + \frac{\sigma_{yy}}{2}\right) = 0,$$

or

$$\left(\frac{\partial^2}{\partial x^2} + \frac{\partial^2}{\partial y^2}\right)(\sigma_{xx} + \sigma_{yy}) = 0. \tag{7.105}$$

This is the equation of compatibility for plane strain or plane stress of an incompressible, viscous substance. It is identical to eq. (7.100), the compatibility equation for the elasticity model. It follows that Airy's stress function satisfies the compatibility equation for the viscosity model, eq. (7.105), so that solutions for stresses in plane strain or plane stress of an elastic body are solutions for the stresses in incompressible, viscous bodies with the same boundary conditions.

Folding of firm, viscous strut member embedded in soft, viscous medium

A consequence of the similar forms of the rheological equations for elastic and viscous materials is that theories of folding of both materials are nearly identical. Thus, the differential equation for the deflection of an elastic member in an elastic medium is

$$BI\frac{d^4v_1}{dx^4} + P\frac{d^2(v_1 + v_0)}{dx^2} + kv_1 = 0, \qquad \text{Elastic}$$

whereas, as we will show, the corresponding differential equation for viscous member in a viscous medium has the form

$$B'I\frac{d^4\dot{v}_1}{dx^4} + P\frac{d^2(v_1 + v_0)}{dx^2} + k'\dot{v}_1 = 0, \qquad \text{Viscous}$$

where B' and k' are new constants and the dot over v_1 refers to differentiation with respect to time. The wavelength of initial deflections that will be markedly amplified to form folds in an elastic member is [eq. (3.11)]

$$L_e = 2\pi T\left[\frac{B}{6B_0}\right]^{1/3},$$

where B is the elastic modulus of the member and B_0 is the elastic modulus of the medium, whereas, as we will show, the corresponding wavelength for a viscous member is

$$L_v = 2\pi T\left[\frac{\eta}{6\eta_0}\right]^{1/3},$$

where η is the viscosity coefficient of the member and η_0 is the viscosity coefficient of the medium. The similarities between these wavelengths are striking. The equations indicate that the wavelength is governed by the ratio of the elastic moduli or viscosity coefficients of the member and the medium.

Bending Resistance of Viscous Member.—An initially deflected viscous member is embedded in a viscous medium of infinite thickness (Fig. 7.7A). The member folds as a result of an axial load, P, and folding is resisted by the stiffness of the member and the constraint of the medium. We will ignore temporarily the constraint of the medium.

The member is initially deflected in a simple sinusoidal form,

$$v_0 = \delta_0 \sin\left(\frac{2\pi x}{L_0}\right), \tag{7.106}$$

where δ_0 is the amplitude and L_0 is the wavelength of the initial deflection. When

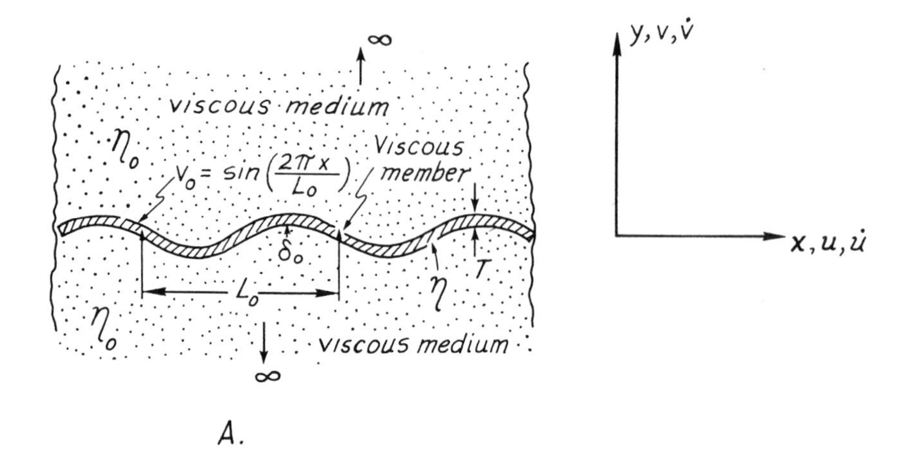

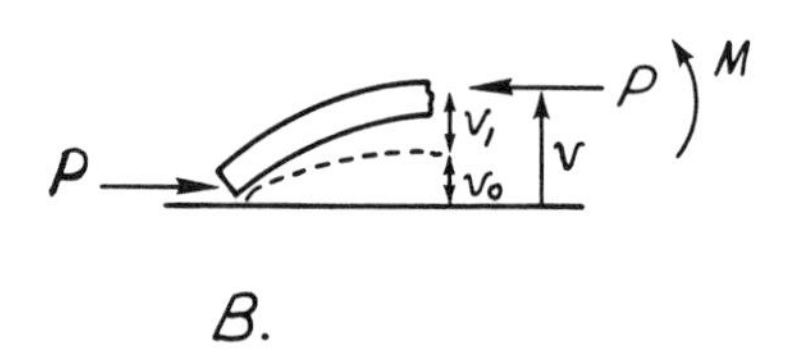

Figure 7.7. Initially deflected, stiff, viscous member embedded in soft, viscous medium.

the axial load is applied to the member, deflection increases so that the member is subjected to a bending moment, M, given by

$$M = -P(v_0 + v_1), \tag{7.107}$$

where v_1 is the additional deflection of the member (Fig. 7.7B).

For an elastic member, the bending moment is

$$M = BI\left(\frac{d^2v_1}{dx^2}\right). \tag{7.108}$$

The problem is to derive a similar expression for a viscous member.

The equations relating stress and strain rate of the member are [eqs. (7.95)–(7.97)]

$$\sigma_{xx} = p + 2\eta\dot{\epsilon}_{xx},$$

$$\sigma_{yy} = p + 2\eta\dot{\epsilon}_{yy},$$

and

$$\sigma_{zz} = p = \frac{\sigma_{xx} + \sigma_{yy} + \sigma_{zz}}{3},$$

because

$$\dot{\epsilon}_{zz} = 0$$

for plane strain. Here z is in the plane of the member, normal to the direction of the axial load, x is parallel to the direction of loading, and y is normal to the plane of the member.

To a first approximation we can assume that the stress normal to the plane of the member is zero. This requires that the viscosity of the medium be small relative to the viscosity of the member. Also, we can assume that the strain in the z-direction is zero. Thus,

$$\sigma_{yy} = 0 \quad \text{and} \quad \sigma_{zz} = p,$$

so that

$$\sigma_{xx} = \frac{\sigma_{xx}}{2} + 2\eta\dot{\epsilon}_{xx},$$

or

$$\sigma_{xx} = 4\eta\dot{\epsilon}_{xx} = A\dot{\epsilon}_{xx}, \tag{7.109}$$

where

$$A = 4\eta\frac{\partial}{\partial t}.$$

For an elastic member,

$$\sigma_{xx} = B\epsilon_{xx},$$

which is identical in form to eq. (7.109).

If the strain rate in the member is proportional to the distance from the neutral plane, the bending moment in a viscous member is

$$M = A\left(I\frac{d^2v_1}{dx^2}\right),$$

where A is substituted for B in eq. (7.108). Thus,

$$M = 4\eta\frac{\partial}{\partial t}\left(I\frac{d^2v_1}{\partial x^2}\right) = 4\eta I\frac{d^3v_1}{\partial x^2\,\partial t} \tag{7.110}$$

and the moment equation becomes

$$4\eta I\frac{\partial^3 v_1}{\partial x^2\,\partial t} + P(v_0 + v_1) = 0. \tag{7.111}$$

Exercise

Eq. (7.111) is the differential equation for the deflection of an isolated, initially deflected viscous member. Solve it and show that, if the initial deflection is a sinusoidal form, $v_0 = \delta_0 \sin(2\pi x/L_0)$, the expression for the total deflection, $v_1 + v_0$, is

$$v = \delta_0 \sin\left(\frac{2\pi x}{L_0}\right) \exp\left[\frac{P}{4\eta I}\left(\frac{L_0}{2\pi}\right)^2 t\right].$$

Resistance of Viscous Medium.—Now we will calculate the resistance of the viscous medium to bending of the viscous member. If the resisting moment of the medium is M, then the differential equation of deflection would be rewritten as

$$M_m = 4\eta I \frac{\partial^3 v_1}{\partial x^2\, \partial t} + P(v_0 + v_1). \tag{7.112}$$

According to an equation that we derived in Chapter 2, the second derivative of the moment with respect to the x-direction is the force per unit of length, q, imposed by the medium on the member:

$$\frac{d^2 M_m}{dx^2} = -q.$$

Equation (7.112) can be rewritten, therefore, as

$$q + 4\eta I \frac{\partial^5 v_1}{\partial x^4\, \partial t} + P \frac{\partial^2}{\partial x^2}(v_1 + v_0) = 0. \tag{7.113}$$

Now we must derive the constraint, q, in terms of the deflection, v_1, of the confining medium. The member imposes a sinusoidal stress distribution on the confining medium, so that the solution for the stress function of the medium is the same as the stress function we derived for the elastic analog of this problem [eq. (7.28)],

$$\Phi = (C_1 + C_2 y)e^{\lambda y} \sin(\lambda x), \tag{7.114}$$

because the compatibility equations for the elastic substance and the incompressible, viscous substance are identical. Here,

$$\lambda = \frac{2\pi}{L_0},$$

and C_3 and C_4 in eq. (7.28) are zero, as was explained in earlier pages.

The corresponding stress components are

$$\sigma_{xx} = \frac{\partial^2\Phi}{\partial y^2} = \sigma_0(1 + \lambda y)e^{\lambda y}\sin(\lambda x),$$

$$\sigma_{yy} = \frac{\partial^2\Phi}{\partial x^2} = \sigma_0(1 - \lambda y)e^{\lambda y}\sin(\lambda x), \tag{7.115}$$

where σ_0 is the amplitude of the sinusoidal stress distribution.

The velocities, $\dot{v}$ and $\dot{u}$, of points in the medium can be calculated by integrating the rheological equations for an incompressible, viscous substance [eqs. (7.95)–(7.97)],

$$\frac{\partial \dot{u}}{\partial x} = \frac{\sigma_{xx} - p}{2\eta_0},$$

$$\frac{\partial \dot{v}}{\partial y} = \frac{\sigma_{yy} - p}{2\eta_0},$$

and

$$\frac{\partial \dot{u}}{\partial y} + \frac{\partial \dot{v}}{\partial x} = \frac{\sigma_{xy}}{\eta_0}.$$

For plane strain, $p = \dfrac{\sigma_{xx} + \sigma_{yy}}{2}$,

so that we can rewrite the rheological equations as

$$\frac{\partial \dot{u}}{\partial x} = \frac{\sigma_{xx} - \sigma_{yy}}{4\eta_0},$$

$$\frac{\partial \dot{v}}{\partial y} = \frac{\sigma_{yy} - \sigma_{xx}}{4\eta_0},$$

and

$$\frac{\partial \dot{u}}{\partial y} + \frac{\partial \dot{v}}{\partial x} = \frac{\sigma_{xy}}{\eta_0}.$$

Substituting the solutions for the stresses, eqs. (7.115), into the rheological equations and integrating,

$$\frac{\partial \dot{u}}{\partial x} = \frac{\sigma_{xx} - \sigma_{yy}}{4\eta_0} = \frac{\sigma_0\lambda y}{2\eta_0} e^{\lambda y}\sin(\lambda x),$$

$$\dot{u} = -\frac{\sigma_0 y}{2\eta_0} e^{\lambda y}\cos(\lambda x) + g(y), \tag{7.116}$$

$$\frac{\partial \dot{v}}{\partial y} = \frac{\sigma_{yy} - \sigma_{xx}}{4\eta_0} = -\frac{\sigma_0\lambda y}{2\eta_0} e^{\lambda y}\sin(\lambda x),$$

$$\dot{v} = -\frac{\sigma_0}{2\eta_0\lambda} e^{\lambda y}(\lambda y - 1)\sin(\lambda x) + f(x). \tag{7.117}$$

We showed, when we derived the analogous equations for the deflections of an elastic medium, that the arbitrary functions, $f(x)$ and $g(y)$, are zero.

Therefore, at the surface of the viscous medium, that is, at $y = 0$, the rate of vertical deflection is, according to eq. (7.117),

$$\dot{v}_{y=0} = \frac{\sigma_0 \sin(\lambda x)}{2\eta_0 \lambda}.$$

The corresponding equation for an elastic medium is nearly identical [eq. (7.52)],

$$v_{y=0} = \frac{2\sigma_0 \sin(\lambda x)}{\lambda E_0}.$$

Now, $\sigma_{yy} = \sigma_0 \sin(\lambda x)$ is the distribution applied at the boundary of the medium, so that

$$\dot{v}_{y=0} = \frac{\sigma_{yy}}{2\eta_0 \lambda}.$$

Further,

$$\sigma_{yy} = q/b,$$

and

$$\lambda = 2\pi/L_0,$$

so that the expression for the rate of deflection is

$$\dot{v}_{y=0} = \frac{qL_0}{4\pi b\eta_0},$$

or

$$q = \frac{4\pi b\eta_0}{L_0}\,\dot{v}.$$

The problem involves the constraint arising from a medium that surrounds the member so that the constraint should be doubled. Also, the medium constrains additional deflection, $\dot{v}_1$. Thus,

$$q = \frac{8b\pi\eta_0\dot{v}_1}{L_0} \tag{7.118}$$

is the constraint of the medium if η_0 is its viscosity. This is equivalent to the expression of constraint derived by Biot (*3*) and it is identical to that for an elastic medium if we let

$$4\eta_0 \frac{\partial}{\partial t} = B_0.$$

Wavelength of Folds.—Substituting the value of the constraint [eq. (7.118)] into the differential equation for the deflection of a viscous member in a viscous medium, eq. (7.113), we have

$$4\eta I \frac{\partial^5 v_1}{\partial x^4 \, \partial t} + P \frac{\partial^2}{\partial x^2} (v_1 + v_0) + \frac{8b\pi\eta_0}{L_0} \frac{\partial v_1}{\partial t} = 0,$$

or (7.113)

$$\frac{4\eta I}{P} \frac{\partial^5 v_1}{\partial x^4 \, \partial t} + \frac{\partial^2 v_1}{\partial x^2} + \frac{8b\pi\eta_0}{PL_0} \frac{\partial v_1}{\partial t} = -\frac{\partial^2 v_0}{\partial x^2}.$$

Equation (7.113) is a nonhomogeneous, partial differential equation. It can be solved by determining a general solution to the homogeneous part, that is, the part to the left of the equal sign, and by adding a particular solution involving the entire differential equation, including the nonhomogeneous part.

First consider the homogeneous part of the differential equation:

$$\frac{4\eta I}{P} \frac{\partial^5 v_1}{\partial x^4 \, \partial t} + \frac{\partial^2 v_1}{\partial x^2} + \frac{8b\pi\eta_0}{PL_0} \frac{\partial v_1}{\partial t} = 0.$$

We can separate variables by letting the displacement be the product of two undetermined functions,

$$v_1 = f(x)g(t),$$

as we did for the problem of internal buckling of a multilayer in chapter 6. Thus,

$$\frac{4\eta I}{P} \frac{d^4 f(x)}{dx^4} \frac{dg(t)}{dt} + g(t) \frac{d^2 f(x)}{dx^2} + \frac{8b\pi\eta_0}{PL_0} f(x) \frac{dg(t)}{dt} = 0.$$

By dividing through by $\dfrac{dg(t)}{dt} \dfrac{d^2 f(x)}{dx^2}$ we separate the variables,

$$\frac{4\eta I}{P} \frac{[d^4 f(x)/dx^4]}{[d^2 f(x)/dx^2]} + \frac{8b\pi\eta_0}{PL_0} \frac{f(x)}{[d^2 f(x)/dx^2]} = \frac{-g(t)}{[dg(t)/dt]} = C_0,$$

where C_0 is a constant. Thereby the problem of solving the homogeneous part of the partial differential equation becomes a problem of solving two homogeneous, total differential equations containing an undefined constant, C_0. The two equations are

$$\frac{4\eta I}{P} \frac{d^4 f(x)}{dx^4} - C_0 \frac{d^2 f(x)}{dx^2} + \frac{8b\pi\eta_0}{PL_0} f(x) = 0 \tag{7.114}$$

and

$$\frac{dg(t)}{dt} + \frac{g(t)}{C_0} = 0. \tag{7.115}$$

We have already derived the solution to eq. (7.114). It is [see eq. (3.29)]:

$$f(x) = (C_1e^{\beta x} + C_2e^{-\beta x}) \cos \alpha x + (C_3e^{\beta x} + C_4e^{-\beta x}) \sin \alpha x,$$

where

$$\alpha = \left[\left(\frac{b\pi\eta_0}{8L_0\eta I}\right)^{1/2} + \frac{C_0P}{16\eta I}\right]^{1/2}$$

and (7.116)

$$\beta = \left[\left(\frac{b\pi\eta_0}{8L_0\eta I}\right)^{1/2} - \frac{C_0P}{16\eta I}\right]^{1/2}.$$

The solution to eq. (7.115) is

$$\int \frac{dg(t)}{g(t)} = -\frac{1}{C_0}\int dt$$

or

$$g(t) = C_5 \exp\left(\frac{t}{C_0}\right). \tag{7.117}$$

There is no need to proceed further with the solution to the homogeneous part of the differential equation. As we showed before, all the constants, $C_1 \cdots C_4$, must be zero in order to satisfy the boundary condition that the deflection, v_1, which is $f(x)g(t)$ plus a particular solution to the nonhomogeneous differential equation, must be zero at $x = L_0$, regardless of the value of the axial load, P. The variables, α and β, depend on the axial load so that the constants, $C_1 \cdots C_4$, would have to vary with P in order to meet the condition that $f(x) = 0$, at $x = L_0$, for all P.

The pertinent solution, therefore, can be obtained by searching for a particular solution to the nonhomogeneous, partial differential equation, eq. (7.113),

$$\frac{4\eta I}{P}\frac{\partial^5 v_1}{\partial x^4\,\partial t} + \frac{\partial^2 v_1}{\partial x^2} + \frac{8b\pi\eta_0}{PL}\frac{\partial v_1}{\partial t} = \delta_0\left(\frac{2\pi}{L_0}\right)^2 \sin\left(\frac{2\pi x}{L_0}\right), \tag{7.118}$$

in which we have substituted the expression in eq. (7.106) for v_0.

In light of the form of the term on the right-hand side of eq. (7.118), and the observation that second derivatives and fourth derivatives of a sine function yield other sine functions, we should try a solution of the form

$$v_1 = g(t) \sin\left(\frac{2\pi x}{L_0}\right), \tag{7.119}$$

where $g(t)$ is some function of time. In the similar problem that we solved in Chapter 4, we were dealing with a total differential equation rather than a partial differential equation. The trial particular solution we used in Chapter 4, therefore, contained an undetermined constant, A, rather than an undetermined function, $g(t)$. The procedures are basically the same, however.

Substituting the trial solution, eq. (7.119), into the nonhomogeneous, partial differential equation, eq. (7.118),

$$\frac{4\eta I}{P}\left(\frac{2\pi}{L_0}\right)^4 \frac{dg(t)}{dt} \sin\left(\frac{2\pi x}{L_0}\right) - \left(\frac{2\pi}{L_0}\right)^2 g(t) \sin\left(\frac{2\pi x}{L_0}\right) + \frac{8b\pi\eta_0}{PL_0}\frac{dg(t)}{dt} \sin\left(\frac{2\pi x}{L_0}\right) = \delta_0\left(\frac{2\pi}{L_0}\right)^2 \sin\left(\frac{2\pi x}{L_0}\right).$$

Dividing through by $(2\pi/L_0)^2 \sin(2\pi x/L_0)$,

$$a\frac{dg(t)}{dt} - g(t) - \delta_0 = 0, \tag{7.120}$$

where

$$a = \frac{4\eta I}{P}\left(\frac{2\pi}{L_0}\right)^2 + \frac{2bL_0\eta_0}{\pi P}.$$

The solution to the homogeneous part of eq. (7.120) is

$$a\int\frac{dg(t)}{g(t)} = \int dt,$$

or

$$a \ln[g(t)] = t + C_1,$$

or

$$g(t) = C_2 \exp(t/a),$$

so that a general solution to the differential equation, eq. (7.120), is

$$g(t) = C_2 \exp(t/a) - \delta_0.$$

The equation for the deflection, v_1, therefore, becomes [eq. (7.119)]

$$v_1 = g(t) \sin\left(\frac{2\pi x}{L_0}\right) = (C_2 e^{t/a} - \delta_0) \sin\left(\frac{2\pi x}{L_0}\right).$$

The value of the constant of integration, C_2, can be determined with the boundary condition that, when time is zero, the deflection, v_1, is zero for all values of x. Thus,

$$v_1 = 0 = (C_2 e^0 - \delta_0),$$

or

$$C_2 = \delta_0.$$

Thus the deflection caused by the axial load is

$$v_1 = \delta_0(e^{t/a} - 1) \sin\left(\frac{2\pi x}{L_0}\right),$$

where

$$a = \frac{4\eta I}{P}\left(\frac{2\pi}{L_0}\right)^2 + \frac{2bL_0\eta_0}{\pi P}. \tag{7.121}$$

The total deflection, v, is v_1 plus the initial deflection, v_0, where

$$v_0 = \delta_0 \sin\left(\frac{2\pi x}{L_0}\right),$$

so that the total deflection is

$$v = v_1 + v_0 = \delta_0\left\{\exp\left[\frac{tP}{4\eta I(2\pi/L_0)^2 + (2bL_0\eta_0/\pi)}\right]\right\} \sin\left(\frac{2\pi x}{L_0}\right). \tag{7.122}$$

The initial irregularities in the member that will be amplified the most by a given axial load, P, will be those for which a, in eq. (7.121), is minimum—that is, the load for which the denominator in the exponent of eq. (7.122) is a minimum. We can calculate this wavelength by differentiating eq. (7.123) with respect to the wavelength and setting the result equal to zero:

$$\frac{da}{dL_0} = 0 = \frac{-8\eta I(2\pi)^2}{L_0^{-3}} + \frac{2b\eta_0}{\pi},$$

so that the initial deflections that will grow the most rapidly, to form the most distinct folds, are those with a wavelength of (*3*)

$$L_v = 2\pi T\left[\frac{\eta}{6\eta_0}\right]^{1/3} \tag{7.123}$$

We showed that, for an elastic member in an elastic medium, the initial deflections that grow most rapidly with increasing axial loads are those with a wavelength of

$$L_e = 2\pi T\left[\frac{B}{6B_0}\right]^{1/3},$$

which is identical in form to that for viscous members in a viscous medium, eq. (7.123).

Similarities and Differences between Elastic and Viscous Folding.— I hope it is clear by now that many aspects of viscosity and elasticity theories are similar. The reason for these similarities is that the rheological equations of viscosity and elasticity are linear, that is, stress and strain or stress and strain rate are proportional. We have shown that the two-dimensional compatibility equations in terms of stresses are identical, that the differential equations of folding deflection, the reaction of confining media, and the predicted fold wavelengths are analogous, and that the forms of folds are identical in viscous and elastic materials.

According to our analyses, folds in layered viscous materials should be identical in form to folds in elastic materials. Both are sinusoidal if amplitudes are small. Even for large amplitudes, the *elastica* should describe the fold form of either elastic or viscous members. This conclusion has far-reaching implications for interpretation of geologic structures. It indicates that we cannot determine whether a rock flowed viscously or deformed elastically when it was bent into sinusoidal folds. It surprised me, and I am sure it has surprised others, that a fluid material, such as tar, would fold into the same form as an elastic material, such as spring steel. But the theory says this is the way it is and, in so far as the assumptions of the theory are fulfilled in tar and spring steel, the conclusion is indisputable.

There are differences between theories of folding of viscous and elastic materials, of course. For example, relatively small axial loads might produce marked folds in layered, viscous materials, if enough time is allowed. Time occurs as an exponential in the deflection equation for a viscous member,

$$v_v = \delta_0 e^{(t/a)} \sin\left(\frac{2\pi x}{L_0}\right), \qquad \text{Viscous}$$

where

$$a = \frac{4\eta I}{P}\left(\frac{2\pi}{L_0}\right)^2 + \frac{4\eta_0 b}{P}\left(\frac{L_0}{2\pi}\right),$$

so that the amount of deflection is highly sensitive to the amount of time involved in folding. The amount of deflection is equally sensitive to the magnitude of the axial load, P. However, shortening and thickening might overshadow buckling of a viscous member if significant amplification of initial deflections in the member requires too much time. Thus, if the constant, a, is large, the viscous member might largely shorten instead of buckle. According to Biot (*3*), folding is insignificant if η/η_0 or B/B_0, the ratios of the viscosity and the elasticity coefficients, respectively, are less than about 100.

On the other hand, the deflection of an elastic member is independent of time and is insignificant unless the axial load approaches the critical load, P_{crit}:

$$v_e = \frac{P_{\text{crit}}/P}{[(P_{\text{crit}}/P) - 1]}\,\delta_0 \sin\left(\frac{2\pi x}{L_0}\right), \qquad \text{Elastic}$$

where

$$P_{\text{crit}} = BI\left(\frac{2\pi}{L_0}\right)^2 + B_0 b\left(\frac{L_0}{2\pi}\right).$$

There is no critical load for buckling of a viscous member.

References cited in Chapter 7

1. Biot, M. A., 1937, "Bending of an Infinite Beam on an Elastic Foundation": *Trans. Amer. Soc. Mechanical Engineers*, V. 59, p. A1–A7.
2. ———, 1954, "Theory of Stress-Strain Relations in Anisotropic Viscoelasticity and Relaxation Phenomena": *Jour. Applied Physics*, V. 25, p. 1385–1419.
3. ———, 1961, "Theory of Folding of Stratified Viscoelastic Media and its Implications in Tectonics and Orogenesis": *Geol. Soc. America Bull.*, V. 72, p. 1595–1632.
4. ———, 1965, *Mechanics of Incremental Deformations*: John Wiley & Sons, Inc., N.Y.
5. Fung, Y. C., 1965, *Foundations of Solid Mechanics*: Prentice-Hall, Inc., Englewood Cliffs, N.Y.
6. Kreyszig, Erwin, 1962, *Advanced Engineering Mathematics*: John Wiley & Sons, Inc., N.Y.
7. Lee, E. H., 1960, "Viscoelastic Stress Analysis": in *Proc. First Symp. on Naval Structural Mechanics*, Stanford Univ., 1958, J. N. Goodier and M. J. Hoff, eds., Pergamon Press, N.Y., p. 456–482.
8. Timoshenko, S., and Goodier, J. N., 1951, *Theory of Elasticity*: McGraw-Hill Book Co., Inc., N.Y.

8

Chapter Sections

Forms of Folds
Behavior of Chert and Shale During Folding
Analysis of Forms
Buckling of Individual Plastic Layers
  Yielding
  Yield Strength
  Formation of Hinge and Straight Limbs
Plastic Instability
Buckling of Viscous-Plastic, Pseudoplastic, and Nonlinear Elastic Strips
Reexamination of Franciscan Folds
Formation of Hinge Lines in Layered Materials
  Experimental Kink Bands
  Analysis of Orientations of Kink Bands
Kink Bands and Concentric Folds
References Cited in Chapter 8

8

Folding of Interbedded Chert and Shale of Franciscan Formation, San Francisco, California

Forms of folds

All the folds we have studied heretofore have had the classic fold form—smooth, periodic waves of various amplitudes. Some folds in cherts of the Franciscan Formation of California are of this classic form but most of them are quite different. Most of the folds have sharp hinges and straight limbs. Figure 8.1, for example, shows folds in steeply dipping layers of interbedded chert and shale near the northern end of the Golden Gate Bridge, San Francisco, California. The general dip of the layering is toward the left but it abruptly shifts at hinges of folds, causing the dip to reverse.

The forms shown in Fig. 8.2 are common in the Franciscan cherts. Two of the chert beds have been chalked so that they would be distinctive in the photograph. The interbedded material is shale. The fold crests in the chert are systematically different from the fold crests in the shale. Almost invariably, the shale layers have increased in thickness in fold crests whereas the chert beds maintain nearly the same thicknesses in fold crests as in fold limbs (*3*).

Two other aspects of the bedding deserve special attention. The first is changes in thickness of chert layers. The chert layers have an irregular, blebby form and they commonly terminate by pinching off with a blunt end. These irregularities in thickness are visible throughout the Franciscan cherts, even where folds are absent, and therefore are not a result of folding (*3*).

Another striking feature is the sharp, angular form of the folds (Figs. 1.1 and 8.1). The limbs of the folds are nearly straight, concentrating the bending in nar-

Figure 8.1. Folds in Franciscan Formation. Resistant beds are chert, soft interbeds are shale. David Pollard is examining the folds.

row, almost sharp, crestal hinges (*3*). In many places the hinges appear to approach circular arcs, with larger radii of curvature on outsides of hinges and smaller radii of curvature on insides of hinges (Fig. 8.2).

The folds are arranged in an orderly way on the scale of an outcrop even though, in places, they are extremely complicated in detail. In outcrops along roadcuts, ranging up to 50 feet in height in some places at the northern end of the Golden Gate Bridge, one can clearly discern "domains" or bands of bedding planes oriented

Figure 8.2. Folds in interbedded chert and shale. Two chert beds were coated with chalk to make them visible in the photograph. White card is 3 by 5 inches in size.

the same way. The domains are most clearly visible when one stands so that light reflects simultaneously from beds with the same orientation. For example, two domains containing beds oriented approximately vertically are visible in Fig. 8.1.

Most of the folds in the Franciscan cherts are asymmetric; one limb is longer than the other. Those shown in the center of Fig. 8.1 are representative. The limbs dipping toward the left in that figure are about three meters long whereas the limbs dipping toward the right are about one-half meter long.

Behavior of chert and shale during folding

The interbedded chert and shale of the Franciscan Formation folded into forms which are quite different from the smooth, periodic forms we have studied and described theoretically for linear elastic and linear viscous materials. Perhaps, then, the Franciscan rocks displayed rheological properties different from those we have studied.

Detailed examination of the folds themselves should help us determine how the chert and shale actually did behave. The relative behaviors of the chert and the shale must have been different, because the shale beds thickened in fold crests whereas the chert beds generally maintained constant thicknesses. Apparently the relatively soft material that is now shale moved along bedding planes toward fold crests, and the relatively stiff material that is now chert maintained rather uniform thicknesses on limbs and in crests of folds. The cherts seem to have dominated the fold pattern. This behavior is indeed relative because, where the shale interbeds are thin, on the order of one-tenth the thicknesses of chert beds, the chert beds as well as the shale beds thickened in fold crests (*3*).

Either the chert beds were softer than they are today or the process of folding was extremely slow. The chert beds are now brittle and highly fractured so that it is difficult to remove a piece of chert larger than a few cubic centimeters in size. The fracturing must have postdated folding, however, because the beds in the fold crests form smooth curves of fractured rock, the pieces fitting together as in a jig-saw puzzle, without the gaping cracks we would expect to see if individual fragments of chert were rotated as units during folding. Also, the individual fragments cannot be bent perceptibly now without fracturing. Thus, if the chert was deformed to form the folds while in its present condition, the deformation must have been slow enough to allow flowage.

Flowage of the chert in its present condition is unlikely, however. Radiolaria, which are siliceous tests of microscopic organisms, appear to be undeformed even in the hinges of the folds, where strain must have been extreme. These delicate tests certainly would have been deformed unless they were suspended in a relatively soft medium. They apparently were more competent than their chert matrix. They are, indeed, flattened and otherwise distorted in narrow bands along contacts between shale beds and chert beds, but, within the chert, even in the intensely strained hinges, they are not.

Thus it would seem that the chert was soft relative to the radiolaria tests and firm relative to the shale. Presumably the chert was a gel and the shale was a soft mud at the time of folding.

Analysis of forms

The sharp crests and straight limbs of the folds in the Franciscan cherts could be results of several quite different causes. It is conceivable that the chert beds behaved as Newtonian viscous layers and that changes in thickness of chert beds caused the deformation to be essentially restricted to narrow hinge zones. We know that the bending resistance of a slightly bent, viscous member is

$$M_R = AI\frac{d^2v}{dx^2},$$

where

$$I = bT^3/12,$$

d^2v/dx^2 is curvature,

$$A = 4\eta\frac{\partial}{\partial t},$$

and the bending resistance of an elastic member is

$$M_R = BI\frac{d^2v}{dx^2},$$

where B is the elastic modulus (see Chapter 7).

For either material the resistance is proportional to the cube of the thickness, T^3, of the member; therefore, a change in thickness of an isolated member can markedly change the curvature of the member. If a member pinched to half its normal thickness, its curvature in the pinched area would increase by a factor of eight in order to maintain a constant bending moment (see ref. *5*).

Seemingly, then, the blebbiness of the chert beds could have determined positions of fold hinges. However, there is no obvious correlation between positions of thin parts of the blebby chert beds of the Franciscan and crests of folds. Chert beds usually are no thinner in fold crests than elsewhere and thin parts of chert beds seem to occur as frequently near the crests and on the limbs of folds as they do in the crests, where we would expect to see them if they controlled positions of crests (*3*). Some other factor must have controlled the shapes of the folds.

A possible explanation for the sharp hinges and straight limbs of the folds in the chert beds is that the cherts behaved as a viscous-plastic or as an elastic-plastic material at the time of folding. Both materials yield plastically when the applied stress reaches their respective yield strengths. Examples of materials that behave roughly as elastic-plastic substances are cardboard and aluminum. We know that a sharp bend forms in a strip of cardboard and in a strip of aluminum if the strips are subjected to large axial loads. I know of no material that behaves as a viscous-plastic substance but, for the reasons discussed below, the chert beds in the Franciscan may have behaved as such a substance.

Another possible explanation for the sharp hinges and straight limbs of the folds is interaction of adjacent chert layers. We know, for example, that where the shale interbeds are thin, the interaction of chert beds in fold hinges caused the chert beds to thicken. Where shale interbeds are thick, the thickening is restricted to the shale. Perhaps, then, the straight limbs are a result of interaction of chert beds.

We will consider each of these possibilities in detail in the remainder of this chapter

Buckling of individual plastic layers

Yielding.—The chert beds seem to have dominated the folding of the interlayered cherts and shales, and for this reason we can perhaps consider separately the behaviors of the firm cherts and the soft shales.

If we manipulate strips of various materials we notice that strips of some materials bend into smooth, sinusoidal forms; examples are rubber and spring steel. Some materials, however, first bend into the smooth form, then, with increased deflection, bend sharply over a narrow zone or hinge. Cardboard used as the backing of writing tablets is a good example of such materials. With a small axial thrust a strip of cardboard deflects into a smooth curve. Suddenly, however, as the load is increased, a narrow band of the cardboard bends sharply, and the smooth curve snaps into a sharp-crested "V."

This phenomenon clearly is related to the behavior of the material in the strip. We can imagine that cardboard, for example, behaves elastically until a certain critical load is reached, after which it behaves differently, fails, and forms a hinge. In some materials, such as lead metal, the phenomenon happens when deflections are slight. In others, such as spring steel, it happens only with extreme deflections. The stress-strain behavior of cardboard is approximately as is shown in Fig. 8.3. Stress is essentially proportional to strain up to a limit, point A in Fig. 8.3, after which strain increases more rapidly with increasing stress. Finally, at some point (B in Fig. 8.3), stress drops off even though strain continues to increase. This phenomenon is called *yielding* of the material, and the stress at which it occurs is the *yield strength* of the material.

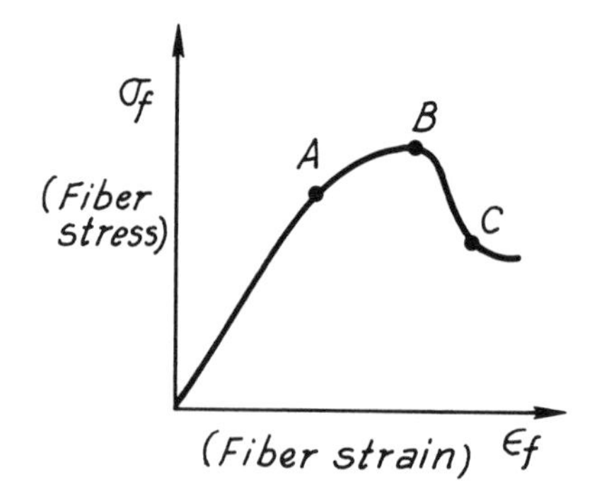

Figure 8.3. Approximate form of stress-strain curve for cardboard.

Now, the amount of fiber strain at the outer edges of a strip is

$$\epsilon_f = T/2\rho,$$

where T is the thickness of the strip and ρ is its radius of curvature. Also, the amount of strain a material can withstand before yielding is proportional to the yield strength and inversely proportional to the elastic modulus of the material. Therefore, we can understand the observation that strips of different materials can withstand different amounts of curvature before they yield; the higher the elastic modulus and the lower the yield strength, the less the curvature at yielding. The ratio of the unconfined compressive strength to the Young's modulus of a ductile material is a good estimate of the maximum fiber strain the material can withstand. The values of the ratios for metals is of the order of 0.5. For a sample of siltstone it was about 0.03; for a claystone, about 0.01; and for Yule marble, about 0.005. Thus, the amount of deflection a strip of rock can withstand before yielding is at least an order of magnitude less than that a strip of steel with the same dimensions can withstand.

Yield Strength.—The simplest rheological model that describes yielding is the *plastic*. According to this model, the shear stress, σ_s, on a body is less than or equal to the shear or yield strength, k, of the plastic:

$$|\sigma_s| \leq k. \tag{8.1}$$

The absolute value symbol is placed around the shear stress symbol in eq. (8.1) because the shear stress can be either positive or negative.

If the material is elastic-plastic, as defined by the Prandtl rheological model, the shear strain is proportional to the shear stress up to the yield strength (see Fig. 1.5). When the shear stress equals the yield strength, the material deforms permanently (Fig. 8.4A).

Strains within thin strips are largely normal strains, however, so that we need to generalize the rheological model. We can readily generalize it by means of Mohr's Circle of stress (Fig. 8.4B). There the yield strength is represented by horizontal dashed lines a distance k from the normal stress axis. The material yields when the state of stress acting on a body is such that the Mohr Circle representing that state is tangential to the dashed lines (Fig. 8.4B).

If we let p be the mean stress, that is, the distance to the center of the Mohr Circle, then the principal stresses, σ_1 and σ_3, are

$$|\sigma_1 - p| \leq k$$

and

$$|\sigma_3 - p| \leq k, \tag{8.2}$$

where

$$p = (\sigma_1 + \sigma_3)/2 = (\sigma_{xx} + \sigma_{yy})/2.$$

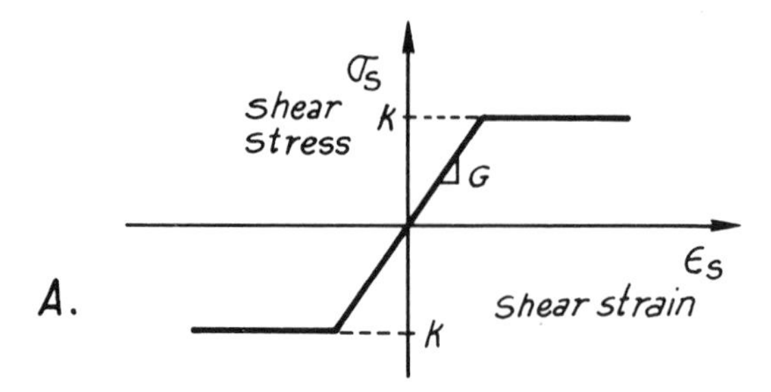

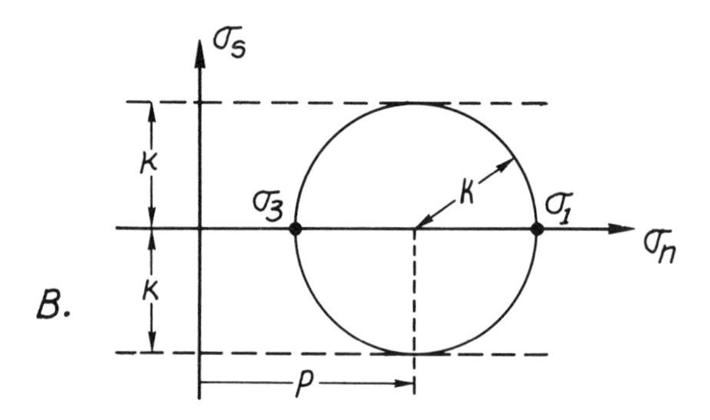

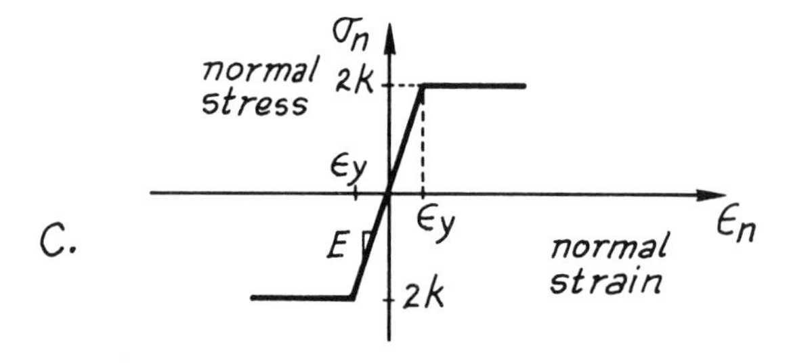

Figure 8.4. Stress-strain relations for elastic-plastic substance.
A. Relation between shear stress and shear strain.
B. Mohr Circle for plastic substance.
C. Relation between normal stress and normal strain.

If the plastic material is deformed under uniaxial compression or tension, as we usually assume in beam and plate theory, the yield criterion, eqs. (8.2), becomes:

Uniaxial Tension

$\sigma_3 = 0,$

$p = \sigma_1/2,$

$$\sigma_1 - \sigma_1/2 \leq k,$$

and

$$\sigma_1 \leq 2k. \tag{8.3}$$

Uniaxial Compression

$$\sigma_1 = 0,$$

$$|\sigma_3| \leq 2k. \tag{8.4}$$

These relations are illustrated in Fig. 8.4C. They show that the yield strength of the simple plastic material in uniaxial tension or compression is twice the yield strength in pure shear. The yield strain in uniaxial compression or tension, as is shown in Fig. 8.4C, is

$$|\epsilon_y| = 2k/E,$$

where E is Young's modulus of the elastic-plastic material.

Formation of Hinge and Straight Limbs.—Now we will consider how a hinge forms during buckling of a strip of elastic-plastic material. Let us assume, as before, that the fiber strains in the bent strip are proportional to the distance from the center of the strip (Fig. 8.5A). We will ignore the fiber strains caused by the axial load itself because they do not affect the general conclusions we will derive (see refs. *3* and *5*).

As the strip begins to bend, the fiber stresses increase in proportion to the strain until the outermost fibers of the strip reach the yield strain, ϵ_y. Then the outermost fibers yield and the stress pattern within the member is nonlinear, as is shown in Fig. 8.5B. As bending continues, the zone of yielding material deepens until it nearly reaches the neutral plane (Fig. 8.5C).

The initial form of a strip buckled by axial loads is sinusoidal,

$$v = C \sin\left(\frac{\pi x}{l}\right),$$

where l is the length of the strip and C is a constant, dependent upon the initial deflection and the axial load, as was explained in Chapter 3. The bending moment is also sinusoidal,

$$M = BI\frac{d^2v}{dx^2} = -CBI\left(\frac{\pi}{l}\right)^2 \sin\left(\frac{\pi x}{l}\right),$$

so that its magnitude is maximum at the midlength of the strip (Fig. 8.6B). The maximum fiber strains are directly proportional to the bending moment,

$$\epsilon_f = T/2\rho = \frac{T}{2}\frac{d^2v}{dx^2} = \frac{M}{2}\left(\frac{T}{BI}\right),$$

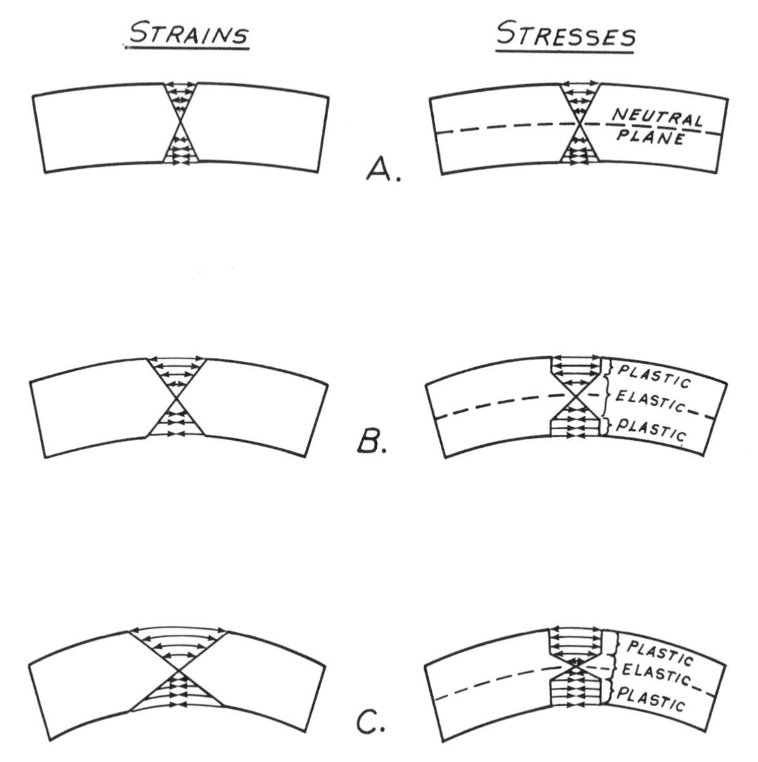

Figure 8.5. Progressive development of plastic zones at top and bottom of elastic-plastic member (after ref. *3*).

so that the maximum fiber strain is at midlength in the strip. It follows, therefore, that the elastic-plastic strip will first begin to yield at midlength and continue to yield there during buckling. This is why a hinge forms and remains during buckling of a strip of elastic-plastic material (Fig. 8.6C).

The change from elastic to plastic deformation in fold hinges is a possible explanation for the straight limbs of some folds, such as those in the Franciscan Formation (Figs. 1.1 and 8.1). The amount of bending of the limbs is a function of the elastic modulus and the yield strength of the member. If the yield strength is high relative to the elastic modulus of a member, such as a strip of aluminum or steel, it will bend markedly before it yields plastically. A thin strip of spring steel can be bent into an O-shape without deforming permanently. When released, the strip will snap back into its original, straight form.

If the yield strength of a member is low relative to its elastic modulus, as, for example, potter's clay, a member will bend very little before it yields plastically. If we bend even slightly a thin roll of potter's clay the roll will maintain the bent shape. The yield strength of potter's clay is so small relative to the elastic modulus

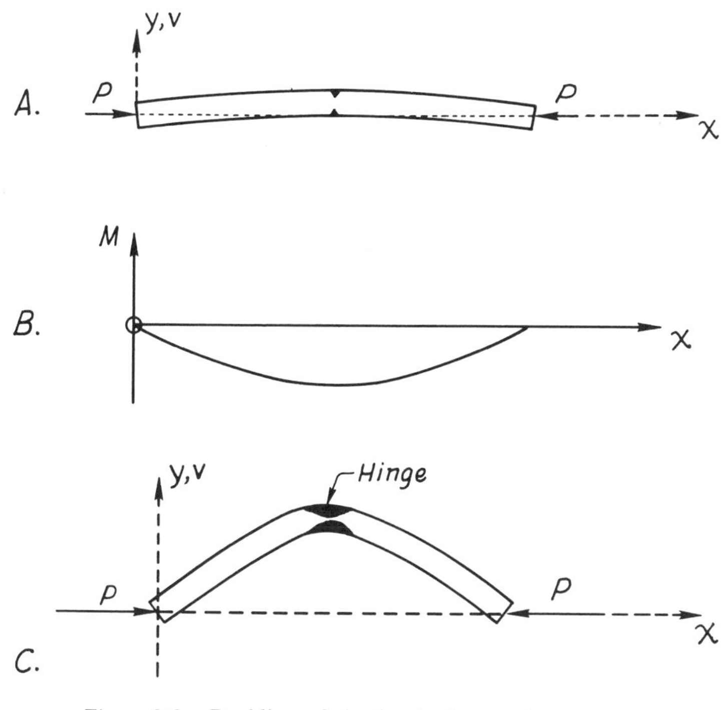

Figure 8.6. Buckling of elastic-plastic member.
A. Sinusoidal deflection of strip.
B. Distribution of bending moments within strip.
C. Yield zone in hinge of buckled strip.

that we would find it difficult to bend the thin roll so little that the roll could snap back into its original shape. When we release the load on the clay, the clay maintains straight limbs and a sharp crest, much as we see in the Franciscan Formation. The yielding behavior of folded rock probably is somewhere between that of steel and that of potter's clay.

Plastic instability

As an initially deflected, elastic-plastic strip is further deflected by axial loads, plastic zones begin to form along the top and bottom of the strip at midlength, and continue to grow as deflection continues. If the strip had a large yield strength and a low elastic modulus, the amplitude, δ, of the wave in the strip would increase indefinitely as the axial load approached the critical load, P_1, of the strip, as is shown by the curve *ABC* in Fig. 8.7. If the yield strength were relatively low, however, plastic zones would begin to form and grow at the crest of the bent strip. The bending resistance of the strip would approach a limiting value, equal to the resis-

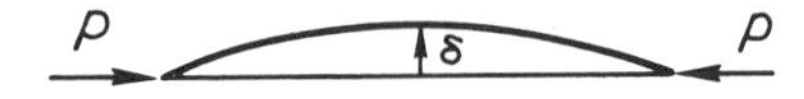

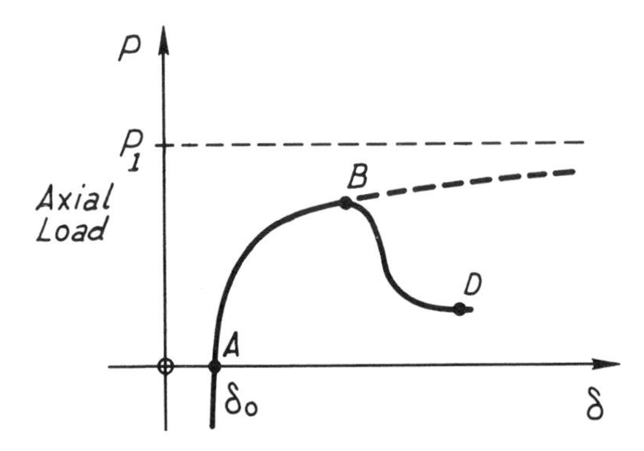

Figure 8.7. Load-deflection curve for initially deflected, elastic-plastic member. Member becomes unstable when axial load reaches value corresponding with point *B*.

tance of the strip when the plastic zones have spread from the edges to the neutral plane of the strip. Then the resisting moment remains constant as the strip continues to deflect. The driving moment, however, continues to increase with increasing deflection, if the axial load is maintained, because the driving moment is equal to the product of the axial load and the deflection.

We can imagine, therefore, that an elastic-plastic strip supporting axial loads becomes *unstable* when the deflection reaches a critical value. The instability can be understood in terms of Fig. 8.7 (*3*). For segment *A-B* of the load-deflection curve, an increase in axial load is required to produce an increase in deflection. The deflecting strip is thus *stable* for this segment of the curve. When the axial load reaches the value corresponding to point *B* on the curve, however, the strip is *unstable*, because further deflection will occur with no further increase in load. Indeed, for deflection to continue under equilibrium conditions, without acceleration, the axial load would have to decrease with increasing deflection, as is shown by curve *B-D* in Fig. 8.7. For this reason, the onset of instability of a strip such as cardboard is expressed as a sudden failure or collapse of the strip (*3*). The limbs of the strip snap back and straighten out, and a narrow hinge forms.

Buckling of viscous-plastic, pseudoplastic, and nonlinear-elastic strips

Adoption of an elastic-plastic model to describe the rheological properties of Franciscan chert would seem to ignore compelling evidence that the chert was soft at the time of folding. As we have shown before, however, the behavior of elastic and viscous materials are analogous, so that we commonly can select either model of material behavior for a problem involving either type of behavior. We will examine the behavior of a viscous-plastic strip thoroughly enough to confirm this conclusion.

The normal-stress, normal-strain relations for the viscous-plastic material are, for uniaxial compression and tension [Fig. 8.8 and eqs. (7.98)],

$$\sigma_n = 4\eta\dot{\epsilon}_n; \qquad \sigma_n \leq 2k. \tag{8.5}$$

When the normal or fiber stress in a viscous-plastic strip reaches the yield strength, $2k$, of the material, the strip deforms plastically. Therefore, if we replace the term "strains" by the term "strain rates" in Fig. 8.5, the diagrams illustrate the development of plastic zones in hinges of viscous-plastic members as well as elastic-plastic members. Thus, development of plastic hinges in viscous-plastic and elastic-plastic members is similar.

The primary difference between viscous-plastic and elastic-plastic buckling of a member is related to the difference between conditions under which plastic deformation takes place. Plastic deformation occurs in the hinges of an elastic-plastic member when the *amount* of deflection of the member reaches a critical value. It occurs in the hinges of a viscous-plastic member when the *rate* of deflection reaches a critical value.

An initially deflected, elastic-plastic member deflects farther by an amount that is related to the magnitude of the axial load, as long as the axial load is below a certain, critical value (Fig. 8.7). Increasing loads are required for increasing deflections. A viscous-elastic member, on the other hand, deflects regardless of the magnitude of the load, and even if the load is constant, as long as the load is compressive. Thus, if the axial load were sufficiently small, the member would deflect viscously,

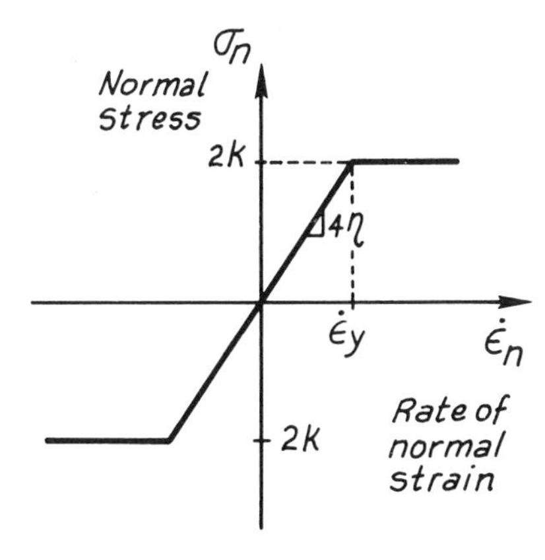

Figure 8.8. Stress, strain-rate relation for viscous-plastic substance.

without plastic deformation, producing a form similar to that of the *elastica*, described in Chapter 4. The driving moment and the fiber stresses in the member are proportional to the axial load and to the deflection, so that the fiber stresses in the outer edges of the viscous-plastic member may equal the yield strength of the material after the member has deflected into a high amplitude.

The primary difference between buckling of elastic-plastic and viscous-plastic members, therefore, is that there is a critical axial load required to cause plastic deformation within elastic-plastic members, whereas there are different amplitudes of deflection required to cause plastic deformation within viscous-plastic members for different values of applied axial load. I know of no way to distinguish between folds formed in rock that behaved as an elastic-plastic substance and folds that formed in rock that behaved as a viscous-plastic substance. The final forms apparently would be the same.

Another type of rheological behavior that could account for the sharp hinges and straight limbs of the folds in the Franciscan chert is *pseudoplastic* behavior. A pseudoplastic material is one in which the rate of shear strain, $\dot{\epsilon}_s$, is exponentially related to the shear stress, σ_s. Thus,

$$\dot{\epsilon}_s = \left(\frac{\sigma_s}{A}\right)^n, \tag{8.6}$$

where A and n are constants. If the exponent, n, is equal to one, eq. (8.6) is the Newtonian viscous model, in which A is the coefficient of viscosity (Fig. 8.9). As

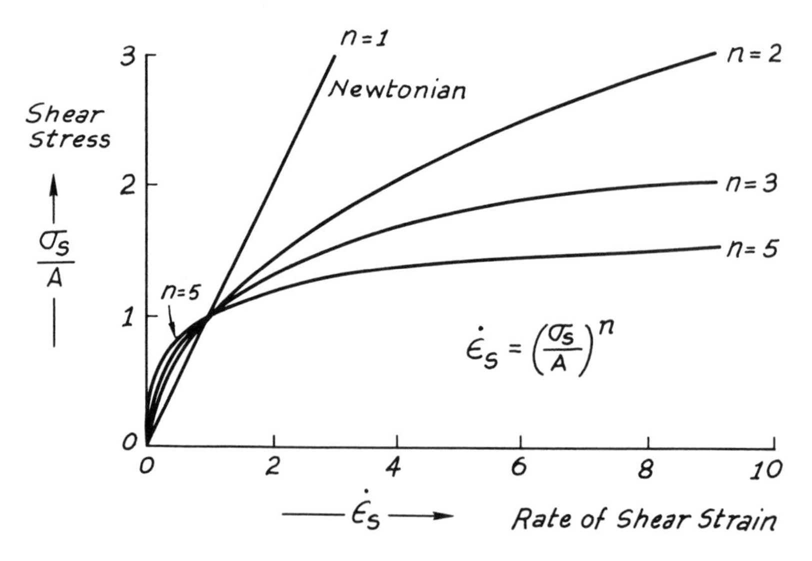

Figure 8.9. Relations between shear stress and rate of shear strain for various pseudoplastic materials. Newtonian viscous substance is special case of pseudoplastic, in which the exponent, n, is one.

we will show in Chapter 14, eq. (8.6) describes quite well the flow of glacial ice if n is approximately equal to four.

The psuedoplastic model for uniaxial compression or tension is

$$\dot{\epsilon}_f = \left(\frac{\sigma_f}{c}\right)^n, \tag{8.7}$$

where

$$C = 2A(2^{1/n}),$$

as we could show by means of Mohr's Circles of stress and strain-rate. Therefore, the resisting moment of a strip composed of pseudoplastic material is

$$M_R = -2\int_0^{T/2} \sigma_f y \, dy.$$

However, the fiber-strain rate is related to the radius of curvature by the relation

$$\dot{\epsilon}_f = \frac{d}{dt}\left(\frac{y}{\rho}\right) = \left(\frac{\sigma_f}{C}\right)^n,$$

so that

$$M_R = -2C\left[\frac{d}{dt}\left(\frac{1}{\rho}\right)\right]^{1/n} \int_0^{T/2} y^{[(1+n)/n]} \, dy,$$

or

$$M_R = -2C\left[\frac{d}{dt}\left(\frac{1}{\rho}\right)\right]^{1/n} \left[\frac{n(T/2)^{[(1+2n)/n]}}{1+2n}\right]. \tag{8.8}$$

As before, T is the thickness, ρ is the radius of curvature, and y is the distance from the neutral plane of the strip [see Chapter 2].

In order to determine the deflection curve of a strip of pseudoplastic material, we equate the resisting moment to the driving moment. The driving moment is the axial load times the deflection of the member, $v_0 + v_1$, where v_0 is the initial deflection and v_1 is the deflection caused by the axial load. Thus,

$$P(v_0 + v_1) + 2C\left[\frac{d}{dt}\left(\frac{1}{\rho}\right)\right]^{1/n} \left[\frac{n(T/2)^{[(1+2n)/n]}}{1+2n}\right] = 0. \tag{8.9}$$

The radius of curvature of the strip is, to a first approximation,

$$\rho = \frac{1}{|(d^2v_1/dx^2)|},$$

so that eq. (8.9) can be rewritten as a differential equation,

$$\left[\frac{\partial}{\partial t}\left(\frac{\partial^2 v_1}{\partial x^2}\right)\right]^{1/n} + \frac{P}{2C}\left[\frac{1+2n}{n(T/2)^{[(1+2n)/n]}}\right](v_0 + v_1) = 0. \tag{8.10}$$

Equation (8.10), the differential equation for small deflections of a pseudoplastic strip, is nonlinear; therefore, it normally cannot be solved exactly. If n is one we can solve it, of course, because then it is equivalent to the equation for the Newtonian member. It turns out that if n is either two or three, the solution is an elliptic integral, similar to the solution for the *elastica* (Chapter 4). If n is four or greater, there appears to be no solution (see ref. *2*). In general, then, we would solve eq. (8.10) numerically, by approximate methods, in order to determine the form of the deflection curve. A computer would greatly aid with the calculations.

Actually, we do not need to solve the equation in order to visualize the deflection forms, however. Let us rewrite eq. (8.10) in the following form:

$$\frac{\partial}{\partial t}\left(\frac{\partial^2 v_1}{\partial x^2}\right) = C_0[P(v_0 + v_1)]^n, \tag{8.11}$$

where C_0 is a constant used to get rid of extraneous parameters. Assume that n is greater than one. With the equation written in this form, it is clear that as the strip deflects, that is, as $v_0 + v_1$ becomes larger, the rate of change of curvature

$$\frac{\partial}{\partial t}\left(\frac{\partial^2 v_1}{\partial x^2}\right)$$

becomes much larger because it is proportional to some power of the deflection. The larger the exponent, n, the more rapidly the curvature changes with increasing deflections. The maximum deflection is at midlength in the strip, so that the curvature increases more rapidly there than it does elsewhere in the member. We can readily imagine, then, that if the exponent is large, the form of the deflecting member will be similar to the "V"-form of a buckled elastic-plastic or viscous-plastic member (Fig. 8.6C). As the exponent becomes small, approaching unity, the deflection form approaches the sinusoidal form of an elastic or a viscous member. Thus the forms of viscous and viscous-plastic strips are end members of the continuum of forms described by the pseudoplastic model.

The elastic analog of the pseudoplastic model can be studied more easily than the pseudoplastic itself, because the elastic analog produces forms that are independent of time. The exponential elastic model is

$$\epsilon_f = \left(\frac{\sigma_f}{D}\right)^n.$$

The resisting bending moment of the material is similar to that of a pseudoplastic, given by eq. (8.8),

$$M_R = -2D\left(\frac{1}{\rho}\right)^{1/n}\left[\frac{n(T/2)^{(1/n+2)}}{1+2n}\right].$$

The driving moment of an initially straight strip is

$$M_D = M_R = Pv,$$

and the curvature for large deflections is

$$\frac{1}{\rho} = \frac{|d^2v/dx^2|}{[1 + (dv/dx)^2]^{3/2}} = \frac{|du/dx|}{[1 + u^2]^{3/2}},$$

where

$$u = dv/dx,$$

so that the differential equation for the deflection of a strip with nonlinear elastic properties is

$$Pv + 2D\left[\frac{|du/dx|}{(1 + u^2)^{3/2}}\right]^{1/n}\left[\frac{n(T/2)^{(1/n+2)}}{1 + 2n}\right] = 0.$$

Elmont Honea, a graduate student at Stanford, solved this latter equation by putting it in the following dimensionless form:

$$|du/dx^*| = (1 + u^2)^{3/2}v^{*n}K,$$

where

$$dv/dx = u,$$

$$x^* = x/S,$$

$$v^* = v/S,$$

and

$$K = \pm\frac{2S}{T}\left[\frac{-PS(+ 2n)}{2Dn(T/2)^2}\right]^n.$$

Here v^* and x^* are dimensionless displacement and distance variables, respectively, and S is the arc length of the strip.

Honea integrated the equation by the Runge-Kutta method of approximate integration (*2*, p. 482), using a computer to perform the calculations. Figure 8.10 shows the forms assumed by nonlinear elastic strips with three different exponents, n, and subjected to different axial loads. Figure 8.10A shows a linear elastic strip, for which $n = 1$. The forms are examples of the *elastica,* which we discussed in conjunction with our examination of ptygmatic folds, in Chapter 4. The curves show the progressive development of the *elastica* as the axial load was increased by increments of 0.2, corresponding with $K = 0.1(0.2)$, $0.3(0.2)$, and $0.5(0.2)$.

Figure 8.10B shows forms assumed by a strip of nonlinear elastic material for which the exponent, n, was 10. The different curves represent the progressive development of forms as the axial load increased by increments of 0.2, as in Fig. 8.10A, but the constant, K, had values of $K = 0.1(0.2)^{10}, 0.3(0.2)^{10} \cdots 0.9(0.2)^{10}$. The curve with the lowest amplitude is similar in form to the *elastica* curve of lowest amplitude, shown in Fig. 8.10A. At high amplitudes, however, the curves in Fig. 8.10B for the nonlinear material have straighter limbs and more curvature concentrated

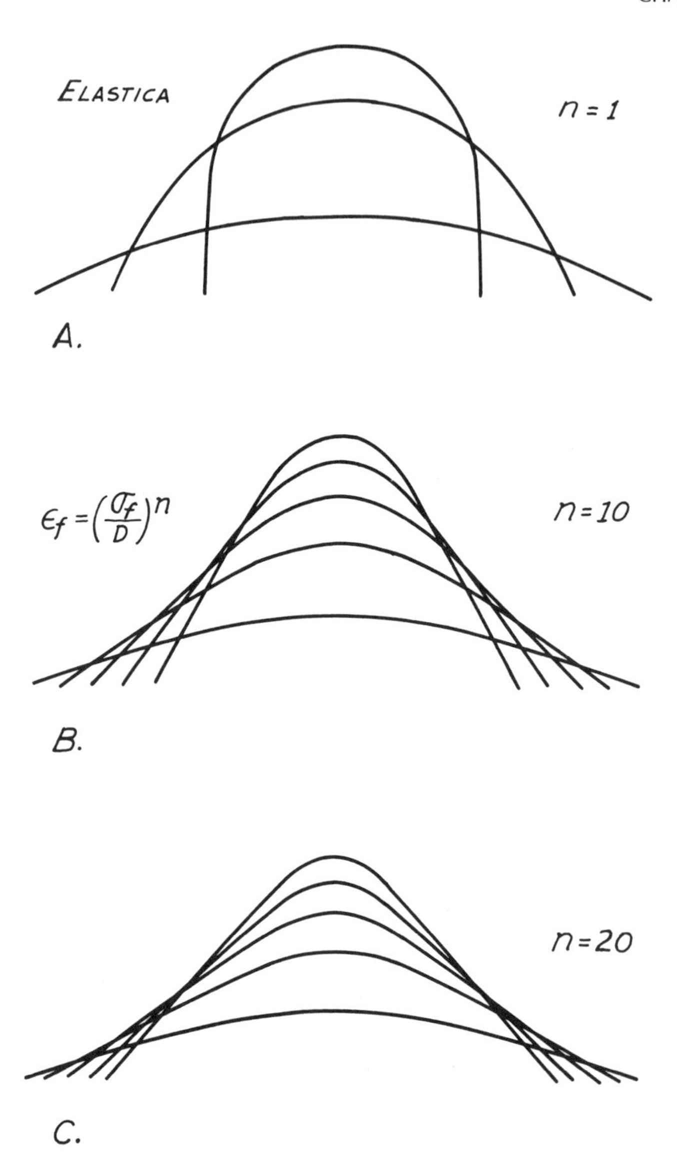

Figure 8.10. Deflection forms of nonlinear elastic strips. As the exponent, *n*, increases, the forms approach that of a buckled plastic strip.

in the crests. The differences in forms become more obvious as the exponent is increased. For example, Fig. 8.10C shows deflection curves for a nonlinear elastic material for which the exponent, *n*, was 20. These curves have markedly straighter limbs and more sharply bent crests.

These deflection forms for different nonlinear elastic materials confirm the conclusion stated earlier that buckled nonlinear elastic strips have forms that are intermediate between those for a linear elastic material (Fig. 8.10A) and those for an elastic-plastic material (Fig. 8.6C). Forms of buckled elastic and plastic strips, therefore, are end members of a spectrum of forms described by nonlinear elastic strips. The same is true for viscous, pseudoplastic, and plastic strips.

Reexamination of Franciscan folds

We have restricted our attention to details of the forms of the sharp hinge and the relatively straight limbs of a fold. Seemingly we can explain the forms in terms of several rheological models. However, if we concentrate overmuch on these details of forms we risk a myopic view of the folds in the Franciscan Formation. When we step farther from the outcrop containing the single fold shown in Fig. 8.2 we see that the fold is merely part of a broader pattern. The pattern is as characteristic of the folded cherts as are the sharp crests and straight limbs (*3*). When we stand about twenty feet from the outcrop we begin to see zones or *domains* (*6*) of parallel, straight layers alternating with domains of layers with a different orientation (Fig. 8.1). If we stand about 50 feet from the outcrop, as shown in Fig. 8.11, the domains dominate the pattern and details of the hinges become invisible. The hinges form the boundaries between the domains.

When we recognize the folded layer as a local expression of a broader pattern of domains we realize that the form of the individual layers must be determined by factors other than their rheologic properties. The layers obviously are not isolated, as we have assumed thus far. Deflection of any layer induces loads on surfaces of neighboring layers, causing regimentation of the layer and its neighbors (*3*).

In order to understand the fold patterns and domains within the Franciscan cherts we need to be able to define boundary conditions of beds containing the domains. Fortunately, Ellen was able to find a quarry where folded cherts of the Franciscan Formation are bounded above and below by relatively undeformed rocks. The quarry is near the north end of the Golden Gate Bridge at San Francisco, California (*3*). Ellen mapped the area around the quarry and showed that the folded sequence of cherts is sandwiched between massive greenstone below and relatively massive chert above. Thus, he postulated that the interbedded chert and shale of the folded sequence were bounded above and below by units that behaved essentially as rigid boundaries during folding, as we postulated for the Carmel Formation, in Chapter 6.

The patterns of folding in the interbedded chert and shale exposed in the quarry are of special interest, therefore, because we can make a good guess about boundary conditions during folding. The patterns are complicated; we will follow only part of Ellen's analysis of them. We can readily analyze a fold pattern within interbedded chert and shale near the overlying, massive cherts. Ellen (1970) has analyzed

Figure 8.11. Domains of similarly oriented chert layers of the Franciscan Formation at the north end of the Golden Gate Bridge, San Francisco, California. Linear boundaries between domains are hinge lines of folds. Stephen Ellen and Bruce Clark are examining the folds.

all the patterns by the methods we will follow and I refer you to his dissertation for the full analysis (*3*).

The folds near the upper boundary of the interbedded chert and shale sequence are shown in Fig. 8.12. Boundaries of some of the chert beds have been traced on the photograph. The general attitude of nonfolded cherts is shown on the right-

Figure 8.12. Folds in interbedded chert and shale of the Franciscan Formation, exposed in quarry near north end of Golden Gate Bridge. Nonfolded strata on right. Box folds of low amplitude on right. Stephen Ellen sitting on talus slope in foreground.

hand side of Fig. 8.12. The beds dip about twenty degrees toward the right. Near the center of the photograph in Fig. 8.12 the dips of the chert beds abruptly reverse over short distances, producing chevron-like folds. The cross-section shown in Fig. 8.12 is nearly at right angles to the axes of the folds. If the cross-section were oblique to the axes, we would see domains such as those shown in Fig. 8.11. The domain pattern that is so common in Franciscan cherts is simply a result of oblique cross-sections of folds similar to those shown in Fig. 8.12.

Formation of hinge lines in layered materials

One feature of the pattern shown in Fig. 8.12 invites special attention. The hinges of chert beds form linear trends on the face of the outcrop. These hinge lines are quite straight and all are inclined at angles ranging from about 25 degrees to 35 degrees with respect to the nonfolded beds shown on the right-hand side of Fig. 8.12. These hinge lines form the domain boundaries that are clearly visible in Figs. 8.1 and 8.11. Thus, whereas the most prominent features at close examination are the sharp crests and straight limbs of the individual folded layers, the most prominent features from a distance are lines of fold hinges, cutting across lines formed by nonfolded beds. Let us focus now on determining the significance of the hinge lines.

Experimental Kink Bands.—During the course of Ellen's investigation of the origin of the folds in the cherts of the Franciscan Formation, he experimented with multilayers with various kinds of physical properties (*3*). His field mapping indicated that the folds in the quarry (Fig. 8.12) formed under boundary conditions of essentially rigid boundaries above and below a packet of folded chert and shale beds. His experiments, therefore, consisted of shortening, edgewise, multilayers bounded above and below by rigid boundaries.

The soft layers in his multilayers were potter's clay. The stiff layers were varied from corrugated cardboard to rubber to another cardboard, used for posters. The corrugated cardboard was about 2 mm thick and was quite compressible. In buckling of isolated strips it yielded at sharp hinges. The other cardboard was about one mm thick, was less compressible, and could be bent into a long, smooth arch before it failed suddenly at a hinge, under axial loading. The rubber strips were about one mm thick and were about as flexible as a wide rubber band or a strip of an automobile inner tube.

We will examine a few of Ellen's experiments because they help us to understand folds in the Franciscan Formation and, it turns out, folds in many other places as well (*3*).

Figure 8.13A shows a stack of black rubber strips interlayered with white potter's clay. The multilayer was about 20 cm by 25 cm in dimensions. The rubber strips were about two cm wide, equal to the width of the plexiglass tank containing them. The clay layers were not as wide as the rubber strips, allowing changes in volume as the multilayer was compressed. Squeezing of the clay between the rubber strips was apparent in later stages of folding (Figs. 8.13E and 8.13F). The multilayer was compressed by means of a piston advancing from left to right in Fig. 8.13.

Figure 8.13B shows deflection and separation of layers near the right-hand edge of the multilayer. The average shortening was 0.015. As shortening continued, to a value of 0.023, a second, complementary deflection occurred in the upper-central part of the multilayer (Fig. 8.13C). Another small wave began to form near the piston, at the upper left. When shortening had reached 0.031, another complementary deflection occurred in the lower-left part of the multilayer, near the piston (Fig. 8.13D). As the multilayer was shortened further, to 0.08 (Fig. 8.13E), another complementary deflection occurred in the lower-left part of the multilayer.

If we connect the points of origin of each of the buckles, as shown in Figs. 8.13D and 8.13E, the resultant lines form an interesting pattern. The pattern formed when we include the buckle near the piston, on the lower left in Fig. 8.13D, is asymmetric, whereas the pattern formed when we include the buckle formed slightly later, Fig. 8.13E, is symmetric. We will discuss this pattern in more detail later, when we discuss Biot's theory. Here it is sufficient to observe that a symmetric pattern seems to be fundamental under the conditions of the experiment. It is worth noting also, here, that the buckles formed successively, from right to left, as though

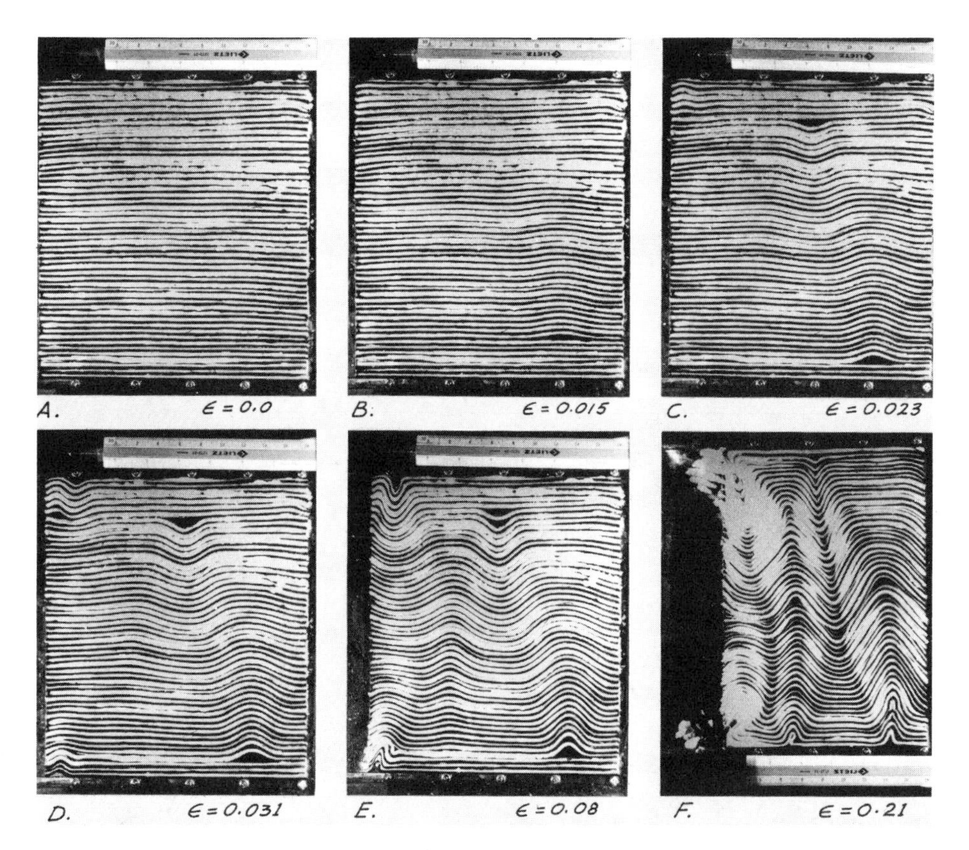

Figure 8.13. Buckling of experimental multilayer consisting of black rubber strips and soft, white clay strips. Multilayer confined rigidly at top, bottom, and right-hand side. Driving piston on left-hand side. Shortening or multilayer indicated at base of photograph. (Photos by Stephen Ellen.)

perhaps the formation of one buckle caused another, complementary buckle to initiate on the opposite side of the multilayer.

Figure 8.13F shows the pattern developed in the multilayer when average axial shortening reached a value of 0.21. Several features of the pattern are of interest, including the straight limbs of the folds, the box-like folds near the boundaries, and the domains. The pattern is similar, in some respects, to the patterns in the Franciscan Formation (Figs. 8.1, 8.2, and 8.12). We can envision certain lines and domains in the experimental folds if we use a great deal of imagination. However, the patterns in the experiments and the patterns in the field have more differences than similarities.

There are at least two aspects of the experiment that we can change in order to improve the correspondence between the patterns. We can change experimental materials and we can change the way we deform the materials. First we will change the method of deforming the multilayer of rubber and clay. The folds shown in Fig. 8.12 are markedly asymmetric so that we should guess that shear played a certain role in their development. Ellen applied shear to experimental multilayers by means of the apparatus shown in Fig. 8.14. The apparatus consists of four arms,

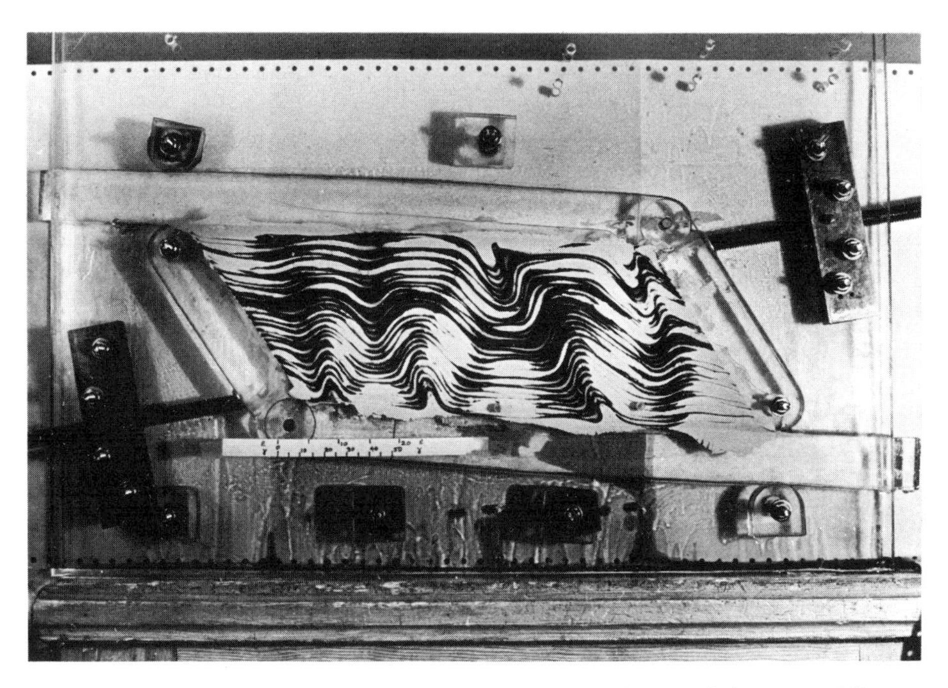

Figure 8.14. Apparatus used by Ellen to simultaneously shear and shorten multilayers. One end of each short plexiglass arm rotates around a point. The other end and each long arm slide as threaded rods are advanced. Multilayer consists of black rubber strips and white, soft clay strips. Shear strain (γ, gamma) was 50 percent and shortening (ϵ, epsilon) was 19 percent at time photograph was taken. (Photo by Stephen Ellen.)

forming a parallelogram, mounted between two sheets of plexiglass, forming sides. Two of the arms rotate about fixed axes, consisting of bolts penetrating the two arms and the two sides of plexiglass (Fig. 8.14). The two long arms slide horizontally and rotate slightly as the two short arms are rotated by threaded rods. The multilayer is subjected to both shortening and shear in the apparatus. The relative amounts of shear and shortening can be adjusted by changing the lengths of the shorter arms; the shorter the short arms, the greater the shear relative to the shortening. The amount of shear is indicated, in percent, by the scale labeled gamma (γ) on the base of the apparatus, and the amount of average normal strain, parallel to the original layers, is designated by epsilon (ϵ).

Figures 8.15 and 8.14 show several stages of folding of black rubber strips and white clay strips during one experiment. Initial waves are visible in the undeformed multilayer, Fig. 8.15A. As the multilayer was shortened, some of the initial waves were amplified and folds developed progressively from right to left in the apparatus (Figs. 8.15A through 8.15D). In spite of the shear strain, which was more than twice the longitudinal strain, the folds were essentially symmetric. Only after amplitudes were considerable did the fold pattern faithfully reflect the shear. For example, the second anticline from the right-hand side of the apparatus became markedly asymmetric between the time the photographs shown in Figs. 8.15D and 8.14 were taken. The right-hand limb of the fold straightened and the left-hand limb overturned near the top and base of the anticline. In addition, the hinge of the fold rotated toward the left. The two anticlines on the left in Fig. 8.14 formed late; therefore, they are nearly symmetric.

The resemblance between the overturned anticlines on the right-hand side of Fig. 8.14 and the folds in the Franciscan Formation, Fig. 8.12, is greater than the resemblance between the folds shown in Fig. 8.13 and those in the Franciscan. One obvious difference, however, is that concentric folds such as those shown in Figs. 8.14 and 8.15 are rarely seen in the Franciscan. This naturally prompts us to try other experimental materials, materials that behave much the way we envisioned when we were studying details of sharp fold hinges and straight limbs, earlier in this chapter. This is what Ellen did. He selected cardboard for the stiff layers and clay for the soft layers. Interbedded strips of white cardboard and red clay are shown in Fig. 8.16. The multilayer was deformed in the apparatus shown in Fig. 8.14.

The forms of the initial deflections of the layers are shown in Fig. 8.16A. There was an irregular pattern of low waves of various amplitudes and wavelengths in the layers before the layers were shortened and sheared. Thus, we might guess that the fold pattern would be roughly sinusoidal, amplifying some of the initial waves, such as we saw in the experiment with rubber strips and clay (Fig. 8.15). This is not what happened at all, however, as is shown in Figs. 8.16B through 8.16D. Instead of a sinusoidal pattern, a kink-like pattern developed in the layers. The long limbs of the kink folds rotated slightly, the amount of rotation being essentially equal to the rotation of the short arms of the apparatus. The short limbs of the kink folds rotated markedly by the time the layers had shortened by about six percent (Fig. 8.16D).

Ellen demonstrated that the pattern shown in Fig. 8.16 is fundamental and not merely a product of a peculiarity of the experiment. He performed another experiment in which a marked sinusoidal wave was induced initially in the multilayer before the multilayer was deformed under the same conditions as those illustrated in Fig. 8.16. The resulting fold pattern was the same as that shown in Fig. 8.16D. The initial wave did not grow, but, instead, was erased and then replaced by a kink band similar to those shown in Fig. 8.16.

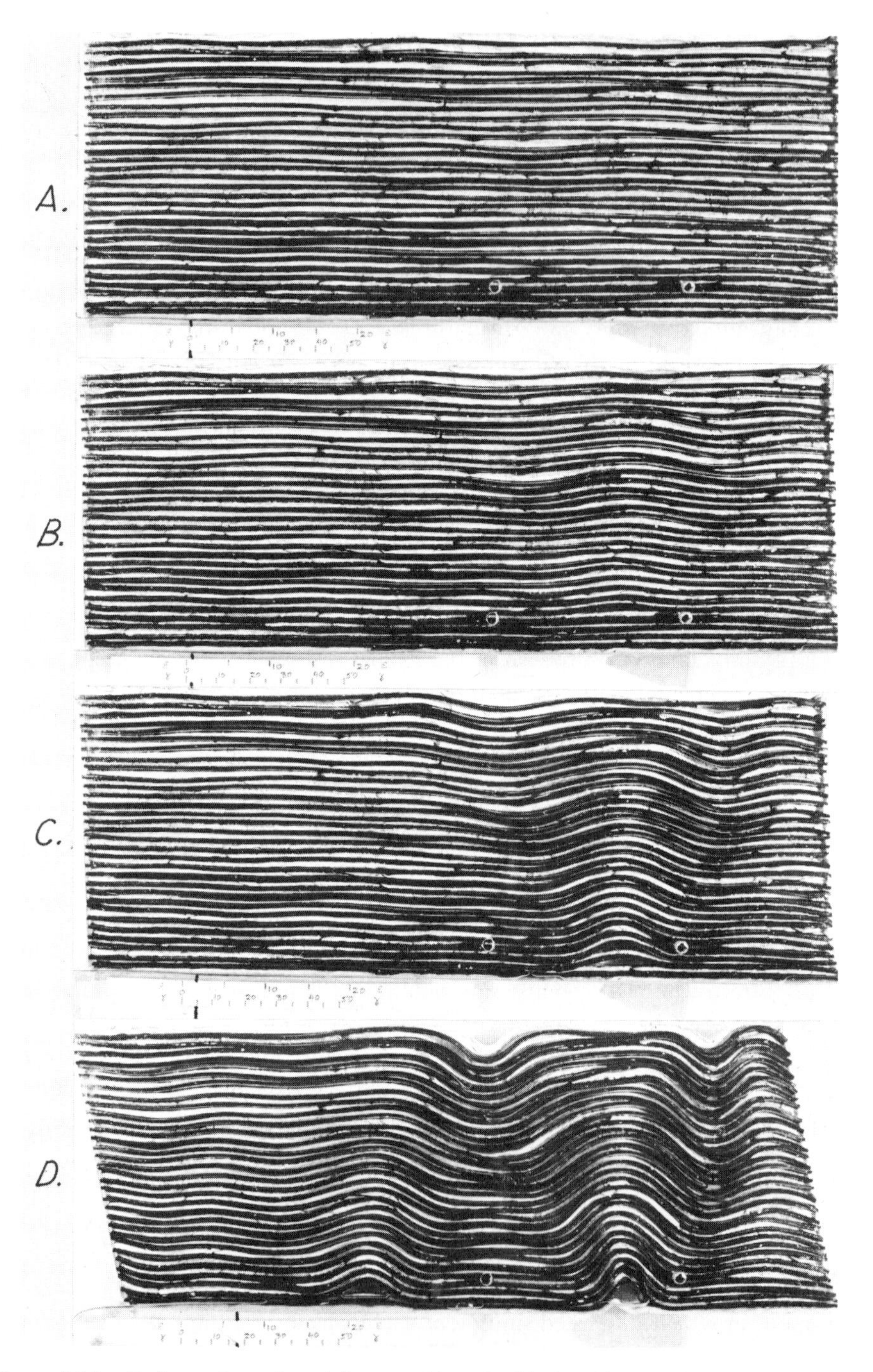

Figure 8.15. Deformation of multilayer of interbedded rubber and clay strips in apparatus shown in Fig. 8.14. Average shear strain (γ, gamma) and shortening (ϵ, epsilon) indicated on scale at bottom of each photograph. (Photos by Stephen Ellen.)

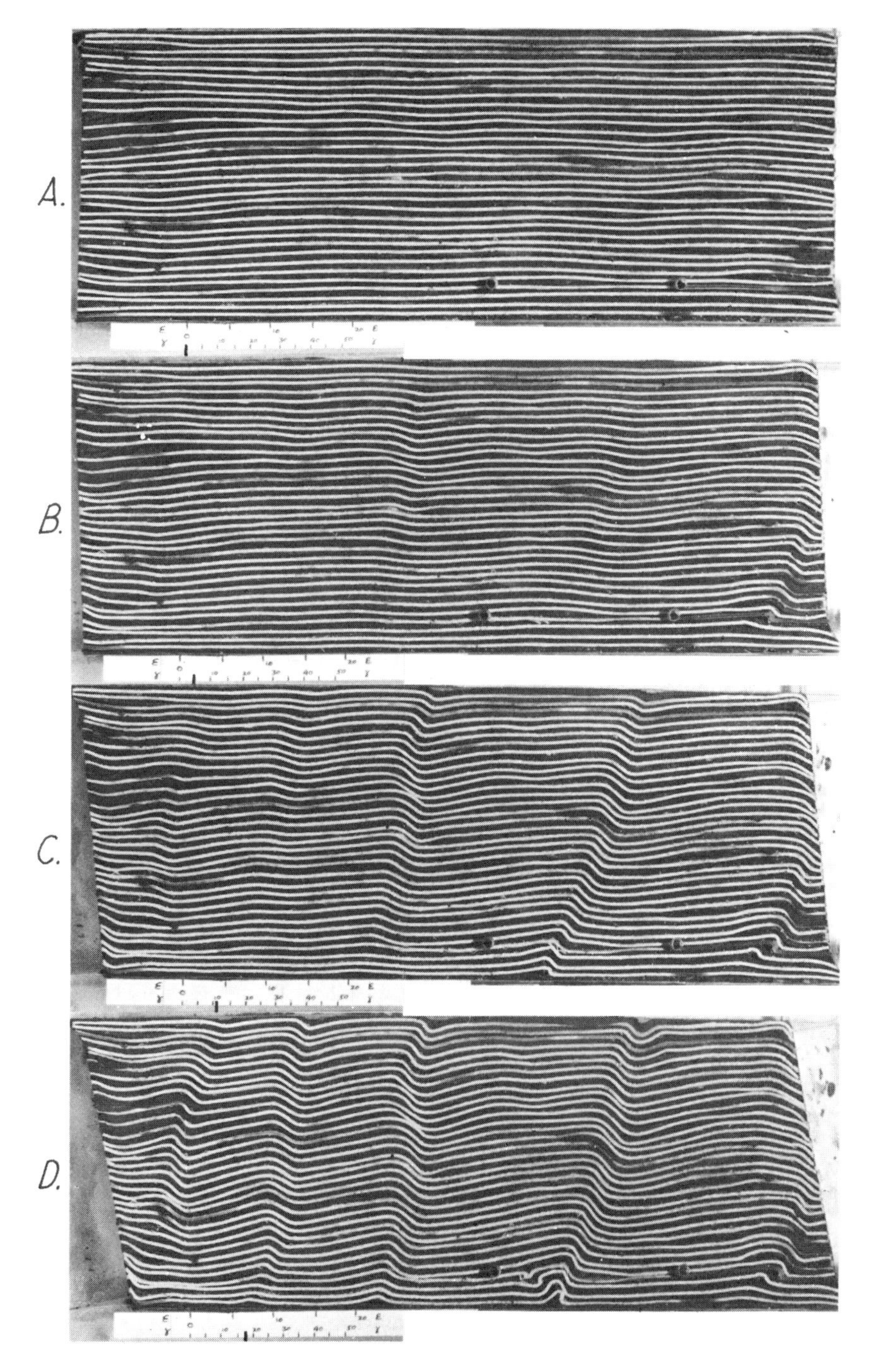

Figure 8.16. Deformation of multilayer of white cardboard strips and dark, soft clay strips in apparatus shown in Fig. 8.14. Shortening (ϵ) and shear strain (γ) indicated at bottom of each photograph (in percent). (Photos by Stephen Ellen.)

The pattern of kink bands shown in Fig. 8.16 confused Ellen and me when we saw it develop. We were perplexed by the sense of motion of the two limbs of each of the kinks. It seemed to suggest shear in the opposite sense to that we knew was applied to the multilayer. If we did not know better, we might suppose that the kink pattern was a result of clockwise shear on the top and bottom of the multilayer rather than the counterclockwise shear that actually was applied. We soon realized, however, that the kinking phenomenon is not related to the shear, but, instead, to axial loading of the layers. That is, kinking is a buckling phenomenon, not a shear phenomenon. The reason the kink bands had the orientation shown in Fig. 8.16 is that the ends of the multilayer were rotated during the experiment. Thus, the boundary conditions of the experiment were much as is shown in Fig. 8.17C.

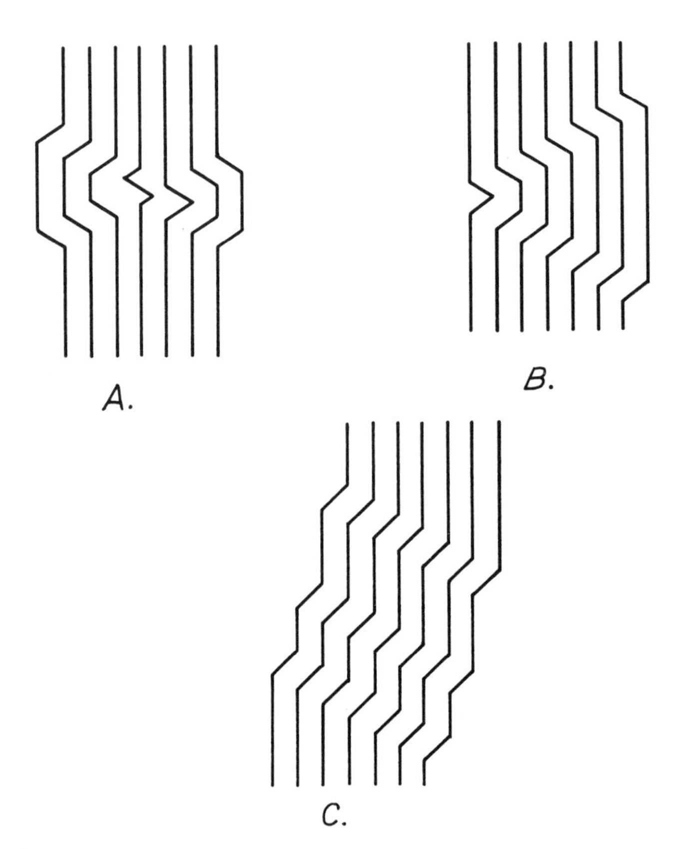

Figure 8.17. Three modes of kinking of a multilayer.

A. Intersecting kink bands. Ends of multilayer do not move laterally.

B. Complementary kink bands. Ends do not move laterally.

C. Kink bands with same orientation. Ends of multilayer must move laterally relative to each other.

Later on in the experiment, after the amplitudes of the kinks had increased, the shear began to become important, straightening and lengthening the short limbs of the kinks shown in Fig. 8.16D and steepening and shortening the long limbs of the kinks.

The development of kink bands with a single orientation, such as those shown in Fig. 8.16, always requires one end of the multilayer to move laterally, normal to bedding, relative to the other end of the multilayer, much as is illustrated in Fig. 8.17C. Conjugate sets of kink bands will form as shown in Figs. 8.17A and 8.17B, when the ends of the multilayer cannot move laterally.

Perhaps the most striking feature of the folds shown in Fig. 8.16 is the appearance of domains, bounded by lines connecting fold hinges. In this respect the pattern that developed in the experimental multilayer, Fig. 8.16, resembles the pattern in the natural multilayer in the Franciscan Formation, Fig. 8.12. A major difference is that the natural folds are more tightly compressed than the experimental folds. Ellen did not attempt to produce experimentally tightly compressed kink folds because they are clearly illustrated in experiments performed by L. E. Weiss (*6*) with decks of cards. I refer you to figs. D, E, and F of plate 17, ref. *6*, for examples of kinks that resemble those in the Franciscan Formation, Fig. 8.12. Essentially, the experiments by Weiss show that kinks such as those in the Franciscan might be a result of kinking followed by shear. The kinking produces forms such as those shown in Figs. 8.16 and 8.18A and the shear causes the short limbs to overturn as is illustrated in Fig. 8.18B. The folds in the Franciscan cherts are strikingly similar to the sheared kinks, shown in Fig. 8.18B.

Analysis of Orientations of Kink Bands.—We have seen that there are certain similarities and certain differences between the buckling of multilayers consisting of interlayered stiff rubber strips and soft clay strips and multilayers consisting of interlayered stiff cardboard strips and soft clay strips. Even though the multilayers were loaded in the same way in the same apparatus, the fold patterns are different. In experiments with rubber strips, the fold forms are rounded (Fig. 8.15D), whereas in experiments with cardboard strips, the fold forms are kinked (8.16D). A similarity is that in both experiments linear features were developed. The linear features crossed the bedding planes obliquely. The linear feature in the case of the cardboard strips was an alignment of hinges. The hinge lines were inclined to undeformed bedding planes at angles ranging between 60 and 70 degrees. The linear features in the case of the rubber strips were lines drawn from the cores of synclines to cores of adjacent anticlines. These lines were inclined to original bedding planes at angles of about 60 degrees (see Fig. 8.15D). Why should these angles be approximately the same? We will discuss this observation in more detail later, after we have discussed a theoretical analysis of buckling of multilayers. It is interesting to note here, however, that Patterson and Weiss (*4*) found that the angles between hinge lines and bedding in experimentally kinked phyllite are about

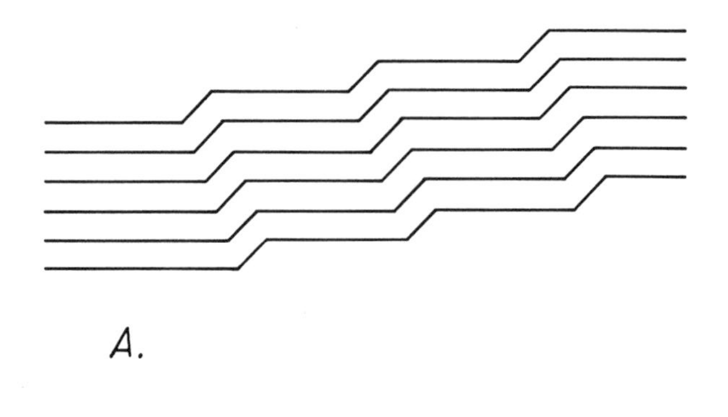

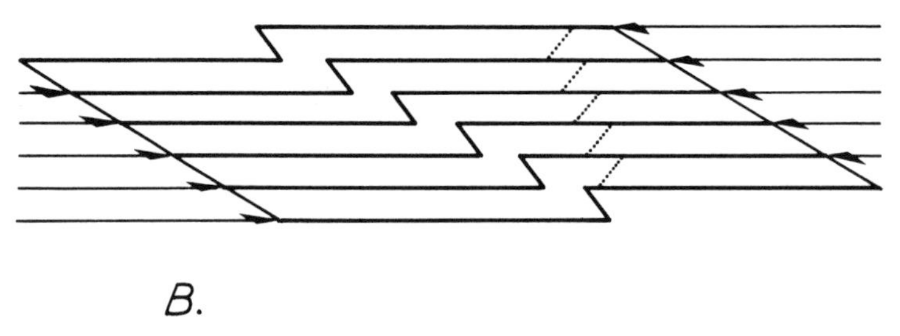

Figure 8.18. Formation of overturned limb of kink fold by layer-parallel shear after deformation caused by formation of original kink bands.

60 degrees, and that Weiss (*6*) found that the angles are about 60 degrees for kinks produced in confined cards, loaded axially. The angles for kinked cards ranged from 58 to 75 degrees, depending on the magnitude of the confinement of the card decks (*6*, p. 313).

A theory that Maurice Biot (*1*) has developed seems to provide insight into the origin of kink bands and the conditions under which kink bands form. The theory will not be developed in detail here because it involves many lengthy manipulations. Biot develops it in his monograph on incremental deformations (*1*). Although another theory, one developed by Ellen, using methods of plasticity theory, is pertinent, it too is complicated; therefore, we will restrict our attention to a few equations developed by Biot.

The following differential equation, derived by Biot (*1*), relates deflections in the x- and y-directions. The equation ignores bending rigidity of the multilayer (*3*).

$$\left(G_i - \frac{S}{2}\right)\frac{\partial^4\psi}{\partial x^4} + 2(2B_i - G_i)\frac{\partial^4\psi}{\partial x^2\,\partial y^2} + \left(G_i + \frac{S}{2}\right)\frac{\partial^4\psi}{\partial y^4} = 0. \tag{8.12}$$

Here ψ (psi) is a displacement function, defined in terms of the displacement, u, in the x-direction and the displacement, v, in the y-direction,

$$u = -\frac{\partial \psi}{\partial y},$$

and

$$v = \frac{\partial \psi}{\partial x}.$$

Also,

$$S = S_{22} - S_{11}$$

is the difference between the principal initial stresses, S_{22} and S_{11}, which are oriented parallel to the y- and x-directions, respectively (*1*).

The two moduli, G_i and B_i, are defined by the relations (*1*)

$$\left.\begin{aligned} G_i &= \frac{tG_1G_2}{t_1G_2 + t_2G_1}, \\ B_i &= \frac{t_1}{t}G_1 + \frac{t_2}{t}G_2, \end{aligned}\right\} \tag{8.13}$$

where t is the thickness of a unit element, t_1 and t_2 are the thicknesses of the stiff and the soft layers, respectively, and G_1 and G_2 are their shear moduli (see Fig. 8.19). The materials of the multilayer are assumed to be incompressible.

We can deduce a great deal about the solution of eq. (8.12) by using D'Alembert's solution (*7*, p. 295),

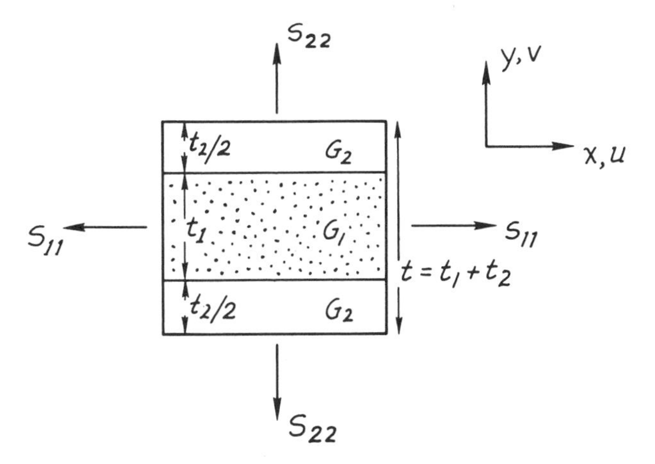

Figure 8.19. Element of multilayer. Stiff layer has shear modulus, G_1, and thickness, t_1. Soft layer has shear modulus, G_2, and thickness, t_2. S_{11} and S_{22} are normal stresses applied to element.

$$\psi = f(x - \mu y), \tag{8.14}$$

where μ (mu) is a constant. If we differentiate eq. (8.14) four times, with respect to x

$$\frac{\partial^4 \psi}{\partial x^4} = f^{\mathrm{IV}}(x - \mu y),$$

and with respect to y

$$\frac{\partial^4 \psi}{\partial y^4} = \mu^4 f^{\mathrm{IV}}(x - \mu y).$$

Therefore, eq. (8.14) is a solution to eq. (8.12) and if we combine the two equations we derive

$$\left(G_i - \frac{S}{2}\right) + 2(2B_i - G_i)\mu^2 + \left(G_i + \frac{S}{2}\right)\mu^4 = 0, \tag{8.15}$$

so that the four solutions for μ are

$$\mu \pm \sqrt{\left\{-\left(\frac{2B_i - G_i}{G_i + S/2}\right) \pm \sqrt{\left[\left(\frac{2B_i - G_i}{G_i + S/2}\right)^2 - \left(\frac{G_i - S/2}{G_i + S/2}\right)\right]}\right\}},$$

or

$$\mu_1 = -\mu_2 = \sqrt{[-m + \sqrt{(m^2 - n)}]},$$

and (8.16)

$$\mu_3 = -\mu_4 = \sqrt{[-m - \sqrt{(m^2 - n)}]},$$

where

$$m = \frac{2B_i - G_i}{G_i + S/2},$$

$$n = \frac{G_i - S/2}{G_i + S/2}. \tag{8.17}$$

The two types of roots Biot considered lead to two distinct types of solutions, presumably representing two distinct types of folding, so that we will consider the two cases in some detail.

In one case,

$$m > 0 \quad \text{and} \quad n < 0.$$

That is, according to eqs. (8.17),

$$2B_i > G_i,$$

and (8.18)

$$S/2 > G_i,$$

which state that the normal modulus, B_i, is greater than twice the shear modulus, G_i, and that the differential axis stress, S, is greater than twice the shear modulus. In this case, according to eqs. (8.16), roots μ_1 and μ_2 are real and μ_3 and μ_4 are imaginary. The solution for the deflection is sinusoidal, according to Biot (*1*, p. 195), and the parameter, μ_1, is defined as

$$\mu_1 = L/2T, \tag{8.19}$$

where L is the wavelength of folds and T is the thickness of the entire multilayer. The solution that yields this value is the same as the solution we derived for folding of a multilayer between two rigid boundaries, Chapter 6. The physical significance of μ_1 is shown in Fig. 8.20A. It defines the slope of the lines connecting the core of a syncline with the cores of adjacent anticlines.

Another, more interesting type of folding is deduced by Biot for situations where all four roots of eq. (8.16) are real. This case, which he calls instability of the second kind, occurs when

$$m < 0 \quad \text{and} \quad m^2 > n > 0,$$

that is, when

$$G_i > 2B_i,$$

$$G_i > S/2, \tag{8.20}$$

and

$$m^2 > n.$$

According to Biot, the values of μ corresponding with a minimum differential axial stress, S_{min}, are

$$\mu_{min} = \pm\sqrt{\left(\frac{G_i - 2B_i}{G_i + S_{min}/2}\right)}, \tag{8.21}$$

where

$$S_{min} = 4\sqrt{(G_i - B_i)B_i}. \tag{8.22}$$

The physical meaning of the direction, μ_{min}, is similar to the meaning of the direction, μ_1, shown in Fig. 8.20A. The physical significance is quite different, however. The value of μ_{min} is the slope of a line, called a *characteristic*. The characteristic direction is similar to the slip line of plasticity theory. A slip line in a plastic may be a line of discontinuity, or fault, across which displacements are discontinuous. Thus, the characteristic direction, given by eq. (8.21), is the direction of a discontinuity within the multilayer. Because a kink band is a discontinuity in a multilayer, it seems reasonable to interpret μ_{min} as the direction of a kink band (*1,3*).

There are several interesting conclusions that we can draw when we interpret

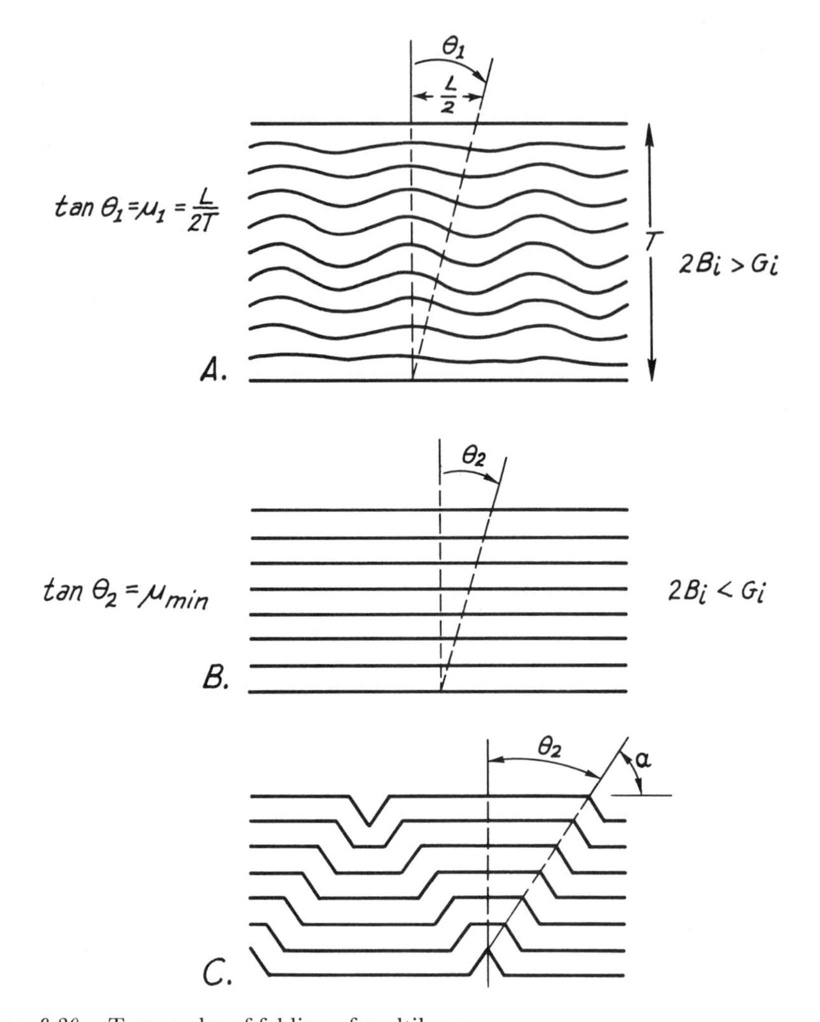

Figure 8.20. Two modes of folding of multilayer.

A. Sinusoidal folds. Dashed lines are characteristic directions. Relatively low shear modulus.

B. Characteristic directions in multilayer with relatively high shear modulus.

C. Kink folds. Characteristic directions are hinge lines.

the kink folds, in the experiments and in the Franciscan Formation, as characteristics. First, if the normal modulus, B_i, of the multilayer is much smaller than the shear modulus, G_i, eq. (8.21), the value of the tangent of the slope of the characteristic is unity, so that

$$\tan \theta_2 = 1 = \mu_{\min}, \qquad \theta_2 = \alpha = 45^\circ. \tag{8.23}$$

That is, the angle between the layers and the characteristic, or kink band, α, is 45°. This is the minimum value for the angle. As the normal modulus increases, the angle θ_2 (Fig. 8.20C) decreases and the angle, α, between the kink band and the bedding increases. As the normal modulus approaches half the shear modulus, that is, as

$$2B_1 \rightarrow G_1$$

[eq. (8.21)], the angle between the bedding and the characteristic directions approaches ninety degrees. When the normal modulus is greater than half the shear modulus [eq. (8.18)], the type of folding is sinusoidal, so that the transition between kinking and sinusoidal folding should occur when

$$2B_1 = G_1.$$

These theoretical conclusions indicate that, where interbedded soft and stiff layers are firmly bonded together and both stiff and soft layers have relatively high resistance to shear, folding of the kink variety should predominate. Where the contacts between the layers are relatively weak, or where the soft layers are weak in shear, however, slippage occurs and folds of the sinusoidal variety should predominate.

The equation Biot derived for the relation between the inclination of characteristics and the relative magnitudes of the shear moduli are in accord with experimental observations of Weiss (*6*). Weiss experimentally deformed paper cards in an apparatus that loaded the cards normally, at right angles to their flat sides, and axially, parallel to their long dimensions. Weiss found that the angle of inclination of kink bands to original layering decreased from about 73 degrees to 58 degrees as the confining load was varied from about 0.3×10^4 N to 1.5×10^4 N (Fig. 8.21). The friction between the cards certainly increased with confinement, so that the increased confinement probably caused the shear resistance, which would appear in our equations as G_1, to increase. Weiss suggested that the probable reason for the break in slope of the curve in Fig. 8.21 at a confinement of about 1.5×10^4 N, is that, at this confinement, shear took place within instead of between cards. The increase of confinement above a load of 1.5×10^4 N had essentially no effect on the angle of kink bands in the experiments of Weiss.

Kink bands and concentric folds

Biot's theory, outlined in preceding pages, indicates that either sinusoidal or kink folding occurs in a multilayer, depending upon relative values of shear and normal moduli of an element of the multilayer. Armed with this information, we will now reexamine the folds observed in the Franciscan Formation and the folds Ellen produced experimentally.

The folds in the Franciscan (Fig. 8.12) apparently are of the kink variety so that the interbedded chert and shale of the Franciscan apparently had a high shear

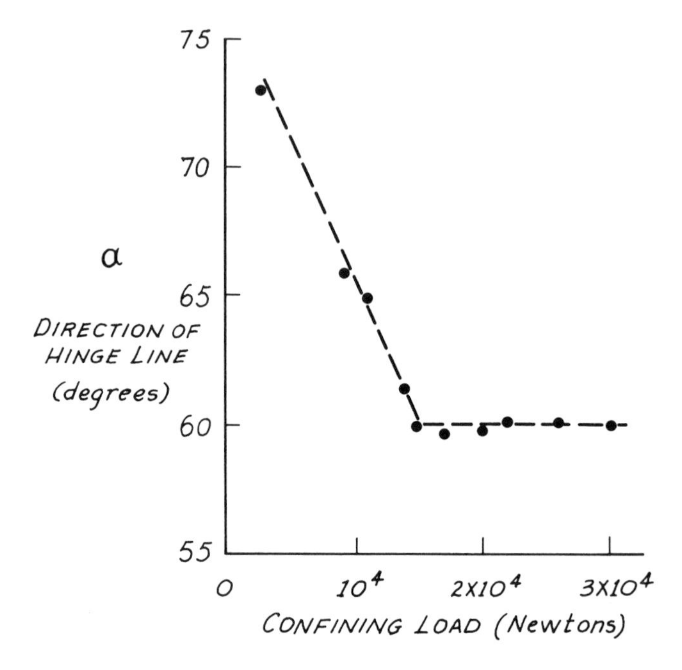

Figure 8.21. Relation between direction of hinge lines of kink bands and confinining loads for experimental multilayers consisting of stacks of cards (after Weiss, ref. *6*, p. 313). See Fig. 8.20C for definition of alpha.

modulus, relative to their normal modulus, at the time of folding [eq. (8.20)]. Further, the folds must have been sheared after they formed, because the hinge lines of the folds are inclined at an angle less than 45 degrees to bedding [eq. (8.23)]. Thus, hinge lines may have been rotated by the shearing process illustrated in Fig. 8.18.

The experiments Ellen performed with cardboard and rubber strips seem to indicate that the chert yielded at the time of kinking, producing the sharp hinges. The rubber strips in his experiments folded into smooth forms, whereas the cardboard, which yielded, folded into angular forms. As was indicated earlier in this chapter, yielding of viscous-plastic or elastic-plastic substances or deformation of a pseudoplastic substance with a large exponent would produce sharp hinges such as the ones in the Franciscan. We cannot, therefore, select a unique rheological model for the behavior of the chert beds solely on the basis of the fold forms. We can state, however, that the chert beds displayed nonlinear behaviors at the time of folding.

Perhaps one of the most interesting features of the experiments Ellen performed is the insight they provide into the process of concentric folding. There are a few natural examples of sinusoidal folding, in which the wavelengths remain constant and in which the folds die out upwards and downwards by diminishing of amplitudes, without changing wavelengths (see Chapter 6). There are many examples in nature of kink folds, in which the bending is concentrated in fold hinges, similar to those in the Franciscan Formation. Probably the majority of natural folds are of the concentric variety, the pure form of which has variable wavelengths and amplitudes at different positions in the fold and consists essentially of parts of concentric arcs, as is shown in Figs. 8.22A and 8.23.

Progressive developments of experimental concentric folds are shown in Figs. 8.13 and 8.15, in which rubber strips are the stiff layers and clay strips are the soft

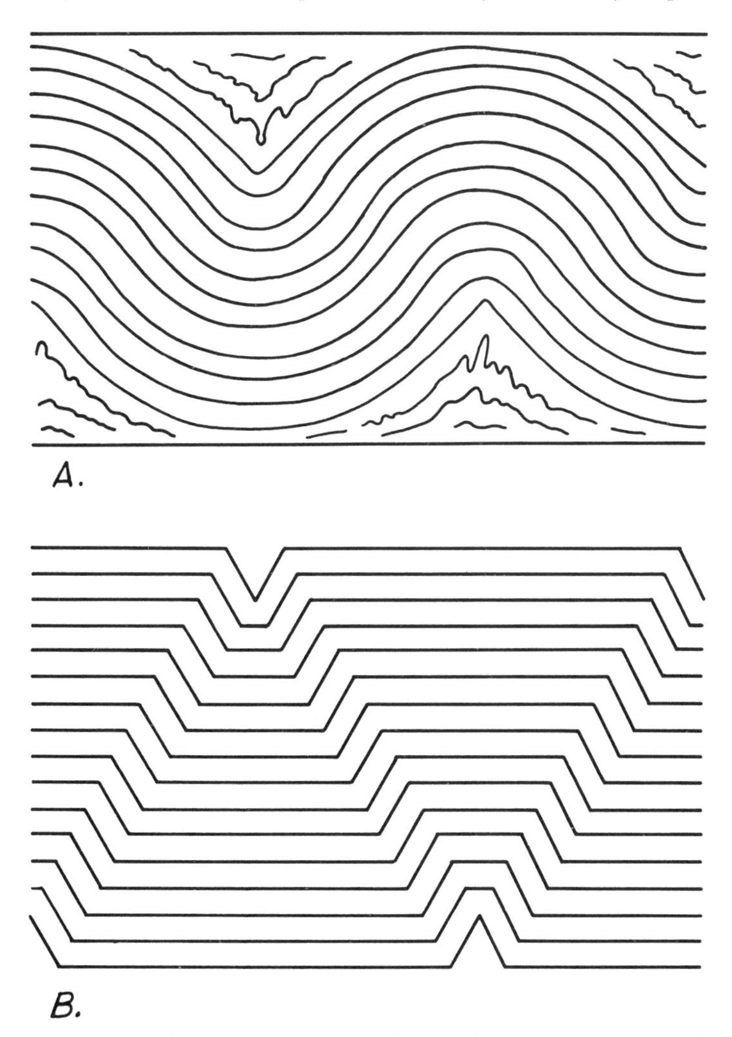

Figure 8.22. Similarities of forms of concentric folds and kink folds.
A. Concentric folds.
B. Kink folds.

Figure 8.23. Concentric fold in amphibolite, Serad Valley, California. (Photo by Stephen Ellen.)

layers. The folds developed progressively as though displacements within the layers propagated across the layers at an oblique angle to bedding, reflected at the rigid boundaries, and propagated across the layers again. This behavior is similar to that of kinks observed by L. E. Weiss, who experimented with cards, and by Stephen Ellen, who experimented with interbedded clay and cardboard (*6,3*).

Another similarity between concentric folds and some kink folds is the gross pattern of deformation, as is illustrated in Figs. 8.22A and 8.22B. Beds in anticlines of concentric folds are smoothly curved upward whereas beds in anticlines of kink folds are abruptly bent upward at fold hinges. Yet, there are similar directions visible in the two types of folds. The experimental folds shown in Fig. 8.13E are intermediate between the ideal concentric forms shown in Fig. 8.22A and the ideal kink forms shown in Fig. 8.22B.

A plausible mechanical explanation for concentric folds can be deduced by comparing the experimental folds in multilayers consisting of interbedded strips of rubber and clay, Fig. 8.15, and of interbedded strips of cardboard and clay, Fig. 8.16. The two experimental multilayers were subjected to identical displacement boundary conditions, as far as Ellen could determine, yet the multilayer of rubber strips formed smooth, concentric forms whereas the multilayer of cardboard strips formed angular, kink forms (*3*). Some of the details of the correspondence between concentric folding and kink folding are being investigated by Ellen. His preliminary results indicate that concentric folds and kink folds are expressions of the same folding processes, but that concentric folds tend to form in natural materials that have high yield strengths, relative to their elastic moduli or viscosity coefficients, and kink folds form in natural materials that have relatively low yield strengths.

References cited in Chapter 8

1. Biot, M. A., 1965, *Mechanics of Incremental Deformations*: John Wiley & Sons, Inc., N.Y.
2. Davis, H. T., 1962, *Introduction to Nonlinear Differential and Integral Equations*: Dover Publications, Inc., N.Y.
3. Ellen, S., (in prep., 1970), "Development of Folds in Interbedded Chert and Shale, Franciscan Formation" (approx. title): Ph. D. Dissertation, Stanford University, Stanford, California.
4. Patterson, M. S., and Weiss, L. E., 1966, "Experimental Deformation and Folding in Phyllite": *Geol. Soc. Amer. Bull.*, V. 77, p. 343–374.
5. Timoshenko, S., and Gere, J. M., 1961, *Theory of Elastic Stability*: McGraw-Hill Book Co., Inc., N.Y.
6. Weiss, L. E., 1969, "Flexural-slip Folding of Foliated Model Materials": In *Proc. Conference on Research in Tectonics*, Baer, A. J., and Morris, D. K., eds., Ottawa: Geological Survey of Canada, GSC Paper, 68-52, p. 294–357.
7. Wylie, C. R., Jr., 1966, *Advanced Engineering Mathematics*: McGraw-Hill Book Co., Inc., N.Y.

PART II

PATTERNS OF FAULTS, JOINTS, AND DIKES

9

Chapter Sections

Faults Near Timber Mountain Caldera
Model of Conditions During Basin and Range Faulting
Anderson's Theory of Faulting
  Stresses on Arbitrary Plane
  Mohr Circle in Two Dimensions
  Orientations of Faults According to Anderson's Theory
Analysis of Fault Pattern Near Timber Mountain Caldera
References Cited in Chapter 9

9

Fault Pattern at Timber Mountain Caldera, Nevada

Faults near Timber Mountain Caldera

Timber Mountain Caldera is in southern Nye County, Nevada, within the Basin and Range Province of the western United States. The area is characterized by Cenozoic normal faults with north-south trends and steep dips (*3*).

Timber Mountain itself is a high area that projects upward in the center of the caldera. The mountain is encircled by the doughnut-shaped depression of the caldera, which collapsed along ring faults during and after an episode of tuff eruption (Fig. 9.1). The ring faults extend to a depth of at least 450 meters, according to Cummings (*3*). The area around Timber Mountain first domed upward and then collapsed as tuff was ejected. Some Basin and Range faults formed at the time of doming, but most of the longest faults became active after caldera collapse (*3*).

The trend of Basin and Range faults in the general area of Timber Mountain is consistently oriented north-south except near the northeast, southeast, and southwest quadrants of the circular caldera, where the faults tend to curve toward the center of the caldera. Faults with a north-south trend are absent within the circular zone of ring faults around Timber Mountain. The fault pattern in Timber Mountain itself is irregular and is distinctively different from the pattern outside the caldera (*3*).

Model of conditions during Basin and Range faulting

In order to explain the convergence of the Basin and Range faults toward the center of subsidence and the lack of Basin and Range faults within the caldera, Cummings (*3*) postulates that the zone of ring faults around the Caldera acted as

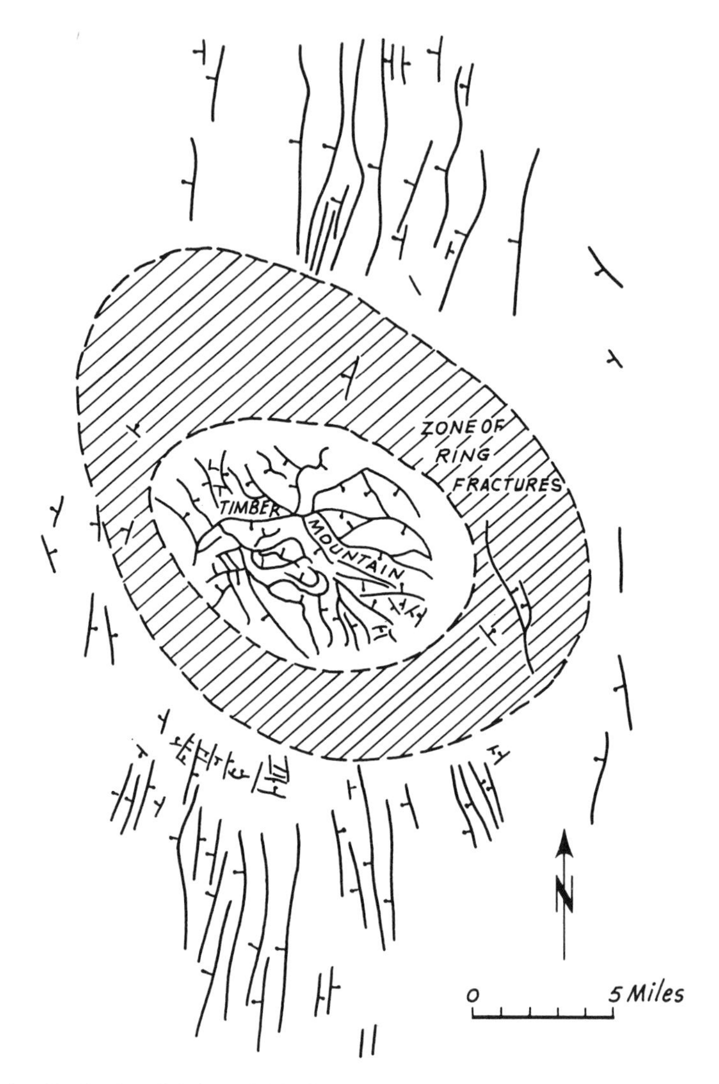

Figure 9.1. Faults in vicinity of Timber Mountain Caldera. Bar and ball on downthrown side of fault (after Cummings, ref. *3*, p. 2788).

a cylindrical surface which decoupled or isolated rocks within the caldera from conditions imposed upon rocks outside the caldera. He assumes that the caldera can be represented by an empty, circular hole. He represents the area surrounding the caldera by a thin elastic body of large areal extent. Finally, he assumes that traces of Basin and Range normal faults somehow correspond with the directions of one of the principal stresses in the elastic body. The stresses and the Basin and Range faults are presumed to be caused by east-west tension in the earth's crust.

Since Cumming's analysis is based on Anderson's theory of faulting, we will examine the fundamentals of this theory before proceeding with the analysis.

Anderson's theory of faulting

Anderson (*1*) was the first geologist to explain orientations of different types of faults in terms of a theory of failure. He stated that "the theory of fault and dyke formation is comparatively simple" (*2*). In fact, he made the theory of fault orientations appear to be so simple that few investigators have questioned the logic on which the theory was based. According to Anderson, it became evident long ago that there are two distinct types of fractures:

> There is, in the first place, tensile fracture, along planes which are normal to the direction of maximum tension in the tested specimen. This can only take place, presumably, when there is at least one direction of tension, but the other type, known as shear fracture, may occur both when there is and when there is not tension, if the stresses in different directions differ sufficiently from one another.... The two classes of rupture are therefore fundamentally different (ref. *2*, pp. 2 and 3).

Anderson explained that Navier and Coulomb developed a theory which predicts that planes of shear rupture do not coincide precisely with planes of maximum shear stress. The angle between the two directions is related to the angle of internal friction of the material.

In a paper entitled "The Dynamics of Faulting," published in 1905 (*1*), Anderson first presented his idea of the relations between the Navier-Coulomb theory of failure and the three types of faults—normal, thrust, and transcurrent:

> It was...by an application of the theory that I succeeded, I believe, for the first time, in explaining the three main classes of faulting. It was also shown in this paper that Navier's principle is necessary to explain the inclination of thrusts and normal faults (ref. *2*, p. 4).

He continued:

> Navier's principle is not only necessary, but sufficient to explain all that is yet certain about the limiting stress conditions, and the directions of rock fracture, in the upper part of the crust (ref. *2*, p. 4).

Stresses on an Arbitrary Plane.—Anderson then proceeded to relate the three types of faults to the three different orientations of principal stress directions with respect to the surface of the earth. Because we will find Mohr's Circle to be most useful for summarizing Anderson's results, we will derive it now.

Figure 9.2A shows a cubic element, of unit breadth, loaded in the horizontal, 3-direction by the minimum principal stress, σ_3, and in the vertical, 1-direction by

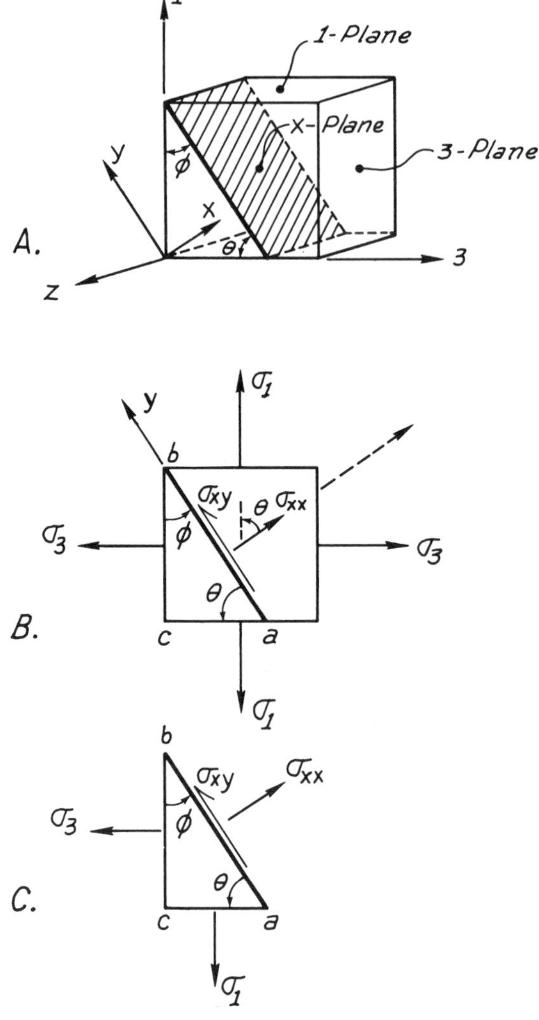

Figure 9.2. Normal and shear stresses on plane inclined to 1- and 3-directions.
A. Square element showing position of arbitrary plane.
B. Plane projected as line on z-plane of element.
C. Free-body diagram of triangular part of z-plane.

the maximum principal stress, σ_1. The x- and y-directions are rotated with respect to the principal directions. By definition of principal stresses, the shear stresses on the faces of the cubic element are zero. The problem is to calculate stresses normal, σ_{xx}, and tangential, σ_{xy}, to an inclined plane containing the y- and z-axes. The plane cuts the 3-axis at an angle θ. The equations we will derive for normal and shear stresses on the x- and y-planes, in terms of the principal stresses and the angle

θ, will define Mohr's Circle. We have used Mohr's Circle several times in preceding discussions but we have not yet examined it thoroughly.

If the square body is in equilibrium, the resultant forces in the 1- and 3-directions must be zero. For convenience, take the thickness of the square body to be unity (1). Then,

$$\left[\sum F_3 = 0\right], \qquad -\sigma_3 cb + ab(\sigma_{xx} \sin\theta - \sigma_{xy}\cos\theta) = 0.$$

But $cb = ab \sin\theta$, so that

$$-\sigma_3 \sin\theta - \sigma_{xy}\cos\theta + \sigma_{xx}\sin\theta = 0. \tag{9.1}$$

Let $l = \cos\theta$ and $m = \sin\theta$ [l and m are called direction cosines]. Then eq. (9.1) becomes

$$-\sigma_3 m + \sigma_{xx} m - \sigma_{xy} l = 0. \tag{9.2}$$

Similarly, for the 1-direction,

$$\left[\sum F_1 = 0\right], \qquad -\sigma_1 ca + \sigma_{xx} ba\cos\theta + \sigma_{xy} ba\sin\theta = 0,$$

or

$$-\sigma_1 l + \sigma_{xx} l + \sigma_{xy} m = 0. \tag{9.3}$$

Now, to find σ_{xx} and σ_{xy} in terms of σ_1 and σ_3, solve eqs. (9.2) and (9.3) simultaneously. Multiply eq. (9.3) by l and eq. (9.2) by m, and add:

$$\sigma_{xx}(l^2 + m^2) = \sigma_3 m^2 + \sigma_1 l^2,$$

or

$$\sigma_{xx} = \frac{\sigma_1 + \sigma_3}{2} + \frac{\sigma_1 - \sigma_3}{2}\cos 2\theta, \tag{9.4}$$

because $l^2 + m^2 = 1$.

Equation (9.4) was derived as follows:

$$\sigma_1 l^2 + \sigma_3 m^2 = \sigma_{xx},$$

and

$$l = \cos\theta, \qquad m = \sin\theta.$$

But

$$\sin^2\theta = \frac{(1 - \cos 2\theta)}{2} = m^2,$$

and

$$\cos^2\theta = \frac{(1 + \cos 2\theta)}{2} = l^2,$$

so that

$$\sigma_1 \frac{(1 + \cos 2\theta)}{2} + \sigma_3 \frac{(1 - \cos 2\theta)}{2} = \sigma_{xx},$$

or

$$\frac{\sigma_1}{2} + \frac{\sigma_1}{2} \cos 2\theta + \frac{\sigma_3}{2} - \frac{\sigma_3}{2} \cos 2\theta = \sigma_{xx},$$

which is eq. (9.4).

In order to derive an expression for σ_{xy}, multiply eq. (9.2) by l and eq. (9.3) by m and subtract the results:

$$\begin{array}{l} -\sigma_3 lm - \sigma_{xy} l^2 + \sigma_{xx} lm = 0 \\ -\sigma_1 lm - \sigma_{xy} m^2 - \sigma_{xx} lm = 0 \\ \hline \sigma_{xy} = (\sigma_1 - \sigma_3) lm \end{array}$$

However,

$$lm = \sin \theta \cos \theta$$

and

$$\sin 2\theta = 2 \sin \theta \cos \theta,$$

so that

$$\sigma_{xy} = \left(\frac{\sigma_1 - \sigma_3}{2}\right) \sin 2\theta. \tag{9.5}$$

Mohr Circle in Two Dimensions.—Equations (9.4) and (9.5) define a circle in polar coordinates. Perhaps this becomes evident if we rewrite them in the following form:

$$\sigma_{xy} = A \sin 2\theta,$$

$$\sigma_{xx} = B + A \cos 2\theta,$$

where A, the radius of the circle, and B, the distance from the origin of coordinates to the center of the circle, are

$$A = \left(\frac{\sigma_1 - \sigma_3}{2}\right)$$

and

$$B = \left(\frac{\sigma_1 + \sigma_3}{2}\right).$$

The circle defined by these equations is called Mohr's Circle, an example of which is shown in Fig. 9.3.

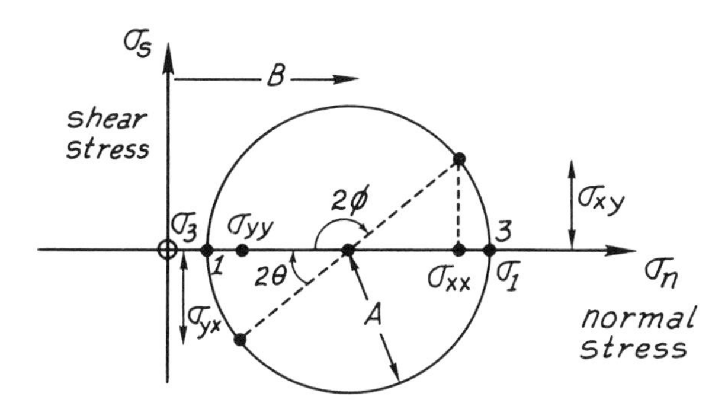

Figure 9.3. Mohr Circle—stress space. Coordinates of points in stress space represent the stresses acting on planes in physical space. The x-plane is the plane normally to which acts σ_{yy}, the normal stress in the y-direction. Angles in stress space are measured between points on a Mohr Circle. Magnitudes of angles in physical space are doubled in stress space.

It is extremely important to remember that the equations we have derived define a circle in *stress space*, not in physical space. That is, the coordinate axes are in terms of stresses, not in terms of distances. The coordinates of a point in stress space (Fig. 9.3) represent the stresses acting on a plane in physical space (Fig. 9.2). Therefore, the x-plane is the one normally to which acts σ_{yy}, the normal stress in the y-direction. The coordinates of point x in Fig. 9.3 represent the stresses acting *on* the x-plane in physical space, that is, σ_{yx} and σ_{yy}. Similarly, the coordinates of point 3 in Fig. 9.3 represent the stresses acting on plane 3 in Fig. 9.2, that is, σ_1. Angle θ is the angle, in physical space, between the y- and the 3-directions and between the x- and 1-directions. These relations between physical space and stress space must be mastered before Mohr's Circle can be applied with confidence.

Exercise

The external stresses on the square in Fig. 9.2 were chosen so that they are *principal stresses.* That is, there are no shear stresses applied to the surfaces of the square body. However, the equations could have been derived with the more general state of stress shown in Figure 9.4. Derive equations for shear stress, σ_s, and normal stress, σ_n, in a logical manner, explaining each step. Bear in mind that the magnitude of σ_{xy} is equal to the magnitude of σ_{yx}. Place your results here:

$$\sigma_s = \tag{9.6}$$

$$\sigma_n = \tag{9.7}$$

Mohr's Circle provides an easy way to derive expressions for principal stresses and maximum shear stresses, and to derive expressions for stresses on a plane with

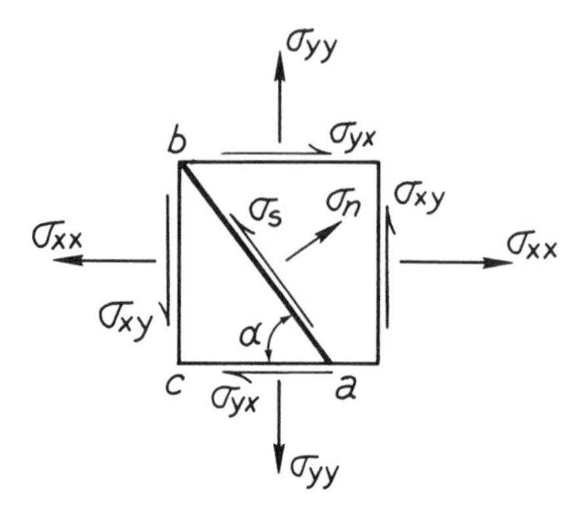

Figure 9.4. Stresses on plane within cubic element supporting external shear and normal stresses.

any orientation. Suppose that a cube supports external shear and normal stresses in the *x*- and *y*-directions (Fig. 9.5A). The state of stress on any plane passing through the cube, paralleling the *z*-direction, which is normal to the page, can be represented on the Mohr Circle shown in Fig. 9.5B. The magnitudes and directions of the principal stresses and the maximum shear stresses can be determined graphically. The procedure is as follows. Establish a coordinate system on graph paper with the abscissa representing normal stresses and the ordinate representing shear stresses (Fig. 9.5B). With an appropriate scale, plot the magnitudes of the normal stresses, σ_{xx} and σ_{yy}, on the abscissa. This established points *A* and *B* in Fig. 9.5B, for example. From point *A* on that figure, draw a vertical line, the length of which represents the magnitude of the shear stress, σ_{xy}. The line is drawn vertically upward if the shear stress on side *cd* of the square shown in Fig. 9.5A acts counterclockwise with respect to the center of the element. The line is drawn downward if the shear has the opposite sense. This operation locates point D, which is on the Mohr Circle. If we now bisect the distance *AB*, we obtain point *C*, the center of the Mohr Circle. The distance *DC*, then, represents the radius, and the complete circle can be drawn as is shown in Fig. 9.5B. The principal stresses are equal in magnitude to *OE* and *OF*. Angle *ECG* in Fig. 9.5B is twice the angle, θ, which is defined in Fig. 9.5A as the angle between the *x*-direction and the direction of the plane normally to which σ_3 acts. Remember that *θ is positive in physical space if it is a counterclockwise angle.* It is the angle between the positive *x*-direction and the direction of the arrow of σ_1 in physical space. In stress space, 2θ is the angle between the point the coordinates of which represent the stresses acting on the *x*-plane $(\sigma_{yy}, \sigma_{yx})$ and the point the coordinates of which represent the minimum principal stress, (σ_3). *In stress space, 2θ is positive if it is a clockwise angle.*

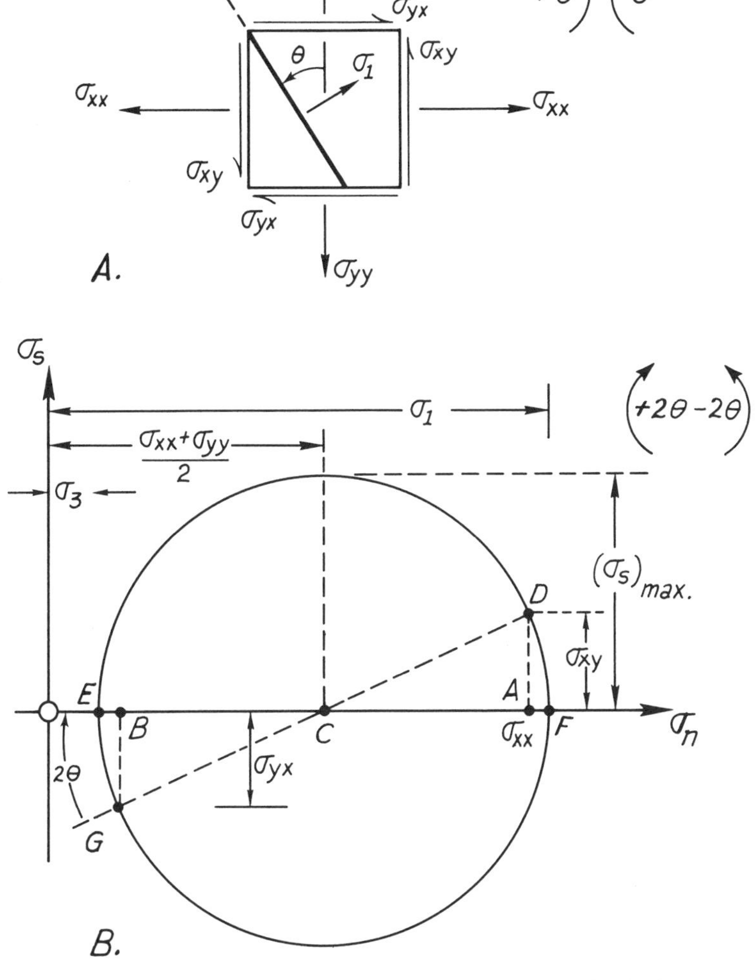

Figure 9.5. Relations among stresses in *x*- and *y*-directions and principal stresses and maximum-shear stresses.

A. Physical space.

B. Stress space.

Exercises

1. Figure 9.6 shows six different states of stress on unit squares. Solve graphically for σ_1, σ_3, and θ.

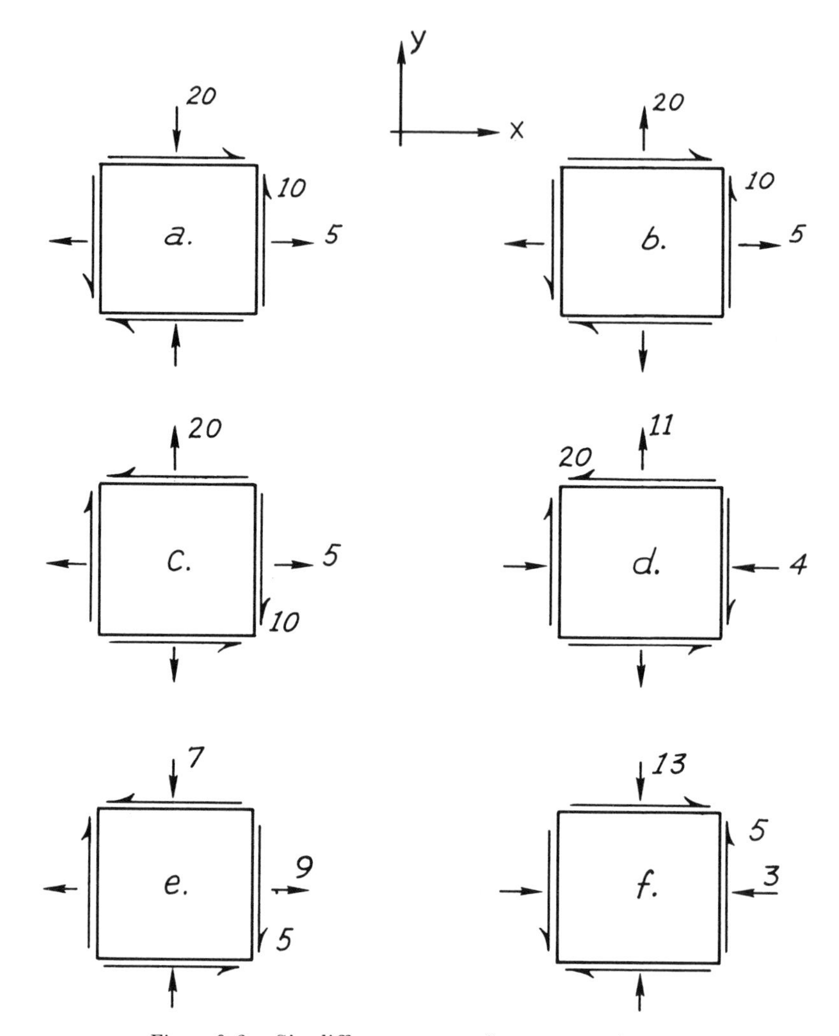

Figure 9.6. Six different states of stress on unit squares.

2. By means of the Mohr Circle in Figure 9.5, show that

$$\sigma_1 = \left(\frac{\sigma_{xx} + \sigma_{yy}}{2}\right) + \tfrac{1}{2}[(\sigma_{xx} - \sigma_{yy})^2 + 4\sigma_{xy}^2]^{1/2},$$

and

$$(\sigma_s)_{\max} = \pm\tfrac{1}{2}[(\sigma_{xx} - \sigma_{yy})^2 + 4\sigma_{xy}^2)]^{1/2}.$$

3. Show that the maximum and minimum normal stresses act at right angles to each other. Show that the planes of maximum shear stress are normal to one

another and bisect the angle between the directions of maximum and minimum normal stresses.

Orientation of Faults According to Anderson's Theory.—Anderson (*2*) selected Coulomb's law to relate shear and normal stresses required for failure of rock. Although Coulomb initially proposed his law to describe the pressure distribution of a soil against a retaining wall, the law has been applied by many investigators to explain the state of stress at failure of rock and concrete.

Coulomb's law is

$$|\sigma_s| \leq C - \sigma_N \tan \phi, \tag{9.8}$$

where C is cohesion, σ_N is the stress normal to the potential plane of failure, and ϕ is the angle of internal friction. Compressive normal stresses are negative. According to eq. (9.8), the shear stress, σ_s, on a plane is less than or equal to the cohesion plus the friction across the plane. If eq. (9.8) is plotted in stress space, Fig. 9.7, we see that Coulomb's law describes two straight lines that slope at an angle ϕ with respect to the normal-stress axis. They intersect the axis of shear stress at C, the cohesion of the material; if the normal stress is zero, the shear strength is equal to the cohesion of the substance.

According to eq. (9.8) and Fig. 9.7, failure is impending if a Mohr Circle, for example the circle of diameter *EO* in Fig. 9.7B, is tangential to the failure envelopes. That circle represents a possible state of stress at failure. On the other hand, if the state of stress in the material were such that the Mohr Circle representing that state does not touch the envelopes, the substance is stable. Such a state of stress is represented by circle *DO* in Fig. 9.7. A Mohr Circle cannot cut across the envelopes because failure occurs if a circle is tangential to them.

Failure is independent of the intermediate principal stress, σ_2, according to the Coulomb criterion.

The coordinates of the points of intersection of a Mohr Circle with the failure envelopes are the values of the shearing and normal stresses acting on the planes of failure, lines *A-A′* and *B-B′* in Figure 9.8A, at the time of failure. *A* and *B* in Fig. 9.8B are such points.

The angle between the directions of the maximum compression, σ_3, and the planes of failure is $45° - \phi/2$, as is indicated in Figs. 9.8A and 9.8B, in accord with the relations between stress space and physical space derived in earlier pages.

Now suppose, as does Anderson (*2*), that one principal stress is normal to the surface of the earth. This is necessary *on* the earth's surface, where the stresses normal and tangential to the surface are zero, but for states of stress below the earth's surface, this assumption usually is invalid. Nevertheless, this is what Anderson assumed, so let us accept his assumption tentatively. If one principal stress is as-

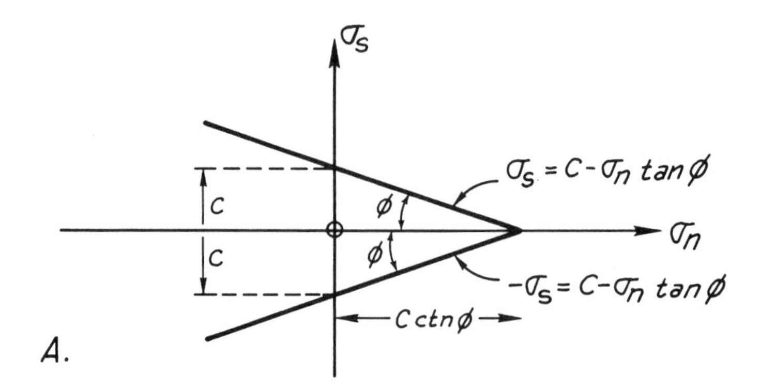

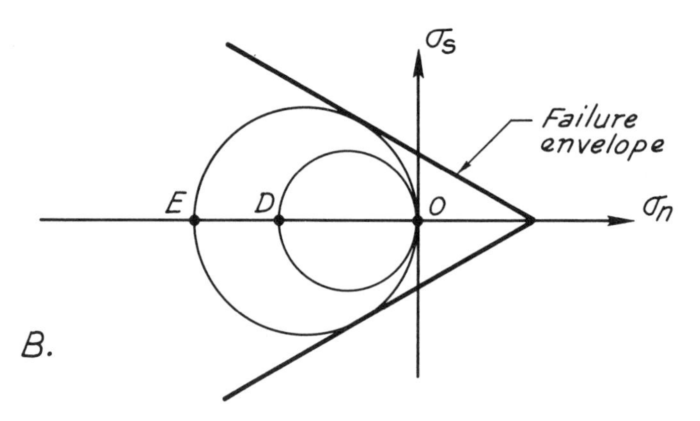

Figure 9.7. Coulomb's law plotted in stress space.

A. Coulomb's law is an equation for two straight lines that slope at an angle of ϕ.

B. Failure is impending in a Coulomb material when the state of stress is such that a Mohr Circle, for example the circle of diameter *EO*, is tangential to the failure envelopes.

sumed to be normal to the earth's surface, the other two principal stresses must be in the plane of the earth's surface.

If the largest compressive stress is vertical, and if the difference between it and the smallest compressive stress satisfies the failure criterion, a normal fault will form, according to Anderson's theory. At the earth's surface, the vertical stress is zero, so that the smallest compressive stress is tensile (Fig. 9.9A). However, this is a special case of the state of stress required for a normal fault. All that is required, in general, is that the largest compressive stress be vertical. If the largest compressive stress is vertical, there are two planes, inclined at 45° to the vertical, along

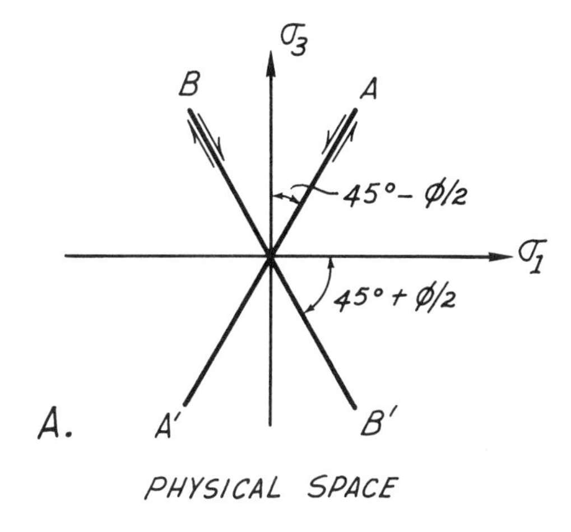

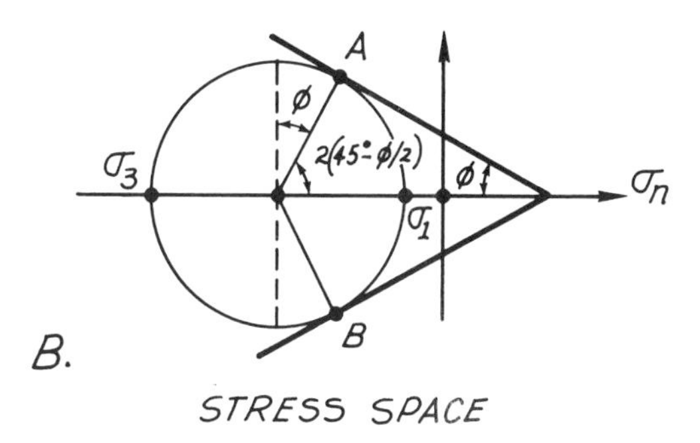

Figure 9.8. Interpretation of Coulomb law. The coordinates of the points of tangential contact of Mohr Circle with failure envelopes are the values of the shear and normal stresses acting on failure planes, A-A' and B-B'. The angle between the direction of maximum compression, σ_3, and the planes of failure is $45° - \phi/2$.

which the tangential stress is maximum. However, as is shown in Fig. 9.9A, normal faults will be inclined at angles of less than 45° to the vertical because of internal friction. Thus, the failure planes are those along which the difference between the applied shear stress and the resisting shear stress is minimum.

Suppose that the largest compressive stress is horizontal. If the material fails, either a transcurrent or a thrust fault will form, according to Anderson's theory. If the smallest compressive stress is vertical, a thrust fault will form (Fig. 9.9B).

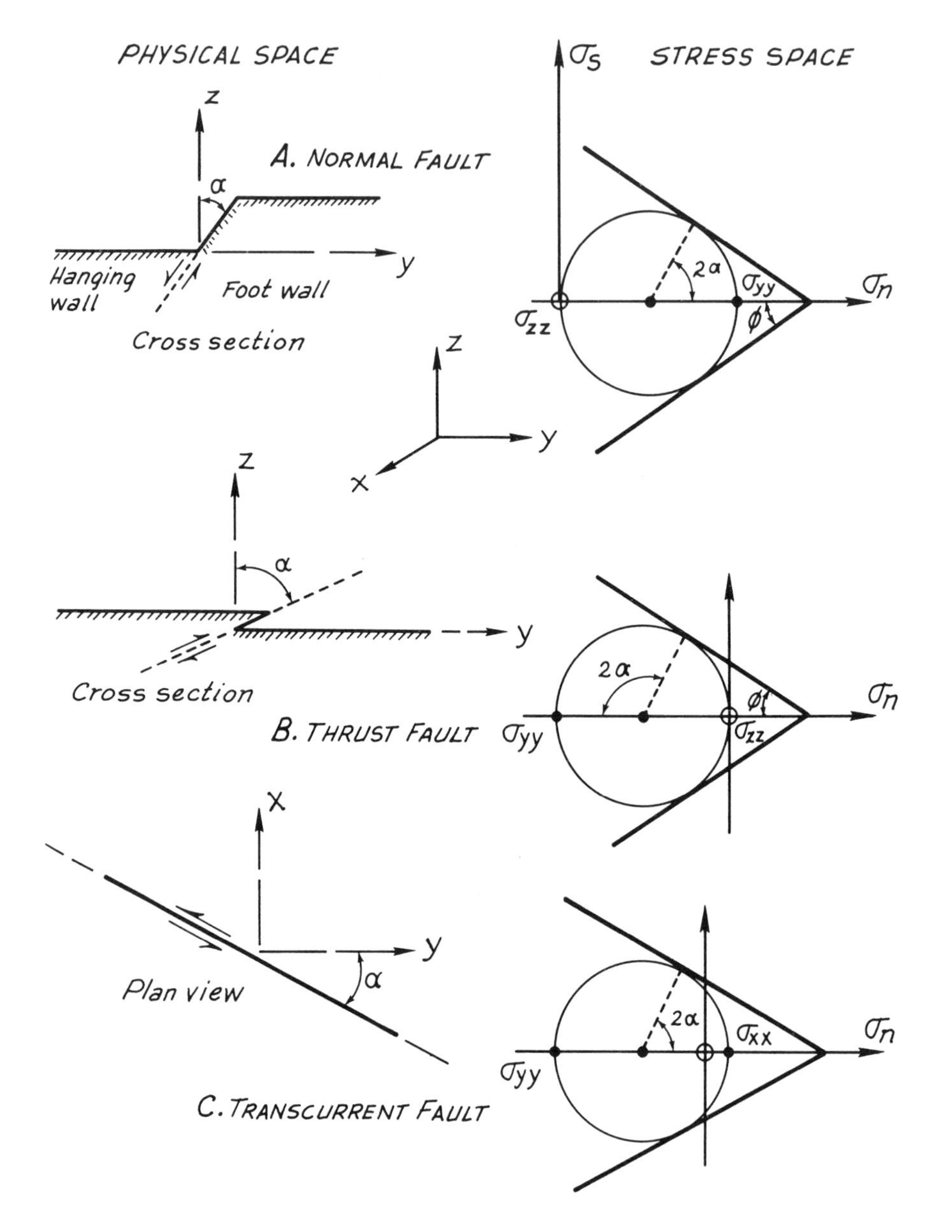

Figure 9.9. Anderson's concept of states of stress at the surface of the earth at the time of faulting.

However, if the smallest compressive stress is horizontal, the stress is relieved by horizontal motion of the material on each side of the fracture (Fig. 9.9C). The fault apparently should be inclined at an angle of less than 45° to the maximum compression.

This is the essence of Anderson's theory of the relations between types of faults and types of states of stress.

Subsequent investigators have extended Anderson's hypothesis and have proposed that joint patterns as well as fault patterns may be directly controlled by regional stress patterns. Thus DeSitter states*:

> Both joints and faults are believed to have a common origin to deformative stresses. . . and must therefore be closely related. The nature of this relation, however, has seldom been ascertained. In general the belief prevails that there exists a gradual transition from joints with no motion along their faces, through joints with small motions, to small faults, and then to large faults (ref. *4*, p. 122).

Because of regularities of many patterns of joints and faults, it has been natural to assume that they somehow reflect patterns of regional stress fields. Anderson presented a plausible basis for the assumption. DeSitter discussed regularities in a fracture pattern in an anticline and a syncline that are side by side (Fig. 9.10A).

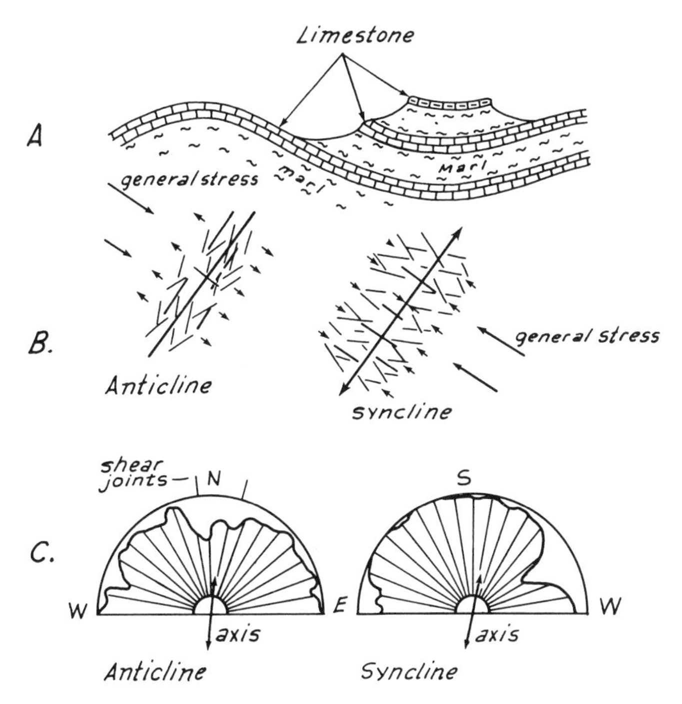

Figure 9.10. Idealized diagram of patterns of small joints and faults in an anticline and syncline that are side by side. (Redrawn from ref. *4*, p. 124.)

* The following passages from L. U. DeSitter, *Structural Geology*, are quoted with the permission of the publisher, McGraw-Hill Book Co., Inc.

He stated that

> . . . a pattern of very small faults, which can be compared to a joint pattern, can be observed (in. . . the crest of the anticline and in the trough of the syncline). The most striking feature of the fault patterns is that both show a double maximum with an intervening angle of 20° in the anticline and 34° in the syncline, but oriented differently in relation to the axis (ref. *4*, p. 123).

DeSitter depended upon Anderson's hypothesis of fault orientations to explain the pattern shown in Fig. 9.10B:

> This theory gives us a ready answer to the question of why sets of differently oriented small faults developed in the anticline and syncline mentioned above. In the anticlinal arch of a competent limestone layer we may expect a local tensional stress in the outer arc perpendicular to the axis. This would therefore be the direction of smallest stress, with the overburden constituting the median stress, and the largest stress parallel to the axis. . . In the syncline we may expect, on the contrary, a local compression in the direction perpendicular to the axis. In both cases a set of shear-joints develops, at an acute angle bisected by the largest stress, but differently oriented in relation to the axis direction. In this very simple case in Algeria we found remarkably good agreement between theory and practice, but in general the agreement is much less obvious and the joint-pattern or fault pattern much more complicated (ref. *4*, p. 125).

Analysis of fault pattern near Timber Mountain Caldera

The Basin and Range faults in the vicinity of Timber Mountain (Fig. 9.1) trend north-south, dip steeply, and are of the normal variety, so that, according to Anderson's theory, the maximum compression was vertical and the minimum compression was horizontal, in the east-west direction. The intermediate stress was in the north-south direction, parallel to the traces of the faults. Since the faults curve toward the center of the caldera, the directions of intermediate normal stress were affected by the existence of the caldera. Thus, according to Anderson's theory, the problem of describing the surface traces of normal faults in the vicinity of the caldera is one of determining the directions, that is, the "trajectories," of the intermediate stress there.

The stresses in the vertical direction need not be considered in our analysis of the fault pattern at Timber Mountain because the vertical stresses do not affect the surface trace of a normal fault, according to Anderson's theory, as long as the vertical stress is the maximum compression. Accordingly, we will follow Cummings (*3*) in treating the earth's surface as an infinite plate and the caldera as a hole in the plate (Fig. 9.11A). We will assume that the plate is subjected to a uniform tension in the east-west direction.

We will not follow in detail the derivation of the stress pattern in an elastic plate containing a circular hole. We will solve a similar problem when we consider

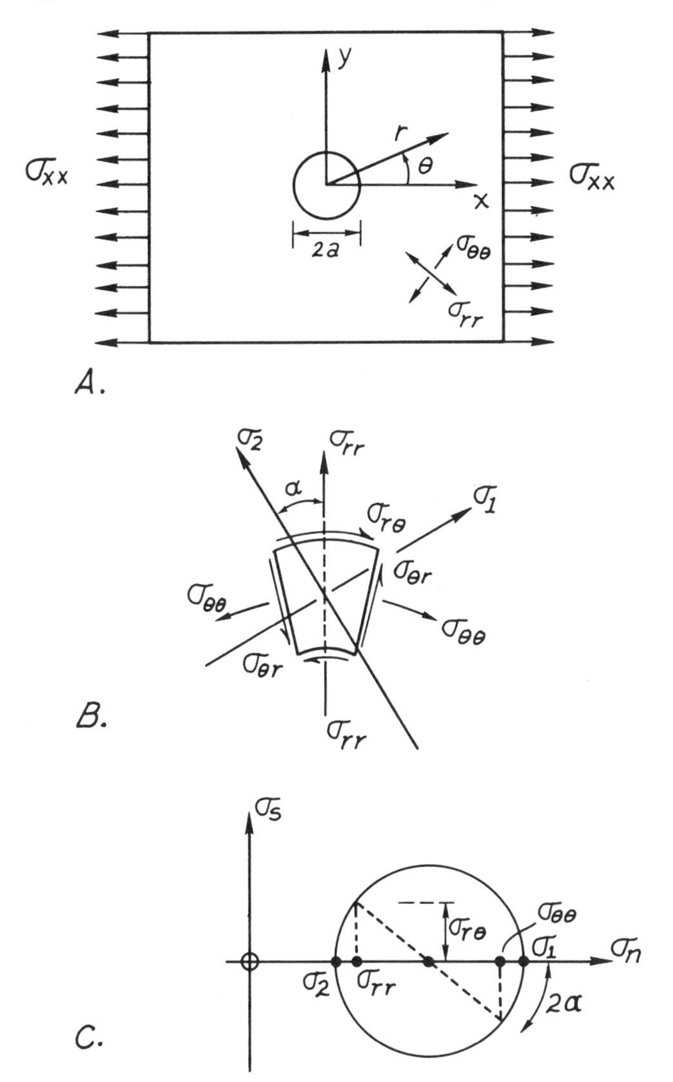

Figure 9.11. Plate containing circular hole and supporting uniaxial stresses.

A. Coordinate system.

B. Small element showing relation between intermediate principal stress direction and radial direction.

C. Mohr Circle for state of stress at point in plate.

the dike patterns at Spanish peaks. According to Timoshenko and Goodier (ref. *5*, p. 80), the stresses in the plate are

$$\sigma_{rr} = \frac{\sigma_{xx}}{2}\left[1 - \left(\frac{a}{r}\right)^2\right] + \frac{\sigma_{xx}}{2}\left[1 + 3\left(\frac{a}{r}\right)^4 - 4\left(\frac{a}{r}\right)^2\right]\cos 2\theta,$$

$$\sigma_{\theta\theta} = \frac{\sigma_{xx}}{2}\left[1 + \left(\frac{a}{r}\right)^2\right] - \frac{\sigma_{xx}}{2}\left[1 + 3\left(\frac{a}{r}\right)^4\right]\cos 2\theta, \quad (9.9)$$

and

$$\sigma_{r\theta} = \frac{-\sigma_{xx}}{2}\left[1 - 3\left(\frac{a}{r}\right)^4 + 2\left(\frac{a}{r}\right)^2\right]\sin 2\theta,$$

where r and θ are polar coordinates, r is radial distance from the center of the circular hole, and θ is the angle between the direction of applied tension, the x-direction, and the radial direction (Fig. 9.11A).

The magnitudes of stresses are of no particular interest to us, however, because the fault traces are correlated with the directions of the intermediate principal stress. The directions can be calculated as follows. Fig. 9.11B shows a small element of the plate, with radial, σ_{rr}, circumferential, $\sigma_{\theta\theta}$, and tangential, $\sigma_{r\theta}$, stresses acting on its surface. The directions of principal stresses σ_1 and σ_2 are also indicated. Let the angle between the direction of the intermediate stress and the radial direction, r, be α (alpha) (Fig. 9.11B). According to the Mohr Circle in Fig. 9.11C, the tangent of 2α is equal to

$$\tan 2\alpha = \frac{\sigma_{r\theta}}{(\sigma_{\theta\theta} - \sigma_{rr})/2}$$

so that

$$\alpha = \tfrac{1}{2}\arctan\left[\frac{2\sigma_{r\theta}}{\sigma_{\theta\theta} - \sigma_{rr}}\right]. \quad (9.10)$$

Thus we can define the directions of the intermediate stress in terms of the stress components, which, according to eqs. (9.9), are determined by the distance from the center of the circular hole and by the angle, θ.

Figure 9.12A shows the directions of the intermediate principal stress at several points around a circular hole. Smooth lines drawn through the short line segments correspond with the directions of the intermediate stress. These smooth lines are the intermediate-stress trajectories. Fig. 9.12B shows the trajectories and Fig. 9.12C shows a map of the actual fault traces at Timber Mountain.

The actual fault traces and the theoretical trajectories correlate remarkably well. Minor discrepancies could be caused by the noncircular form and finite depth of the actual caldera. The circumferential faults outside the caldera are not predicted by the theory, but these faults may have formed during doming or collapse of the caldera (*3*).

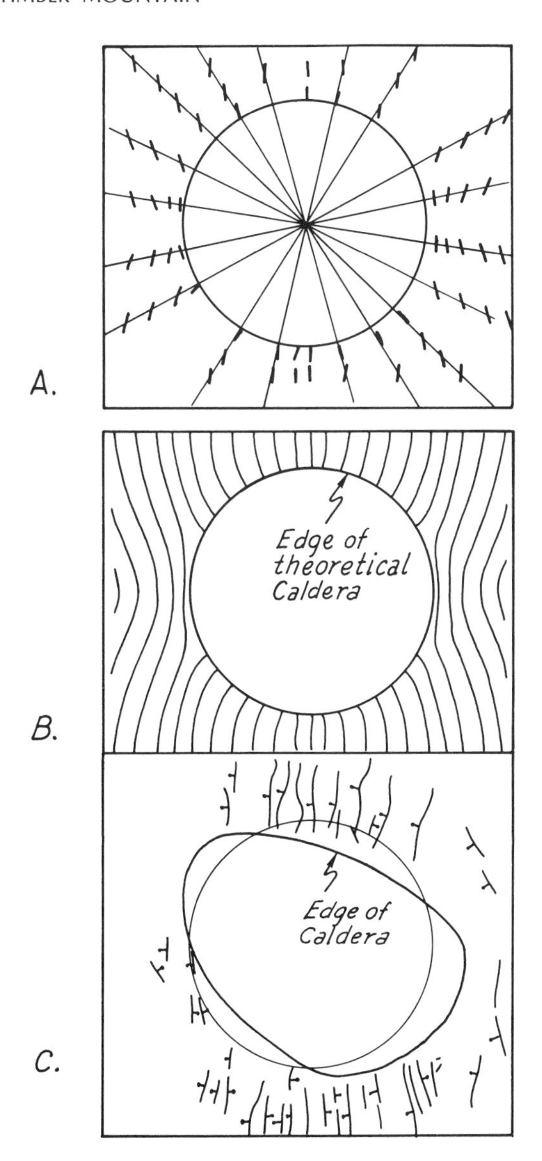

Figure 9.12. Comparison of trajectories of intermediate stress with fault pattern near Timber Mountain Caldera. (Redrawn from ref. *3*, pp. 2792 and 2793.)

A. Directions of intermediate principal stress at several points around theoretical caldera.

B. Smooth curves drawn through short lines shown in *A*.

C. Actual fault pattern around caldera.

References cited in Chapter 9

1. Anderson, E. M., 1905, "The Dynamics of Faulting": *Trans. Edin. Geol. Soc.*, V. 8, pt. 3, p. 387.
2. ———, 1942, *The Dynamics of Faulting and Dyke Formation with Application to Britain*: Oliver and Boyd, Edinburgh, London.
3. Cummings, D., 1968, "Mechanical Analysis of the Effect of the Timber Mountain Caldera on Basin and Range Faults": *Jour. Geophys. Research*, V. 73, p. 2787–2794.
4. DeSitter, L. U., 1956, *Structural Geology*: McGraw-Hill Book Co., Inc., N.Y.
5. Timoshenko, S., and Goodier, J. N., 1951, *Theory of Elasticity*: McGraw-Hill Book Co., Inc., N.Y.

10

Chapter Sections

Sheet Structure
Fletcher Quarry
  Characteristics of Sheet Fractures in Fletcher Quarry
    Lengths of Sheet Fractures
    Terminations of Sheet Fractures
    Thicknesses of Sheets
  Microfractures in Chelmsford Granite
Causes of Sheet Fracturing
Griffith Theory of Fracture
  Stress Concentrations Around Cavities and Cracks
    Circular Hole
    Elliptic Hole
  Strain Energy and Surface Energy
Analysis of Lengths of Sheet Fractures in Fletcher Quarry
  Stress-Strain Relations for Initially Stressed Elastic Body
  Relation Between Depth and Length of Sheet Fractures
  Buckling of Sheets Near Ground Surface
  Possible Origin of Rift Microcracks
Analysis of Thicknesses of Sheets in Fletcher Quarry
Theoretical Interlude: Growth of Fractures in Brittle-Elastic Substances
  Griffith Substance
  Tension
  Compression
References Cited in Chapter 10

10

Formation of Sheet Structure in Granite

Sheet structure

A common feature of many bodies of granitic rocks is *sheet structure*, which divides the rocks into lenses, plates, "sheets," or "beds" by means of jointlike fractures. Most sheet fractures are broadly curved and others are essentially flat; all tend to be parallel to the overlying ground surface. Thus, topographic forms, such as domes, are wrapped by sheets like the outer layers of an onion (*7*). Sheet structure is especially well exposed in granite quarries (*6,15*), in glaciated terrain such as New England and the Sierra Nevada (*7*), and on many rounded topographic eminences, such as Stone Mountain, Georgia, and Enchanted Rock, Texas.

The origins of sheet structure are enigmatic. Whereas most joints are curviplanar or conchoidal fractures, oriented at nearly any angle from horizontal to vertical, sheet fractures commonly are flat. The spacing between sheet fractures generally increases with increasing depth. Indeed, the increase in spacing with depth is so regular in parts of New England that Jahns (*15*) was able to estimate depth of glacial erosion by measuring thicknesses of sheets in quarries and hillsides.

Sheet structure within granitic rocks is independent of structures formed early in the history of the rocks (*6,15*). It is independent of the original shape, transecting contacts of main intrusive bodies and roof pendants (*15*). Primary structures such as pegmatites, aplites and other late dikes, mineralogic layering, and mineral fabrics are all transected by sheet fractures in some places (*15*).

Sheet fractures terminate laterally by gradually thinning until they disappear or by intersecting other structures. They commonly die out against mica bands or against more steeply dipping joints, faults, or shear zones (*15*).

According to Jahns (*15*), who examined sheet structure throughout New

England, parts of some granitic bodies that contain many steeply dipping joints appear to lack sheets entirely, and the thicknesses and degree of perfection of sheeting varies with texture and mineralogy of the granitic rock.

Fletcher Quarry

We will examine the sheet structure in a quarry in New England in order to focus more clearly on the peculiarities of the structure. The Fletcher Quarry, near Westford, Massachusetts, has been selected because Jahns (*15*, and personal communication) has studied it for many years and I have examined the sheet fractures and microcracks in some detail.

Fletcher Quarry is a big hole (Fig. 10.1). It is roughly rectangular in plan, perhaps 150 to 200 meters long and 100 to 150 meters wide. Its depth ranges up to 75 or 80 meters although most of the quarry floor is about 50 meters deep.

The different characters of the sides of the quarry reflect the different eras of quarrying practice. The southeastern side is terraced or benched, much as the Bingham open-pit mine in Utah, showing the oldest quarrying technique, which used natural fractures in the rock and a minimum of splitting to free blocks of the sheets (Fig. 10.1). The northwestern side of the quarry is vertical for most of its height, as are the northeastern and southeastern sides. The northwestern side is scalloped in plan, each scallop representing about half a core hole, four feet in diameter, used to free the granite blocks during more recent quarry operations (see left side of Fig. 10.2). The most recent quarrying techniques produce the smooth vertical faces of the northeastern and southwestern sides of the quarry. Examples of these faces are visible in the middle distance of Figs. 10.1 and 10.2. The more modern techniques involve cutting by means of torches and wire saws. An example of a smooth face cut by wire is shown in Fig. 10.3.

Characteristics of Sheet Fractures in Fletcher Quarry.—Sheets on the northwestern side of the quarry range in thickness from a few centimeters to a meter near the top of the quarry, as visible in the upper right in Fig. 10.2. They are one to three meters thick at mid-depth, as shown in the center of Fig. 10.2, near where two workmen are standing, and three to ten meters in the lower part of the quarry barely visible in the lower left in Fig. 10.2. The lower part of the quarry is visible in the lower part of Fig. 10.1, a view across the quarry toward the south, from the northern corner of the quarry. The sheets dip gently toward the south, off the southern flank of a hill in which the quarry has been developed (*15*).

The sheet fractures here, as elsewhere, are independent of primary structures within the granite. Mineralogic layering, which, presumably, is primary, is nearly vertical in the part of the quarry shown in Fig. 10.2, nearly horizontal in part of the distant wire-sawed face shown in the upper right in Fig. 10.1, and nearly vertical

Figure 10.1. View south across Fletcher Quarry. Deepest part of quarry is below crane in center of photo. Sheets thin from about ten meters at base of quarry to less than one meter at top of quarry, visible in upper left.

on the face of the quarry wall shown in the upper middle and upper left in Fig. 10.1. In addition, long dimensions of feldspar laths and mica books are nearly vertical throughout the quarry. Thus, the horizontal sheet fractures within Fletcher Quarry intersect mineralogic layering and foliation planes at angles ranging from zero to ninety degrees. Fig. 10.3 shows a subhorizontal sheet fracture transecting

Figure 10.2. Northwestern side of Fletcher Quarry, Mass. Sheets increase in thickness with increasing depths in quarry.

mineralogic layering which dips steeply toward the left, on a northeastern sawcut face of Fletcher Quarry.

Lengths of sheet fractures.—The narrow cracks that form the sheet fractures in Fletcher Quarry average about 2 mm in thickness and they range from hairline cracks to gaping gashes up to a centimeter wide. Individual fractures extend

Figure 10.3. Sheet fracture on wire-sawed face in Fletcher Quarry. Fracture is 1.5 mm wide and cuts across mineralogic layering, which dips about 80 degrees to left.

laterally from a few meters to tens of meters (*15*). For example, the separation across the sheet fracture shown in Fig. 10.3 is 1.5 mm and the fracture extends laterally for at least 12 meters. It terminates about three meters to the left of the view, where it becomes parallel to a sheet fracture a few centimeters above it. These fractures are at a depth of about 30 meters below the original ground surface.

The sheet fractures seem to be of different lengths in different parts of Fletcher Quarry. The longest fracture is at midheight on the northwest side of the quarry; it is at least 70 meters long. Many of the fractures are relatively short near the ground surface, averaging perhaps three to five meters in length. There the sheets are bounded by curved and flat fractures that separate the granite into lenses and thin plates (see sheets in upper left of Fig. 10.1). The sheet fractures tend to be flat but short near the bottom of the quarry.

Lengths of sheet fractures were measured on photographs taken of vertical faces on three sides of Fletcher Quarry. The ends of the sheet fractures were determined in the field. The data are scattered but two conclusions seem to be warranted:

1. The sheet fractures tend to be longer on the northwestern (grain) faces than on the northeastern and southwestern (hardway) faces of the quarry. The outlines of two sheets are clearly exposed on the floor of the quarry. Both are roughly elliptic in plan, with a ratio of major axis to minor axis of about two. The ratio of lengths of longest sheet fractures exposed on grain and hardway faces is also about two.
2. The lengths of sheet fractures seem to be maximal at a depth of about 30 meters below the original ground surface and seem to decrease with depth at depths greater than about 30 meters. There are long, flat fractures near the ground surface but they are much less abundant than the short, curviplanar fractures (see upper left in Fig. 10.1).

Terminations of sheet fractures.—Sheet fractures terminate in several ways. Some fractures narrow laterally and gradually fade and disappear, distant from other fractures. Others terminate a few centimeters beneath or above an adjacent, parallel sheet fracture. For example, Fig. 10.4 shows three sheet fractures separated by a few centimeters near midheight in the photograph of a vertical wire-sawed face. The parallel, slightly arched lines near the top and bottom of the photo are not fractures; rather, they are marks made when wire-saw blades were changed. The lowest sheet fracture enters the view from the left and dies out in a series of en-echelon cracks that step downward, away from the other two fractures.

The middle sheet fracture is about one meter long. It dies out in en-echelon cracks at its right-hand end and dies out by narrowing at its left-hand end. The upper sheet fracture terminates by narrowing about one-half a meter to the left of the view and about one meter to the right of the view. Feather-like cracks, such as those visible above the upper sheet fracture, occur in many parts of Fletcher Quarry but I do not know their origin. They are about three centimeters long and are subparallel with the sawcut face, perhaps dipping ten degrees with respect to the face. They occur in several places where there is a single sheet fracture, and therefore are not necessarily caused by interference of sheet fractures.

In some places, sheet fractures curve and die out against older cracks. This type of termination is especially common near the ground surface, where sheets

Figure 10.4. Terminations of sheet fractures deep within Fletcher Quarry. Lowest sheet fracture enters view from left and dies out in series of en-echelon cracks. Middle fracture is about one meter long. It dies out in en-echelon cracks at right-hand end and by narrowing at left-hand end. Upper fracture dies out beyond view. Feather-like cracks above upper fracture are common in the quarry but their origin is unknown.

are characteristically lens-shaped. One example of this type of termination is shown in Fig. 10.5, which is a photograph of a vertical face near the original ground surface over Fletcher Quarry.

In one place, deep within the quarry, however, two sheet fractures crosscut

Figure 10.5. Curved sheet fracture terminates by intersecting flat fracture at high angle near top of Fletcher Quarry.

each other at an intersection of about 20 degrees. One crack follows the trend of the other crack for a distance of about two centimeters and then continues on its original orientation for about one meter.

Thicknesses of sheets.—In the Fletcher Quarry, sheets range from a fraction of a meter to about ten meters in thickness. They generally thicken with depth, as can be confirmed by examination of Fig. 10.2. The increase in thickness with depth is irregular in detail, however. In some places there may be a sheet of three meters' thickness in a zone of sheets predominantly less than a meter thick. Conversely, a zone of thick sheets may be disrupted by two or three thin sheets.

Jahns (*15*) measured thicknesses and depths of 35 sheets exposed on the southeastern face of Fletcher Quarry. I measured about an equal number of sheets on the southwestern and northwestern faces. The data are plotted in Fig. 10.6; they indicate that the sheets tend to thicken with increasing depth and that the increase in sheet thickness increases with depth, so that we can well imagine that at some depth there are no sheet fractures at all.

Also plotted in Fig. 10.6 are average data that Jahns (*15*) collected in more than 100 quarries and 44 natural exposures of Chelmsford granite, the granite quarried in Fletcher Quarry. He excluded data from quarries and natural exposures in the

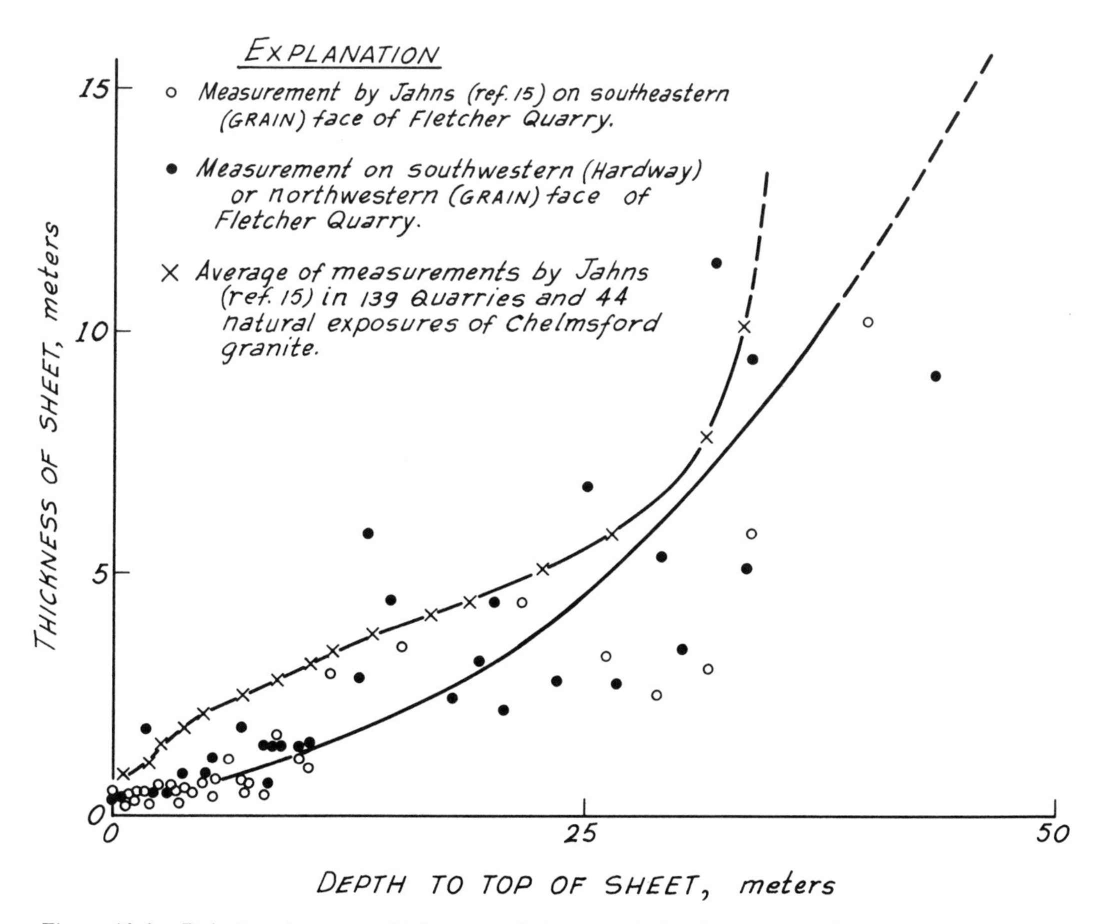

Figure 10.6. Relations between thicknesses of sheets and depths to tops of sheets in Chelmsford granite.

south and southeast sides of hills, where relative movement of glacial ice excavated much rock by plucking.

Both curves in Fig. 10.6 substantiate the impression one has as he looks over Fletcher Quarry—thin sheets near the ground surface and progressively thicker sheets at depth. The curve representing the average of many measurements in Chelmsford granite emphasizes the rapid increase in thickness of sheets at depths greater than about 30 meters below the ground surface.

Microfractures in Chelmsford Granite.—The Chelmsford granite quarried in Fletcher Quarry splits more easily on some planes than on others. The plane of easiest splitting is called the *rift*, which is roughly horizontal in the quarry. A plane of slightly more difficult splitting is called the *grain*, which, in Fletcher Quarry, is vertical and NE–SW. A plane of more difficult splitting and oriented at right angles to the rift and grain is called the *hardway*. Thus, quarrymen identify rift, grain, and hardway of granite in order of ease of splitting (*6*).

An experienced quarryman can identify rift, grain, and hardway with his eyes closed. He recognizes them by their "feel." I am told that the rift forms the smoothest split surface, the grain forms a slightly rougher and more abrasive surface, and the hardway forms the roughest of the three surfaces. A quarryman can even distinguish surfaces of sheets from surfaces of rift, grain, and hardway, because sheet surfaces are very smooth and some places are almost polished.

The stonecutter uses the three directions of splitting to control the shape of the stone he is working.

Field mapping of orientations of microfractures, which range from about $\frac{1}{2}$ to 5 mm and average about 2 mm in length, suggests that the rift and grain of the granite in Fletcher Quarry are parallel to directions of relatively intense microfracturing; more microcracks seem to be oriented parallel to rift and grain than to other directions in the granite. The rift is most easily identified by observation of microcracks but the grain is sometimes difficult to recognize in the field.

The validity of these field observations is indicated by a study of thin sections of the granite by S. D. Peng (personal communication). Peng classified microcracks according to the type of mineral with which they are associated and according to the position of the crack relative to the grains. Thus, *transgranular* cracks cut across mineral grains, and *intergranular* cracks are between mineral grains. Peng determined orientations and lengths of the various types of cracks. Figure 10.7 is a stereonet, showing orientations of the microcracks in one sample of Chelmsford granite. The lines are contours of numbers of cracks with various orientations. The normals to the cracks are plotted relative to the normals to the rift (R), grain (G), and hardway (H). Thus the plane of the hardway is vertical and parallel to the long dimension of the page, the rift is vertical and parallel to the short dimension of the page, and the grain is horizontal in the figure.

According to Fig. 10.7, most of the microcracks are parallel or subparallel to

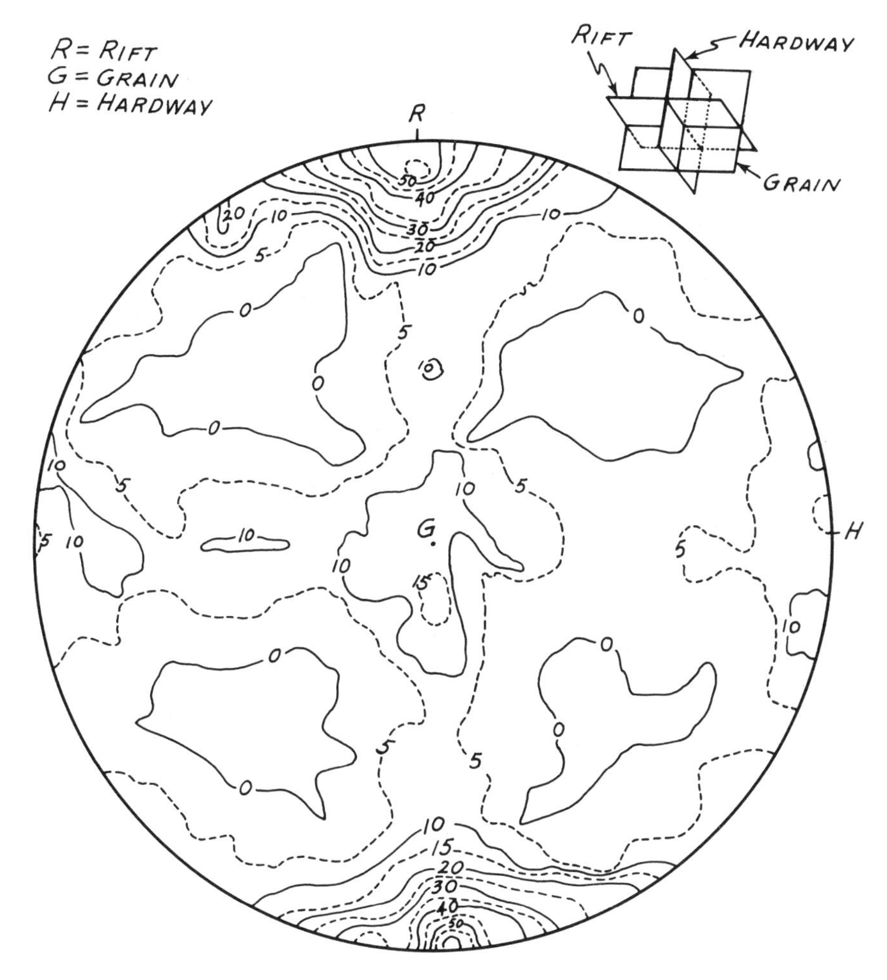

Figure 10.7. Stereonet plot of poles of microcracks in a sample of Chelmsford Granite. Most of the microcracks are nearly parallel with the rift. (Prepared by S. D. Peng, Stanford University.)

the rift. A few cracks are parallel to the hardway and a moderate number are parallel to the grain. Most of the cracks are associated with quartz. According to Peng, about 45 percent of the microcracks in the samples he studied are intergranular, between quartz grains; 20 percent are transgranular, in quartz grains; and 35 percent are transgranular, in feldspar grains. Further, the average intergranular cracks are about the same size as the quartz grains, the average transgranular cracks in quartz are larger than the quartz grains, and the average transgranular cracks in feldspar are shorter than the feldspar grains.

Peng found that the cleavage planes of mica tend to be subparallel to the grain and that twin planes of feldspar grains in the granite appear to be randomly oriented (Peng, personal comm.).

Thus, the Chelmsford granite in Fletcher Quarry is markedly anisotropic with respect to orientations of microfractures and mica. Rift and grain, the two directions of easiest splitting, are correlated with preferred orientations of microfractures. The rift is horizontal, so that slightly tilted sheet fractures in the quarry are nearly parallel with the rift, whereas curved sheet fractures near the original ground surface cut sharply across the rift.

Causes of sheet fracturing

Sheet fracturing seems to be related to energy stored within rock. Granite quarrymen are intimately acquainted with this energy or "pressure," which causes sudden splitting or sheeting of granite freshly exposed in quarry floors or causes scaling of conchoidal flakes from floors or sides of quarries. Sometimes the strain energy is expressed by the formation of diagonal cracks connecting partly completed sheets. Commonly it is expressed by lateral expansion of sheets as they are freed at two ends by channeling. Other examples of manifestations of stored energy are cited in papers by Dale (*6*) and Jahns (*15*).

Sheeting has been attributed to other causes, including local or regional compressive stresses imposed after crystallization, insolation, and mechanical action of fire, frost, and vegetation. Jahns (*15*) examines each of these alternative causes and concludes that they are of minor significance in New England granites. He feels that sheet structure and the manifestations of compressive strain in granite quarries are "compatible with an origin involving dilatancy through removal of overlying load" (*15*, p. 81).

Our old friend G. K. Gilbert (*7*) seems to have recognized clearly that sheets are an expression of relief of primary confining pressure. He examined sheets that parallel topographic forms in the area of the present Yosemite Park, in the Sierra Nevada of California. At the time Gilbert concerned himself with sheet structure, there were two theories of its origin. According to one theory, the curved sheets were original structures in the granite, and domes, such as Half Dome in Yosemite Park, were caused by sculpture controlled by pre-existing sheets. Alternatively, the curved sheet structures were thought to have originated after the domes formed, with the sheet structure in some way caused by undefined reaction from the surface (*7*).

Gilbert stated that he became convinced, after examining sheet structure at several places in Yosemite Valley (*7*), that the sheets postdate the topography. One argument he presented is that the sheets are limited to shallow depths. He noted that the sheet fractures extend to a depth of about 50 feet on Half Dome, and about

100 feet on another dome. If the sheets were original, Gilbert reasoned, we would expect the sheets to continue indefinitely downward and inward in domes. Another observation in favor of the theory that the sheets are secondary is that they are parallel to valley sides so that they form anticlines on ridges and synclines in valleys. Gilbert reminds us that anticlines in folded strata, such as in the Appalachian Mountains, can form either valleys or ridges, or even be discordant with respect to topography.

In order to determine the cause of sheeting, Gilbert (*7*) contemplated familiar manifestations of contraction and expansion. The effect of contraction is illustrated by drying of mud and cooling of lava. Cooling and drying begin at the surface and progress inside, causing cracks to form normally to the surface and to separate the materials into columns. On the other hand, effects of expansion are illustrated by exfoliation of rocks when they are rapidly heated, as during a forest fire. Flakes of rock spall off along partings nearly parallel to the surface of the rock. Thus Gilbert concluded that sheet fractures in granite are more nearly like exfoliation produced by fire, which is ascribed to expansion (*7*).

The most likely reason for expansion in granites, according to Gilbert, is unloading. In his words (*7*, p. 32 and 33):

> When the granite came into existence by the cooling of the parent magma it was buried under a deep cover of older rock. Because of that cover it was subject to compressive stress, and that compressive stress was of course balanced by internal expansive stress competent to cause actual expansion if the external pressure were removed. As in course of time the load was in fact gradually removed, the compressive stress was diminished and the expansive stress became operative.... The parts of the granite successively exposed at the surface were in a condition of potential expansion, or tensile strain, and that strain would be relieved by the separation of layers through the development of division planes approximately parallel to the surface.

He also pointed out that the effect of cooling and contraction of granite magma would be to counteract the expansion caused by removal of overburden.

Sheet fracturing probably results from enlargement of small cracks and flaws that exist in the granite prior to sheeting. Our theoretical analysis of sheeting begins with a discussion of Griffith's failure criterion, which describes conditions under which small flaws, such as microcracks, propagate in brittle-elastic materials. Then we will develop a theory to account for the fact that granite forms under high pressures and is highly strained at the earth's surface. The theory is nothing more than a quantitative statement of the concepts established by Gilbert (*7*) and Jahns (*15*). Finally, we will combine these theories to explain, tentatively, several common features of sheet fractures.

Griffith theory of fracture

In the preceding chapter we have examined two cases that support the common belief that patterns of large-scale fractures reflect patterns of large-scale stress fields. Nobody, however, has explained how the patterns are related. Even though there are fractures at some scale in every outcrop, very little is understood about how they form, grow, branch, and join to produce faults, joints, or sheet fractures.

In fact, there is at least one fundamental fallacy in using any failure criterion to predict orientations of fractures within bodies, or even to predict the stresses at which a body of rock will fail by a throughgoing fracture. Once a fracture has begun, the state of stress changes markedly in its vicinity. The stresses that determine the directions of propagation of a crack are those in the vicinity of the crack, along its surface, so that the influence of the crack itself on the stresses in its environment can be highly important (*4,10,20*).

Suppose that we select Coulomb's law as the failure criterion of a brittle-elastic body. Then, according to the law, if we increase the stresses on the body, the body will begin to fracture somewhere. It will begin to crack near the point where the failure criterion is first satisfied. But as soon as the crack forms, the stresses in its vicinity are markedly changed, and Coulomb's law is inadequate to predict further fracturing of the material. We cannot predict the stresses near the crack merely with knowledge of the stresses at the boundary of the material. To predict the orientation of a throughgoing fracture, or failure plane, we must know the direction of propagation of a crack. This is the big hangup with failure criteria: at best, they predict the direction of initiation of a crack.

Another problem with the Coulomb failure criterion is that it predicts that the angle between the direction of maximum compression and the fracture plane is a constant, equal to $\pm(45° - \phi/2)$. According to this criterion, failure is by shear. However, we infer from tests on rock samples, from study of columnar joints and mud cracks, and from other lines of evidence that planes of fractures can form parallel to two principal stress directions. Joints that formed during folding commonly are parallel to fold axes and normal to bedding. Thus, they seem to have formed parallel to two principal stress axes.

Another problem with the Coulomb criterion and with most other theories regarding the strength of brittle solids is the wide discrepancy of several orders of magnitude between the strength of a solid in practice and the strength of a solid predicted by calculating the stress required to separate two adjoining layers of atoms (*10,19*). This difficulty was ingeniously overcome by Griffith (*8,9*), who postulated that materials contain small cracks, flaws, and other discontinuities around which stress is concentrated when the materials are deformed. As a result of the stress concentrations, a material can fail when relatively small stresses are applied to its surface.

Griffith's investigation was founded on theoretical work by Inglis (*12*), who calculated stresses around an elliptical hole in a plate. Inglis (*12*, p. 219) stated that "...by making one axis very small the stresses due to the existence of a fine straight crack can be investigated"; the solution for the stresses is exact so that it is applicable to the extreme limits of form which an ellipse can assume. Inglis recognized the marked effect of flaws on the strength of brittle materials:

> The destructive influence of a crack is a matter of common knowledge; and is particularly pronounced in the case of brittle nonductile materials. This influence is turned into useful account in the process of glass cutting. A fine scratch made on the surface produces such a local weakness to tension that a fracture along the line of the scratch can be brought about by applied forces which produce in the rest of the plate quite insignificant stresses (ref. *12*, p. 219).

Stress Concentrations Around Cavities and Cracks.—A hole in a stressed elastic body has a marked effect on the state of stress in its vicinity. The stresses that were carried by the material missing from the hole are supported by surrounding material (Fig. 10.8). They are greater near the hole than away from it. Because of the hole in the material, the stresses in the vicinity of the hole are greater than the stresses applied to the boundaries of the material. Thus, we say that the hole causes *stress concentration* (*17*) in its neighborhood. We will define stress concentration as the ratio of the stress at any point in the medium to the uniform stresses applied at the boundary of the body.

Circular hole.—For example, if the hole is circular, the stress concentration in the medium is as shown in Fig. 10.9. Figure 10.9A shows a plate with a hole at its center. The plate is uniformly loaded along one boundary by a stress, σ_{yy}. Figures 10.9B and 10.9C show relations between stress concentration and distance from the edge of the circular hole, as determined by eqs. (9.9) (*17*). The variation of stress concentration with vertical distance, that is, with the y-direction, is shown in Fig. 10.9B. We see that the stress in the radial direction, σ_{rr}, is zero at the boundary of the hole. It is very nearly equal to the value of the applied stress at a distance of $r = 3a$, that is, at a distance of about three radii. The tangential stress, $\sigma_{\theta\theta}$, approaches zero at a distance of about three radii. Its concentration is -1 at the boundary of the hole, so that if the applied stress is compressive, the tangential stress is tensile at the top of the hole. The magnitude of the tangential stress is equal to the magnitude of the applied stress.

Now let us examine the variation of stress concentration with distance in the horizontal direction (Fig. 10.9C). The radial stress, σ_{rr}, is zero at the boundary of the hole. It increases to about $\sigma_{yy}/2$ at a distance of one and one-half radii from the center of the hole. At about three radii it approaches zero; it approaches the magnitude of the horizontal applied stress, σ_{xx}, which is zero at the boundary of the

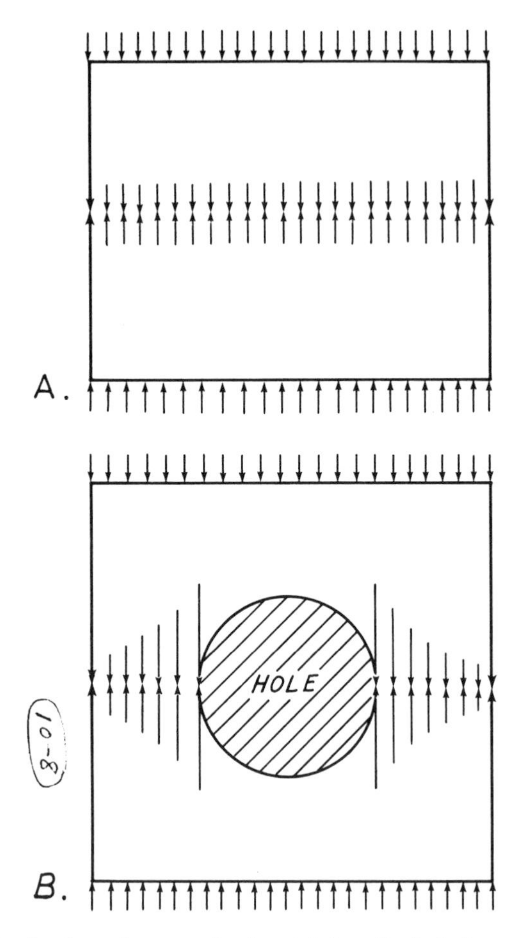

Figure 10.8. Redistribution of stresses in the vicinity of a hole in a plate.

A. In a flawless plate the stresses acting across a plane through the center of the plate are the same as the stresses applied at the boundary of the plate.

B. A hole in the plate markedly affects the stress distribution on the central plane. The stresses that were carried by the material missing in the hole are supported by material surrounding the hole. They are greater near the hole than away from it.

plate. The tangential stress, $\sigma_{\theta\theta}$, is equal to three times the applied stress, σ_{yy}, at the boundary of the hole. It decreases to the value of the applied stress at about three radii. Therefore, at the side of the hole, the magnitude of the tangential stress is three times the magnitude of the applied stress. If the applied stress is compressive, the tangential stress is compressive. If the applied stress is tensile, the tangential stress is also tensile, but it is three times as intense (*17*).

Our analysis of the stresses in the vicinity of a hole in an elastic plate indicates

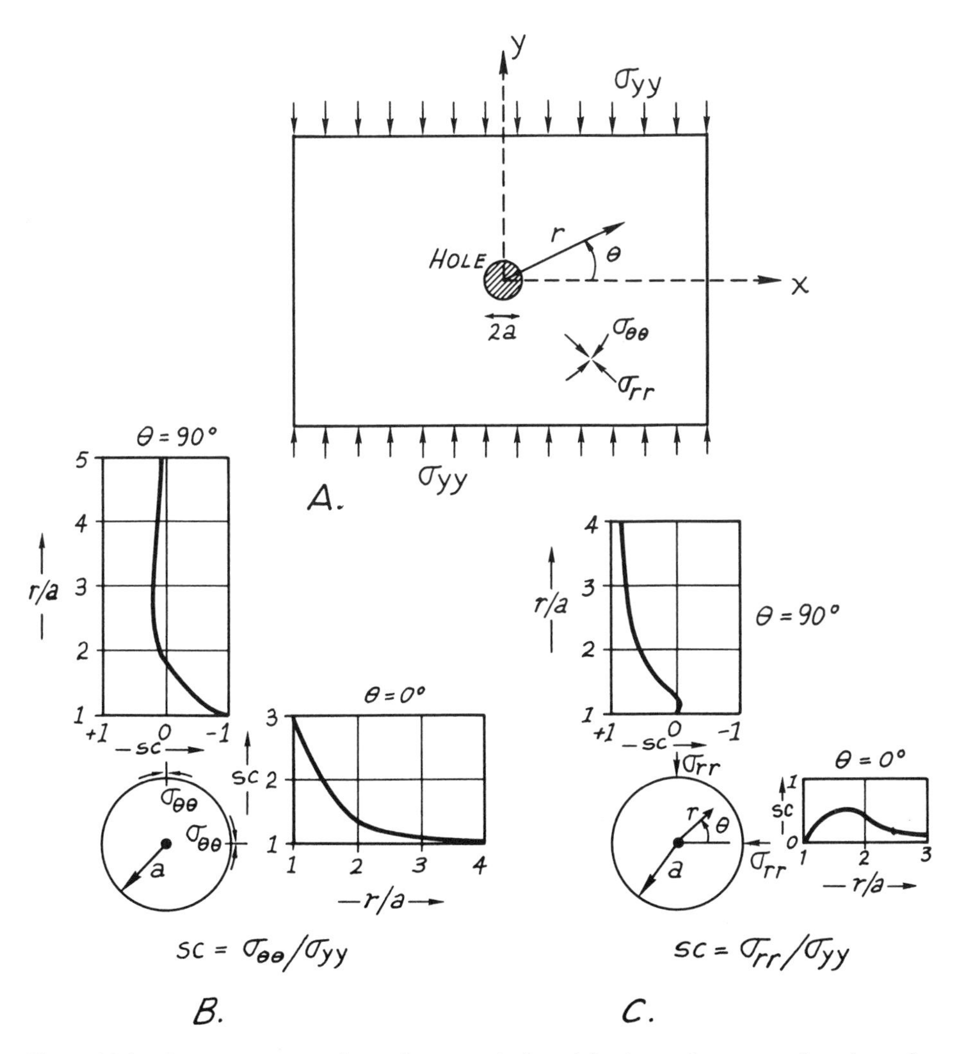

Figure 10.9. Stress concentrations along vertical and horizontal axes passing through center of circular hole in uniaxially-loaded plate.

A. Plate containing circular hole of radius *a*.

B. Variation of concentration of circumferential stress, $\sigma_{\theta\theta}$, with distance from center of hole.

C. Variation of concentration of radial stress, σ_{rr}, with distance from center of hole.

(Redrawn from figure in ref. *17*.)

that the stress distribution in a plate is markedly affected by the hole only within an area of about three radii from the center of the hole, and that a hole can be a very effective stress concentrator.

Elliptic hole.—Circular holes in materials are rare. Flaws that cause failure of materials commonly are believed to be cracks, which are far from being circular in cross-section. Griffith (*8,9*) adopted the hypothesis of Inglis (*12*) that the cross-section of an elongate fracture can be represented by a compressed ellipse in which the long axis has the same length as the crack. Suppose that an elliptic hole is oriented in an elastic material so that its major or minor axis is parallel to the maximum compression applied at the boundary of the material. Then the stress tangential to the surface of the ellipse is (*14,17,23*):

$$\sigma_\beta = \frac{2ac(P + Q) + (Q - P)[(a + c)^2 \cos 2\beta - (c^2 - a^2)]}{a^2 + c^2 - (c^2 - a^2) \cos 2\beta}, \tag{10.1}$$

where β is an elliptic coordinate, σ_β is the normal stress acting parallel to the surface of the elliptic hole, $2a$ and $2c$ are, respectively, the width and height of the ellipse, and Q and P are principal stresses (see Fig. 10.10). If $\beta = 0$, eq. (10.1) is the expression for the tangential stress at the point of intersection of the horizontal axis and the periphery of the hole. If $\beta = 90°$, eq. (10.1) expresses the stresses at the vertical axis of the hole (Fig. 10.10A).

Figure 10.10 shows concentrations of circumferential normal stresses around elliptic holes with major-axes ratios of $\frac{1}{2}$ and 2 and with various boundary loadings. M is the ratio of the horizontal to the vertical applied stresses (*17*). For uniaxial loading, $M = 0$; for biaxial loading in which $P = 3Q$, $M = \frac{1}{3}$; and for hydrostatic loading, $M = 1$.

If the horizontal applied stress, Q, is zero, the tangential stress along the vertical axis, that is, at $\beta = 90°$, is, from eq. (10.1) and Fig. 10.10,

$$\sigma_\beta = -P.$$

Along the horizontal axis, that is, at $\beta = 0°$,

$$\sigma_\beta = P\left(1 + \frac{2c}{a}\right). \tag{10.2}$$

For a circular hole, the maximum stress concentration is +3 (Fig. 10.9). But for an elliptic hole, with a width-to-height ratio of 2, the maximum stress concentration is +5, if the long axis of the ellipse is normal to the direction of applied axial loading (Fig. 10.10). Therefore, if a uniaxial tensile stress is applied to the plate shown in Fig. 10.10, the maximum stress at the periphery of the hole will be magnified five times. It is apparent from eq. (10.2) that, as the ellipse approaches the form of a crack, the width, a, becomes very small and the tangential stress at the tip of the crack becomes very large. Thus, a narrow slit can be a highly effective stress concentrator (see ref. *17,21*).

Strain Energy and Surface Energy.—The growth of fractures in an elastic body decreases the energy stored in the body. If the reduction of energy is equal

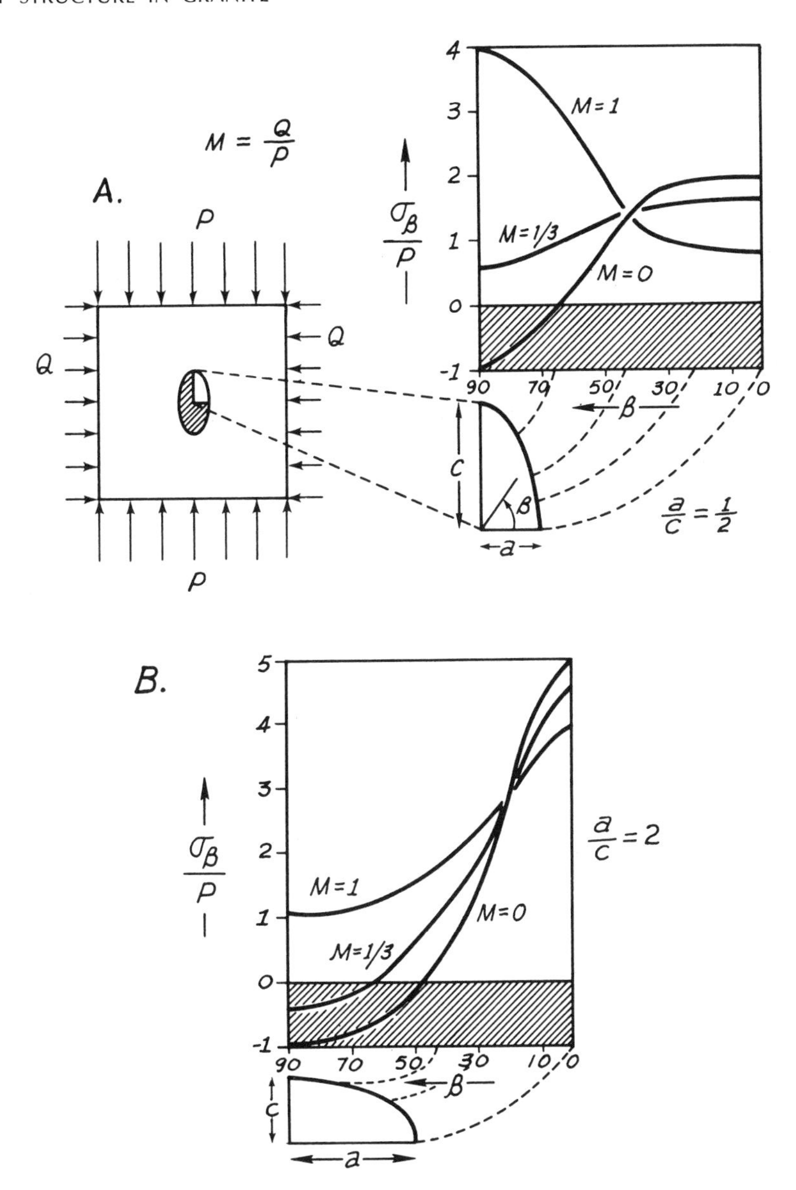

Figure 10.10. Concentration of stresses along surfaces of elliptic holes with width-to-height ratios of 1/2 and 2. (Redrawn from figure in ref. *17.*)

to or greater than the surface energy of newly formed crack surfaces, the crack will propagate spontaneously. This is the essence of Griffith's theory of failure. Let us examine it in some detail.

If an element of elastic material is loaded uniaxially, the element deforms and the load does work on the element. Figure 10.11A shows an element supporting a force, $F_y = \sigma_{yy}\,\delta x\,\delta z$. In response to the force the element elongates an amount, $\epsilon_{yy}\,\delta y$ (Fig. 10.11B). The work, W, done by the force on a body is the integral of the force times the increment of displacement,

$$W = \int_0^v F_y\,dv,$$

so that the work done on the small elastic element is (*14,23*)

$$\delta W = \int_0^{\epsilon_{yy}} (\sigma_{yy}\,\delta x\,\delta z)\,\delta y\,d\epsilon_{yy} = \delta x\,\delta y\,\delta z \int_0^{\epsilon_{yy}} (E\epsilon_{yy})\,d\epsilon_{yy}$$

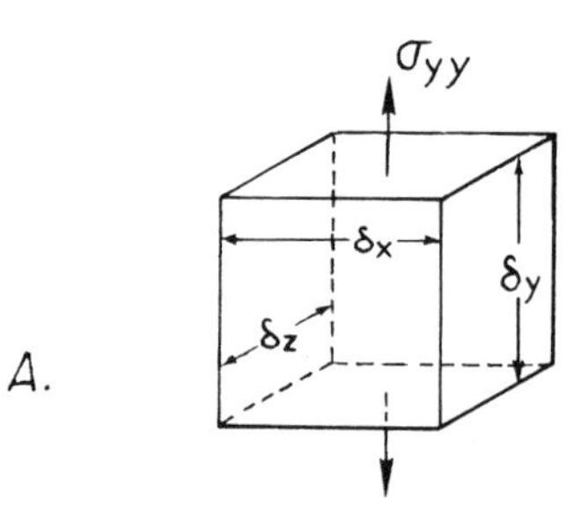

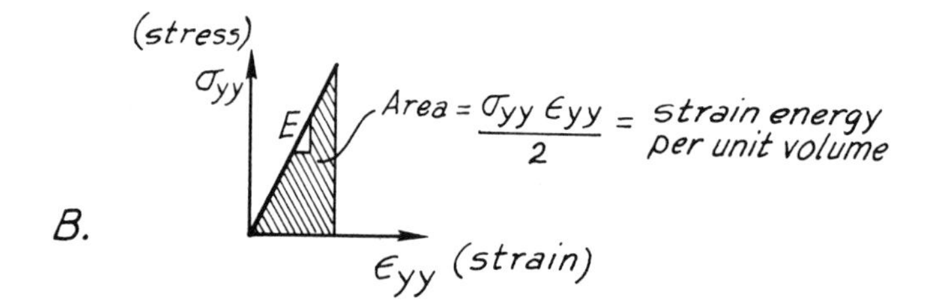

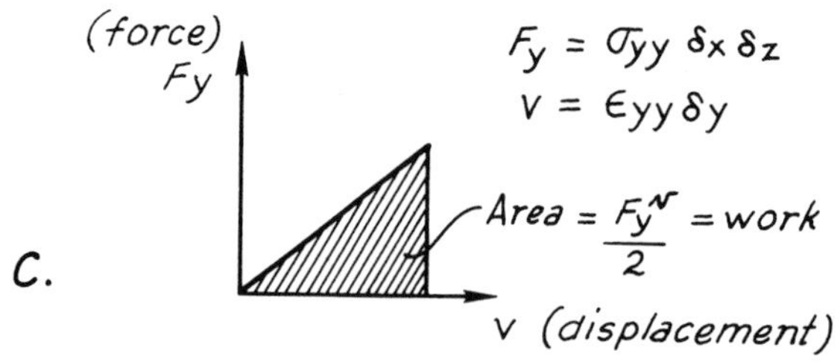

Figure 10.11. Work performed on elastic element and strain energy stored in element.

A. Element supporting uniaxial stress.

B. Strain energy is defined as area beneath stress-strain curve of elastic substance times volume of material.

C. Work done on element is integral of force times displacement, or the area under the force-displacement curve.

or

$$\delta W = \delta x\ \delta y\ \delta z\left(\frac{E\epsilon_{yy}^2}{2}\right) = \delta x\ \delta y\ \delta z\left(\frac{\sigma_{yy}\epsilon_{yy}}{2}\right).$$

The relation between force and displacement of a boundary of an elastic element is a straight line, so that the work done on the element is proportional to the triangular area beneath the force-displacement curve, Fig. 10.11C (*23*). The area and, therefore, the work is

$$\delta W = \delta x\ \delta y\ \delta z\ \frac{\sigma_{yy}\epsilon_{yy}}{2}.$$

When an elastic body such as a cylinder of rock or steel is extended, part of the work performed on the body is converted into heat and part is converted into permanent deformation of the body. Both of these components of energy or work are quite small, however, if the loads are small (*14*). Most of the work is stored in such bodies as potential energy, so that if the load is released, the body springs back to its original shape. This potential energy is called *strain energy*. The strain energy per unit of volume stored in a body is defined as the area beneath the stress-strain curve (Fig. 10.11B). Thus, strain energy, V, is

$$V = (\tfrac{1}{2})\sigma_{yy}\epsilon_{yy}\ \delta x\ \delta y\ \delta z.$$

The same type of analysis we have made for an element loaded uniaxially can be made for an element supporting a full complement of shear and normal stresses. The total strain energy, V, in this case is the sum of the components of strain energy produced individually by the stress components (*23*, p. 147):

$$V = (\tfrac{1}{2})[\sigma_{xx}\epsilon_{xx} + \sigma_{yy}\epsilon_{yy} + \sigma_{zz}\epsilon_{zz} + \sigma_{xy}\epsilon_{xy} + \sigma_{xz}\epsilon_{xz} + \sigma_{yz}\epsilon_{yz}]\ \delta x\ \delta y\ \delta z. \tag{10.3}$$

In order to explain the low tensile strength of elastic materials that contain flaws, Griffith (*8,9*) assumed that the flaws can be represented by elongate elliptical holes. He considered the change in strain energy stored in a plate containing a growing elliptical hole. Suppose that a plate is loaded uniformly along two opposing edges so that it is loaded uniaxially, and that it is constrained on those edges so that the boundaries cannot displace (Figs. 10.12A and 10.12B). If a small elliptic hole is formed in the plate (Fig. 10.12C), the stresses applied to the rigidly fixed edges of the plate are reduced. The plate containing a hole, in effect, has a lower elastic modulus than the plate containing no flaws; if the modulus is decreased and the displacement, or strain, is unchanged, the stress must be decreased. As a result of the reduction of stress, the strain energy stored in the plate is reduced. We will not calculate this reduction of strain energy, but, according to Timoshenko and Goodier (*23*, p. 161) and to Jaeger and Cook (*14*, p. 313), it is

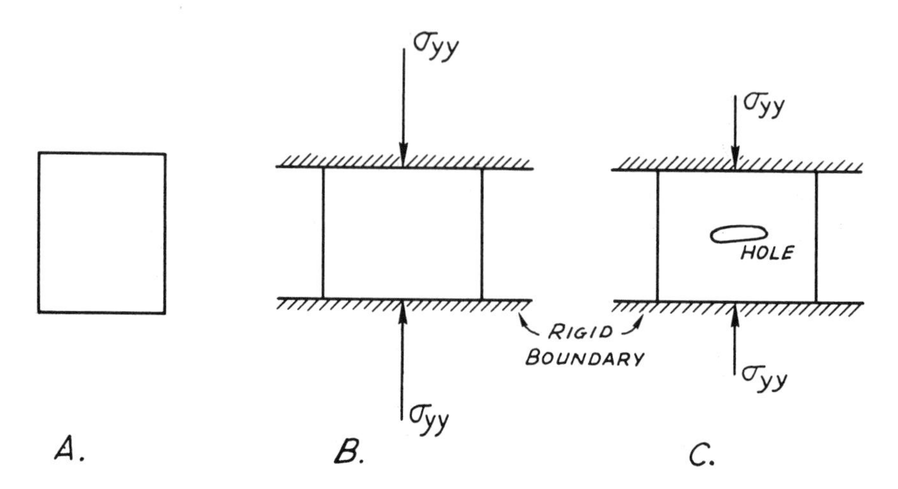

Figure 10.12. Effect of elliptic hole on strain energy stored within elastic plate.
A. Unloaded, undeformed plate.
B. Plate loaded by rigid boundaries.
C. Elliptic hole cut in loaded plate. Boundaries are rigid so that edges of plate do not displace. Stresses applied to boundaries are decreased so that strain energy stored in plate is decreased.

$$\delta V = \frac{\pi c^2(\sigma_{yy})^2 b}{E}. \tag{10.4}$$

Now, if the elliptic crack lengthens, the strain energy is further changed by the amount

$$\frac{d(\delta V)}{dc} = \frac{2\pi c(\sigma_{yy})^2 b}{E}, \tag{10.5}$$

where dc is the change in length of the crack and σ_{yy} is the tensile stress applied to the edges of the plate required to cause the crack to propagate.

Lengthening of the crack causes an increase of surface energy associated with the rupture of bonds. Let this energy be (*14*)

$$dE_s = 4bT(dc),$$

where T is a surface energy per unit of area, similar to surface tension of liquids.

The condition that the crack will propagate spontaneously, according to Griffith, is that the increase in surface energy of the crack with change in length of the crack is equal to the decrease of strain energy stored in the elastic body. Thus,

$$\frac{dE_s}{dc} = 4Tb = \frac{2\pi c(\sigma_{yy})^2 b}{E},$$

or

$$(\sigma_{yy})_{\text{crit}} = \left(\frac{2TE}{\pi c}\right)^{1/2}, \tag{10.6}$$

is the critical tensile stress, applied to the edges of the plate, required to cause the elliptic crack to propagate spontaneously. The greater the length, c, of the original crack, the lower is the tensile stress required to cause it to propagate spontaneously (*14*, p. 321).

Analysis of lengths of sheet fractures in Fletcher Quarry

We have seen that the length of the longest sheet fracture at each depth in Fletcher Quarry, near Westford, Massachusetts, seems to decrease with increasing depths in the quarry. Also, there are many short, curved sheet fractures near the ground surface. Further, there are abundant microfractures in the Chelmsford granite of the quarry, and many of the microfractures are essentially horizontal. Now we will explore a possible explanation of these phenomena.

Stress-Strain Relations for Initially Stressed Elastic Body.—In all our analyses of elastic deformation, we have been concerned with bodies which, we implicitly assumed, were under zero strain at the ground surface, where stresses are zero by convention. However, many rocks, including granites, formed at great depths, where they were subjected to great pressures and elevated temperatures. These rocks are "out of equilibrium" at the ground surface, with respect to both the temperature and the pressure under which they formed. One expression of the "disequilibrium" of granite is that it gradually weathers, or changes its mineralogy, tending toward chemical equilibrium. Another expression of the "disequilibrium," according to Jahns (*15*) and to Gilbert (*7*), is sheeting and other types of fracturing. Our analysis will be restricted to effects of pressure changes but it would not be appreciably changed if we included effects of temperature changes.

We know that granitic rocks exposed at the ground surface are in environments quite different from those in which they were formed. It seems reasonable to assume that granitic rocks are strained at the ground surface even though the stresses acting on them might be negligible there. Also, it seems reasonable to assume that granitic rocks were under zero strain in the environment in which they solidified from a melt. There the crystals presumably fitted together perfectly because they grew into spaces between other crystals in the partly crystallized magma. Thus, let us take zero strain in granite to be the state of strain at the time it cooled and solidified.

The usual stress-strain relations for an elastic body that formed at the earth's surface are three equations of the form [eqs. (5.31)]

$$\epsilon_{xx} = \frac{1}{E}[\sigma_{xx} - \nu(\sigma_{yy} + \sigma_{zz})]$$

and three equations of the form [eqs. (5.38)]

$$\epsilon_{xy} = \frac{1}{G}\sigma_{xy}.$$

If we solve eqs. (5.31) for normal stresses rather than normal strains, we derive the following alternate statement of the three stress-strain relations:

$$\left.\begin{aligned} \sigma_{xx} &= 2G\epsilon_{xx} + \lambda(\epsilon_{xx} + \epsilon_{yy} + \epsilon_{zz}) \\ \sigma_{yy} &= 2G\epsilon_{yy} + \lambda(\epsilon_{xx} + \epsilon_{yy} + \epsilon_{zz}) \\ \sigma_{zz} &= 2G\epsilon_{zz} + \lambda(\epsilon_{xx} + \epsilon_{yy} + \epsilon_{zz}) \end{aligned}\right\} \tag{10.7}$$

where λ and G are Lamé's constants,

$$\left.\begin{aligned} \lambda &= \frac{E\nu}{(1+\nu)(1-2\nu)}, \\ &\text{and} \\ G &= \frac{E}{2(1+\nu)}. \end{aligned}\right\} \tag{10.8}$$

Suppose, however, that a different elastic body solidified under *initial normal stresses*, p_{xx}, p_{yy}, and p_{zz}, and *initial shear stresses*, p_{xy}, p_{xz}, and p_{yz}. Where the applied external stresses, σ_{xx}, σ_{yy}, and σ_{zz}, are zero, for example at the ground surface, the initially stressed body would expand by an amount, in the x-direction,

$$\epsilon'_{xx} = -\frac{1}{E}[p_{xx} - \nu(p_{yy} + p_{zz})].$$

Similarly it would shear by an amount,

$$\epsilon'_{xy} = -\frac{1}{G}p_{xy}.$$

The total strain, which we shall designate by an asterisk, ϵ^*, of such an initially stressed body will be the sum of the strain, ϵ, caused by the surface stresses, σ_{xx}, σ_{xy}, etc., plus the strain, ϵ', caused by the initial stresses, $p_{xx}, p_{yy}, \ldots, p_{xy}$, etc. Thus, the total normal strain in the x-direction would be

$$\epsilon^*_{xx} = \epsilon'_{xx} + \epsilon_{xx} = \frac{1}{E}[\sigma_{xx} - p_{xx} - \nu(\sigma_{yy} - p_{yy} + \sigma_{zz} - p_{zz})].$$

Therefore, the usual stress-strain relations for elastic bodies would be replaced by the following equations for *initially stressed* bodies:

$$\left.\begin{aligned}\sigma_{xx} - p_{xx} &= 2G\epsilon^*_{xx} + \lambda(\epsilon^*_{xx} + \epsilon^*_{yy} + \epsilon^*_{zz})\\ \sigma_{yy} - p_{yy} &= 2G\epsilon^*_{yy} + \lambda(\epsilon^*_{xx} + \epsilon^*_{yy} + \epsilon^*_{zz})\\ \sigma_{zz} - p_{zz} &= 2G\epsilon^*_{zz} + \lambda(\epsilon^*_{xx} + \epsilon^*_{yy} + \epsilon^*_{zz})\end{aligned}\right\} \tag{10.9}$$

Similarly, the stress-strain relations for shear would be

$$\left.\begin{aligned}\sigma_{xy} - p_{xy} &= G\epsilon^*_{xy}\\ \sigma_{xz} - p_{xz} &= G\epsilon^*_{xz}\\ \sigma_{yz} - p_{yz} &= G\epsilon^*_{yz}\end{aligned}\right\} \tag{10.10}$$

If an element of elastic material that formed under initial stresses at depth is taken from the ground surface, where the external stresses are zero, to a depth where the external stresses are equal to the magnitude of the initial stresses, the strains are zero, in accord with our fundamental assumption.

Relation Between Depth and Length of Sheet Fractures.—Suppose that a granite body solidified and cooled at a great depth, D, and at this depth was subjected to hydrostatic pressure, $p_{xx} = p_{yy} = p_{zz} = -\gamma D$, and $p_{xy} = p_{xz} = p_{yx} = 0$, where γ is the unit weight of the overburden. Then suppose that the ground surface was lowered by erosion until the granitic body was exposed at the ground surface. Finally, we shall assume plane strain, so that, during erosion, there were no horizontal deformations; thus the horizontal strains, ϵ^*_{xx} and ϵ^*_{yy}, are zero, even at the ground surface where the vertical strain is maximum. Under these conditions, the normal stress-strain relations, eqs. (10.9), reduce to

$$\sigma_{xx} - p_{zz} = \lambda\epsilon^*_{zz} = \sigma_{yy} - p_{zz} \tag{10.11}$$

and

$$\sigma_{zz} - p_{zz} = 2G\epsilon^*_{zz} + \lambda\epsilon^*_{zz} = (2G + \lambda)\epsilon^*_{zz}. \tag{10.12}$$

The vertical normal stress, σ_{zz}, is proportional to the depth below the ground surface, so that

$$\sigma_{zz} = -\gamma z. \tag{10.13}$$

Combining eqs. (10.11), (10.12), and (10.13), and solving for σ_{xx},

$$\sigma_{xx} - p_{zz} = \lambda\epsilon^*_{zz} = \left(\frac{\lambda}{2G + \lambda}\right)(\sigma_{zz} - p_{zz})$$

or

$$\sigma_{xx} = \gamma\left[\frac{\lambda}{2G + \lambda}(D - z) - D\right]. \tag{10.14}$$

Equations (10.13) and (10.14) are relations for the vertical, σ_{zz}, and horizontal, σ_{xx}, stresses in the granite body.

The next step in our analysis of the growth of sheet fractures is to inquire about a failure criterion for the granite. We showed earlier that Chelmsford granite contains abundant microfractures. It seems natural to suppose that these behave as Griffith cracks and that they propagate under certain conditions to form sheet fractures.

If we assume that the cracks propagate when the stress at the crack tip reaches a critical value, we can presumably use eq. (10.1) as the criterion of fracture propagation,

$$\sigma_\beta = \frac{2ac(P + Q) + (Q - P)[(a + c)^2 \cos 2\beta - (c^2 - a^2)]}{a^2 + c^2 - (c^2 - a^2) \cos 2\beta},$$

where c is the half-length of the fracture, a is its half-width, $P = \sigma_{xx}$, and $Q = \sigma_{zz}$. The elliptic angle, β, is zero or 180° at each tip of the fracture, and therefore the tangential stress at the tips of the fracture is

$$\sigma_\beta = Q\left(1 + 2\frac{c}{a}\right) - P = \sigma_{zz}\left(1 + 2\frac{c}{a}\right) - \sigma_{xx}. \tag{10.15}$$

Substituting eqs. (10.13) and (10.14) for σ_{xx} and σ_{zz} into eq. (10.15),

$$\sigma_\beta = -\gamma z\left(1 + 2\frac{c}{a}\right) - \gamma\left[\frac{\lambda}{2G + \lambda}(D - z) - D\right].$$

Rearranging,

$$\frac{\sigma_\beta}{\gamma D} = \left(1 - \frac{z}{D}\right)\left(1 - \frac{\lambda}{2G + \lambda}\right) - 2\frac{z}{D}\frac{c}{a},$$

or

$$\frac{c}{a} = \frac{1}{2}\left[\left(\frac{D}{z} - 1\right)\left(1 - \frac{\lambda}{2G + \lambda}\right) - \frac{\sigma_\beta}{\gamma D}\left(\frac{D}{z}\right)\right],$$

or

$$\frac{c}{a} = \frac{1}{2}\left[\left(\frac{D}{z} - 1\right)\left(\frac{1 - 2\nu}{1 + \nu}\right) + \frac{\sigma_\beta}{p_{zz}}\left(\frac{D}{z}\right)\right], \tag{10.16}$$

where ν is Poisson's ratio.

Equation (10.16) is plotted in Fig. (10.13), using a value of 0.2 for Poisson's ratio, and using various ratios of tensile strength, σ_β, and initial stress p_{zz}. According to our solution and the graphs in Fig. 10.13, the length, $2c$, of sheet fractures increases very rapidly near the ground surface, that is, at depths less than about $\frac{1}{50}$ of the depth at which the granite formed. At shallow depths the sheet fractures become very large; theoretically, they become boundlessly large at the ground surface.

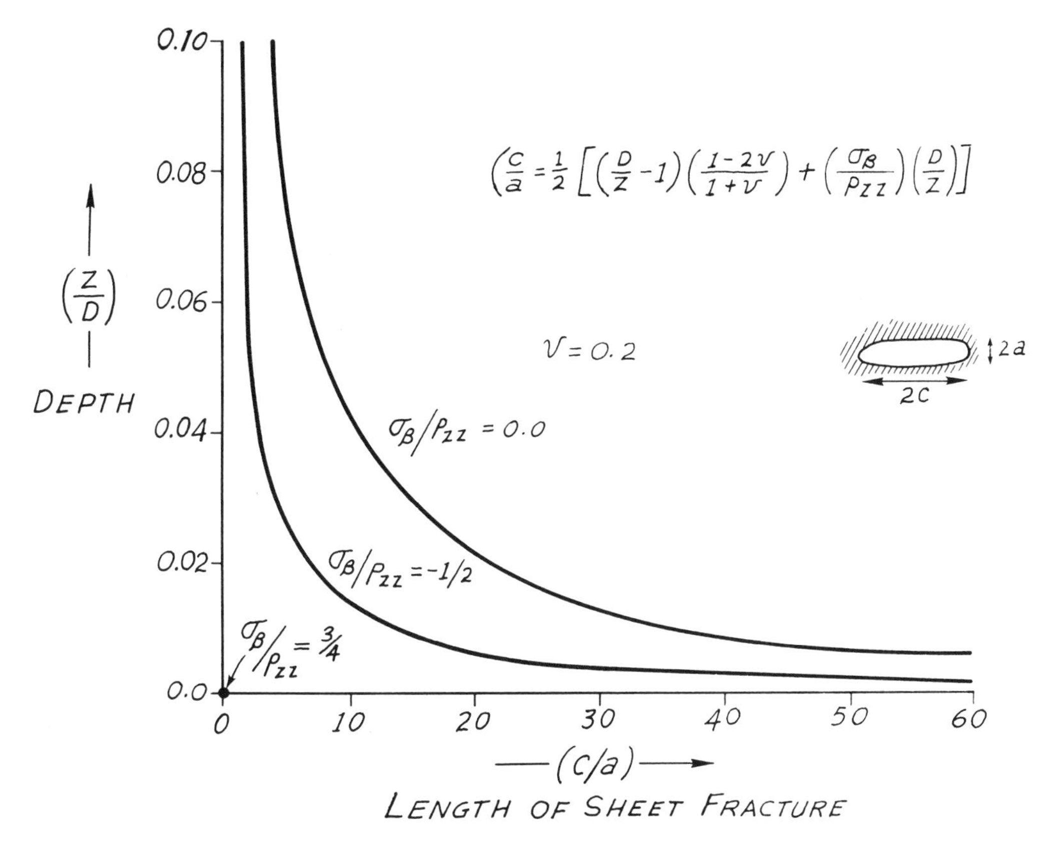

Figure 10.13. Theoretical relation between length of sheet fracture and depth below surface of earth.

The relation between length of sheet fractures and depth below the ground surface depends upon the magnitude of the initial pressure, p_{zz}, of the granite body, relative to the magnitude of the inherent tensile strength, σ_β, of the granite. Unfortunately, it is difficult to estimate the inherent tensile strength, so that it is difficult to estimate the initial pressure, p_{zz}, of the granite. According to Jaeger and Cook (*14*, p. 323), the strength is related to the surface energy, T_0, of the material,

$$\sigma_\beta = \left(\frac{4T_0E}{\rho}\right)^{1/2},$$

where ρ is the radius of curvature of the tip of the sheet fracture. This equation provides minor help, however, because it is difficult to determine meaningful values of surface energy for rock. In fact, the energy consumed by plastic deformation at the end of a propagating crack usually is several times as great as the surface energy of the crack surfaces, so that surface energy probably is not the most important factor anyway (*22*). In spite of these problems, however, we can state that the initial pressure in the granite must be at least 1.3 times the inherent tensile strength of the granite in order for the sheet fractures to form at all by the mechanism we are proposing. The greater the initial pressure, the larger will be the sheet fractures at a given depth.

Measurements on various faces of Fletcher Quarry indicate that the subhorizontal sheet fractures exposed on quarry faces parallel to the grain, that is, the NW and SE faces, tend to be longer than those exposed on faces parallel to the hardway, that is, the NE and SW faces of the quarry. Indeed, the longest sheet fractures visible at a given depth on the NW faces are about twice as long as the longest sheet fractures visible on the SE faces of the quarry. These observations correlate with stress measurements made in the bottom of the quarry by Hooker and Johnson (*11*), who indicated that the ratio of principal horizontal stresses is about two.

According to the two-dimensional theory of stress at the edge of an elliptic crack, eq. (10.15), the stress at the crack tip is

$$\sigma_\beta = Q\left(1 + 2\frac{c}{a}\right) - P,$$

where Q is vertical and P is horizontal. This same equation can be used, as a first approximation, for comparing stresses on ends of principal axes of three-dimensional cracks (*21*).

Let c_1 be the lengths of sheet fractures exposed on the NW face of Fletcher Quarry, parallel to the grain, and c_2 be the lengths of sheet fractures exposed on the SE face of the quarry, parallel to the hardway. Let P_1 be oriented parallel to the grain and P_2 be parallel to the hardway. Then we have

$$\sigma_\beta = Q\left(1 + 2\frac{c_1}{a}\right) - P_1$$

and

$$\sigma_\beta = Q\left(1 + 2\frac{c_2}{a}\right) - P_2.$$

σ_β, Q, and a are the same for the two directions. To a first approximation,

$$2\frac{c_1}{a} \gg 1 \ll 2\frac{c_2}{a}$$

and

$$P_1 \gg \sigma_\beta \ll P_2,$$

so that, to a first approximation,

$$\frac{c_1}{c_2} \approx \frac{P_1}{P_2}.$$

Thus, the ratio of the principal axes of the sheet fractures, which are crudely elliptic in plan, should be roughly proportional to the ratio of the horizontal stresses that caused the fractures to propagate. This is the case for the present stresses in Fletcher Quarry, where *(11)*,

$$\frac{P_1}{P_2} = \frac{-4{,}530 \text{ psi}}{-2.343 \text{ psi}} \approx 1.9$$

and

$$\frac{c_1}{c_2} \approx 2,$$

as indicated above. Also, the maximum compression is oriented N53°E, that is, roughly parallel to the grain in Fletcher Quarry *(11)*.

Buckling of Sheets near Ground Surface.—The relation we have derived for length of sheet fractures agrees in a general way with the measurements of longest sheet fractures at different depths in Fletcher Quarry. Below a depth of about 20 meters, the length of the largest sheet fracture at each depth decreases with increasing depths. Above a depth of about 20 meters in the quarry, however, most of the sheet fractures are much shorter than they should be, according to our theory, anyway. The sheet fractures appear to become shorter and shorter near the ground surface, rather than become boundlessly large as predicted by the theory.

Perhaps the sheet fractures near the ground surface propagate by a different mechanism than do the sheet fractures at depth. The fractures near the surface characteristically are curved and commonly terminate by intersecting other fractures at high angles (see Fig. 10.5). The fractures are essentially horizontal at depth in the quarry, where I found only a few fractures that terminated by intersecting other fractures at high angles. These appeared to be fresh and probably formed during quarrying.

The mechanism of fracture propagation near the ground surface probably is related to the thicknesses of the sheets. The sheets commonly are less than a meter thick near the ground surface and they are up to about 10 meters thick near the base of the quarry (see Fig. 10.6). Perhaps the thin sheets buckle. The critical axial stress for buckling of an unconfined, thin elastic plate is

$$\sigma_{\text{crit}} = \frac{Bt^2}{12}\left(\frac{\pi}{l}\right)^2,$$

where t is the thickness and l is the length of the plate. This equation presumably would apply to buckling of sheets near the surface of the quarry. The thickness and lateral confinement of sheets decrease with decreasing depths, but the axial stress,

σ_{xx}, remains moderately large. Therefore, we can imagine that sheets could buckle and bend near the ground surface. We know from previous analyses that buckling of layers involves compression and extension on opposite sides of the layers. The extension would tend to cause sheet fractures to turn toward surfaces of sheets rather than propagate parallel to other sheet fractures. Perhaps, therefore, a sheet fracture near the ground surface propagates until it is long enough to allow buckling of the sheet overlying or underlying it. At that point the fracture curves upward or downward, depending upon whether the sheet bends upward or downward. Most of the fractures should curve upward because upward deflections generally will meet less resistance than downward deflections. Thus, we can understand the curved, short fractures near the ground surface, and thus we can understand the deviation of lengths of sheet fractures from the lengths we would predict from our theory, which does not consider buckling or bending (Fig. 10.14).

Possible Origin of Rift Microcracks.—An interesting feature of our theory of sheet fracturing is that it may provide an explanation for another common feature of granitic rock, such as the Chelmsford granite at Fletcher Quarry. Microcracks that parallel the rift of Chelmsford granite are strongly oriented parallel to the horizon (Jahns, personal comm.). The sheet fractures may curve and even become vertical in some places near the ground surface, but the rift microfractures are horizontal. This suggests that the rift cracks predate the sheet structure. Jahns (personal comm.) suggests that the rift microcracks are horizontal because they form when the granite is at depths where effects of variations of surface topography are minimal on the stresses that cause the microfractures to form.

Our theory of sheet propagation is compatible with the conclusion by Jahns. It indicates that sheet fractures should be very short, less than 25 times the width of the fractures, at depths greater than $\frac{1}{50}$ of the depth at which the vertical stress is equal to the initial pressure in the granite (Figs. 10.13 and 10.14). These small sheet fractures may be the horizontal rift microcracks so commonly observed in granitic rocks.

Analysis of thicknesses of sheets in Fletcher Quarry

Certainly one of the most striking features of sheet fractures in Fletcher Quarry is that they are close together near the ground surface and rather widely spaced near the bottom of the quarry (Fig. 10.2). As we noted earlier (Fig. 10.6), the thicknesses of sheets definitely increase with increasing depths in the quarry. The theory of initial stresses may account for the observed variations of sheet thicknesses.

The basic premise with which we attempt to explain the observed variations of sheet thickness is that part of the strain energy induced by unloading of the granite is transformed into surface energy and plastic work along newly formed sheet fractures, as the granite is brought toward the ground surface. According to this premise, there is a critical strain energy, above which sheet fractures form.

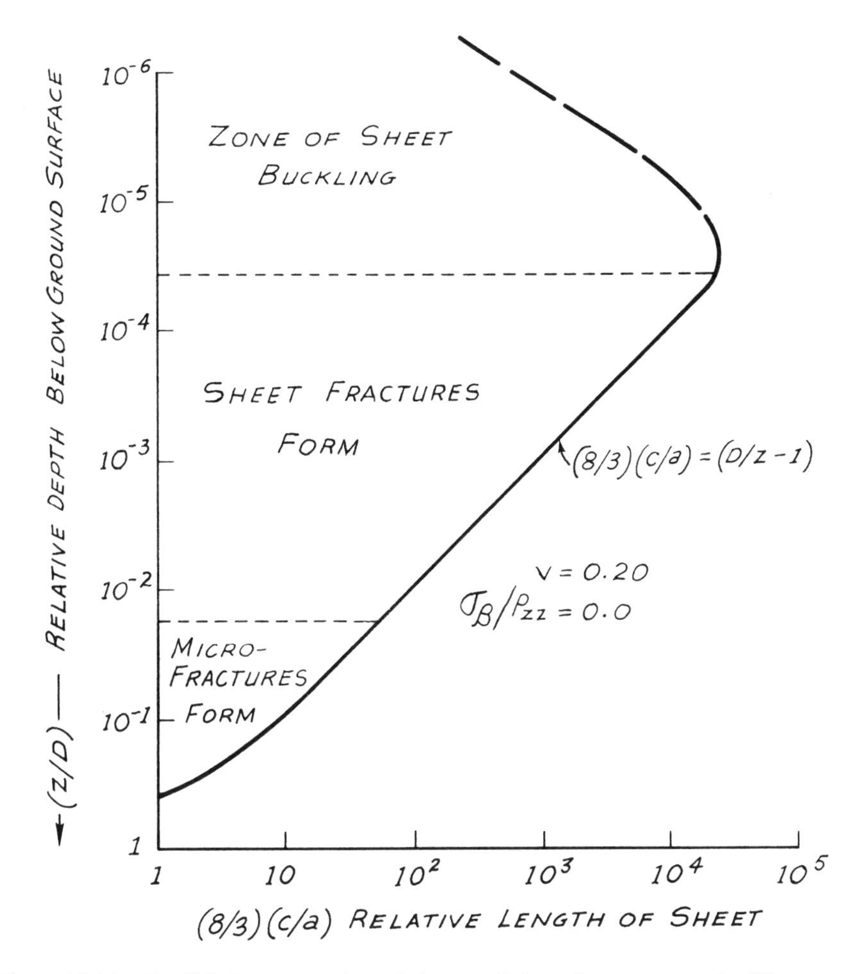

Figure 10.14. Possible interpretation of theory of sheet-fracture growth. Theory seems to predict that microfractures will form at depth, below depth of formation of sheet fractures.

Let the part of the energy be V_0. The surface energy per unit of area, plus the plastic work per unit of area, is T, so that the surface energy and plastic work are $2T\,\delta x\,\delta y$ per sheet fracture, where δx is the length and δy is the breadth of an element of area of the fracture surface. Part of the strain energy is converted into surface energy along the sheet fracture and into plastic work neár the crack tip, so that

$$V_0 = 2nT\,\delta x\,\delta y, \tag{10.17}$$

where n is the number of cracks in the unit element. Thus, part of the strain energy, V_0, in a small element of granite is converted into surface energy and plastic work which equal the surface energy and plastic work per crack in the element times the number of cracks in the element.

The average spacing between sheet cracks in the element is approximately equal to the height of the element divided by the number of cracks. Thus, the spacing can be derived from eq. (10.17) by differentiation with respect to depth, z,

$$\frac{dV_0}{dz} = 2T\,\delta x\,\delta y\,\frac{dn}{dz},$$

so that the spacing, or thickness of sheets, t, is

$$t = \frac{dz}{dn} = \frac{2T\,\delta x\,\delta y}{dV_0/dz}. \tag{10.18}$$

Now we need to derive an expression for the change in strain energy with depth, dV_0/dz.

In earlier pages we defined the strain energy per unit of volume as the area under the stress-strain curve for uniaxial strain. The problem we are considering is one of uniaxial strain because we are assuming that horizontal strains are zero. The stress-strain equation for the initially stressed material we are considering is [eq. (10.9)]

$$\sigma_{zz} - p_{zz} = (\lambda + 2G)\epsilon^*_{zz}. \tag{10.19}$$

The strain, ϵ^*_{zz}, is zero at the depth of solidification of the granite. The stress-strain relation for initially stressed granite is shown in Fig. 10.15A. The total strain energy stored per unit volume of granite is shown in Fig. 10.15A, as the area beneath the stress-strain curve, which is

$$\frac{\delta V}{\delta x\,\delta y\,\delta z} = \tfrac{1}{2}(\sigma_{zz} - p_{zz})\epsilon^*_{zz},$$

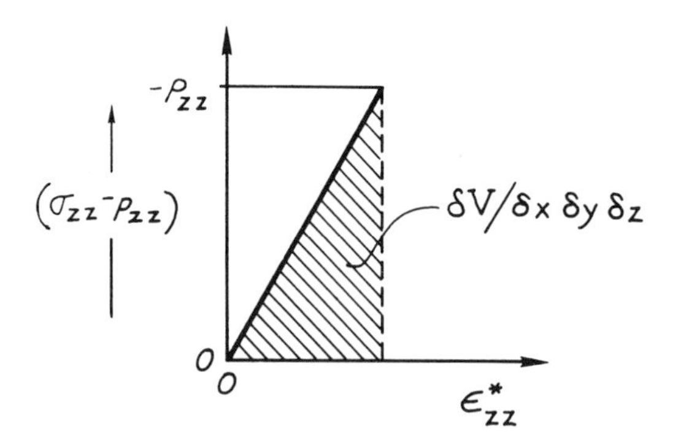

Figure 10.15. Strain energy per unit of volume for initially stressed elastic material.

so that the change of strain energy with change in depth is

$$\frac{dV}{dz} = \tfrac{1}{2}\,\delta x\,\delta y(\sigma_{zz} - p_{zz})\epsilon_{zz}^{*}.$$

Our previous analysis suggests that sheet fractures do not form below a certain depth, D_0, in granite. Below this depth only microcracks should exist. Therefore, let us assume that all the strain energy induced in granite above a depth of D_0 is transformed into surface energy along sheet fractures. In this case the expression for the change in strain energy with depth would be (Fig. 10.15B)

$$\frac{dV_0}{dz} = \tfrac{1}{2}(\sigma_{zz} + \gamma D_0)(\epsilon_{zz}^{*})\,\delta x\,\delta y. \tag{10.20}$$

Combining eqs. (10.19) and (10.20), the change in strain energy is

$$\frac{dV_0}{dz} = \tfrac{1}{2}(\sigma_{zz} + \gamma D_0)^2\left(\frac{1}{2G + \lambda}\right)\delta x\,\delta y. \tag{10.21}$$

Therefore, substituting eq. (10.21) into eq. (10.18). we have an expression for the thickness of sheets,

$$t = \frac{2T\,\delta x\,\delta y}{dV_0/dz} = \frac{4T(2G + \lambda)}{(\sigma_{zz} + \gamma D_0)^2}. \tag{10.22}$$

However, $\sigma_{zz} = -\gamma z$, so that

$$t = \frac{4T}{\gamma^2}(2G + \lambda)\frac{1}{(D_0 - z)^2},$$

or

$$t = \left(\frac{2}{\gamma D_0}\right)^2 T(2G + \lambda)\frac{1}{[1 - (z/D_0)]^2}.$$

Let

$$t' = \frac{t}{(2/\gamma D_0)^2 T(2G + \lambda)}.$$

Then,

$$t' = \frac{1}{[1 - (z/D_0)]^2}. \tag{10.23}$$

The relation between sheet thickness and depth is shown in Fig. 10.16, which shows that the thickness should be a minimum near the ground surface, where z approaches zero. There $t' = 1$. The sheets should increase in thickness rapidly with depth, approaching infinity as the critical depth, D_0, for sheeting is approached. The form of this theoretical relation between sheet thickness and depth corresponds remarkably well with field observations (Fig. 10.6). This can be illustrated by super-

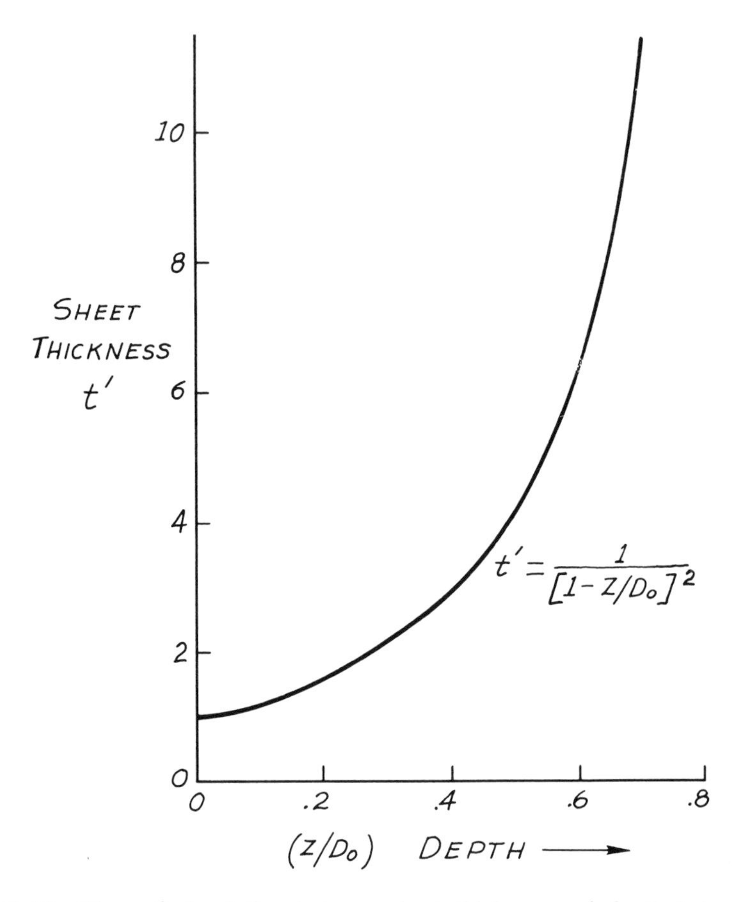

Figure 10.16. Theoretical relation between sheet thickness and depth below ground surface.

imposing a theoretical curve on a graph of the data collected by Jahns (*15*) (Fig. 10.6) in many quarries in Chelmsford granite (Fig. 10.11), using the data to estimate the constants in eq. (10.22). Measurements of the thicknesses of sheets near the ground surface are the most different from what the theoretical curve suggests. The actual sheets are thinner than we would predict with the theory. Perhaps the deviation is caused by near-surface buckling of the thin sheets.

Theoretical Interlude: Growth of fractures in brittle-elastic substances

Griffith Substance.—We have examined the stresses around a crack, which Inglis (*12*) represented by an elliptic hole. But to derive a failure criterion we must state something about the distribution of cracks in the solid and the effects of various

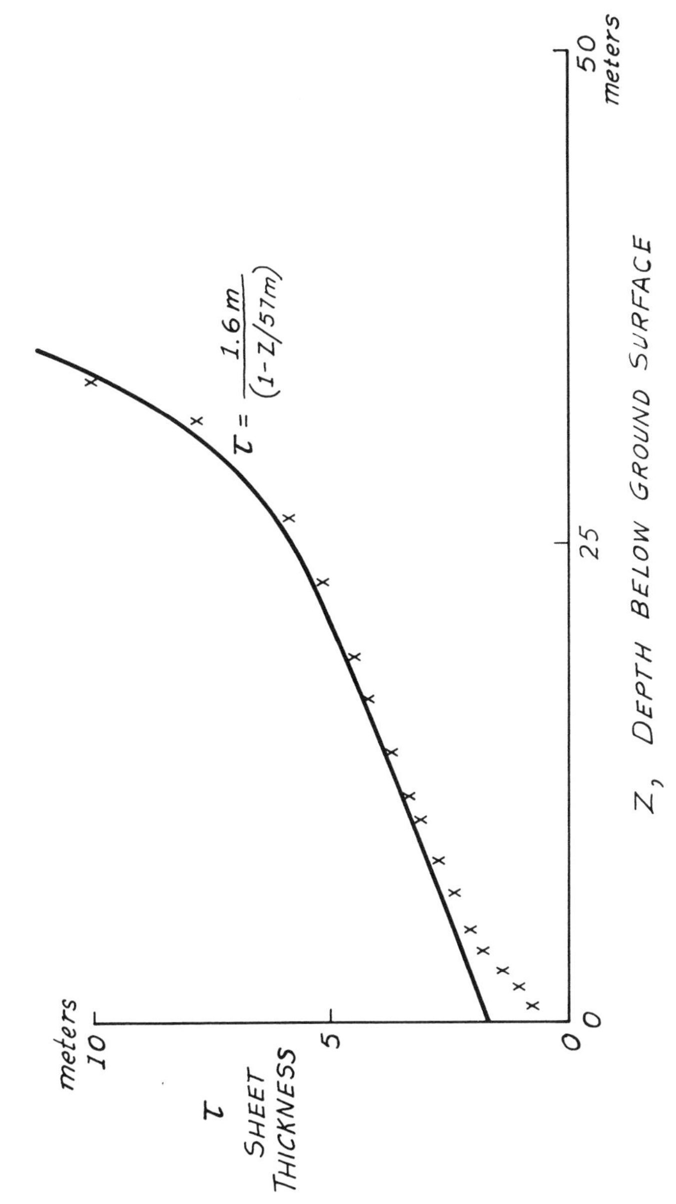

Figure 10.17. Comparison of theoretical relation between sheet thickness and depth below ground surface with average measurements made by Jahns (ref. *15*) in many quarries and natural exposures in Chelmsford granite.

kinds of loading—simple tension, compression, shear, and combinations of these. The Griffith substance (*8,9*) is considered to contain a large number of small cracks oriented at random, but sufficiently far apart so that the maximum stress around each crack is unaffected by stresses induced by the presence of other cracks. Based on the rapid decrease in stress concentration with distance from circular hole in a plate, Fig. 10.9, we can infer that interference of stress fields probably is small if the distances between cracks are several crack lengths.

Tension.—If such a Griffith substance is in uniaxial tension, we can predict approximately the state of stress at failure and the direction of propagation of a large fracture. The most highly stressed cracks are oriented with their long axes normal to the direction of applied tension, are longest, and have the highest length-to-width ratios, $2c/a$. The part of the surface of the crack that is most highly stressed is the crack tip, and therefore the crack will propagate in the direction of its long axis; the crack will propagate in its plane. If the crack is not deflected by an inhomogeneity in the material, the orientation of the final megascopic fracture will be parallel to the initial fracture.

According to Jaeger and Cook (*14*), if the following inequality,

$$3\sigma_1 > -\sigma_3, \tag{10.24}$$

is satisfied, the maximum tension along the crack surface is at the crack tip. Here σ_3 is the confining pressure. Therefore, if the inequality (10.24) is satisfied, we can predict the orientation of the final fracture; it will be parallel to the crack with the highest tensile stress along its surface. Also, if ineq. (10.24) is satisfied, the tensile strength of the material is independent of the magnitude of the confining pressure.

If we designate the tensile strength of the Griffith substance by T_0, we can rewrite ineq. (10.24) as

$$-\sigma_3 < 3T_0. \tag{10.25}$$

Therefore, if the magnitude of the confining pressure is less than three times the uniaxial tensile strength of the substance, the substance will fail along macroscopic fractures that are normal to the direction of applied tension.

Compression.—Unfortunately, if ineq. (10.25) is unsatisfied, we are unable to predict orientations of macroscopic fractures with Griffith's criterion. If the criterion is satisfied we know that there is tension somewhere along the surface of a Griffith crack and that the tension is sufficiently great for the crack to grow. But the maximum tension is not at the very tip of the crack (*2,3,4,10,14,20*).

The exact position of the maximum tension depends upon the state of stress. Figure 10.18 shows the relation between the angle, ψ, between a Griffith crack and the σ_1 direction, and the angle, β, between the σ_1 direction and the normal to the surface of a Griffith crack where tension is maximum. The relation is plotted for

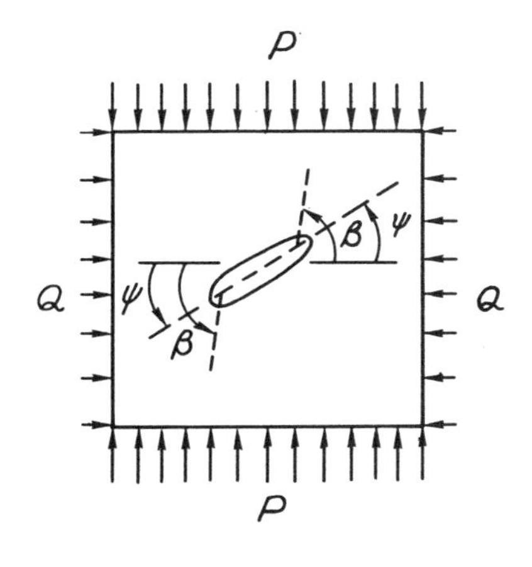

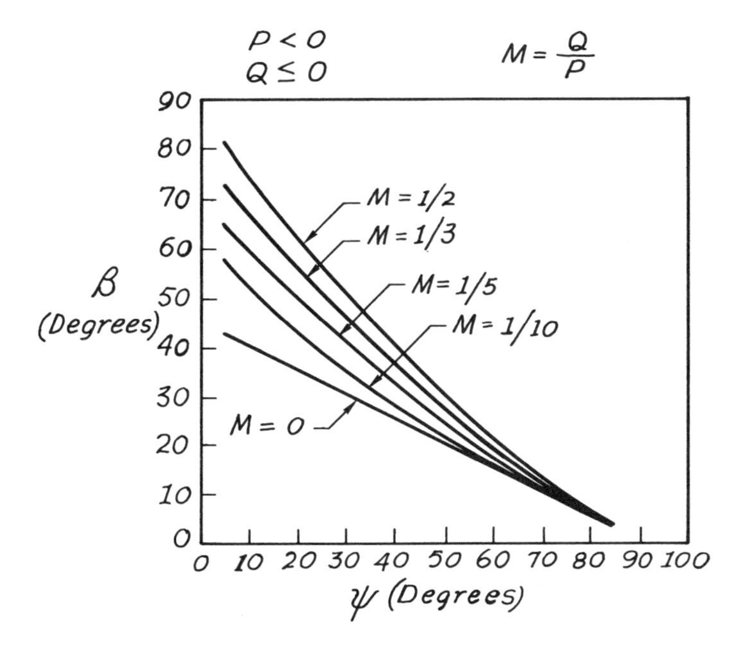

Figure 10.18. Relation between orientation of Griffith crack and position of maximum tension on the surface of crack. Beta, β, is the angle between the Q direction and the normal to the surface of the Griffith crack, where tension is maximum. Psi, ψ, is the angle between the long axis of the Griffith crack and the Q direction (after Paul and Gangal, ref. *20*).

several ratios, σ_3/σ_1, of the minimum and maximum principal stresses in the plane of the fracture. If the crack grows normal to its surface, where tension is greatest, it will grow out of the plane of the initial Griffith crack (Fig. 10.19A). This was

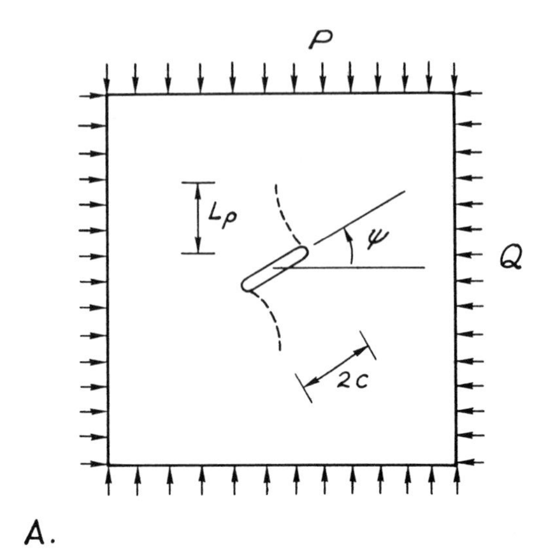

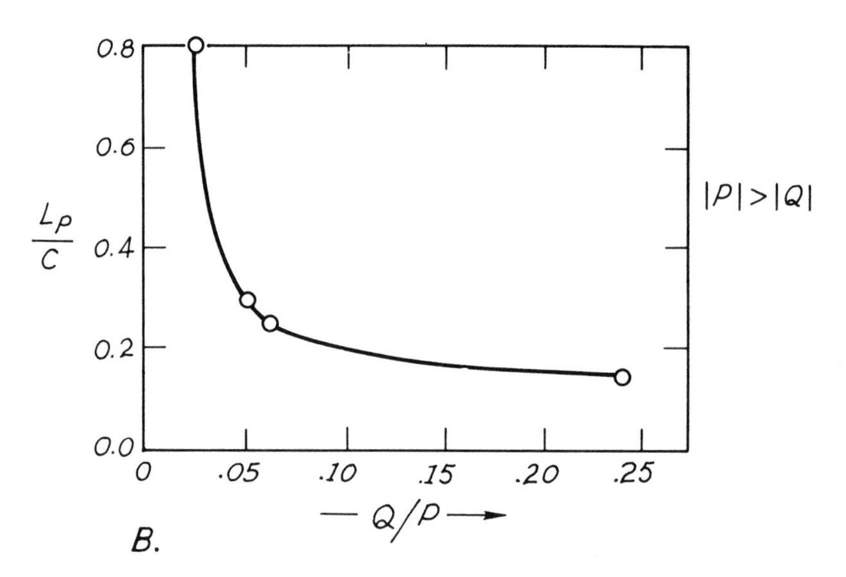

Figure 10.19. Growth of an elliptical crack.

A. Dashed lines indicate paths of propagation of cracks from surface of ellipse. The cracks originated near, but not at, the top of the ellipse (after ref. *10*).

B. Relation between length of stable crack and ratio of applied stresses (after ref. *10*).

verified experimentally by Bombolakis (*2*) and by Hoek (*10*). They found that holes in plates of glass and plexiglass grew in size by emitting cracks that extended along curved paths until the emitted cracks became approximately parallel to the direction of maximum compression, σ_3. According to Hoek (*10*), the distances of propagation of the emitted cracks is controlled by the lengths of the original elliptic holes and the magnitudes of the principal stresses (Fig. 10.19B).

For uniaxial compression, the tensile stress at the tip of an emitted crack that is parallel to the direction of applied loading is independent of the shape of the crack (compare, for example, Figs. 10.10A and 10.10B for $M = 0$; also, see ref. *23*). Crack propagation would continue to the edge of the material under uniform uniaxial compression if the applied stress were equal to the theoretical tensile strength of the material. Thus, in uniaxial compression, the macroscopic fracture should be parallel to the direction of the applied compression unless the material fails first by some other mechanism, such as the one described below.

According to Hoek's (*10*) experimental study of fracture propagation under conditions of biaxial compression, propagating cracks curve toward, but may stop before they become parallel to, the direction of the maximum compressive stress applied to the boundary of the body. This behavior can be understood in terms of Fig. 10.10. We see that if the ratio of principal stresses is greater than about $\frac{1}{10}$, there is compression at the tip of an elliptic crack with a width-to-height ratio of 2. A crack presumably will not grow if its surface is entirely in compression, so that under certain states of stress the crack will cease propagating before it becomes parallel to the direction of maximum compression.

With the concepts we have now developed, we can piece together a theory of one way that rocks could fail in shear. The largest, most favorably oriented cracks would begin to grow first (Fig. 10.20A) as the magnitudes of the stresses are increased. When these cracks become immobilized, shorter and less favorably oriented cracks begin to grow. All of them curve toward the direction of maximum compression (Fig. 10.20B, C, and D) (*10*). The specimen has now reached a stage where it is essentially a compact granular solid. It consists of a mass of particles that fit together nearly perfectly, as pieces of a jigsaw puzzle, because they are bounded by fractures.

Now we can begin to understand why Coulomb's law, which was invented to describe failure of soil and other granular solids, for which it works fairly well, sometimes predicts approximately the orientations of families of joints and faults (*1*). Some rocks, under some conditions of loading, will granulate by growth of internal cracks and flaws. Once the rock has granulated by growth of internal cracks, it should deform much as a soil, for which Coulomb proposed his law.

It is becoming increasingly clear that gross failure of rock, including the formation of faults and joints, involves several factors. Sometimes fracturing is dominated

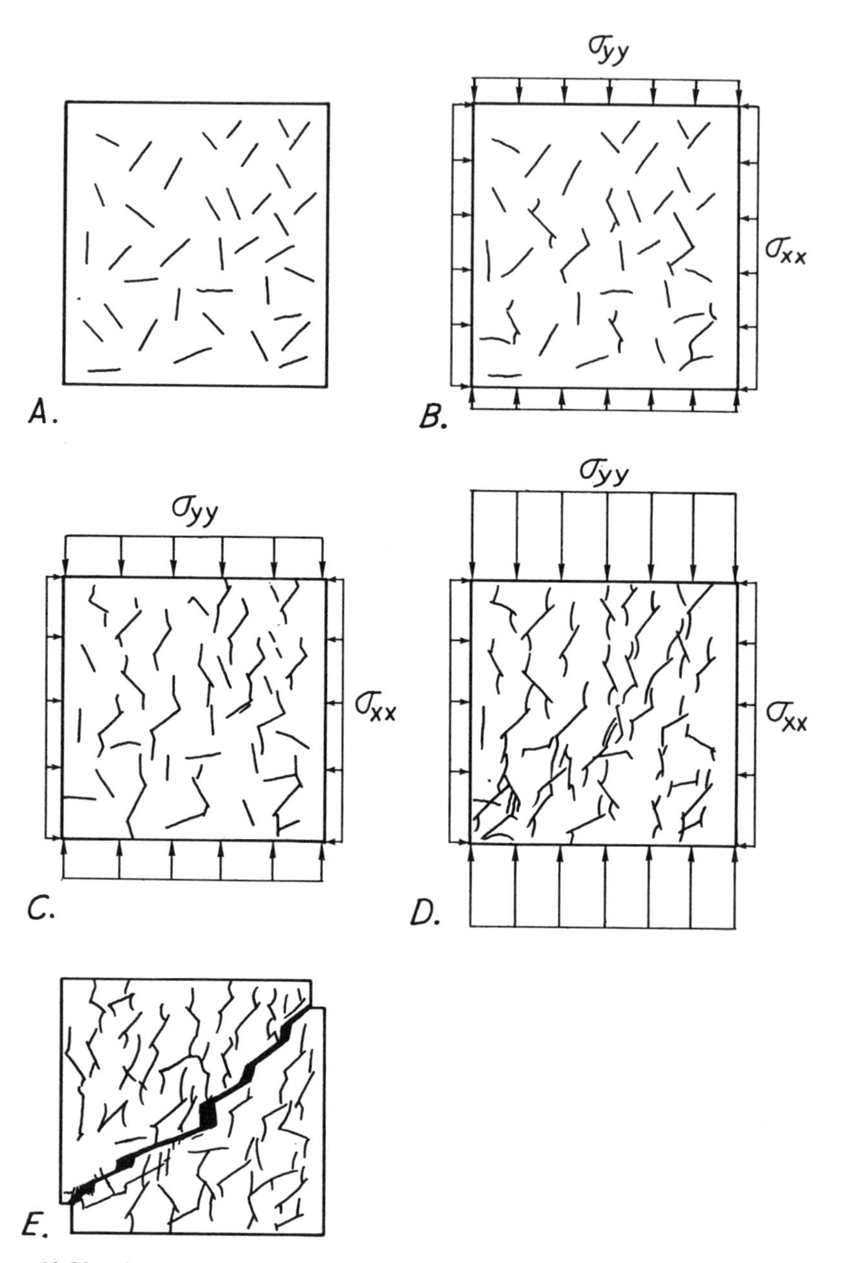

Figure 10.20. A concept of how a large-scale fracture develops in a brittle-elastic substance.

A. Griffith substance containing many cracks.

B. Under load the most favorably oriented cracks and the longest cracks begin to propagate in the direction of the largest compressive stress.

C. As the load increases, shorter and less favorably oriented cracks begin to grow.

D. The specimen now consists essentially of a granular solid.

E. A major fracture forms by coalescence of many small fractures.

by one, sometimes by a combination of factors. The theory of faulting proposed by Anderson (*1*) is easy to comprehend but it probably applies to very few actual fractures. Initial stress or strain of rocks probably was an important factor during the formation of many joints. Exfoliation, "spheroidal weathering," and spalling, as well as sheeting, probably are related to initial stresses, but initial stresses alone cannot account for all these features (*5,19,16,24*).

One factor which we have ignored is displacement. The growth of many megascopic fractures, including faults, is associated with relative displacements of bodies. Indeed, different kinds of faults are classified according to types of displacements along them (*1*). The possible types of displacement of parts of bodies should have a significant effect on the form of megascopic fractures that form in them (*18*). An example of megascopic fracture being controlled by displacement boundary conditions is the formation of "shear fractures" in uniaxially-loaded rock specimens. Fig. 10.21 shows three cores of Chelmsford granite. The left-hand core was not loaded. The middle core was loaded until it failed by shear predominantly along a single plane, involving lateral movement of the top half of the core with respect to

Figure 10.21. Effect of displacement boundary conditions on fracture mode. Left-hand specimen not loaded. Middle specimen loaded under conditions that allowed one end of specimen to move laterally with respect to other end; result, "simple shear" failure. Right-hand specimen loaded under conditions that restricted lateral movement of ends of rock core; result, "pure shear" failure.

the bottom half. The right-hand core failed along cone-shaped fractures, involving only vertical movement of the cones and lateral movement of the rock surrounding the cones. Thus, the middle core failed in "simple shear" whereas the right-hand core failed by "pure shear." S. D. Peng (personal comm.) loaded a series of cores the ends of which were lubricated with teflon sheets and silicon grease, and another series of cores the ends of which were unlubricated and rested against steel platens. Most of the cores with lubricated ends failed predominantly by "simple shear," as did the middle core in Fig. 10.21, and most of the cores with unlubricated ends failed by "pure shear," as did the right-hand core in Fig. 10.21. The lubrication allowed the ends of cores to slip horizontally relative to each other. Thus, the displacement boundary conditions can be an important factor in determining the type of gross failure of rock cores and they must be equally important in nature.

Meaningful analyses of fracture patterns will have to include consideration of each of the factors we have discussed as well as others we have not recognized. The factors can be determined only by careful field study. Most important, however, we should bear in mind the fact that failure criteria, alone, usually will not predict orientations of megascopic fractures.

References cited in Chapter 10

1. Anderson, E. M., 1942, *The Dynamics of Faulting and Dyke Formation with Application to Britain*: Oliver and Boyd, Edinburgh, London.
2. Bombolakis, E. G., 1964, "Photoelastic Investigation of Brittle Crack Growth Within a Field of Uniaxial Compression": *Tectonophysics*, V. 4, p. 343–351.
3. Brace, W. R., 1960, "An Extension of the Griffith Theory of Fracture to Rocks": *Jour. Geophys. Research*, V. 65, p. 3477–3480.
4. Brace, W. R., and Bombolakis, E. G., 1963, "A Note on Brittle Crack Growth in Compression": *Jour. Geophys. Research*, V. 68, p. 3790–3813.
5. Cottrell, A. H., 1965, "Mechanism of Fracture": in *Fracture, Proc. 1st Tewksbury Symposium*; Butterworth Inc. Washington, D.C.
6. Dale, T. N., 1923, "The Commercial Granites of New England": *U.S. Geol. Survey Bulletin*: 738.
7. Gilbert, G. K., 1904, "Domes and Dome Structure of the High Sierra": *Bull. Geol. Soc. America*, V. 15, p. 29–36.
8. Griffith, A. A., 1920, "The Phenomena of Rupture and Flow in Solids": *Phil. Trans. Roy. Soc. London*, Ser. A, V. 221, p. 163–198.
9. ———, 1924, "Theory of Rupture": *1st Internat. Cong. Applied Mech., Proc.*, Delft.
10. Hoek, E., 1965, "Rock Fracture Under Static Conditions": *Nat. Mech. Engrg. Res. Inst., Council for Sci. and Indust. Research*, Rept. MEG 383, Pretoria, South Africa.
11. Hooker, V. E., and Johnson, C. F., 1969, "Near-Surface Horizontal Stresses Including the Effects of Rock Anisotropy": *U.S. Bur. Mines*, Rept. Invest. 7224.
12. Inglis, C. E., 1913, "Stresses in a Plate Due to the Presence of Cracks and Sharp Corners": *Inst. Naval Architects, Trans.*, V. 55, p. 1, London.

13. Jaeger, J. C., 1967, "Brittle Fracture of Rocks": in *Failure and Breakage of Rock; Proc. 8th Symp. Rock Mech., Amer. Inst. Mining Engrs.*, Inc., N.Y., p. 3–57.
14. Jaeger, J. C., and Cook, N. G. W., 1969, *Fundamentals of Rock Mechanics*: Methuen and Co., Ltd., London.
15. Jahns, R. H., 1943, "Sheet Structure in Granites: Its Origin and Use as a Measure of Glacial Erosion in New England": *Jour. Geology*, V. 51, p. 71–98.
16. McClintock, F. A., and Walsh, J. B., 1962, "Friction on Griffith Cracks in Rocks Under Pressure": *Proc. Nat. Congr. Appl. Mech., 4th.*, Berkeley, p. 1015–1021.
17. Obert, L., Duvall, W. I., and Merrill, R., 1960, "Design of Underground Openings in Competent Rocks": *U.S. Bur. Mines, Bull.* 587.
18. Oertel, G., 1965, "The Mechanism of Faulting in Clay Experiments": *Tectonophysics*, V. 2, p. 343–393.
19. Orowan, E., 1949, "Fracture and Strength of Solids": *Rept. Prog. Physics*, V. 12, p. 185–232.
20. Paul, B., and Gangal, M., 1967, "Initial and Subsequent Fracture Curves for Biaxial Compression of Brittle Materials": in *Failure and Breakage of Rock, Proc. 8th Symp. Rock Mech., Amer. Inst. Mining Engrs., Inc.*, N.Y., p. 113–141.
21. Sack, R. A., 1946, "Extension of Griffith's Theory of Rupture to Three Dimensions": *Proc. Phys. Soc.*, London, V. 58, p. 729–736.
22. Tetelman, A. S., and McEvily, A. J., Jr., 1967, *Fracture of Structural Materials*: John Wiley and Sons, Inc., N.Y.
23. Timoshenko, S., and Goodier, J. N., 1951, *Theory of Elasticity*: McGraw-Hill Book Co., Inc., N.Y.
24. Walsh, J. B., and Brace, W. F., 1964, "A Fracture Criterion for Brittle Anisotropic Rock": *Jour. Geophys. Research*, V. 69, p. 3449–3456.

11

Chapter Sections

Spanish Peaks Region
Dike Mountain
  Stock and Radial Dike Swarm
  Possible Causes of Dike Pattern
  Theoretical Relation of Dike Pattern to Stock
    Introduction to problem
    Equations of equilibrium in polar coordinates
    Compatibility equation in polar coordinates
    Stresses in infinite plate containing pressurized hole
  Comparison of Theoretical and Actual Patterns of Dikes at Dike Mountain
East and West Spanish Peaks
  Geologic Setting
  Odé's Approach
  Method of Images
  Two Circular Pressurized Holes in Infinite Plate
  Comparison of Theoretical and Actual Patterns of Dikes at Spanish Peaks
References Cited in Chapter 11
Other References on Fracturing, Diking, Jointing and Faulting

11

Dike Patterns at Spanish Peaks, Colorado

Spanish Peaks region

The Spanish Peaks region of south-central Colorado is blessed with intrusive rocks of nearly every variety geologists have imagined: dikes, sills, stocks, plugs, laccoliths, and sole injections; granites, syenites, diorites, and gabbros (4). Certainly the most impressive form of intrusive in the Spanish Peaks region, however, is the dike. Dikes are so conspicuous there that F. V. Hayden noticed them in 1869 during one of his whirlwind tours which formed part of the famous Hayden Survey of the western United States (4). The dikes were studied by R. C. Hills (2) at about the turn of the century and later by Adolf Knopf (5), whose detailed petrographic descriptions of the rock types led to his famous discussion of the "lamprophyre concept."

The Spanish Peaks area is a high plain, gradually rising westward and punctuated here and there by large igneous masses that project above the plain to form Dike Mountain and the Spanish Peaks themselves. Long walls and ridges formed by dikes radiate much as spokes of wheels around hubs formed by Spanish Peaks and Dike Mountain.

The widths of the dikes range from one foot to about 100 feet. Some of them extend for 14 miles. They usually fill joints and in some places several dikes fill the same joint. Most of the dikes are vertical or nearly so (*4*).

The regional structure of the Spanish Peaks area is dominated by La Veta syncline, an asymmetric trough trending roughly north-south and passing beneath West Peak and Dike Mountain (Fig. 11.1). The eastern side of the synclinal trough dips gently and the western side dips steeply and is overturned in a few places. Immediately west of the trough are the Sangre de Cristo Mountains, which have been thrusted eastward along faults that generally parallel the axis of La Veta syncline (Fig. 11.1). Normal faults are uncommon in the area (*3,4*).

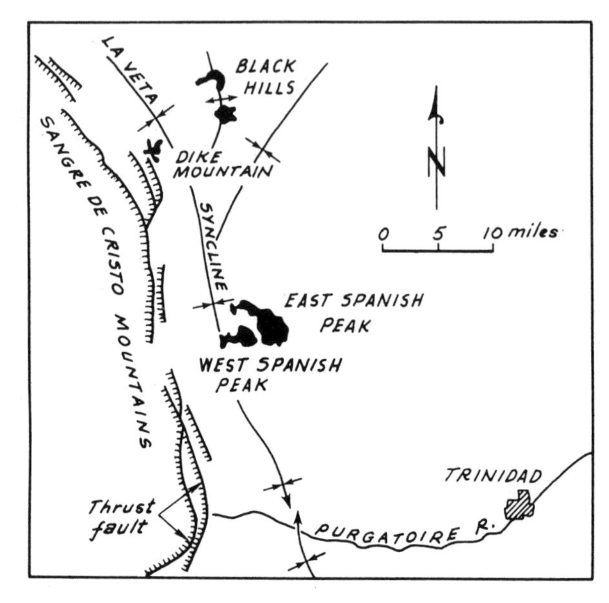

Figure 11.1. Spanish Peaks area, Colorado (after ref. *3*, p. 580).

Conglomerates and sandstones in the Spanish Peaks region are highly jointed. The dominant joint system strikes roughly east, normal to the axis of La Veta syncline, but other systems have strikes ranging around the compass. The highly jointed rocks occur all around Spanish Peaks for distances up to ten miles east of the peaks and three miles west of the peaks. Older shales and limestones, which are exposed on the limbs of La Veta syncline, beyond the limits of the sandstone outcrops, display no well-developed joints (*3,4*).

The dikes associated with the Spanish Peaks and with Dike Mountain are essentially restricted to the jointed sandstone; few dikes invaded the shales (*3,4,5*).

Dike Mountain

Stock and Radial Dike Swarm.—Dike Mountain is a stock of syenodiorite porphyry which intruded Eocene sandstones near the axis of La Veta syncline (Fig. 11.1). It has the irregular form of an ink blot and is $1\frac{1}{4}$ miles long from east to west and about one mile long from north to south (Fig. 11.2). According to Ross Johnson (*3*), there is no evidence of metamorphism or doming by intrusion of the Dike Mountain stock.

Associated with the stock is a swarm of radial dikes which radiate from a center within the stock. A few dikes actually join the rock within the stock (*3,4*). The dikes on the western side of the mountain are shorter than those on the eastern side, extending about five miles east and $2\frac{1}{2}$ miles west of the stock.

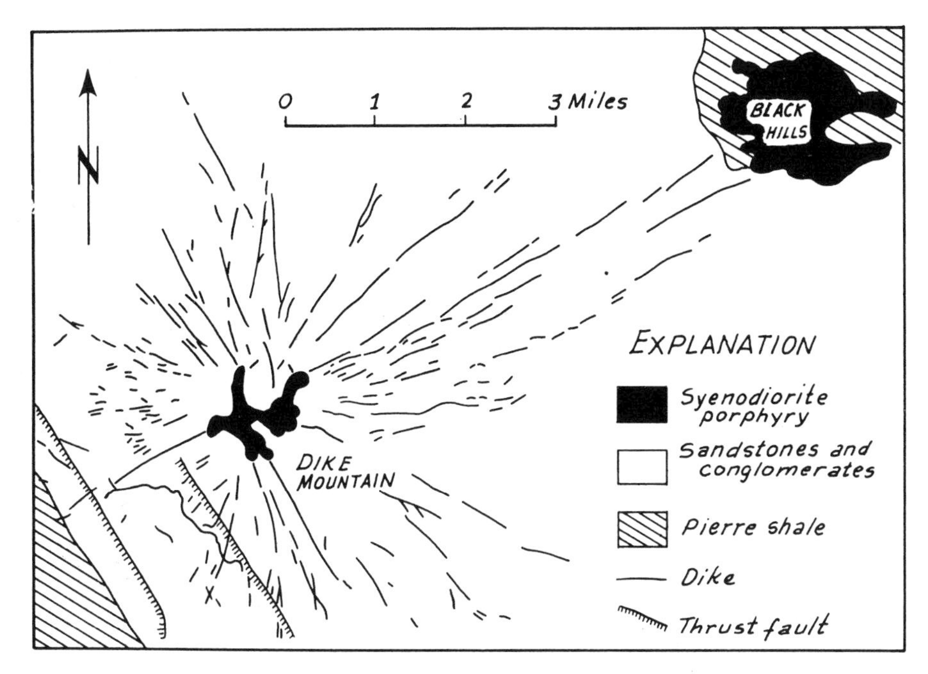

Figure 11.2. Dike swarm around Dike Mountain, Spanish Peaks area, Colorado (after ref. *3*, p. 587).

The composition of the dikes is similar to that of the stock, and for that reason Ross Johnson (*3*), who has been studying rocks of the Spanish Peaks area in detail, postulates that the dikes and the stock represent a single episode of intrusive activity, involving the same magma.

Possible Causes of Radial Dike Pattern at Dike Mountain.—About as many causes have been suggested for the dike patterns as there have been geologists studying them. R. C. Hills (*2*) suggested that the dikes occupy radial fractures produced by doming of the sandstones during intrusion of the stocks. Ross Johnson (*4*) however, reports that the sandstones were not domed. Adolf Knopf (*5*) mentioned the possibility that the dike magmas moved into joints. He suggested that the joints formed normally to the axis of La Veta syncline, as a result of tension developed by folding, prior to dike intrusion.

Johnson (*3,4*) also believed that the joints provided channels for the dikes and that the joints predated intrusion. He imagined that the joints formed throughout the region as a result of "intermittent orogenic stresses of varying directions during the folding of the La Veta syncline" (ref. *4*, p. 37). The joint pattern is complex, according to Johnson, because each of the several stress fields imposed on the rocks of the area created new families of shear and tension joints.

His explanation of the relation between the joints and the dikes is not particularly clear to me but I believe that the following is a reasonably accurate interpretation. The dike pattern, he reasoned, is related to the stocks, to the low viscosity of the magma, and to the preexisting joints. Apparently he imagined the viscosity of the magma that formed the stock of Dike Mountain to be so high that the magma was not able readily to insinuate itself into joints. Thus, this magma did not form many dikes. However, he suggested, the pressure exerted by intrusion of the stock affected the pattern of later dikes. The dikes themselves formed from a less viscous magma which moved along joints oriented normally to surfaces of equal pressure, which were oriented concentrically with respect to the stock. Thus, concentric, equal-pressure surfaces induced by the stock caused the dikes to be crudely radial; joints oriented concentrically with respect to the stock were closed by the pressure in the stock, whereas radial joints were able to open relatively easily under the pressure exerted by the magma in the dikes.

Johnson's (*3,4*) analysis of the dike pattern is partly dependent upon Odé's (*6*) theoretical study of the dike pattern at West Spanish Peak. According to Odé, as we see in more detail in later pages, the dikes surrounding the Spanish Peaks are parallel to trajectories of maximum compressive stresses, largely caused by pressure in the stock of West Peak.

Later on we will follow in some detail Odé's theoretical analysis of the patterns, partly because it provides an excellent introduction to problems involving polar coordinates. We should, however, bear in mind the point made by Johnson (*3,4*) that regional stress patterns, such as those considered by Odé, strongly influenced which sets of joints were selected as channels of dike intrusion but the channels themselves almost certainly existed before intrusion.

Theoretical Relation of Dike Pattern to Stock.—*Introduction to problem.*— We will follow Odé's general approach by assuming that the dikes surrounding Dike Mountain parallel a radial stress pattern that existed at the time the dikes were intruded (*6*). Thus, the dikes follow trajectories of radial stresses. We will show that the trajectories of principal stresses surrounding a pressurized circular hole in a plate of infinite extent are radial and concentric with respect to the hole.

We will idealize the actual, irregular outline of the stock at Dike Mountain by a circular hole. The stress pattern in the plate at distances of one or two diameters from the hole is, indeed, unaffected by the shape of the hole, according to the principle of Saint-Venant (*7*). Near the hole the pattern is complex, but at moderate distances it is quite simple.

Another simplification we will employ is the assumption that sedimentary rocks around Dike Mountain behaved as an elastic plate of infinite horizontal extent. Our conclusions would be unchanged if we assumed that the rocks were infinitely thick and that they were penetrated by a vertical circular tube. Consequently, this simplification does not place our analysis in jeopardy of being oversimplified, at least in terms of the plan shape of the dike complex. If we were interested in the shape of the complex in vertical cross-section, we might have to use a different conceptual model.

Finally, we will assume that the sedimentary rocks are essentially homogeneous, isotropic, and linearly elastic materials on the scale with which we are dealing. This assumption allows us to use the powerful tools of elasticity theory.

Equations of equilibrium in polar coordinates.—In discussing the stresses in bodies subjected to point loads or to uniform loading in a circular hole, it is advantageous to write the differential equations of equilibrium in terms of polar coordinates.

Consider the equilibrium of a small element cut out of the body (Fig. 11.3).

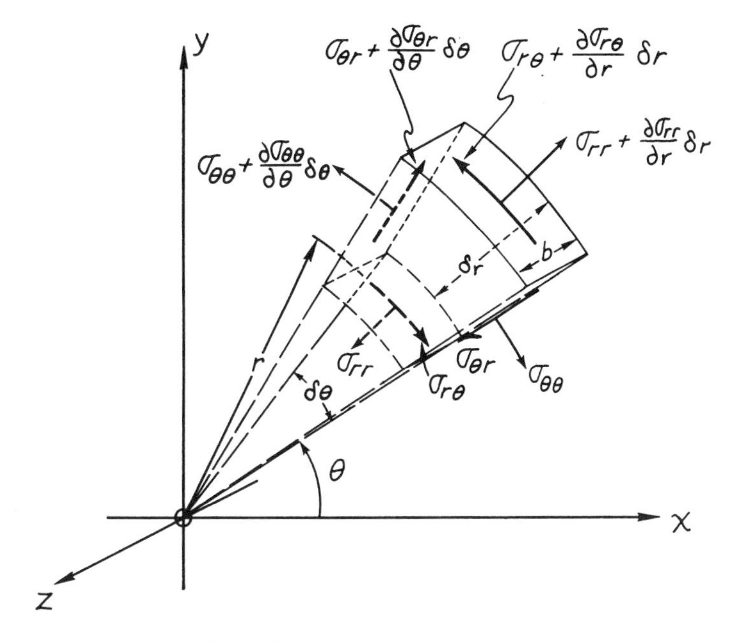

Figure 11.3. Stress element in polar coordinates.

The position of the end of the element closest to the origin, *O*, is defined by a radial distance, *r*, and an angle, θ, between the *r*- and *x*-directions. The normal stress in the radial direction is designated by σ_{rr} and the circumferential stress, acting in the θ-direction, is $\sigma_{\theta\theta}$. The shear stresses are $\sigma_{r\theta}$ and $\sigma_{\theta r}$, the first subscript indicating the normal to the plane on which the stress acts. There is a gradient in the stresses across the width and height of the element so that the stresses on opposite faces of the element are different (Fig. 11.3).

The summations of forces in the *r*- and θ-directions must be zero for the element to be in equilibrium; therefore, $\sum F_r = 0$ and $\sum F_\theta = 0$. The areas of faces $ABDC$ and $EFGH$ are the same, $\delta r b$, where b is the thickness of the element. The area of the curved face $DCGH$ is $(r + \delta r)\,\delta\theta b$ and the area of the curved face $ABEF$ is $r\,\delta\theta b$. Multiplying the stress components by the appropriate areas, to convert the loads into forces, and summing the forces in the radial direction,

$$\left[\sum F_r = 0\right],$$

$$R\left(r + \frac{\delta r}{2}\right)\delta\theta\,\delta r b + \left(\sigma_{rr} + \frac{\partial\sigma_{rr}}{\partial r}\,\delta r\right)(r + \delta r)\,\delta\theta b$$
$$- r\sigma_{rr}\,\delta\theta b + \left(\sigma_{\theta r} + \frac{\partial\sigma_{\theta r}}{\partial\theta}\,\delta\theta\right)\delta r b - \sigma_{\theta r}\,\delta r b$$
$$- \left(\sigma_{\theta\theta} + \frac{\partial\sigma_{\theta\theta}}{\partial\theta}\,\delta\theta\right)\delta r b \sin(\delta\theta) - \sigma_{\theta\theta}\,\delta r b \sin(\delta\theta) = 0, \quad (11.1)$$

where R is the body force in the radial direction. The last two components of force in eq. (11.1) are added to the forces in the radial direction because the circumferential stresses are not normal to the radial direction. We can simplify eq. (11.1) because some of the terms are much smaller than others. Thus,

$\delta r/2$ is much smaller than r, to which it is added, and

$\delta\theta \approx \sin(\delta\theta)$

if $\delta\theta$ is very small, so that in the limit eq. (11.1) reduces to

$$\frac{\partial\sigma_{rr}}{\partial r} + \frac{1}{r}\frac{\partial\sigma_{r\theta}}{\partial\theta} + \frac{\sigma_{rr} - \sigma_{\theta\theta}}{r} + R = 0, \quad (11.2)$$

where $\sigma_{r\theta}$ was substituted for the shear stress of the same magnitude, $\sigma_{\theta r}$. Equation (11.2) is the differential equation of equilibrium for the radial direction.

When forces are summed in the θ-direction, we find that

$$\frac{1}{r}\frac{\partial\sigma_{\theta\theta}}{\partial\theta} + \frac{\partial\sigma_{r\theta}}{\partial r} + 2\frac{\sigma_{r\theta}}{r} + \Omega = 0, \quad (11.3)$$

where Ω is the body force in the θ-direction.

Exercise

Derive eq. (11.3).

Compatibility equation in polar coordinates (7).—From here on we will assume that effects of body forces are negligible compared with effects of surface forces. Thus, let R and Ω be zero in eqs. (11.2) and (11.3). Then the equations of equilibrium are satisfied by a stress function, Φ, where the stress function is related to the components of stress by the relations (7)

$$\left.\begin{aligned}\sigma_{rr} &= \frac{1}{r}\frac{\partial\Phi}{\partial r} + \frac{1}{r^2}\frac{\partial^2\Phi}{\partial\theta^2},\\ \sigma_{\theta\theta} &= \frac{\partial^2\Phi}{\partial r^2},\\ \text{and}\qquad &\\ \sigma_{r\theta} &= \frac{1}{r^2}\frac{\partial\Phi}{\partial\theta} - \frac{1}{r}\frac{\partial^2\Phi}{\partial r\,\partial\theta} = -\frac{\partial}{\partial r}\left(\frac{1}{r}\frac{\partial\Phi}{\partial\theta}\right).\end{aligned}\right\} \tag{11.4}$$

Equations (11.4) could be derived by the method illustrated below.

There are many solutions for the stress function, Φ, that would satisfy eqs. (11.4) and the boundary conditions, but the correct solution must satisfy the compatibility equation as well. In terms of cartesian coordinates, the compatibility equation is

$$\left(\frac{\partial^2}{\partial x^2} + \frac{\partial^2}{\partial y^2}\right)\left(\frac{\partial^2\Phi}{\partial x^2} + \frac{\partial^2\Phi}{\partial y^2}\right) = 0. \tag{11.5}$$

The problem of deriving a compatibility equation for polar coordinates is one of replacing the second derivatives, $\partial^2\Phi/\partial x^2$ and $\partial^2\Phi/\partial y^2$, with expressions involving polar coordinates.

By reviewing our calculus, we find that the chain rule tells us that

$$\frac{\partial\Phi}{\partial x} = \frac{\partial\Phi}{\partial r}\frac{\partial r}{\partial x} + \frac{\partial\Phi}{\partial\theta}\frac{\partial\theta}{\partial x}. \tag{11.6}$$

We need an expression for the second derivative, however, so we differentiate eq. (11.6) with respect to x:

$$\begin{aligned}\frac{\partial^2\Phi}{\partial x^2} &= \frac{\partial}{\partial x}\left(\frac{\partial\Phi}{\partial r}\frac{\partial r}{\partial x}\right) + \frac{\partial}{\partial x}\left(\frac{\partial\Phi}{\partial\theta}\frac{\partial\theta}{\partial x}\right)\\ &= \left[\frac{\partial}{\partial x}\left(\frac{\partial\Phi}{\partial r}\right)\right]\frac{\partial r}{\partial x} + \frac{\partial^2 r}{\partial x^2}\frac{\partial\Phi}{\partial r} + \left[\frac{\partial}{\partial x}\left(\frac{\partial\Phi}{\partial\theta}\right)\right]\frac{\partial\theta}{\partial x} + \frac{\partial\Phi}{\partial\theta}\frac{\partial^2\Phi}{\partial x^2}.\end{aligned} \tag{11.7}$$

By using the chain rule once more, we can perform the differentiations indicated in brackets in eq. (11.7):

$$\begin{aligned}\frac{\partial}{\partial x}\left(\frac{\partial\Phi}{\partial r}\right) &= \frac{\partial(\partial\Phi/\partial r)}{\partial r}\frac{\partial r}{\partial x} + \frac{\partial(\partial\Phi/\partial r)}{\partial\theta}\frac{\partial\theta}{\partial x}\\ &= \left(\frac{\partial^2\Phi}{\partial r^2}\right)\frac{\partial r}{\partial x} + \frac{\partial^2\Phi}{\partial r\,\partial\theta}\frac{\partial\theta}{\partial x}.\end{aligned}$$

Similarly,

$$\frac{\partial}{\partial x}\left(\frac{\partial\Phi}{\partial\theta}\right) = \frac{\partial^2\Phi}{\partial\theta^2}\frac{\partial\theta}{\partial x} + \frac{\partial^2\Phi}{\partial r\,\partial\theta}\frac{\partial r}{\partial x}.$$

The next step in the derivation is to get rid of the terms involving the cartesian coordinate, x. The relations between cartesian coordinates and polar coordinates are, according to Fig. 11.4,

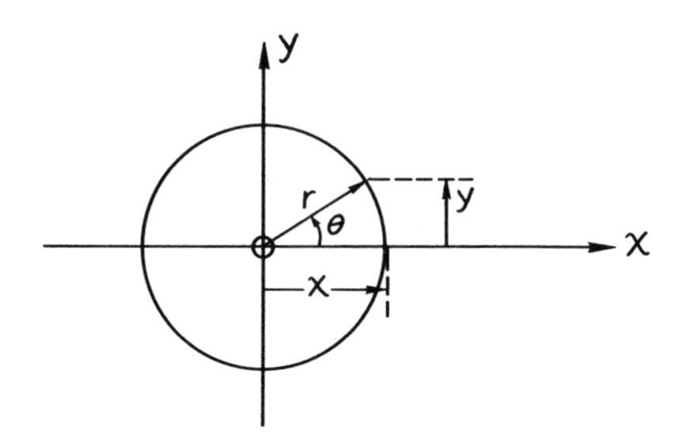

Figure 11.4. Relation between polar and cartesian coordinates.

$$r = \sqrt{(x^2 + y^2)},$$

and

$$\theta = \tan^{-1}(y/x).$$

Differentiating r with respect to x,

$$\frac{\partial r}{\partial x} = \frac{x}{\sqrt{(x^2 + y^2)}} = \frac{x}{r}.$$

Similarly,

$$\frac{\partial \theta}{\partial x} = \frac{1}{1 + (y/x)^2} \frac{-y}{x^2} = \frac{-y}{r^2}.$$

Differentiating once more,

$$\frac{\partial^2 r}{\partial x^2} = \frac{r - x(r/x)}{r^2} = \frac{1}{r} - \frac{x^2}{r^3} = \frac{y^2}{r^3},$$

and (11.8)

$$\frac{\partial^2 \theta}{\partial x^2} = -y \frac{-2}{r^3} \frac{\partial r}{\partial x} = \frac{2xy}{r^4}.$$

Finally, if we substitute the relations $(x/r) = \cos\theta$ and $(y/r) = \sin\theta$ into eqs. (11.8) and then substitute the resulting eqs. (11.8) into eq. (11.7), we find that the second derivative of phi with respect to x is

$$\frac{\partial^2 \Phi}{\partial x^2} = \frac{\partial^2 \Phi}{\partial r^2} \cos^2\theta - \frac{2}{r} \frac{\partial^2 \Phi}{\partial r\, \partial \theta} \sin\theta \cos\theta + \frac{1}{r^2} \frac{\partial^2 \Phi}{\partial \theta^2} \sin^2\theta$$

$$+ \frac{1}{r}\frac{\partial \Phi}{\partial r}\sin^2\theta + \frac{2}{r^2}\frac{\partial \Phi}{\partial \theta}\sin\theta\cos\theta. \qquad (11.9)$$

Similarly for the y-direction,

$$\frac{\partial^2 \Phi}{\partial y^2} = \frac{\partial^2 \Phi}{\partial r^2}\sin^2\theta + \frac{2}{r}\frac{\partial^2 \Phi}{\partial \theta\,\partial r}\sin\theta\cos\theta + \frac{\partial \Phi}{\partial r}\frac{\cos^2\theta}{r}$$

$$- 2\frac{\partial \Phi}{\partial \theta}\frac{\sin\theta\cos\theta}{r^2} + \frac{\partial^2 \Phi}{\partial \theta^2}\frac{\cos^2\theta}{r^2}. \qquad (11.10)$$

If we add eqs. (11.9) and (11.10), we have an expression for one of the operators of the compatibility equation, eq. (11.5):

$$\left(\frac{\partial^2}{\partial x^2} + \frac{\partial^2}{\partial y^2}\right)\Phi = \frac{\partial^2 \Phi}{\partial r^2} + \frac{1}{r}\frac{\partial \Phi}{\partial r} + \frac{1}{r^2}\frac{\partial^2 \Phi}{\partial \theta^2} = \nabla^2\Phi. \qquad (11.11)$$

Exercise

Derive eqs. (11.10) and (11.11).

Now we can write the compatibility equation in terms of polar coordinates:

$$\left(\frac{\partial^2}{\partial r^2} + \frac{1}{r}\frac{\partial}{\partial r} + \frac{1}{r^2}\frac{\partial^2}{\partial \theta^2}\right)\left(\frac{\partial^2 \Phi}{\partial \theta^2} + \frac{1}{r}\frac{\partial \Phi}{\partial r} + \frac{1}{r^2}\frac{\partial^2 \Phi}{\partial \theta^2}\right) = 0. \qquad (11.12)$$

If we use the operator defined by eq. (11.11), we can write the compatibility equation in shorthand forms:

$$\nabla^2\nabla^2\Phi = \nabla^4\Phi = 0. \qquad (11.13)$$

Stresses in infinite plate containing pressurized hole.—Now that we have written the compatibility equation in a form that corresponds with the polar geometry of the problem we wish to solve, we can proceed to find a stress function that satisfies the boundary conditions for the problem. According to the conditions we have assumed to have existed at the time of formation of the dike pattern at Dike Mountain, the magma in the stock of Dike Mountain exerted a pressure, p, at the boundary of the plug, with a radius of r_0. Also, the effect of the pressure in the plug diminishes to zero at large distances from the plug. Thus,

$$\left.\begin{aligned} &\sigma_{rr} = \sigma_{\theta\theta} = 0, \qquad \text{as } r \to \infty \\ &\text{and} \\ &\sigma_{rr} = p, \qquad \text{at } r = r_0 \end{aligned}\right\} \qquad (11.14)$$

are the boundary conditions our solution must satisfy. We can calculate the components of stress from the stress function by using eqs. (11.4).

The stress applied to the edge of the circular hole is a uniform pressure (Fig. 11.5). We can imagine, therefore, that the stresses at any radial distance, r, are the same, independent of position around the hole. The stresses may vary with radial

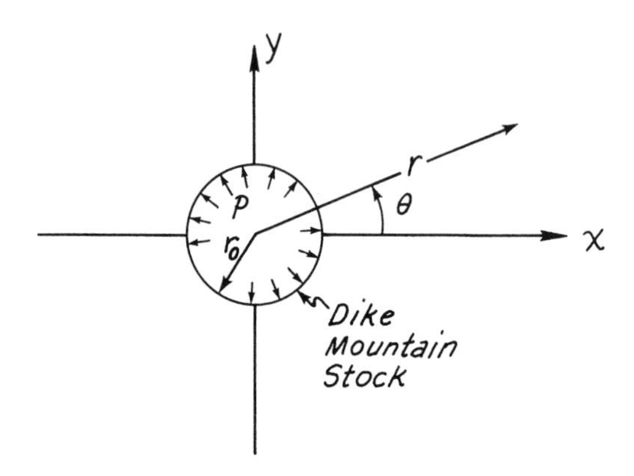

Figure 11.5. Dike Mountain stock represented by circular hole in large sheet of elastic material. Pressure, p, exerted by magma in stock on edges of circular hole.

distance from the hole but, by symmetry, they should be independent of θ. It follows, by eqs. (11.4), that the stress function is independent of θ and that the compatibility equation becomes an ordinary differential equation [θ = constant in eq. (11.12)]:

$$\left(\frac{d^2}{dr^2} + \frac{1}{r}\frac{d}{dr}\right)\left(\frac{d^2\Phi}{dr^2} + \frac{1}{r}\frac{d\Phi}{dr}\right) = \frac{d^4\Phi}{dr^4} + \frac{2}{r}\frac{d^3\Phi}{dr^3} - \frac{1}{r^2}\frac{d^2\Phi}{dr^2} + \frac{1}{r^3}\frac{d\phi}{dr} = 0. \quad (11.15)$$

This differential equation is similar in form to those we have solved in preceding chapters. A significant difference is that the independent variable, r, appears in coefficients of the derivatives. We can convert the equation to the standard form that we have solved before, however, by substituting a new variable for r. Thus, let (7)

$$r = e^t,$$

which is equivalent to

$$t = \ln r,$$

be the relationship between the new variable, t, and r. Here ln means "natural logarithm of."

The derivatives of t and Φ are:

$$dt = dr/r,$$

$$\frac{d\Phi}{dr} = \frac{D}{r} = \frac{1}{r}\frac{d\Phi}{dt},$$

where $D = d\Phi/dt$, for shorthand. The second derivative can be calculated as follows:

$$\frac{d}{dr}\left(\frac{1}{r}\frac{d\Phi}{dt}\right) = \frac{1}{r}\frac{d}{dr}\left(\frac{d\Phi}{dt}\right) - \frac{1}{r^2}\frac{d\Phi}{dt}.$$

However,

$$\frac{1}{r}\frac{d}{dr}\left(\frac{d\Phi}{dt}\right) = \frac{1}{r}\frac{d^2\Phi}{dt^2}\frac{dt}{dr} = \frac{1}{r^2}\frac{d^2\Phi}{dt^2},$$

because

$$\frac{d}{dr} = \frac{1}{r}\frac{d}{dt}.$$

Thus,

$$\frac{d^2\Phi}{dr^2} = \frac{1}{r^2}(D^2 - D).$$

Similarly, the third and fourth derivatives are

$$\frac{d^3\Phi}{dr^3} = \frac{1}{r^3}(D^3 - 3D^2 + 2D),$$

and

$$\frac{d^4\Phi}{dr^4} = \frac{1}{r^4}(D^4 - 6D^3 + 11D^2 - 6D).$$

Although these expressions for derivatives in terms of the new variable appear to be worse than the differential equation with which we started, they actually simplify the equation. Substituting the derivatives into eq. (11.15),

$$\frac{1}{r^4}(D^4 - 6D^3 + 11D^2 - 6D) + \frac{2}{r^4}(D^3 - 3D^2 + 2D) - \frac{1}{r^4}(D^2 - D) + \frac{1}{r^4}D = 0.$$

Gathering terms and multiplying through by r^4, we find that the differential equation becomes a fourth-order equation with constant coefficients:

$$(D^4 - 4D^3 + 4D^2) = 0, \tag{11.16}$$

where D^4, for example, is shorthand for the fourth derivative of Φ with respect to t.

Now the equation is in a form that we can solve readily with the trial solution,

$$\Phi = e^{mt}, \tag{11.17}$$

where m is a constant. Substituting eq. (11.17) into eq. (11.16),

$$D = d\Phi/dt = me^{mt}$$

. . . .

. . . .

$$D^4 = d^4\Phi/dt^4 = m^4e^{mt},$$

so that eq. (11.16) becomes

$$(m^4 - 4m^3 + 4m^2)e^{mt} = 0,$$

or

$$m^4 - 4m^3 + 4m^2 = 0.$$

Factoring,

$$m^2(m - 2)^2 = m^4 - 4m^3 + 4m^2 = 0, \tag{11.18}$$

so that the roots are

$$m = 0 \quad \text{and} \quad m = 2.$$

It follows that

$$\Phi = e^{mt} = Ae^{0t} = A$$

and

$$\Phi = Be^{2t}$$

are solutions of eq. (11.16), where A and B are constants. These are independent, particular solutions of eq. (11.16), and the general solution would appear to be

$$\Phi = A + Be^{2t}.$$

However, we know that the solution of a fourth-order differential equation must have four arbitrary constants. Ours has only two, so that we know there are other possible solutions to the differential equation. This difficulty arose because there were only two distinct roots in the characteristic equation, eq. (11.18). The general solution to eq. (11.16) turns out to be (see a textbook on applied mathematics or on ordinary differential equations):

$$\Phi = (A + Ct) + (B + Et)e^{2t}, \tag{11.19}$$

where A, B, C, and E are arbitrary constants. Note that the general solution is derived by replacing A by $A + Ct$ and B by $B + Et$.

In order to determine what our solution means, we need to dispose of the new variable, t, and perform the appropriate differentiations on Φ in order to derive expressions for the stresses. Thus,

$$r = e^t \quad \text{or} \quad t = \ln r,$$

so that

$$\Phi = A + C \ln (r) + Br^2 + Er^2 \ln (r).$$

Also,

$$\left.\begin{aligned}\sigma_{rr} &= \frac{1}{r}\frac{\partial\Phi}{\partial r} = \frac{C}{r^2} + 2B + E(1 + 2\ln r),\\ \sigma_{\theta\theta} &= \frac{\partial^2\Phi}{\partial r^2} = \frac{-C}{r^2} + 2B + E(3 + 2\ln r),\\ \text{and}\quad \sigma_{\theta r} &= 0.\end{aligned}\right\} \tag{11.20}$$

The next step in the analysis is to evaluate the arbitrary constants, B, C, and E. According to the boundary conditions, eqs. (11.14),

$$\sigma_{rr} = \sigma_{\theta\theta} = 0 \qquad \text{as } r \to \infty,$$

so that constants B and E in eqs. (11.20) must be zero. Also, according to eqs. (11.14),

$$\sigma_{rr} = p \qquad \text{at } r = r_0.$$

If the pressure is tensile, p is a positive number. If the pressure is compressive, p is negative. From the first of eqs. (11.20),

$$\sigma_{rr} = p = C/r_0^2$$

so that

$$C = pr_0^2,$$

and the stress components, eqs. (11.20), become

$$\left.\begin{aligned}\sigma_{rr} &= p\left(\frac{r_0}{r}\right)^2,\\ \sigma_{\theta\theta} &= -p\left(\frac{r_0}{r}\right)^2, \qquad r \geq r_0\\ \text{and}\quad \sigma_{r\theta} &= 0.\end{aligned}\right\} \tag{11.21}$$

Comparison of Theoretical and Actual Patterns of Dikes at Dike Mountain.—According to our solution for the stresses in the plate, the radial, σ_{rr}, and the circumferential, $\sigma_{\theta\theta}$, normal stresses are of equal magnitude but of opposite sign. Further, the stresses diminish rapidly, in proportion to one over the square of the distance from the edge of the circular hole. The maximum normal stresses are around the edge of the hole, where they are equal in magnitude to the pressure in the hole. Our solution indicates that the circumferential normal stress is tensile, which is somewhat disturbing in light of the preexisting joints in the sandstone. However, we have ignored regional compressive stresses in the rock surrounding the stock. Equal lateral pressures exerted on the rocks regionally could be superimposed upon our solution without changing the orientations of the principal stresses. This superposition could change the circumferential stress to zero or even make it compressive.

The radial and circumferential normal stresses are principal stresses because the shear stresses are zero on the edges of the polar element (Fig. 11.3). Accordingly, the principal stress trajectories are radial and concentric. The radial direction is the direction of maximum compression and the tangential direction is the direction of minimum compression. Following Ross Johnson's (*3,4*) analysis, we would propose that preexisting joints oriented roughly concentrically would be held closed by the radial compression, whereas joints with a radial orientation, relative to the stock, could be opened more easily by the wedging action of magma. In this sense, therefore, our solution predicts the gross radial form of the dike pattern surrounding Dike Mountain (Fig. 11.6).

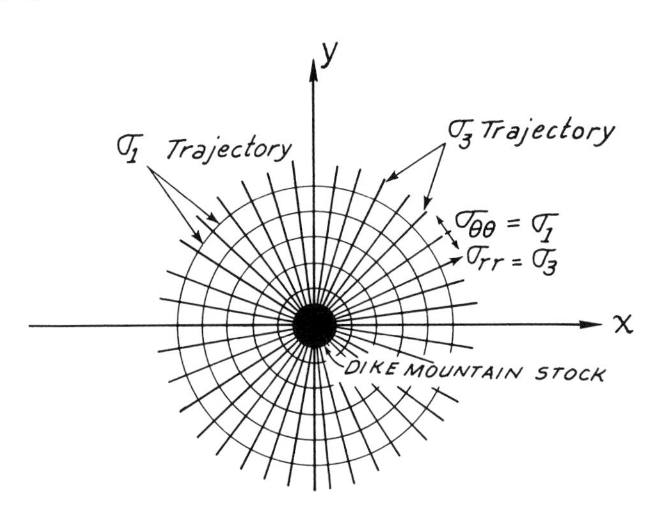

Figure 11.6. Theoretical pattern of dikes around Dike Mountain. Dikes should be radial, following trajectories of radial stresses.

The actual dike pattern at Dike Mountain is different from the theoretical pattern. The dikes on the eastern side of Dike Mountain are much longer than those on the southwestern side (Fig. 11.2). Odé (*6*) considered similar aspects of the dike pattern at Spanish Peaks; we will therefore defer this problem for the moment.

East and West Spanish Peaks

Geologic Setting.—Dike Mountain is actually one of the minor topographic forms in the Spanish Peaks area. The landscape is dominated by the Spanish Peaks themselves. East Peak rises about 6500 feet and West Peak about 7500 feet above the surrounding plain, which is itself at an elevation of 6000 feet (*4*).

The dikes radiating from East and West Peaks are on a grander scale also. They extend for about twelve miles east of the peaks.

According to Ross Johnson (*3*), East Spanish Peak is a large stock roughly circular in plan and composed mostly of granite porphyry. The sedimentary rocks on the south and west sides of the stock were domed by intrusion of the granite porphyry. West Spanish Peak is a stock composed of several varieties of syenodiorite. The intrusions metamorphosed surrounding sedimentary rocks but did not dome or otherwise seriously deform them.

The dike swarm surrounding the two Spanish Peaks converges on West Spanish Peak with few exceptions—some dikes converge on East Peak and a few others converge on points elsewhere (Fig. 11.7). The dikes extend much farther east than west of the peaks. The outline of the dike swarm forms an ellipse with the long axis trending roughly east-west and with one focus on West Peak. Many of the dikes bend around to an east-west course even though they converge on the northeast or southeast quadrants of West Peak. The dikes are most highly concentrated to the west of the peaks. They are short there and they terminate several hundred feet east of the shales and limestones exposed on the western side of La Veta syncline (Fig. 11.7).

The radial dikes range in composition from basalt and gabbro through granite. The gabbro and basalt dikes are short and are relatively distant from West Peak. Most of the dikes are syenodiorite porphyry and occur on all sides of West Peak. According to Johnson (*3*, p. 584), a swarm of parallel dikes, trending east-west, is younger than the swarm associated with West Peak. The parallel dikes are omitted from Fig. 11.7 for simplicity (see Johnson, ref. *4*, plate 1).

Johnson (*4*, p. 36) determined relative ages of dike magma, primarily by studying dike intersections. He suggests that the sequence of intrusion is as follows (abridged) in the vicinity of the Spanish Peaks:

4. (f) Granite (youngest)
   (e) Granodiorite
   (d) Syenite
   (c) Syenodiorite
   (b) Diorite
   (a) Gabbro
3. Radial dike swarm of West Spanish Peak
2. Syenodiorite stock of West Spanish Peak
1. Granite stock of East Spanish Peak (oldest)

These rocks were intruded as separate pulses, according to Johnson (*4*), in late Eocene time, after thrusting and formation of La Veta syncline. Also, as is indicated above, the types of magmas that formed the radial dike swarms are much more diverse than the types of magmas that formed the stocks in East and West Spanish Peaks. He concludes, for several additional reasons, that the dikes were emplaced

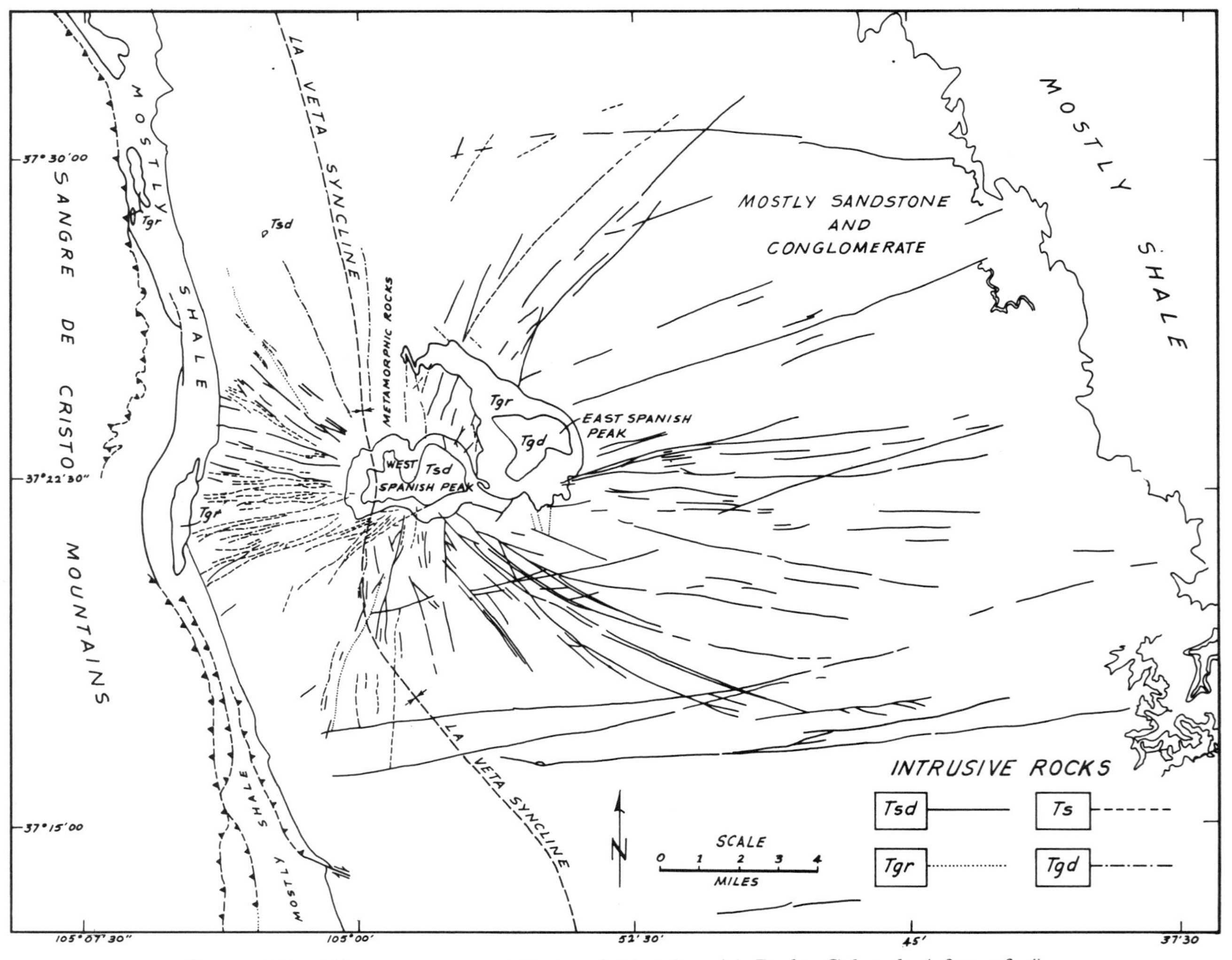

Figure 11.7. Dike swarm around East and West Spanish Peaks, Colorado (after ref. 4).

from sources below, during at least seven episodes of intrusion, rather than being injected radially from West or East Spanish Peak (*4*).

Odé's Approach.—Helmar Odé showed that the gross pattern of dikes surrounding Spanish Peaks can be duplicated theoretically by superimposing a rectilinear, regional stress field upon a radial stress field produced by pressure in a circular hole (*6*). His analysis of the dike pattern is in terms of stress, so that a fundamental assumption is that the gross form of the dike swarm was determined by the stress state of the rocks at the time of intrusion. Thus his statement (ref. *6*, p. 569): "The symmetry of the dike pattern indicates that the total stress field must be symmetric in the same manner." As Johnson (*3,4*) has indicated, however, the pressure in the stock at West Peak almost certainly did not produce the fractures that the dikes intruded. Rather, the assumption is that the pattern of stresses due to the pressure in the stock and to the regional stress pattern influenced the paths of the propagating dikes.

We will assume that the rock surrounding Spanish Peaks was grossly homogeneous, isotropic, and linearly elastic. Further, we will assume that strains are infinitesimal. Three additional assumptions will allow us to compute the stress field throughout the area:

1. The magma rose to the surface through a vertical circular hole;
2. The Sangre de Cristo Mountains to the west acted as a rigid boundary; and
3. The problem can be treated as one of plane stress.

Therefore, the rock behaved as a semi-infinite elastic plate pierced by a circular hole.

Method of Images.—We are supposing that the mountain front of the Sangre de Cristo Mountains, to the west of Spanish Peaks (Fig. 11.7), remained fixed in space so that it behaved as a rigid boundary. However, to calculate the stresses around the stock we would need to know the stresses along the mountain front. But we cannot know them before we solve for the stresses elsewhere. We can extricate ourselves from this dilemma with a technique called the *method of images*, or the method of reflection. With this technique, we reflect the problem region in the vicinity of the stocks across the rigid front so that, in effect, there are now two stocks, equidistant from the rigid front (Fig. 11.8). By this manipulation, the condition of no lateral displacement across the mountain front is satisfied and the stresses can be calculated throughout the problem region, including places along the rigid front, because there are no boundaries with unknown stress.

We have already computed the stress distribution caused by the pressure in one hole in an infinite plate. We can superimpose the stress distributions from two such holes by choosing a suitable origin of coordinates. The method should become clear as we proceed.

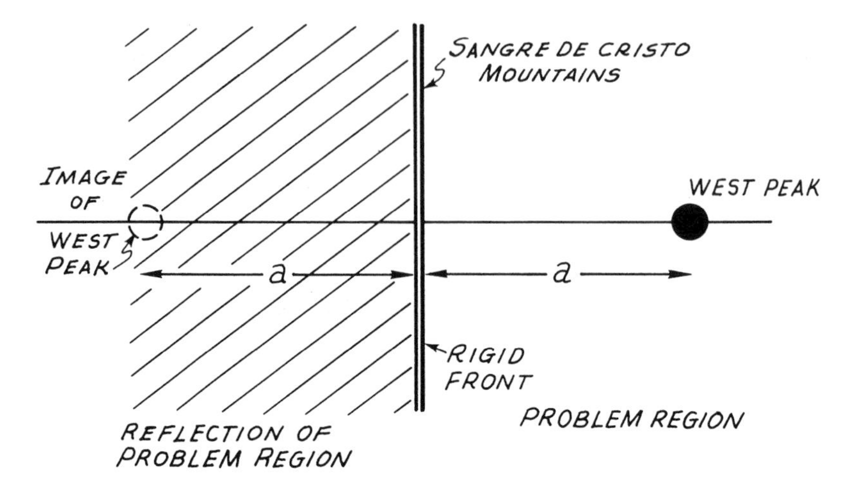

Figure 11.8. Simulation of effect of mountain front by method of images. Problem region in vicinity of West Peak is reflected across rigid front so that, in effect, there are two West Peaks, equidistant from the rigid front.

Two Circular Pressurized Holes in an Infinite Plate.—Suppose that there are two circular holes of radius r_0 in a plate of infinite dimensions. The holes are separated by a distance $2a$ (Fig. 11.9). We want to derive a stress function that will satisfy the condition of zero displacement in the direction normal to a line that is equidistant from the holes and normal to the straight line that connects the centers of the holes. This line corresponds to the rigid mountain front that Odé postulated.

If the pressures in the two holes are equal, there will be no displacement normal to the rigid front. For example, consider the point where a horizontal line, passing through the center of the holes, intersects the rigid front. The magnitude of the radial

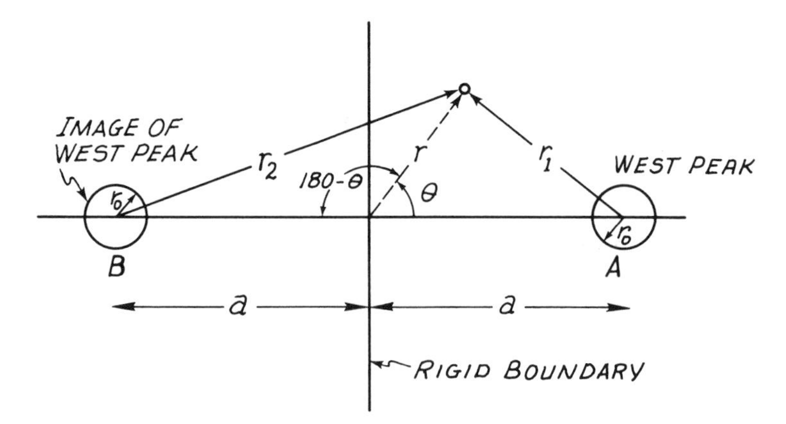

Figure 11.9. Coordinate system for two circular holes in plate of infinite extent.

stress arising from the pressure in the hole at B is the same as that from the pressure in the hole at A. But the stresses are of opposite directions, so that their effects cancel. That is, one tends to displace the front to the right and the other tends to displace the front an equal amount to the left.

Let r_1 be the distance from the hole on the right to some point and let r_2 be the distance from the hole on the left to the same point (Fig. 11.9). Then the solution for the stress function takes the form

$$\Phi = A_1 \ln (r_1) + A_2 \ln (r_2) = A_0 \ln (r_1 r_2), \tag{11.22}$$

if the pressures in the two holes are compressive. We have simply added two solutions of the same type as those that were derived for a single hole in a plate [eq. (11.21)],

$$\Phi = C \ln (r). \tag{11.23}$$

Thus, to derive eq. (11.22), we have *superimposed* two solutions of the type (11.23).

If the magma in one hole were in tension and the magma in the other hole were in compression, the solution would be

$$\Phi = A_1 \ln (r_1) - A_2 \ln (r_2) = A_0 \ln (r_1/r_2). \tag{11.24}$$

We will accept the stress function given in eq. (7.8) because we have assumed that the mountain front behaves as a rigid boundary. Take the origin of coordinates at half the distance between the two sources, that is, at the position of the rigid front (Fig. 11.9).

According to Fig. 11.9 and to the cosine law,

$$r_1^2 = r^2 + a^2 - 2ar \cos \theta,$$

$$r_2^2 = r^2 + a^2 + 2ar \cos \theta,$$

where $\cos (180° - \theta) = -\cos \theta$.

If we define $2\xi^2$ as $a^2 + r^2$, for brevity, the equation for the stress function, eq. (11.22), becomes

$$\Phi = A_0 \ln [(2\xi^2 - 2ar \cos \theta)(2\xi^2 + 2ar \cos \theta)],$$

or

$$\Phi = A_0 \ln [4\xi^4 - 4(ar)^2 \cos^2 \theta]. \tag{11.25}$$

Performing the differentiations indicated by eqs. (11.4), the stress components at any point, (r, θ), are (*6*, p. 571):

$$\left.\begin{aligned} \sigma_{rr} = -\sigma_{\theta\theta} &= \frac{A_0}{[\xi^4 - (ar)^2 \cos^2 \theta]} [\xi^6 - \xi^4 a^2 \sin^2 \theta \\ &\qquad - (\xi ar)^2 \cos^2 \theta - a^4 r^2 \cos^2 \theta \sin^2 \theta], \\ \sigma_{r\theta} &= \frac{A_0 a^2 \sin \theta \cos \theta}{[\xi^4 - (ar)^2 \cos^2 \theta]} [\xi^4 - 2(\xi r)^2 + (ar)^2 \cos^2 \theta]. \end{aligned}\right\} \tag{11.26}$$

Comparison of Theoretical and Actual Dike Patterns at Spanish Peaks.—Odé postulated that the dikes adjacent to Spanish Peaks tended to follow directions of maximum compression in the rock at the time the dikes formed (*6*). The question is, then, do eqs. (11.26) describe a pattern of principal stress directions that is similar to the pattern of dikes? To answer this question, we need to plot trajectories of principal stresses.

Stress trajectories are lines that are parallel to directions of principal stresses at each point within a problem region. The principal stresses are normal to each other at each point, so that principal stress trajectories may be either one of two families of lines that intersect at right angles.

Figure 11.10A shows the directions of the principal stresses at some point in the vicinity of West Peak. The stresses acting on an infinitesimal element are shown on an enlargement of part of Fig. 11.10A. There we see that the direction of σ_3 is inclined to the direction of σ_{rr} by an angle, α. We can relate the angle, α, to the magnitudes of the stresses in the radial and tangential directions with a Mohr diagram (Fig. 11.10B). The coordinates of point r in the Mohr diagram are the stresses acting on the r-plane in physical space. These stresses are $\sigma_{\theta\theta}$ and $\sigma_{\theta r}$. The angle, 2α, is twice the angle between the r-direction and the direction in which σ_3 acts, in accord with the definition we gave to α in Fig. 11.10A.

According to the Mohr Circle in Fig. 11.10B,

$$\tan 2\alpha = \frac{\sigma_{r\theta}}{[(\sigma_{\theta\theta} - \sigma_{rr})/2]} = \frac{\sigma_{r\theta}}{\sigma_{rr}}, \tag{11.27}$$

because $\sigma_{rr} = -\sigma_{\theta\theta}$, according to eqs. (11.26).

In order to calculate orientations of the principal stress trajectories, we substitute eqs. (11.26) into eq. (11.27). Thus,

$$\tan 2\alpha = \frac{a^2 \sin\theta \cos\theta[\xi^4 - 2(\xi r)^2 + (ar)^2 \cos^2\theta]}{[\xi^6 - \xi^4 a^2 \sin^2\theta - (\xi ar)^2 \cos^2\theta - a^4 r^2 \cos^2\theta]}, \tag{11.28}$$

where

$$\xi^2 = (a^2 + r^2)/2.$$

Let us calculate one value for α. Suppose we want to calculate α at a distance of a from the mountain front. Then, we can show that eq. (11.28) reduces to

$$\tan 2\alpha = \sin\theta \cos\theta \tan^2\theta.$$

Now, suppose that $\theta = 45°$. Then,

$$\tan 2\alpha = -0.495,$$

$$2\alpha = -26° \quad \text{or} \quad 180° - 26°,$$

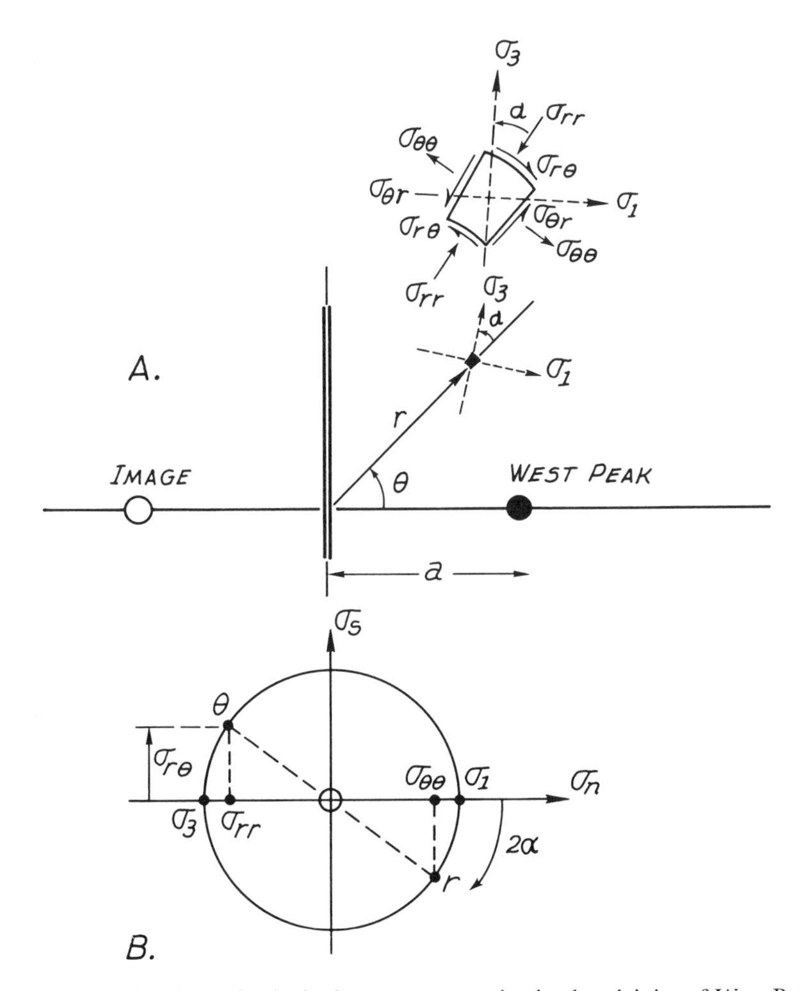

Figure 11.10. Directions of principal stresses at a point in the vicinity of West Peak.

A. Physical space, showing stresses acting on infinitesimal element that is a distance r, θ, from origin of coordinates.

B. Stress space. Relation between angle, α, and stresses at point shown in *A*, above.

and

$$\alpha = -13^\circ \quad \text{or} \quad +77^\circ.$$

The ambiguity is a property of the definitions of trigonometric functions. However, -13° and $+77^\circ$ define two directions normal to each other so that they represent both directions of the principal stresses. The problem is to determine which direction corresponds with which principal stress. To make this decision we need more information about the problem. This information is available in Fig. 11.11A. If the pressure in West Peak is compressive, we would guess that the direction of maximum compression, σ_3, would be nearly radial with respect to West Peak. Therefore, we would select 77° for the direction of σ_3 and −13° for the direction of σ_1 (Fig. 11.11A).

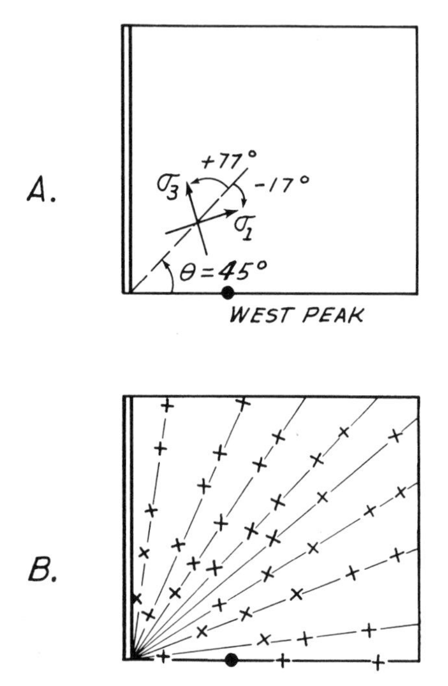

Figure 11.11. Directions of principal stresses at several points within problem region.
A. Principal stress directions at a point with coordinates $r = a$, $\theta = 45°$.
B. Directions of principal stresses at several places along radii from origin of coordinates.

Figure 11.11B shows directions of principal stresses at a large number of points near West Peak. Figure 11.12A shows the principal stress trajectories, which are smooth curves drawn through the points in Fig. 11.11B.

Thus, the method of constructing fields of stress trajectories consists of finding the orientations (α) of the trajectories at many points within a problem region and then drawing smooth curves through the points (e.g., see ref. *1*). The smooth curves are principal stress trajectories, which coincide with directions of principal stresses at every point in the problem region.

Stress magnitudes are not indicated by a field of stress trajectories. However, the magnitudes can be derived with the aid of the Mohr Circle in Fig. 11.10B:

$$\sigma_1 = \sqrt{(\sigma_{\theta\theta}^2 + \sigma_{r\theta}^2)},$$

and

$$\sigma_3 = -\sqrt{(\sigma_{\theta\theta}^2 + \sigma_{r\theta}^2)}.$$

Figures 11.12A and 11.12B show, respectively, fields of principal stress trajectories for two sources having equal pressures and two sources having opposite pressures.

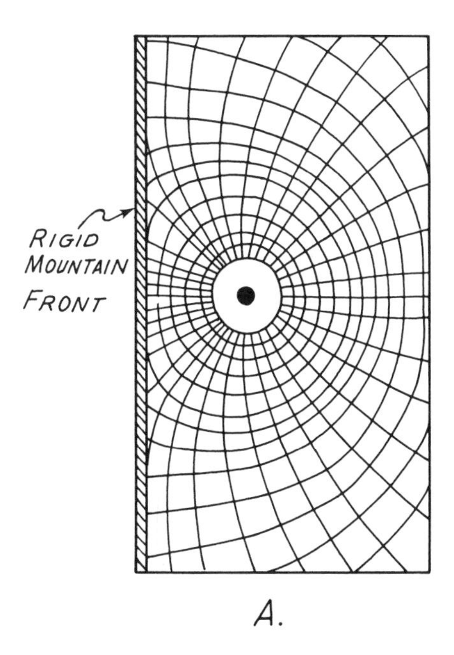

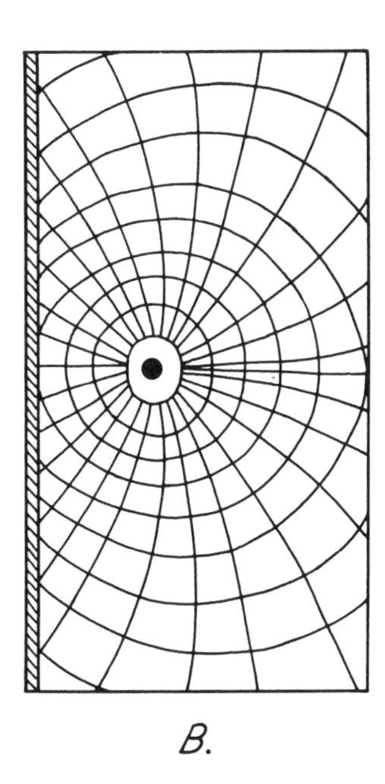

Figure 11.12. Trajectories of principal stresses for different magma pressures in image of West Peak.

A. Two holes with equal compressive pressures. The trajectories are smooth curves drawn through the points in Fig. 11.11B.

B. Trajectories for one hole under compressive pressure and the other hole under tensile pressure (after ref. *6*).

As pointed out by Odé, neither of these patterns corresponds closely to the pattern of dikes at Spanish Peaks (Fig. 11.7) (*6*). To get a better fit for the trajectories, Odé superimposed a constant regional stress field onto the radial fields. Because of the symmetry of the dike pattern along the direction N75°E, Odé suggested that the regional stress pattern should be symmetrical with respect to this direction (*6*).

Odé proposed a stress function of the following form for this regional field:

$$\Phi = (Bx^2/2) + (Cy^2/2), \tag{11.29}$$

which is the stress function for a constant stress, C, in the x-direction and a constant stress, B, in the y-direction. However,

$$x = r \cos \theta$$

and

$$y = r \sin \theta,$$

so that the stress function can be written as

$$\Phi = (B/2)r^2 \cos^2 \theta + (C/2)r^2 \sin^2 \theta,$$

and the components of stress are,

$$\left.\begin{aligned} \sigma_{rr} &= B \sin^2 \theta + C \cos^2 \theta, \\ \sigma_{\theta\theta} &= B \cos^2 \theta + C \sin^2 \theta, \\ \sigma_{r\theta} &= [(B - C)/2] \sin 2\theta. \end{aligned}\right\} \tag{11.30}$$

The resultant stress pattern can be determined by adding the regional stresses, eqs. (11.30), to the local stresses, eqs. (11.26). Figure 11.13 shows the results of superimposing the stress systems. The figure was constructed on the assumption that

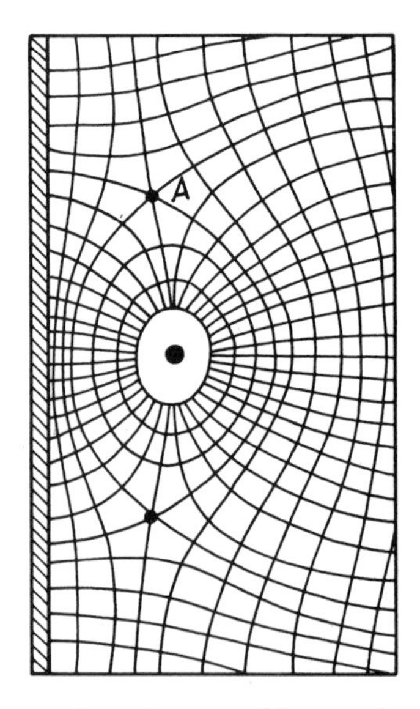

Figure 11.13. Principal stress trajectories caused by equal compressive pressures in West Peak and image of West Peak and by a superimposed regional stress field (after ref. *6*).

$\sigma_{xx} = 2\sigma_{yy}$ for the regional stresses and that $B = A_0/a^2$ and $C = 2A_0/a^2$, where a is the distance from West Peak to the mountain front. The normal stresses at point A in Fig. 11.13 must be hydrostatic, for only then can more than two trajectories cross at a point.

The pattern of stress trajectories shown in Fig. 11.13 is remarkably similar to the pattern of dikes surrounding West Peak (Fig. 11.7). The theoretical pattern could be adjusted better to fit the field data by altering the values of A_0, B, and C in eqs. (11.30) and (11.26). A larger value of the regional stress in the y-direction, for example, would tend to limit the area affected by the pressure from the stock at West Peak, so that radial dikes would be compressed into a rather compact bundle trending in the east-west direction. Perhaps most important, however, is the fact that the real dike pattern can be approximately duplicated theoretically by the superposition of two very simple stress fields.

References cited in Chapter 11

1. Frocht, M. M., 1941, *Photoelasticity*: John Wiley & Sons, Inc., N.Y. (2 vols.).
2. Hills, R. C., 1901, "Description of the Spanish Peaks Quadrangle, Colorado": *U.S. Geological Survey, Geol. Atlas*, Folio 71.
3. Johnson, R. B., 1961, "Patterns and the Origin of Radial Dike Swarms Associated with West Peak and Dike Mountain, South-Central Colorado": *Geol. Soc. America, Bull.*, V. 72, p. 579–590.
4. ———, 1968, "Geology of the Igneous Rocks of the Spanish Peaks Region, Colorado": *U.S. Geol. Survey Prof. Paper* 594-G.
5. Knopf, Adolph, 1936, "Igneous Geology of the Spanish Peaks Region, Colorado": *Geol. Soc. America Bull.*, V. 47, p. 1727–1784.
6. Odé, Helmar, 1957, "Mechanical Analysis of the Dike Pattern of the Spanish Peaks Area, Colorado": *Geol. Soc. America Bull.*, V. 68, p. 567–576.
7. Timoshenko, S., and Goodier, J. N., 1951, *Theory of elasticity*: McGraw-Hill Book Co., Inc., N.Y.

Other references on fracturing, faulting, jointing, and diking

Addinall, E., and Hackett, P., 1964, "Tensile Failure of Rock-Like Materials": *Proc. Sixth Symp. Rock Mech.*, Univ. Missouri, Rolla, Missouri, p. 315–338.

Aso, Kazuo, 1966, *Phenomena Involved in Pre-Splitting by Blasting*: Ph.D. Thesis, Stanford University, Stanford, Calif.

Balk, R., 1937, "Structural Behavior of Igneous Rocks": *Geol. Soc. Amer. Memoir* 5.

Biggs, W. D., 1960, *The Brittle Fracture of Steel*: MacDonald and Evans, Ltd., London.

Birch, Francis, et al., 1942, "Handbook of Physical Constants": *Geol. Soc. Amer. Spec. Paper* 36.

Bombolakis, E. G., 1964, "Photoelastic Investigation of Brittle Crack Growth Within a Field of Uniaxial Compression": *Tectonophysics*, V. 4, p. 343–351.

Brace, W. R., 1960, "An Extension of the Griffith Theory of Fracture to Rocks": *Jour. Geophys. Research*, V. 65, p. 3477–3480.

———, 1961, "Dependence of Fracture Strength of Rocks on Grain Size": *Penn. State Univ. Mineral Expt. Stat. Bull.* No. 76, p. 99–103.

———, and Bombolakis, E. G., 1963, "A Note on Brittle Crack Growth in Compression": *Jour. Geophys. Research*, V. 68, p. 3709–3713.

———, 1964, "Brittle Fracture of Rocks": In *State of Stress in the Earth's Crust*, ed. by W. R. Judd, p. 110–178.

———, 1965, "Some New Measurements of Linear Compressibility of Rocks": *Jour. Geophys. Research*, V. 70, p. 391–398.

———, and Paulding, B. W., Jr., and Scholz, C., 1966, "Dilatancy in the Fracture of Crystalline Rocks": *Jour. Geophys. Research*, V. 71, p. 3939–3953.

Chapman, C. A., 1958, "Control of Jointing by Topography": *Amer. Jour. Sci.*, V. 66, p. 552–558.

Davis, S. N., and Moore, G. W., 1965, "Semidiurnal Movement Along a Bedrock Joint in Wool Hollow Cave, California": *Nat. Speleological Soc. Bull.*, V. 27, p. 133–142.

DeSitter, L. U., 1956, *Structural Geology*: McGraw-Hill Book Co., Inc. N.Y.

Diller, J. S., 1889, "Sandstone Dikes": *Geol. Soc. Amer. Bull.*, V. 1, p. 411–442.

Donath, F. A., 1962, "Analysis of Basin-Range Structure, South-Central Oregon": *Geol. Soc. Amer. Bull.*, V. 73, p. 1–16.

———, 1963, "Strength Variation and Deformational Behavior of Anisotropic Rock": In *Internatl. Conf. State of Stress in Earth's Crust*, The Rand Corp., Santa Monica, Calif.

Firman, R. J., 1960, "The Relationship Between Joint and Fault Patterns In the Eskdale Granite (Cumberland) and the Adjacent Borrowdale Volcanic Series": *Jour. Geol. Soc. London*, V. 116, p. 317–347.

Foster, R. J., 1958, "The Teansaway Dike Swarm of Central Washington": *Amer. Journ. Sci.*, V. 256, p. 644–653.

Gilbert, G. K., 1882, "Post-Glacial Joints": *Amer. Jour. Sci.*, *3rd Ser.*, V. 23, p. 25–27.

———, 1882, "On the Origin of Jointed Structure": *Amer. Jour. Sci.*, *3rd Ser.*, V. 24, p. 50–53.

Goguel, J., 1962, *Tectonics*: W. H. Freeman and Co., San Francisco.

Gretener, P. E., 1965, "Can the State of Stress be Determined From Hydraulic Fracturing Data?" *Jour. Geophys. Research*, V. 70, p. 6205–6212.

Griggs, D. T., and Handin, J. H., 1960, "Observations on Fracture and a Hypothesis of Earthquakes": In *Geol. Soc. Amer. Memoir 79*, p. 357–364.

Harms, J. C., 1965, "Sandstone Dikes in Relation to Laramide Faults and Stress Distribution in the Southern Front Range, Colorado": *Geol. Soc. Amer. Bull.*, V. 76, p. 981–1002.

Hill, P. A., 1965, "Curviplanar (radial, bow-tie, festoon) and Concentric Jointing in Jurassic Dolerite, Mersey Bluff, Tasmania": *Jour. Geol.*, V. 73, p. 255–270.

Hodgson, R. A., 1961, "Classification of Structures on Joint Surfaces": *Amer. Jour. Sci.*, V. 259, p. 493–502.

———, "Reconnaissance of Jointing in Bright Angle Area, Grand Canyon, Arizona": *Amer. Assoc. Petrol. Geol. Bull.*, V. 45, p. 95–97.

———, "Regional Study of Jointing in Comb Ridge-Navajo Mountain Area, Arizona and Utah": *Amer. Assoc. Petrol. Geol. Bull.*, V. 45, p. 1–38.

Howard, J. H., 1960, "Structural Development of the Williams Range Thrust, Colorado": *Geol. Soc. Amer. Bull.*, V. 77, p. 1247–1264.

Hubbert, M. K., 1951, "Mechanical Basis for Certain Familiar Geologic Structures": *Geol. Soc. Amer. Bull.*, V. 62, p. 355–372.

———, and Willis, D. G., 1957, "Mechanics of Hydraulic Fracturing": *AIME, Trans.*, V. 210, p. 153–166.

———, and Rubey, W. W., 1959, "Role of Fluid Pressure in Mechanics of Overthrust Faulting, Part I: Mechanics of Fluid-Filled Porous Solids and Its Application to Overthrust Faulting": *Geol. Soc. Amer. Bull.*, V. 70, p. 115–206.

Irwin, G. R., 1958, Fracture: *Handbuch der Physik.*, V. 6, Flugge, S., Ed., p. 551–591, Springer, Berlin.

Jaeger, J. C., 1961, "The Cooling of Irregularly Shaped Igneous Bodies": *Amer. Jour. Sci.*, V. 259, p. 721–734.

———, 1962, *Elasticity, Fracture, and Flow*, 2nd Ed.: Methuen and Co., Ltd., London.

———, 1963, "Extension Failure in Rocks Subject to Fluid Pressures": *Jour. Geophys. Research*, V. 68, p. 6066–6067.

Kelley, V. C., and Clinton, J. J., 1960, "Fracture Systems and Tectonic Elements of the Colorado Plateau": *Univ. New Mexico Pub. in Geol.*, No. 6.

Kolsky, H., 1965, "Fracture and Cavitation in Glassy Materials": *Soc. Glass Tech. Trans.*, V. 39, p. 394–403.

Lattman, L. H., and Segovia, A. V., 1961, "Analysis of Fracture Trace Pattern of Adak and Kagalaska Islands, Alaska": *Amer. Assoc. Petrol. Geol. Bull.*, V. 45, p. 249–263.

Longwell, C. R., 1945, "Low-Angle Normal Faults in the Basin-and-Range Province": *Trans. Amer. Geophys. Union*, V. 26, p. 107–118.

Lu, P. H., and Scheidegger, A. E., 1965, "An Intensive Local Application of Lensen's Isallo Stress Theory to the Sturgeon Lake South Area of Alberta": *Bull. Canadian Petrol. Geol.*, V. 13, p. 389–396.

Mollard, J. D., 1957, "A Study of Aerial Mosaics in Southern Saskatchewan and Manitoba": *Oil in Canada*, Aug. 5.

Muehlberger, W. R., 1961, "Conjugate Joint Sets of Small Dihedral Angle": *Jour. Geol.*, V. 69, p. 211–219.

Murgatroyd, J. B., 1942, "The Significance of Surface Marks on Fractured Glass": *Jour. Soc. Glass Tech.*, V. 26, p. 155–171.

Murrell, S. A. F., 1958, "The Strength of Coal Under Triaxial Compression": In *Mechanical Properties of Non-Metallic Brittle Solids*, Interscience, N.Y., p. 123–146.

Nadai, A., 1963, *Theory of Flow and Fracture of Solids*, Vol. II: McGraw-Hill Book Co., Inc., N.Y.

Odé, Helmar, 1960, "Faulting as a Velocity Discontinuity in Plastic Deformation": In *Geol. Soc. Amer. Memoir 79*, p. 292–311.

Orowen, E., 1960, "Mechanism of Seismic Faulting": *Geol. Soc. Amer. Memoir 79*, p. 323–345.

Parker, J. M., 1942, "Regional Systematic Jointing in Slightly Deformed Sedimentary Rocks": *Geol. Soc. Amer. Bull.*, V. 53, p. 381–408.

Peterson, G. L., 1966, "Structural Interpretation of Sandstone Dikes, Northwest Sacramento Valley, California": *Geol. Soc. Amer. Bull.*, V. 77, p. 833–842.

Preston, F. W., 1926, "The Spalling of Bricks": *Amer. Ceram. Soc. Jour.*, p. 654–658.

Price, N. J., 1959, "Mechanics of Jointing in Rocks": *Geol. Mag.*, V. 96, p. 149–167.

Raggatt, H. G., 1954, "Markings on Joint Surfaces at Anglesea, Victoria": *Amer. Assoc. Petrol. Geol. Bull.*, V. 38, p. 1808–1809.

Roberts, J. C., 1961, "Feather Fracture, and the Mechanics of Rock-Jointing": *Amer. Jour. Sci.*, V. 259, p. 481–492.

Savin, G. N., 1961, *Stress Concentration Around Holes*: Pergamon Press, N.Y.

Scheidegger, A. E., "Geometrical Significance of Isallo Stress": *New Zealand Jour. Geol. and Geophys.*, V. 6, p. 221–227.

Secor, D. T., 1965, "Role of Fluid Pressure in Jointing": *Amer. Jour. Sci.*, V. 263, p. 633–646.

Seldenrath, I. T. R., and Cramberg, I. J., 1958, "Stress-Strain Relations and Breakage of Rocks": In *Mechanical Properties of Non-Metallic Brittle Materials*, Butterworth's Scientific Pubs., London, p. 79–105.

Shotton, F. W., 1965, "Normal Faulting in British Pleistocene Deposits": *Quart. J. Geol. Soc. London*, V. 121, p. 419–434.

Silverman, I. K., 1957, "Behavior of Materials and Theories of Failure": *Colo. School of Mines*, Quarterly, V. 52, p. 3–18.

Smalley, I. J., 1966, "Contraction Crack Networks in Basalt Flows": *Geol. Mag.*, V. 103, p. 110–114.

Solomon, M., and Hill, P. A., 1962, "Rib and Hackle Marks on Joint Faces at Renison Bell, Tasmania: A preliminary note": *Jour. Geol.*, V. 70, p. 493–496.

Spencer, E. W., 1959, "Geologic Evolution of the Beartooth Mountains, Montana and Wyoming, Part II, Fracture Patterns": *Geol. Soc. Amer. Bull.*, V. 70, p. 467–508.

Spry, Alan, 1962, "The Origin of Columnar Jointing, Particularly in Basalt Flows": *Jour. Geol. Soc. Australia*, V. 8, p. 191–216.

Tandahand, S., and Hartman, H. L., 1962, "Investigation of Dynamic Failure by High-Speed Photography": *Proc. Fifth Symp. Rock Mech.*, Pergamon Press, N.Y., p. 1–32.

Trump, R. P., and Patnode, H. W., 1963, "Laboratory Demonstration of the Failure of Materials Under Uniform Loads": *Proc. Fifth Symp. Rock Mech.*, Pergamon Press, N.Y., p. 661–666.

Turner, F. J., and Weiss, L. E., 1963, *Structural Analysis of Metamorphic Tectonites*: McGraw-Hill Book Co., N.Y.

Umbgrove, J. H. F., 1947, *The Pulse of the Earth*, 2nd Ed.: The Hague, Marinns Nijhoff., p. 294–319.

Volkov, S. D., 1962, *Statistical Strength Theory*: Gordon and Breach, Science Publishers, N.Y.

Walsh, J. B., 1965, "The Effect of Cracks on the Compressibility of Rock": *Jour. Geophys. Research*, V. 70, p. 381–389.

———, and Brace, W. F., 1964, "A Fracture Criterion for Brittle Anisotropic Rock": *Jour. Geophys. Research*, V. 69, p. 3449–3456.

Wuerker, R. G., 1956, "Annotated Tables of Strength and Elastic Properties of Rocks": *Petroleum Branch, AIME.*

———, 1959, "Influence of Stress Rate and Other Factors on the Strength and Elastic Properties of Rocks": *Colo. Sch. Mines Quart.*, V. 5., No. 3.

PART III

FLOW OF ICE, LAVA, AND DEBRIS

12

Chapter Sections

Introduction
Features of Debris-Flow Deposits
Movement of Debris Flows
  Debris Flows at Wrightwood, California
  Debris Flow at Parker, Arizona
Similarities among Debris-Flow, Lava-Flow, and Glacial Deposits
Shape of Longitudinal Profile of Snout
  Sand-Wedge Experiment
  Theoretical Profile of Snout of Plastic Material
  Comparison of Theoretical and Actual Snouts
References Cited in Chapter 12.

12

Formation of Debris-Flow Deposits

Introduction

Debris flow is a process by means of which granular solids, sometimes mixed with relatively minor amounts of entrained water and air, move readily on low slopes. By means of a natural form of this process, tremendous volumes of the products of rock weathering and minor percentages of miscellaneous items such as wood, tin cans, road graders, and stray cattle are transported for miles down canyons, over surfaces of alluvial fans, and over numerous relatively gentle slopes. By themselves, the granular solids in debris flows are relatively immobile so that they flow only on steep slopes, but addition of small proportions of water and air transforms certain granular solids into highly mobile slurries (*4,5,6*).

Debris has been considered to behave much as water by several investigators, who attribute the different flow rates of debris and water to their markedly different viscosities. For example, Sharp and Nobles (*10*) and Curry (*1*) estimated the "viscosity" of two moving debris flows to be about 10^4 poises, or one million times that of water.

Such estimates of viscosities do give us some concept of the sluggishness of debris flows, but they may be misleading in several respects; probably the most important differences between debris flow and water flow cannot be attributed to markedly different viscosities (*4*). A striking feature not explainable in terms of simple viscous behavior is the characteristically blunt terminations or steep outer margins of debris deposits and debris flows (Fig. 12.1). The longitudinal profile of the blunt front of a debris flow is strikingly similar to the profile of the meniscus at the edge of a drop of water on a flat, dry surface. Now, if a purely viscous substance were poured onto a flat, dry surface, the substance would flow indefinitely, or until the layer of the substance were very thin. Thus, the meniscus at the edge of a water drop and the blunt termination of a debris flow are results of imperfections of the

Figure 12.1. Profile of snout of debris-flow deposit near Cris Wicht Camp, Panamint Mountains, California. Deposit about three feet thick.

viscosities of these two substances. The meniscus is attributed to surface tension of water and the blunt termination is attributed to finite strength of slurries (*4*).

Another problem inadvertently introduced by assuming debris to be a simple viscous substance is inability to explain why debris flows can transport coarse particles, such as large boulders and blocks. There is no question that slurries in the form of debris flows can transport large boulders—this has been reported by all observers of debris flows—but there is some disagreement concerning the mechanism of transport. Some investigators maintain that large boulders must be supported by turbulence within a viscous debris flow. Others, however, who have studied debris-flow deposits in the field and who have observed natural and experimental flows conclude that the flow is primarily laminar. A mechanism of transport that is proposed for debris flowing laminarly is the combination of high density and high strength of debris (*4,6*); basically, the strength and density of debris retards or prevents coarse particles from sinking.

Features of debris-flow deposits

Some general features of debris flow and debris-flow deposits are shown, schematically, in Fig. 12.2, by means of a longitudinal section, a plan view, and cross-sections of a single debris-flow *arm* in motion at an instant of time. One or more arms may represent a single episode of debris-flow activity (*4*).

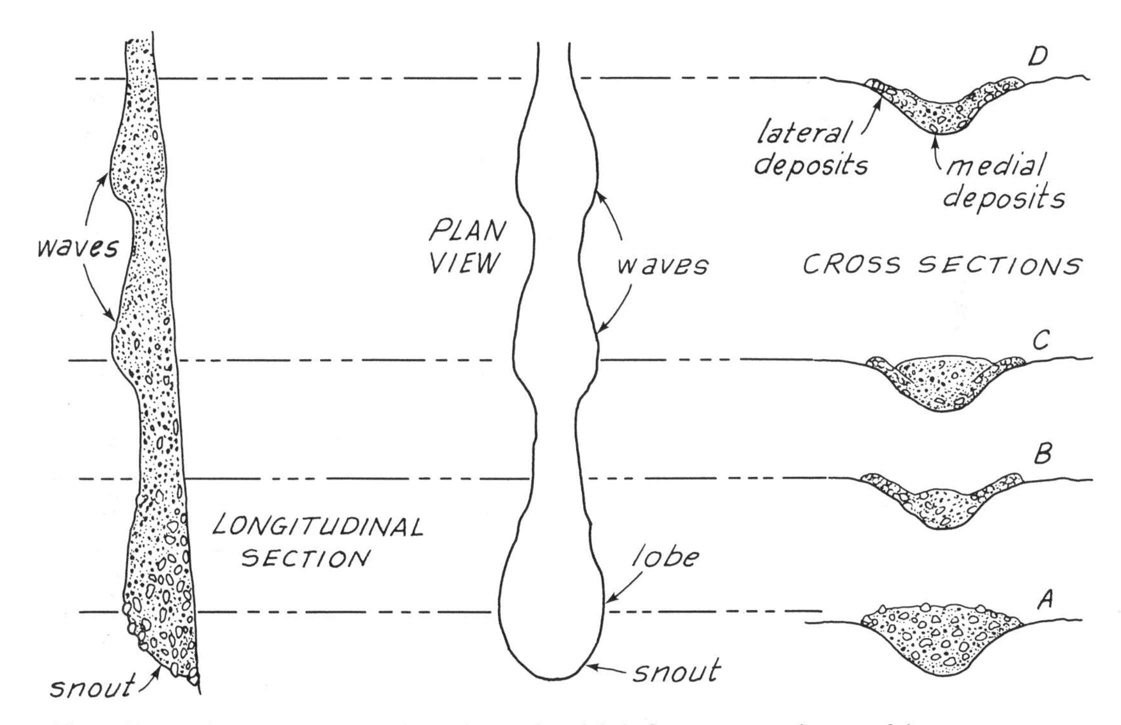

Figure 12.2. Idealized representations of a moving debris-flow arm at an instant of time.

Many debris flows can be pictured as large waves of solid and fluid material flowing more or less steadily through a channel, with superimposed smaller waves traveling at velocities higher than those of the debris flow itself. The *snout*, or steep, frontal region, and succeeding *waves* appear in the longitudinal section. The waves are expressed in plan view by local widening of the flow if the channel sides slope outward.

Each arm of the typical debris-flow deposit comprises a *lobe*, in reference to the overall plan form of the front of the deposit, and *lateral deposits*, that can be distinguished relative to a *channel* in which the debris has moved. Lateral deposits are on the sides of a channel or immediately adjacent to it.

Most of these features are preserved in debris-flow deposits along the lower edge of a steep talus slope, near Klare Springs, in Titus Canyon, Death Valley National Monument (*4*) (Fig. 12.3). Perhaps eight to ten debris waves stopped in the complex of channels on the talus slope before they reached the "dumping grounds" in the bottom of Titus Canyon, in the foreground in Fig. 12.3. One of the debris waves is in the large channel on the left side of the photograph. It has a sharp front with a snout that is about five feet high. The wave is lobate in plan and its longitudinal profile is similar in form to the profile of a falling drop of water, cut in half by a

Figure 12.3. Complex debris-flow deposits on steep talus slope near Klare Springs, Death Valley National Monument, California. *L.D.* means lateral deposit.

plane passing through its bulbous front and tapered tail, and placed on its side in the channel.

On each side of the channel shown in Fig. 12.3 are lateral deposits that record the passage of several earlier waves. In cross-sections normal to channel axes, the lateral deposits appear as a series of sharp-crested wedges of debris perched on the insides of the channels. The lateral deposits within the channels represent combined deposits of several debris waves. For example, the wave preserved in the large channel shown on the left side of Fig. 12.3 deposited a lateral deposit that is visible on the right side of the channel (point #*I* in Fig. 12.3). The passage of at least three larger waves is recorded by the lateral deposits in front of the wave (point #*II* in Fig. 12.3).

The same features can be recognized in debris-flow deposits that are several orders of magnitude larger or smaller than the debris wave in the channel at Klare Springs. Shelton (ref. *11*, fig. 151), for example, shows a photograph of a mammoth debris-flow deposit in southern Idaho. The distal end of the deposit is a lobe, perhaps 2000 feet wide, which can be traced sourceward up a channel bordered by faintly defined lateral deposits. The snout of the lobe rises about 50 feet abruptly above the valley floor. Apparently this debris-flow deposit represents one giant wave of debris that began as a landslide and continued moving as a debris flow down the channel to its present site.

Deposits representing the opposite extreme of the size spectrum of debris flow are shown in Fig. 12.4. These tiny debris-flow deposits formed on freshly excavated embankments of the California Aqueduct, near Lost Hills, California. The embankment is in unconsolidated sands, gravels, and muds which are lenticular bodies deposited near the shoreline of the extinct Tulare Lake. Some of the sand lenses seeped water for several days after the embankment was excavated and the seeping water apparently mobilized part of the loose sand, which flowed down the steep embankment, forming long, narrow arms of sand. The lateral deposits and the bulbous shape of the lobes in these small debris-flow deposits are disproportionate, compared with features of larger deposits, such as those shown in Fig. 12.3 and the one described by Shelton (*11*). Nevertheless, the same *types* of features are recognizable in deposits of all three sizes, presumably indicating that the same types of processes are active on all three scales. It is these processes that we need to understand.

Movement of debris flows

Debris Flows at Wrightwood, California.—Some graduate students and I were most fortunate in being able to see debris flows in action on 20 May 1969 near the town of Wrightwood, in southern California. The winter of 1969 was unusually wet so that unusually thick snow packs accumulated in the San Bernardino Mountains, the source area of the Wrightwood flows. Rapid melting of the snow saturated

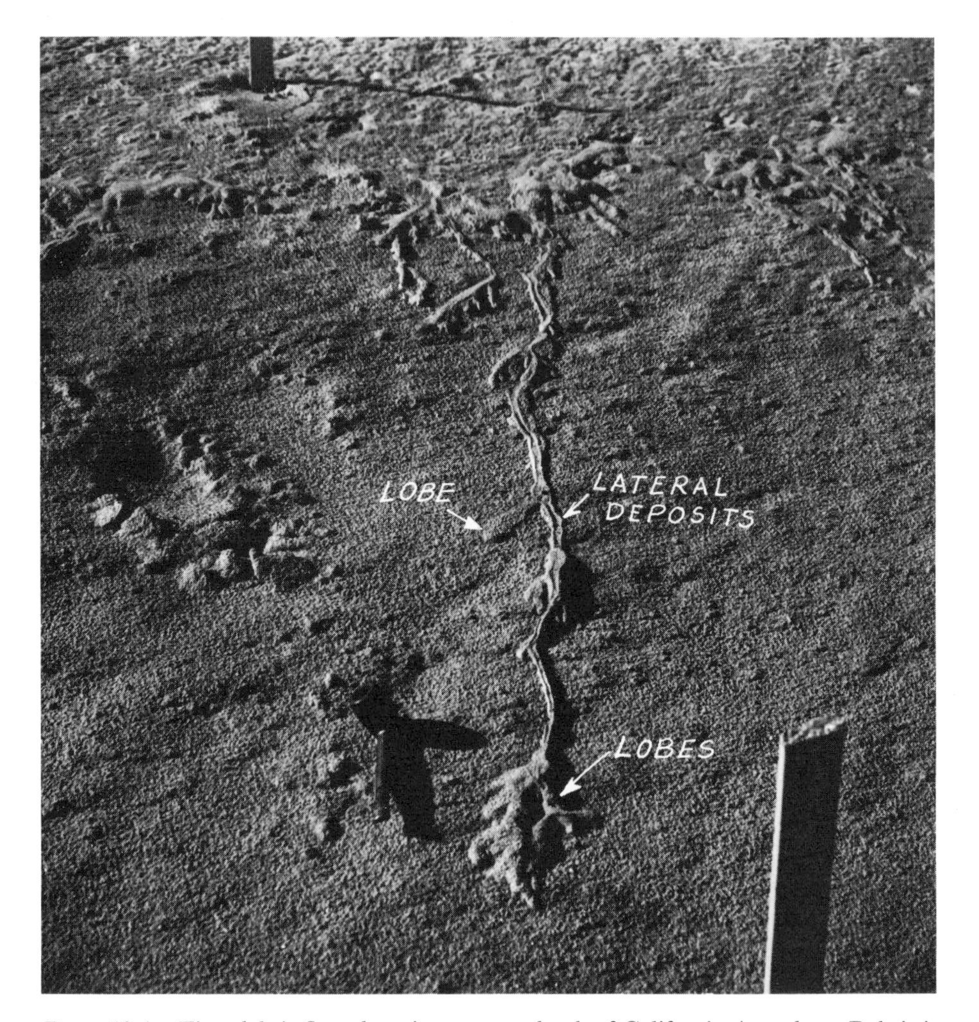

Figure 12.4. Tiny debris-flow deposits on steep bank of California Aqueduct. Debris is sand mobilized by water seeping from bank shown near top of photograph.

debris in a large landslide in the source area, causing masses of the landslide to break off, mix with water, and move in waves of debris down the channels (*6*). The spring thaw produced debris flows for about ten days in 1969.

At about 8:00 in the morning of the 20th we found the channel down which the flows were moving. The flows appeared in the channel about every ten to twenty minutes so that we were able to see many within a few hours. In the desert areas, in contrast, a person is fortunate to see even one debris flow, because the periodicity of flows in a given canyon is on the order of 30 to 100 years and because they usually occur during the heat of the summer.

The flows at Wrightwood were of various sizes and types, but they all displayed the same flow cycle (*6*). A stream of muddy melt water, a few inches deep and three or four feet wide, flowed nearly continuously in the channel. About every ten to twenty minutes, the muddy water gradually became deeper and carried noticeably more and more sediment. A few moments after we noticed the deepening of the muddy water, we would begin to hear a low rumble, signaling the appearance of another debris wave. Then the muddy water usually dwindled to a trickle, apparently when the debris wave became arrested temporarily at a constriction in the channel. Suddenly the debris wave would appear around a bend in the channel, about ten meters upstream (Fig. 12.5A). The rumble, the clanking of boulders tumbling over the snout, and the sloshing and slopping of the mud drowned the sound of our gleeful shouts as we attempted to alert the whole world to the remarkable event unfolding before our eyes.

The front of the typical debris wave consisted largely of boulders, usually six inches to two feet but occasionally three or four feet in diameter (Fig. 12.5A). The foremost boulders sometimes tumbled over the front and were pushed along the channel but they usually seemed to move along more or less together, slowly shifting position. The boulders visible on top of the flow moved ahead and tumbled over the front as though they were on top of the tread of a caterpillar tractor. The snout and the top of the flow, extending ten to thirty feet behind the snout, typically were armoured with boulders (Figs. 12.5A and 12.5B).

Behind the bouldery front, the surfaces of the flows exposed more of the finer-grained mud and the boulders appeared to be more widely separated than they were in the frontal region (Fig. 12.5C). For example, Fig. 12.6A shows a high, oblique view of a debris flow moving from right to left in a channel about eight feet wide. The boulders are quite widely separated and much interstitial mud is visible.

The part of the debris wave that contained boulders commonly was several tens of feet and sometimes may have been 100 meters long. Gradually, the number of boulders decreased and the debris became charged with pebble-sized fragments such as we see on the far side of the channel shown in Fig. 12.6B. Following this debris was material that appeared to have a lower percentage of pebbles, such as we see in the center of the channel in Fig. 12.6B. At the time this photograph was taken, the older, pebbly debris on the far side of the channel was being remobilized by the younger, finer-grained debris in the foreground. Small surface waves are visible in the finer debris.

Gradually, the flow of finer-grained debris became more and more diluted with water until it returned to its normal condition as muddy water (Fig. 12.6C).

Observation of the flows at Wrightwood and of experimental flows seems to indicate that debris flows in laminar flow rather than in turbulent flow (*4,6*).

It is only natural that early concepts of debris flow compared debris flow with water flow, considering the relatively low flow rates of debris to be caused by the

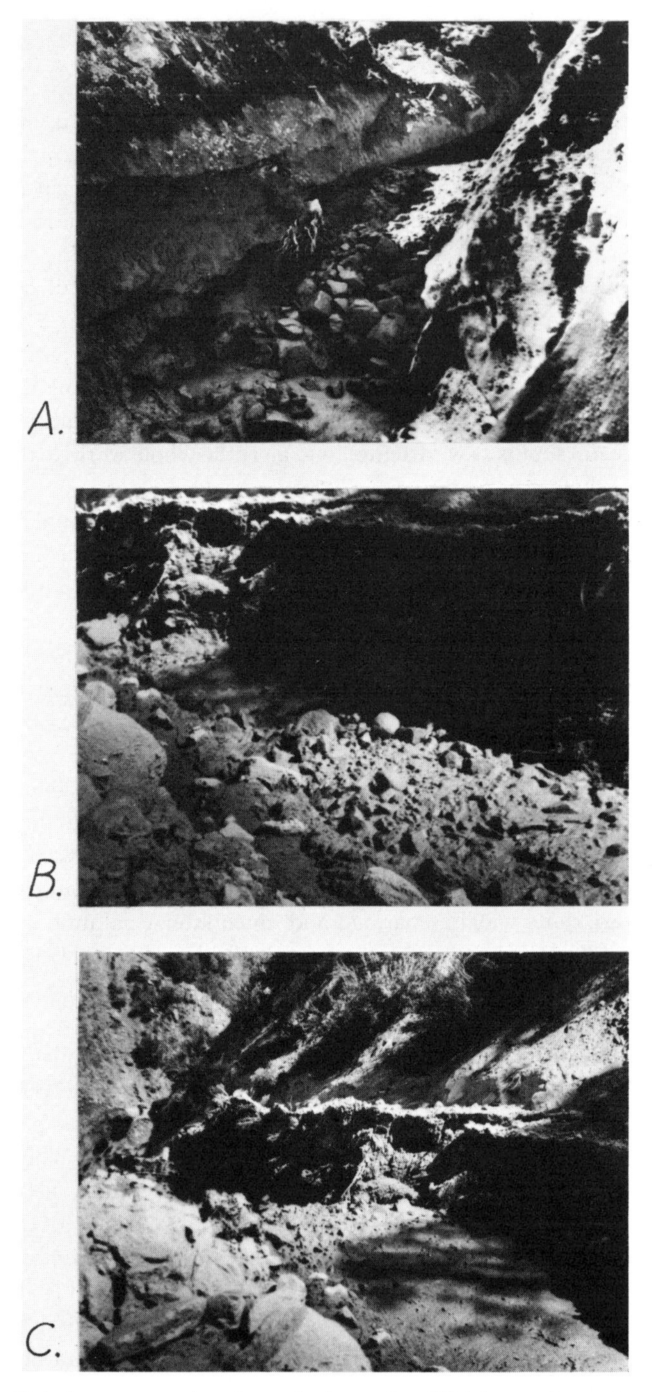

Figure 12.5. Debris wave moving through channel at Wrightwood, southern California. Boulders are concentrated in front of flow. They become more widely separated ten to fifteen feet in back of snout.

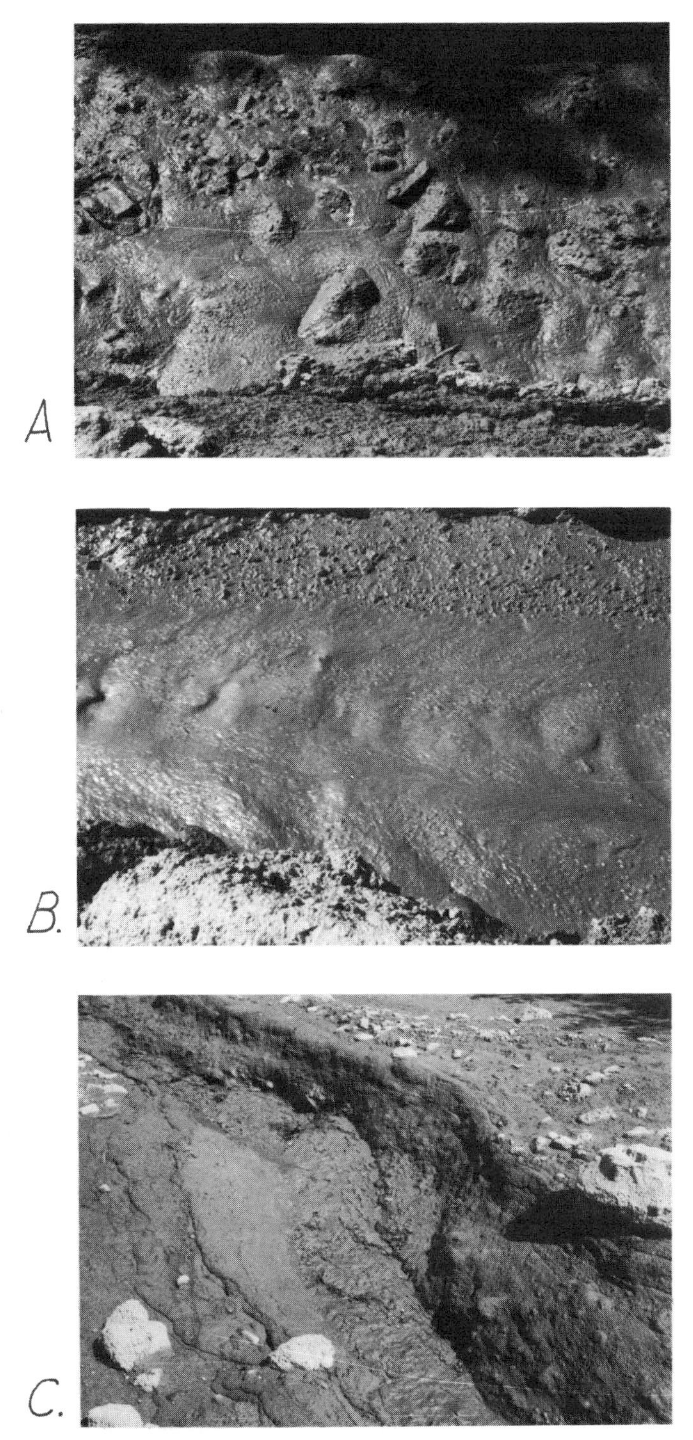

Figure 12.6. Three phases of flow cycle at Wrightwood, southern California.

high viscosity of debris (*1,10*). Following this reasoning, it was only natural to assume that debris flows must be highly turbulent in order to suspend large blocks and boulders. After all, experience with flow of water has indicated that turbulence is required to suspend sediment.

This seems to be carrying the analogy between water flow and debris flow much too far, however. We observed the Wrightwood flows with particular attention to the determination of whether flow is laminar or turbulent, and in addition we have performed experiments at the U.S. Geological Survey Slurry Laboratory (*5,6*) in Menlo Park, California. Some of our experiments were with kaolin-water slurries flowing at constant rates in a semicircular channel of 4-inch diameter. Under the conditions of our experiments, these slurries flow in a laminar manner until clay contents are less than about ten percent by weight; the granular solids in most natural debris flows comprise 60 to 90 percent of the weight of the flowing materials (refs. *3,4,6,10*). The kaolin clay seems to have a marked damping effect on turbulent eddies, which are characteristic of turbulent flow. Thus, when we artificially induced an eddy into the flow, the eddy quickly damped out and the flow resumed its laminar mode.

We noted similar phenomena in the Wrightwood debris flows. An obvious and significant difference between the Wrightwood flows shown in Figs. 12.6B and 12.6C is the character of the flow surfaces. Whereas the muddy water, shown in Fig. 12.6C, has a choppy and irregular surface, the fine-grained debris, shown in Fig. 12.6B, has a smooth, broadly wavy surface. The choppiness is characteristic of shear turbulence and the smooth surface is characteristic of laminar flow. Our observations of the flows at Wrightwood (*6*) further strengthen our conclusion, based on experiments with kaolinite-water slurries, that a small amount of suspended clay and silt is adequate to dampen eddies to a great extent and to reduce greatly the tendency toward turbulent flow. Only the muddy water at Wrightwood flowed turbulently. The debris flows moved in a laminar fashion. Indeed, the debris flows churned and rotated when they flowed over small waterfalls and around constrictions in the channel, but only in this sense did they move in a turbulent manner. When the churning debris reached a relatively straight-walled segment of the channel, the churning entirely disappeared, indicating that the eddies were temporary.

There are other lines of evidence indicating that debris flows normally are in a state of laminar flow (*4*). For example, debris flows handle gently the objects they are transporting. Large boulders and fragile clasts such as fractured boulders and blocks of brittle shale, wood fragments, and cans of all sizes retain their respective identities during debris flow. Many fractured boulders have been carried for miles as parts of recent debris flows, and evidently remained almost intact during their journeys (Fig. 12.7). Yet, some of them were so fragile that they have broken into piles of angular fragments since they came to rest only a few years or decades ago, and the pieces in each pile can be fitted back together as though they were pieces of a giant jigsaw puzzle (*4*).

Figure 12.7. Large block transported by debris flow on east side of Avawatz Mountains, California. Block apparently fell apart after reaching its present location, several miles from source (shown at the right).

Turbulent flow seems to be restricted to the last, fluid phases of normal episodes of debris-flow activity and, perhaps, to unusual flows of gigantic proportions that we have not observed in modern times.

Debris Flows at Parker, Arizona.—Jahns (*3*) described a debris flow that was larger than any of those we saw at Wrightwood. Its mode of flow apparently was similar to that of the smaller ones, however, as is indicated by his description (ref. *3*, p. 12 and 13):

> The flood was an awesome sight. A dark reddish-brown mass of water-lubricated debris moved—very slowly, it seemed—down the last tortuous part of the canyon. It formed a curving, but extremely steep wall, which must have been about 35 feet high at the point where it burst from the narrow mouth of the canyon. Masses of rock more than 30 feet in maximum dimension cascaded down the front and quickly disappeared from view beneath its base....
>
> It was impossible to follow the initial flow as it spread out along the valley floor, but inspection a short time later indicated that it had come to

a halt about a mile from the canyon mouth. Evidently it had fanned out somewhat and had stopped when enough water had soaked into the dry sand and ground of the valley floor to reduce internal lubrication materially. In this, its final position, the front was very steep and about 15 feet in average height. The top of the flow, which dried within a matter of hours, was studded with boulders, most of them several feet in diameter. It was also marked by large wrinkles, generally six inches to more than two feet high, that lay essentially parallel to the margin of the flow.

Similarities among debris-flow, lava-flow, and glacial deposits

Perhaps the most powerful tool we have to decipher the record of geologic processes is analogy. G. K. Gilbert used analogies extensively to analyze problems he studied. By analogy, we compare products of an unknown process with products of a process we think we understand. Thus, we thought we were able to understand many features of folded rocks by contemplating behaviors of elastic metal rods and plates. We thought we could understand the gross form of the radial dike pattern at Dike Mountain by contemplating trajectories of principal stresses around a pressurized hole in a large elastic plate. In each of these examples, the analogy takes us part way toward the understanding of a process.

There seem to be certain analogies among the processes of lava flow, debris flow, and glacier flow. Glaciers, lava flows, and debris flows involve movement of quite different materials—ice, molten rock, and a mixture of granular solids and water—yet, the way they flow and the forms of deposits they leave to record their activities are remarkably similar. For example, all three processes produce similar lateral ridges of debris on the insides of channels or immediately adjacent to channels. If these ridges were formed by glaciers or lava flows they are called moraines (*8*), and if they were produced by debris flows they are called lateral deposits or levees (*4,10*).

There is a group of congealed lava flows on the northwestern flank of Mount Shasta in northern California which are grossly similar in form to the debris-flow deposits we have discussed (see Fig. 65, ref. *11*). The flows spread out in arm-like lobes, terminated at fronts that rise abruptly above the surroundings, just like debris-flow deposits. Also, the flows on Mount Shasta left lateral ridges, and a wave of lava is frozen in the channel midway between source and terminus, similar to the debris-flow wave shown in Fig. 12.3. The similarities in gross forms of termini of glaciers and snouts of debris flows, lateral moraines of glaciers and lateral deposits of debris flows, and lobate forms and surface wrinkles of glaciers and debris flows are well known.

Konrad Krauskopf, who saw the eruption of Paricutín volcano in Mexico, during the winter of 1945–46, recognized certain similarities between lava flow and glacier flow. In his words (ref. *8*, p. 1278):

> Lava movement is strikingly similar to glacier movement. Like a glacier, a flow carries debris on its top, drops the debris in front of its moving tip, and partly overrides it. Like a glacier a flow develops crescentic ridges on its surface which show the differential movement between center and edges. Also like a glacier a flow builds along its sides well-developed lateral moraines, consisting of loose blocks pushed aside as the flow moves.

I think that we can *tentatively* conclude, on the strength of these analogies, that the flows of lava, debris, and ice are similar. Detailed study is obviously required, however, to determine the degree of similarity and to assess the sources of similarities that we might recognize after further study.

The analogies I have recognized between flow of glaciers and flow of debris lead me to suggest, in Chapter 15, an explanation for the causes of the characteristic U-shape of valleys that have been visited by glaciers. Debris-flow valleys tend to be U-shaped, and we can understand why. The same explanation might be valid for glacial valleys if, indeed, the flow of glacial ice and the flow of debris are crudely similar.

Shape of longitudinal profile of snout

All debris-flow deposits I have seen have sharp breaks in slope at their margins (Fig. 12.1). The form of the sharp break is similar to the form of the meniscus at the edge of a drop of water on a flat, dry surface. The form is clearly visible in the profile of small lobes of debris poured on a table (Fig. 12.8A). The debris shown in Fig. 12.8 consisted of fine-grained material collected from the Surprise Canyon alluvial fan in Panamint Valley, California. About 16 percent of the weight of the debris was water that had been added (*4*). The strength of the debris was about 800 dn/cm^2. The debris flowed to the position shown in Fig. 12.8A as it was poured onto the surface of the tilted table; the debris spread radially as an ever-expanding lobe. The snout of the lobe has a sharp break in slope next to the steep front. The longitudinal profile in back of the snout is nearly straight, or slightly convex upward.

The profiles of this lobe and of the debris-flow deposit shown in Fig. 12.1 can be compared with the profile of a debris wedge in two-dimensional flow. Fig. 12.8B shows a cross-section of a debris wedge that was built by adding debris to its apex, causing flow laterally between two plates of glass that were six inches apart. The wedge was composed of fine-grained material admixed with about 14 weight percent of water (*4*). The debris had a strength of about 2700 dn/cm^2 and a density of about 2.19 gm/cc. The profile of the wedge of stronger debris was more strongly convex than that of the lobe of weaker debris; therefore, convexity seems to be a function of the strength of the material.

Sand-Wedge Experiment.—I saw the same general snout form in a sand cone in a Massachusetts gravel pit, where slightly damp and cohesive gravel and sand

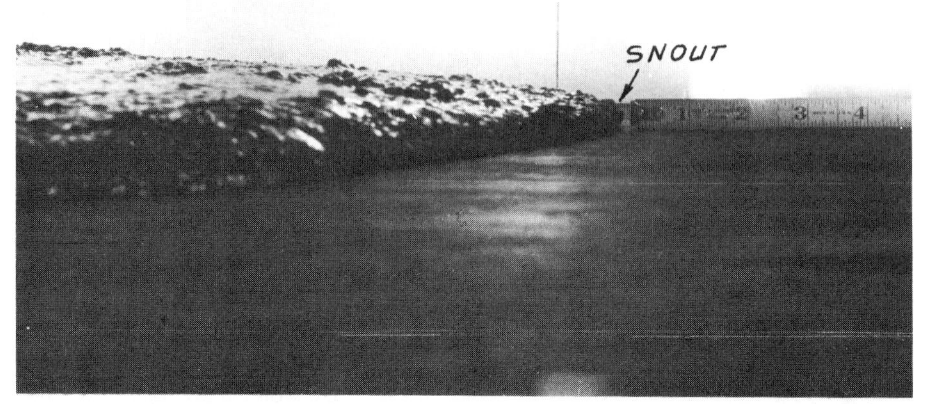

A. DEBRIS LOBE

B. DEBRIS WEDGE

Figure 12.8. Snouts of debris lobe and debris wedge.

A. Lobe formed by pouring fine-grained debris onto tilted table.

B. Wedge formed between plates of glass, 6 inches apart, by adding debris at apex.

were separated by screening and were fed to the apices of conical storage piles by means of a conveyor belt. The sand cone was about 15 feet in height at the time I saw it. Its sides were steep, sloping at perhaps 30 to 40 degrees, and they were quite straight. Near the base of the cone, however, was a steep front, perhaps four or five inches high. The cone surface intersected the ground approximately at a right angle, not at the slope angle of the upper part of the cone. Thus the outer slopes of the cone were compound.

In order to investigate the internal deformation of the sand and the cause of the compound slope of the cone, I built sand wedges and added sand to their apices, causing flow laterally between two plates of glass that were 15 inches apart (*4*). When the wedge so constructed was composed of dry sand, its slopes were simple, intersecting a flat base at the slope angle of the upper part of the wedge; the angle of slope presumably was the angle of repose of dry sand. However, when a small amount of water was admixed with the free-flowing sand, and a wedge was formed by pouring the damp sand onto the apex of a wedge formed between plates of glass, the compound slope formed (Fig. 12.9C).

A moist sand was used for the wedge shown in the photographs in Fig. 12.9. Successive layers were formed by adding, successively, white sand and sand stained with red dye. A white layer was being constructed by adding material to the wedge apex when the photographs of Fig. 12.9 were taken. Figures 12.9B and 12.9C show the progressive development of the white layer as more white sand was added to the apex, and finally, Fig. 12.9D shows the completed layer (*4*).

Between the times the photographs in Figs. 12.9A and 12.9B were taken, a layer of sand moved downward, toward the right, the moving region being bounded by a surface of approximately the shape shown in Fig. 12.10A. The boundary conditions of the moving layer can be deduced approximately by studying the photographs. First, it can be observed that the sand near the apex of the wedge moved downward, normally to the wedge surface, so that there the largest compressive stress must be normal to the surface. This stress is necessarily zero, so that the normal stress acting parallel to the surface must be tensile. In the snout region, however, the surface of the sand wedge moved upward, so that the stress parallel to the surface must be compressive there. Somewhere between these two areas there must be a region, or at least a point, where the stress parallel to the surface is equal to the stress normal to the surface, which is zero.

The representation of a simple plastic substance on a Mohr diagram might be helpful in visualizing the necessity of these conclusions. The component of normal stress that is tangential to the cone surface near the base of the cone is shown at point 1 in Fig. 12.10B, and the circle of diameter $2k$ represents the state of stress required for flowage there. This is a requirement of the basic element of the plastic model—if a state of stress is represented by a circle within the two limiting envelopes, such as circle OD, there can be no failure, whereas failure is impending if the state

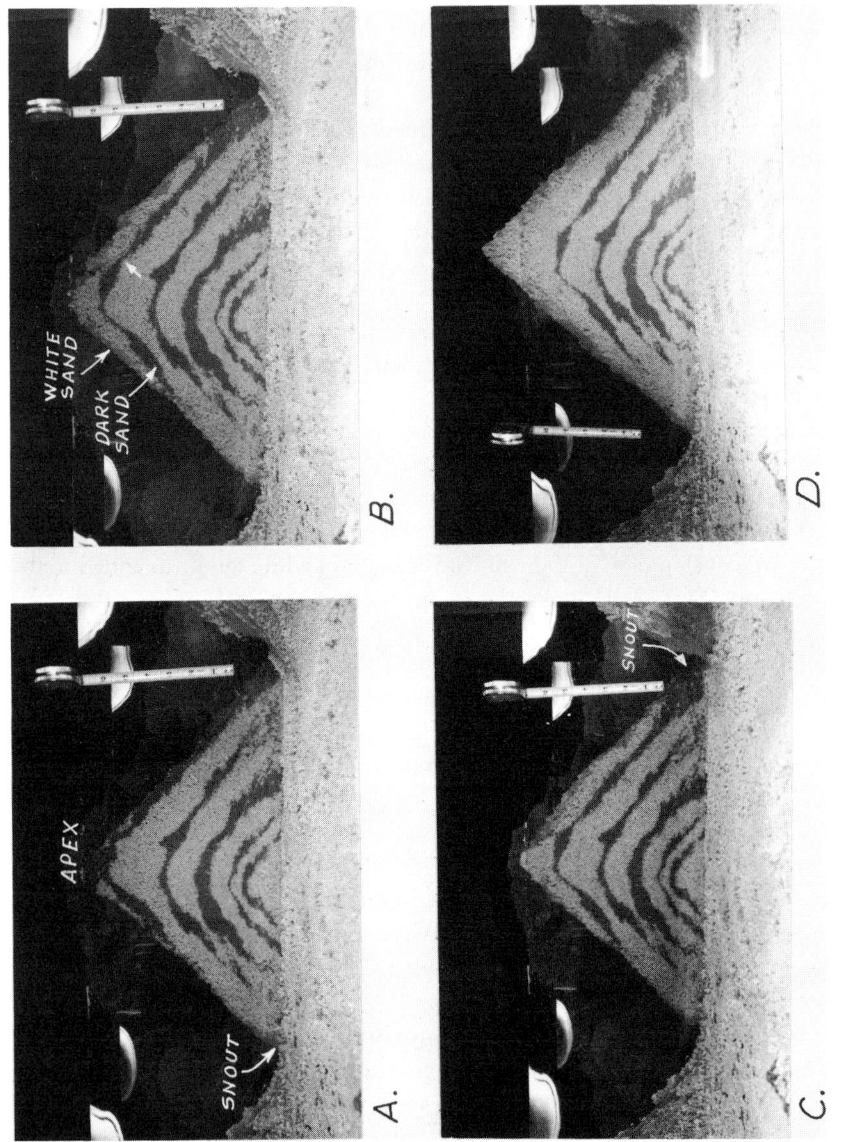

Figure 12.9. Development of layer of white sand on wedge formed between plates of glass. All flowage takes place within two inches of surface of wedge.

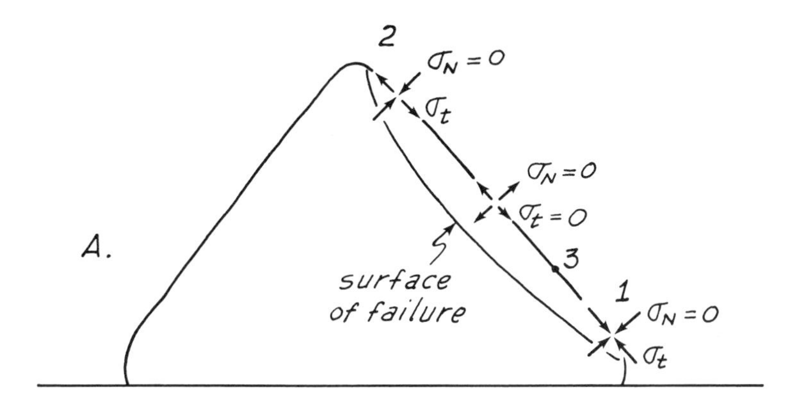

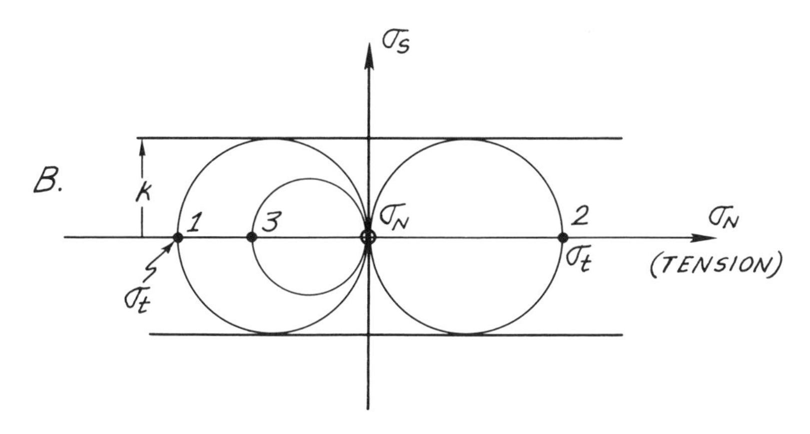

Figure 12.10. State of stress along surface of sand wedge.

of stress is represented by a circle that is tangential to the limiting envelopes. Now, there must be some point along the surface of the sand wedge where the stresses both normal and parallel to the surface are zero, because there must be a region where the tangential surface stresses are intermediate between the two extreme stresses required to cause failure (points 1 and 2, Fig. 12.10B). Point 3 represents a possible value of the normal stress acting tangentially to the surface in this area, and in this case failure is not impending because the circle of this diameter does not intersect the failure envelopes. This prompts the conclusion that part of the layer of sliding sand must have behaved rigidly (*4*).

Theoretical Profile of Snout of Plastic Material.—Regardless of the state of stress along the outer surface of a debris cone or wedge, such as the one shown in preceding figures, the outer surface must be a principal stress trajectory. This is a necessary conclusion because the external stresses acting normally to the surface are zero and there are no shear stresses externally applied to the surface. If substantial parts of snout regions are in a state of critical equilibrium, the shape of the snout should depend upon the physical properties of the material in the snout.

The necessity of such a profile of critical equilibrium can be explained in terms of a familiar experiment. Suppose that a rectangular mass of damp sand rests against a piston (Fig. 12.11A) in a squeezing box of the type King Hubbert (*2*) used for his demonstration of normal and thrust faults in sand. We know from Hubbert's experiments that if the piston is driven forward, the damp sand will fail along an inclined slip plane, such as *A-C* in Fig. 12.11A. The force necessary for thrusting the sand wedge can be calculated, but for the present considerations it is sufficient to rely upon intuition. If the experimental apparatus is tilted, as is shown in Fig. 12.11B, the amount of force required to thrust the sand is less than that required to thrust the sand shown in Fig. 12.11A. This is because gravity is "helping" when the apparatus is tilted. The important point is that, as the slope angle increases, the

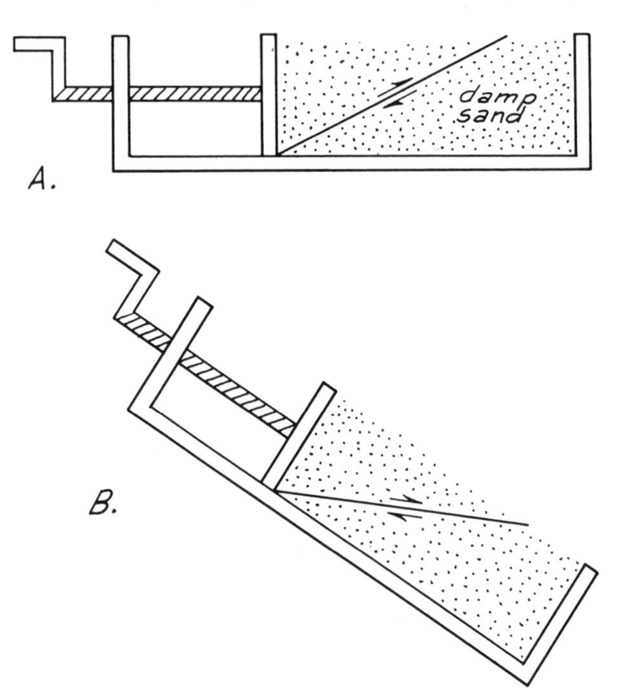

Figure 12.11. Thrusting of damp sand in Hubbert's squeezing machine. When machine is inclined, load required for thrusting is decreased.

force necessary to thrust the sand decreases. Thus, there should be a curved surface such that, if the yield criterion is satisfied by an external force at one point along the surface, the criterion should be satisfied at *every* point along the surface.

The problem is to find the form of this curved surface which corresponds with the form of the snout of debris-flow deposits (Fig. 12.1).

In order to derive an expression for the form of the curved parts of the snouts of debris flows and sand wedges, we will assume that debris is a simple plastic substance. According to the model of the simple plastic, if the maximum shear stress is equal to the yield strength at any point within the plastic, either the plastic is flowing or flow is impending at that point. In general, the relation

$$|\sigma_s| \leq k \tag{12.1}$$

is satisfied everywhere within such a material. Here k is the yield strength of the plastic.

The plastic material along the surface of a debris-flow snout is in critical equilibrium, which means that the yield criterion is satisfied all along the surface. The normal stress acting normally to the surface of the snout, σ_{ns}, is zero, so that the normal stress acting parallel to the surface, σ_{ps}, is twice the shear strength of the debris. We have assumed that debris behaves as a rigid plastic, which requires that the normal stress acting parallel to the surface is

$$\sigma_{ps} = -2k. \tag{12.2}$$

The normal stresses σ_{ns} and σ_{ps} are principal stresses so that the surface of the snout is a principal-stress trajectory; the stress normal to the surface is zero and there are no shear stresses applied *to* the surface. Thus the stresses are as indicated in Fig. 12.12B. The problem of determining an expression for the form of the surface, therefore, is that of deriving an equation for the form of principal-stress trajectories in plastic substances.

We will not follow in detail the derivation of the differential equations for principal-stress trajectories because the derivation is tedious and similar to that of the shear-stress trajectories, described in detail in Chapter 13. The method of deriving principal-stress trajectories in a rigid plastic is described in ref. (*4*).

Let θ be the slope at some point along the surface of a snout of rigid-plastic material (Fig. 12.13). Then, the differential equation describing the change in slope angle with change in height, y, of the snout is

$$\frac{d\theta}{dy} = -\frac{\gamma_d}{2k}. \tag{12.3}$$

Integrating,

$$\theta = \theta_0 - \frac{\gamma_d}{2k}y, \tag{12.4}$$

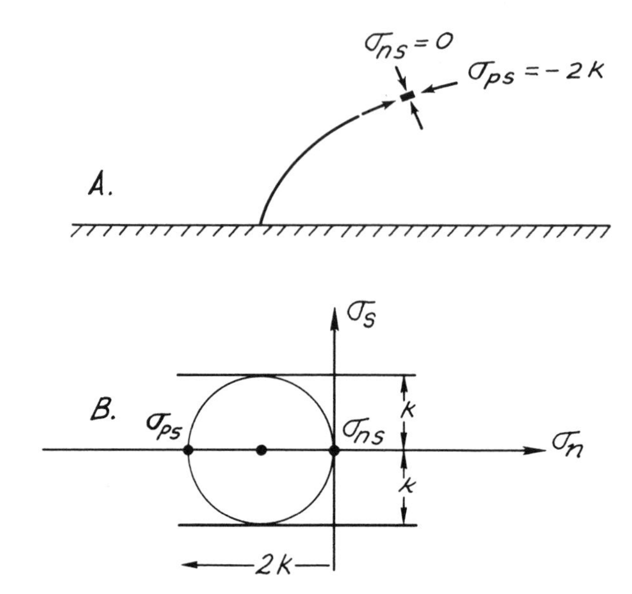

Figure 12.12. Stresses acting on element at surface of plastic in critical equilibrium.

where θ_0 is the value of θ at $y = 0$. Eq. (12.4) gives the slope angle of the snout surface as a function of y. We see that the slope angle decreases linearly with y, so that it approaches zero when

$$y \to \frac{2k\theta_0}{\gamma_d}. \tag{12.5}$$

Theta, of course, is in radians, not degrees.

We can express the form of the snout in x- and y-coordinates by using the definition of the slope of the snout. If θ is the slope angle of the snout, the tangent of the slope angle is the derivative of y with respect to x. Thus,

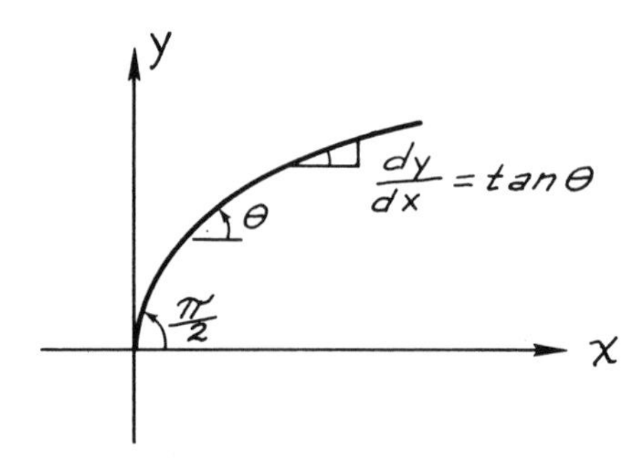

Figure 12.13. Coordinate system of surface of snout.

$$\frac{dy}{dx} = \tan\theta = \tan\left(\theta_0 - \frac{\gamma_d}{2k}y\right). \tag{12.6}$$

Now,

$$\int \frac{dz}{\tan z} = \ln(\sin z) + C_1,$$

so that if we let $z = \theta_0 - (\gamma_d/2k)y$, then

$$dz = -\frac{\gamma_d}{2k}\,dy,$$

and eq. (12.6) can be integrated to yield

$$\int \frac{dz}{-(\gamma_d/2k)\tan(z)} = \int dx = x = -\frac{2k}{\gamma_d}\ln\left[\sin\left(\theta_0 - \frac{\gamma_d}{2k}y\right)\right] + C_1. \tag{12.7}$$

We can evaluate the constants, θ_0 and C_1, in terms of the sketch of the snout shown in Fig. 12.13A. At $x = 0$ and $y = 0$, let the snout be vertical, that is, let $\theta_0 = 90° = \pi/2$. Then eq. (12.7) is satisfied if the second constant, C_1, is zero. Thus, for these conditions, eq. (12.7) becomes (*4*)

$$x = -\frac{2k}{\gamma_d}\ln\left[\cos\left(\frac{\gamma_d}{2k}y\right)\right], \tag{12.8}$$

because $\sin(90° - z) = \cos z$.

Equation (12.8) is plotted in dimensionless form in Fig. 12.14. It describes a bulbous form that is strikingly similar to the profiles of snouts of many debris-flow deposits and solidified lava flows, and of some landslides (e.g., compare Figs. 12.1 and 12.14).

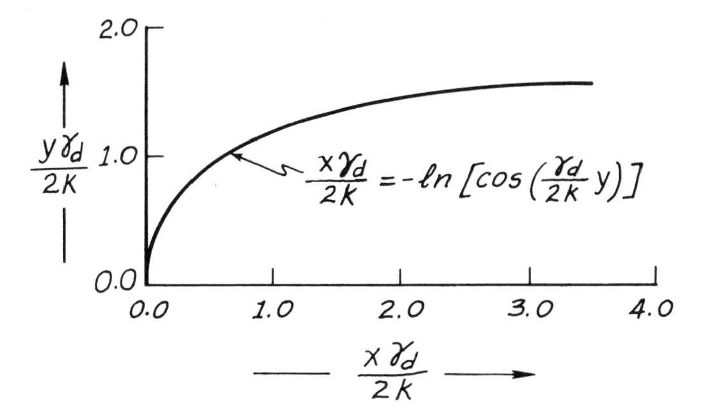

Figure 12.14. Theoretical profile of snout of rigid plastic.

The size of the snout in a rigid plastic, according to our derivations, depends upon the density and shear strength of the plastic. The greater the shear strength and the less the density of the plastic, the larger is the snout. Our solution, however, does not indicate how far we should extend the profile of the snout. It does tell us that, if we extend the back of the snout to infinity, the slope approaches zero and the height approaches a limiting value of $\pi k/\gamma_d$. This value is slightly larger than the critical height of a vertical snout, which is $2k/\gamma_d$ (see ref. *12*).

Comparison of Theoretical and Actual Snouts.—It is interesting and enlightening to use the knowledge we have gained in order to re-examine more carefully what we saw happening in the sand wedge built between two plates of glass by adding moist sand to its apex (Fig. 12.9). Between the times the photographs in Figs. 12.9B and 12.9C were taken, a mass of sand moved downward, toward the right. The cross-sectional shape of the mass of sand is shown, approximately, in Fig. 12.15A. Near the apex of the sand wedge, the mass of sand was sheared along several curved planes, which appear approximately as is shown in Fig. 12.15A. Similarly, at the base of the right-hand side of the wedge, the mass was sheared along several planes within the dark-colored sand. Between these zones of shear, however, the sand mass moved approximately as a rigid body. Thus the mass of moving sand was disturbed internally at each end, but it was undisturbed in the middle. The profile is curved in the snout region whereas it is nearly straight in the central region of the mass of sand.

The shape of the profile in the snout region should be described by the equation we derived, eq. (12.8), in so far as the sand behaves as a simple plastic substance. The thickness of the layer of sand that moved as a rigid mass, behind the snout, should be described approximately by an equation that we shall derive later on,

$$T_c = \frac{k}{\gamma_d \sin \delta}. \tag{12.9}$$

This equation cannot be strictly applicable to the sand wedge, however, because we assumed when we derived it that the slope, and therefore the layer of sand, extended an infinite distance. The layer of rigidly sliding sand is definitely finite in length and it is the only conceivable source of the push that causes sand in the snout to flow, so that either the thickness or the slope angle is slightly greater in the sand wedge than in an infinite sheet. There was no obvious change in the thickness of the layers of rigidly sliding sand in the sand wedges, but the slopes of the sand wedges definitely decreased as the wedges were built up. For example, Fig. 12.16 shows successive profiles of the wedge shown in Fig. 12.9. The slopes of the sides of the wedges decreased from about 51 degrees to 47 degrees as the sand wedge was built up from 4.5 inches to 9 inches in height. When the wedge was about 11 inches high its slope was still about 47 degrees, so that the slope seemed to be approaching a limiting value as the size of the wedge was increased (*4*).These observations sug-

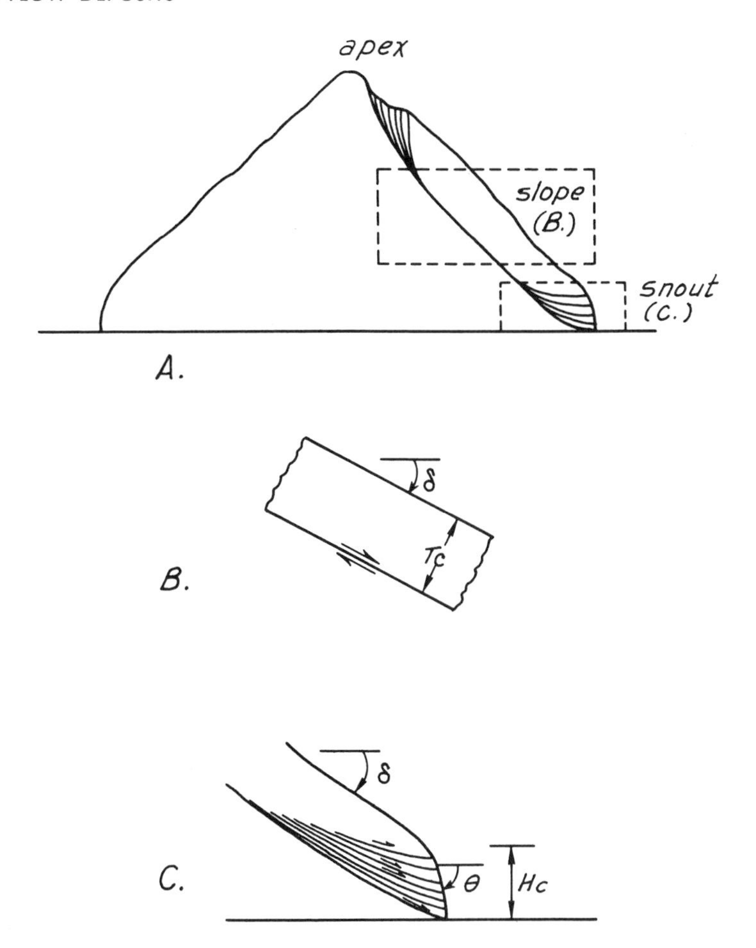

Figure 12.15. Types of deformation within sand wedge.
A. Form of layer that slides.
B. Central part of layer slides along single plane.
C. Snout region deforms along several slip lines.

gest that we can reasonably represent the sliding layer by an infinitely long layer of sand moving down a slope. Apparently, when the sand wedge exceeds about 10 inches in height, the moving layer behaves essentially as though it were infinitely long. This conclusion was verified by building one sand wedge to a height of about two feet, and taking photographs at several stages to compare profiles. The limiting slope for the sand used in the experiment was found to be about 45 degrees. The angle of slope of the dry sand was about 35 degrees.

On the basis of these observations, I think that we can reasonably approximate the flow in snout regions as a combination of a rigid layer, of thickness T_c [eq. (12.9)],

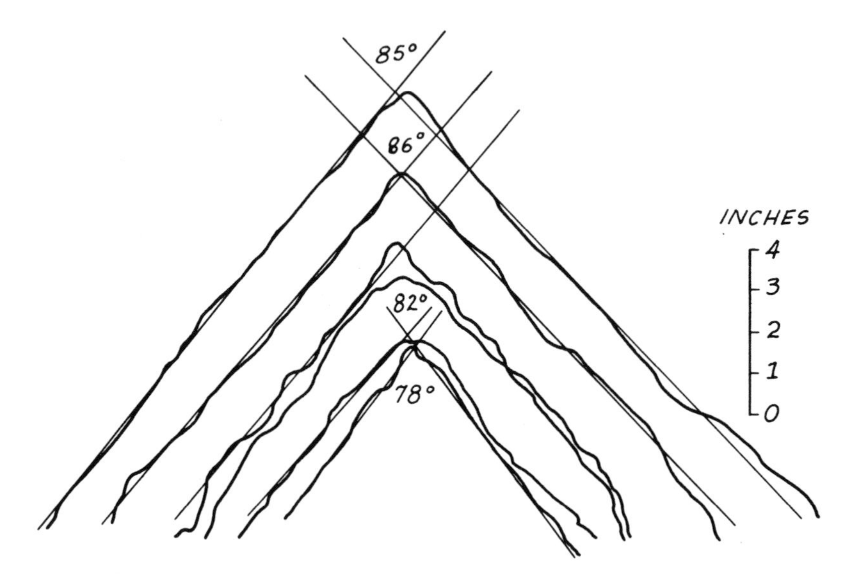

Figure 12.16. Successive profiles of a sand wedge.

$$T_c = \frac{k}{\gamma_d \sin \delta},$$

and a snout height [eq. (12.4)]

$$y = -\left(\theta - \frac{\pi}{2}\right)\frac{2k}{\gamma_d} \tag{12.10}$$

which depends upon the slope angle, θ, of the snout. Where the snout and the straight back join, the slope angles, θ and δ, must be equal, so that where $y = H_s$, the height of the.snout is

$$H_s = \left(\frac{\pi}{2} - \delta\right)\frac{2k}{\gamma_d}. \tag{12.11}$$

H_s is the vertical distance from the surface on which the plastic is flowing to the place where the uniform slope behind the snout breaks off to form the curved snout itself (Fig. 12.15C).

We have shown that the general forms of snouts of sand cones, lava flows, debris flows, and landslides are similar to the form of the snout of a rigid-plastic substance in critical equilibrium. Further, we should be able to estimate the yield strength of the various materials by measuring the heights of their snouts. Then, if H_s is the snout height, the yield strength of the material is

$$k = \frac{H_s \gamma_d}{\pi[1 - (\delta/90)]}, \tag{12.12}$$

where δ, measured in degrees, is the angle of inclination of the uniform slope behind the snout.

Presumably we should be able to estimate the strength of debris by measuring the height of a snout at the periphery of a debris-flow deposit. For example, Fig. 12.8B shows snouts of a debris wedge produced experimentally by pouring debris from above so that the debris flowed laterally between parallel plates of glass. The snout on the left-hand side is about 3.5 cm high and has a minimum slope of about 25 degrees.

Thus,

$$H_s = 3.5 \text{ cm},$$

$$\delta = 25 \text{ degrees}.$$

The debris had a density of about 2.2 gm/cc, so that according to eq. (12.12),

$$k = \frac{H_s \gamma_d}{\pi[1 - (\delta/90)]} \approx \frac{3.5(2.2)(980)}{\pi[1 - (25/90)]} \approx 3 \times 10^3 \text{ dn/cm}^2.$$

According to measurements with a cylindrical viscometer (*4*), the shear strength of the slurry used for the debris wedges was about 2.5×10^3 dn/cm², so that the theory seems to check out.

However, snout heights depend upon factors we have ignored. For example, the properties of lava are presumably very sensitive to the temperature of the lava, and, as a result, the strength estimated by the method outlined here would probably be the strength of relatively cool lava at the periphery of the flow. Similarly, the strength of debris in a debris flow can be extremely sensitive to changes in the moisture content of the debris. The strength of one sample of fine-grained debris from a deposit in Panamint Valley, California, for example, changed from about 10^3 dn/cm² to 1.5×10^4 dn/cm², that is, more than an order of magnitude, when the moisture content of the debris lost about two percent of its weight, from about 16 to 14 (*4*). We could reasonably assume, therefore, that the strength of debris in the snout of a debris flow can be much greater than the strength of the debris in the interior of the flow because water could be lost more rapidly from the snout to the dry surface of the alluvial fan than from the interior of the flow. Perhaps we can conclude that strengths estimated by measuring snout heights will tend to be high values for the debris.

Another problem that arises when we attempt to relate the snout heights and strengths of various materials is that certain materials seem to have a type of resistance to shear that is not accounted for in theory by a simple plastic substance. For example, it might bother other people, as it bothers me, to consider most sand or even fine-grained debris to be a simple plastic substance with a shear strength, k. Perhaps the Coulomb model, used in soil mechanics, better describes the strength of such materials. The Coulomb model can be written as

$$|\sigma_s| \leq C - \sigma_n \tan \phi, \tag{12.13}$$

where tensile normal stresses are positive, and where

C = cohesion,

σ_n = normal stress on shear plane,

and

$\tan \phi$ = coefficient of friction.

Proceeding as we did when we derived the critical thickness of the simple plastic substance, we could show (*4*) that the critical thickness of a Coulomb substance on a very long slope is

$$T_c = \frac{C}{\gamma_d \cos \delta(\tan \delta - \tan \phi)}; \qquad \delta > \phi, \tag{12.14}$$

which can be compared with the critical thickness of a simple plastic,

$$T_c = \frac{k}{\gamma_d \sin \delta}. \tag{12.15}$$

Similarly, the height of a snout in a Coulomb material would be

$$H_s = \frac{\pi C}{\gamma_d}\left(1 - \frac{\delta}{90}\right) \tan\left(\frac{\pi}{4} + \frac{\phi}{2}\right), \tag{12.16}$$

where δ is in degrees. This equation, also, can be compared with the equation for the snout height of a simple plastic,

$$H_s = \frac{\pi k}{\gamma_d}\left(1 - \frac{\delta}{90}\right). \tag{12.17}$$

We see in each of these equations that if the angle of internal friction, ϕ, is zero, the equations for the Coulomb plastic reduce to the equations for a simple plastic. We also see that, in order to determine the Coulomb properties of a material, such as debris, we need two independent measurements, because there are two strength constants, C and ϕ. The necessary measurements cannot be made on a single snout. Measurements of two different sets of parameters of a debris-flow deposit are needed to determine the Coulomb friction and cohesion of the debris. For example, the thickness and surface slope of a wide debris-flow deposit might be one set of measurements and the size and density of the largest boulders transported by a debris flow, or the height and slope of the snout, might be the other set of measurements.

References cited in Chapter 12

1. Curry, R. R., 1966, "Observation of Alpine Mudflows in the Tenmile Range, Central Colorado": *Geol. Soc. America. Bull.*, V. 77, p. 771–776.

2. Hubbert, M. K., 1951, "Mechanical Basis for Certain Familiar Geologic Structures": *Geol. Soc. America Bull.*, V. 62, p. 355–372.
3. Jahns, R. H., 1949, "Desert Floods": *Eng. and Sci. Monthly*, Calif. Inst. Tech. May, p. 10–14.
4. Johnson, A. M., 1965, "A Model for Debris Flow": *Ph.D. dissertation*, The Pennsylvania State University, Univ. Park, Penna.
5. ———, and Hampton, M. A., 1968, "Subaerial and Subaqueous Flow of Slurries": Ann. Progress Report to U.S. Geol. Survey, Branner Library, Stanford University, Stanford, Calif.
6. ———, and Hampton, M. A., 1969, "Subaerial and Subaqueous Flow of Slurries": Final Report to U.S. Geol. Survey, Branner Library, Stanford University, Stanford, Calif.
7. Johnson, W., and Mellor, P. B., 1962, *Plasticity for Mechanical Engineers*: D. Van Nostrand Co., Ltd., London.
8. Krauskopf, K. B., 1948, "Lava Movement at Paricutín Volcano, Mexico": *Geol. Soc. America Bull.*, V. 59, p. 1267–1284.
9. Prager, W., and Hodge, P. G., 1951, *Perfectly Plastic Solids*: John Wiley & Sons, Inc., N.Y.
10. Sharp, R. P., and Nobles, L. H., 1953, "Mudflow of 1941 at Wrightwood, Southern California": *Geol. Soc. America Bull.*, V. 64, p. 547–560.
11. Shelton, J. S., 1966, *Geology Illustrated*: W. H. Freeman & Co., San Francisco, Calif.
12. Sokolovski, V. V., 1965, *Statics of Granular Media*: Pergamon Press, N.Y.

13

Chapter Sections

Mechanisms of Transport
Settling of Sphere in Plastic Substance
Simplification of Problem
Equations Describing Plane Strain of a Rigid-Plastic Substance
Equations for stresses
Equations of equilibrium
Equations for velocities within rigid plastic
Bearing Capacity of Rigid Plastic
Relation Between Depth of Penetration of Sphere and Strength of Plastic
Modifications to account for three-dimensional conditions and buoyancy of plastic
Comparison of theory and experiment
Transport of Blocks and Boulders on Surprise Canyon Alluvial Fan
References Cited in Chapter 13
Other References on Soil Mechanics and Plasticity Theory

13

Transport of Boulders and Blocks by Debris Flow

Mechanisms of transport

A characteristic of debris-flow deposits which distinguishes them from stream-laid sediment is the appearance of random distribution of large particles set in a matrix of finer-grained debris (*6*). In this respect they resemble glacial till. Yet, when we examine debris-flow deposits in detail, we begin to see evidence of the sorting of coarse clasts (*6,8*). The coarse clasts tend to be concentrated at the edges, in the snouts and in the channels of debris-flow arms. A characteristic of the dozens of flows observed at Wrightwood (*8*), California, is that boulders were concentrated in the frontal and medial portions of flows (e.g., see Fig. 12.5).

The sorting of coarse clasts by debris flow, indeed even the transport of coarse clasts, seems to be related to several aspects of the process of debris flow. One factor is the relative densities of the boulders being transported and of the matrix of the debris. The density of the debris that formed the deposits on the Surprise Canyon alluvial fan (*6*) in 1917 ranged from about 2.1 to 2.4 gm/cc. The boulders transported by this flow had densities ranging from about 2.5 to 2.6. Thus, boulders completely submerged in the debris flow nearly floated.

Another factor, one which has not been analyzed adequately in terms of debris properties, is *lift*. Goldsmith and Mason (*4*) have shown that spheres suspended in a viscous fluid, flowing axially in a circular tube, are subjected to lift. The lift causes the spheres to migrate radially inward or outward, depending on the relative densities of the spheres and fluid and upon the position of the spheres within the tube.

One possible cause of the sorting and transport of coarse clasts in debris flows has been analyzed in detail by Bagnold (*1,2*). The mechanism Bagnold describes is also an appealing explanation for the sorting in tongues of dry debris moving on talus slopes (see ref. *8*). The basic concept of Bagnold's theory is that, where flowing granular solids are bounded on one side by a rigid boundary, grains can move more readily parallel to the boundary or away from the boundary than toward the boundary. This is particularly evident for particles moving along the rigid boundary. The attempt of grains to move toward the boundary is resisted by forces exerted on the grains by the boundary. The values of these forces, per unit area, is Bagnold's "dispersive pressure" (*1*).

The force, or change in momentum per unit of time, exerted by the boundary on a particle, in the direction normal to the boundary, is (*1*)

$$F_d = 2mU \cos \alpha, \tag{13.1}$$

where m is the mass of the particle, U is the velocity, parallel to the rigid boundary, of a particle with respect to a particle below, and α is the angle between the plane of impact of the particle with a particle below and the direction of movement of the particle. The angle of the plane of impact, α, is related to the packing of the particles and seems to be essentially equivalent to the angle of internal friction of the granular solids (*1*).

When there are many particles moving generally parallel to a rigid boundary, the dispersive force acting on any individual particle can be imagined to be an average force for a succession of glancing collisions among grains. The velocity, U, in this case can be imagined to be the relative velocity of grains. The relative velocity is proportional to the radius of the grains and to the velocity gradient, dU/dy, so that

$$U \propto R(dU/dy). \tag{13.2}$$

Equation (13.1), therefore, can be rewritten as

$$F_d = C_1 mR(dU/dy) \cos \alpha, \tag{13.3}$$

where C_1 is a constant.

The dispersive force in dry sand flows on talus cones is resisted by the weight of each grain of sand. The weight of each grain is proportional to its mass, so that

$$W_t \propto m.$$

Therefore, we can write the ratio of the driving force, F_d, to the resisting force, W_t, as

$$\frac{F_d}{W_t} = C_2 R(dU/dy) \cos \alpha. \tag{13.4}$$

In the case of particles suspended in a viscous fluid, we would replace the weight by an effective, submerged weight (*1*).

The dispersive force acting on a particle, according to eq. (13.4), is proportional to the shear rate, dU/dy, suggesting that when particles in a dry flow are sheared together, the larger particles should drift toward the zone of least shear rate, that is, toward the upper surface and the periphery of dry flows. This, of course, is exactly what is observed.

Bagnold's dispersive force mechanism is a logical explanation for the sorting and transport of large clasts in a mass containing many interacting clasts.

The ability of debris flows to transport coarse clasts is related to at least one more factor—the yield strength of slurries (*6,7,8*). The strength of debris retards the sinking of boulders into the debris, whether moving or stationary. An observation made while conducting experiments which were designed to relate the strength of slurries to the transport of spherical particles bears on the validity of the statement that slurries have finite strength (*7*). This statement is questionable because of the similarities between the meniscus of water, which is explained in terms of surface tension of water, and the shape of the profile of debris-flow snouts, which we have explained in Chapter 12 in terms of strength of debris.

The experiments were made with wooden spheres ranging from 1.6 to 2.6 cm in diameter and ranging in density from 1.09 to 2.28 gm/cc. The spheres were placed in a rectangular tank of slurry, consisting of water, quartz sand, and kaolin clay (*7*). The densities of the slurries ranged from 1.2 to 2.0 gm/cc. As predicted by the theory that we will develop subsequently, the depth to which the spheres settled into the slurries is related to the strengths of the slurries. But during the experiments, I noticed that slurry adjacent to each sphere was displaced slightly downward and outward, much like the profile of the meniscus that forms around a pin that is supported by surface tension of water. A question that came to mind, therefore, is whether the spheres were supported by surface tension between the spheres and the slurries. In order to answer this question, the following experiment was made: A sphere was placed on the slurry and it settled until about half of it projected above the surface of the slurry. Thus, the sphere floated. Then the sphere was pushed about one centimeter below the surface of the slurry and released. It neither sank to the bottom nor bobbed back to the surface again. It stayed put. The strength of the slurry apparently prevented the sphere from rising to the surface or from sinking farther. If the sphere had been supported by surface tension, it would have behaved much as a pin does when it is placed on water: If the pin is placed gently on the water surface, surface tension prevents it from getting completely wet, and therefore the pin floats. If the pin is pushed below the water surface, however, it sinks to the bottom of the container of water.

Settling of sphere in plastic substance

Simplification of Problem.—The problem of calculating the upward force a plastic substance exerts on a sphere settling into it is most difficult. As far as I know, it has not been solved. We can derive an approximate solution, however, if we introduce some simplifying assumptions (*7*).

As a spherical particle sinks into a plastic substance, it displaces the plastic laterally within certain zones on each side of the sphere. These zones probably appear much as is shown in Fig. 13.1A. The zone of displacement is larger than the radius of the sphere. In order for the sphere to sink into the plastic, the sphere must apply sufficient pressure to the plastic to cause the plastic to yield not only along the surface of the sphere, but everywhere within the zone of flow. Buoyancy of the plastic is a second type of resistance to sinking of the sphere. The buoyancy is equal to the weight of the displaced plastic.

The upward pressure exerted by the plastic material on the submerged part of the sphere can be calculated, approximately, by ignoring the submerged part of the sphere, so that the sphere has a flat side (Fig. 13.1B). Then the solution for the pressure required to drive a smooth, flat punch, with a circular cross-section, into the plastic can be used to calculate the upward pressure exerted by the plastic on the sphere (Fig. 13.1C). This problem has been solved (*10,13*), but the solution is

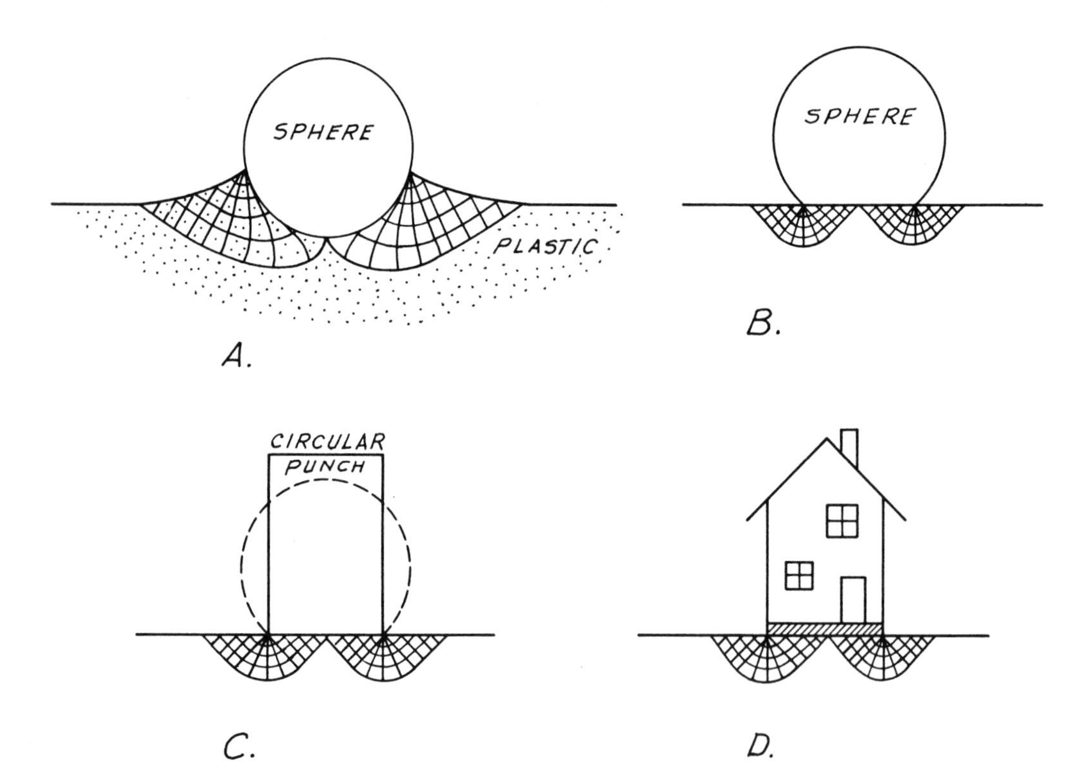

Figure 13.1. Zones of deformation of rigid plastic beneath various objects.

A. Approximate pattern of slip lines beneath sphere.
B. Slip-line pattern beneath flat-sided sphere.
C. Slip-line pattern beneath circular punch.
D. Slip-line pattern beneath building with slab foundation.

difficult to follow. The problem is greatly simplified if we make it two-dimensional by assuming that the punch is rectangular in cross-section, with one dimension much greater than the other. We will compare our two-dimensional solution with the three-dimensional solution and show that the solutions do not differ greatly. Posed as a two-dimensional problem, the problem of settling of a sphere in a dense, plastic substance is identical to the problem of bearing capacity of a simple plastic soil loaded by a slab foundation (*11,16*) (Fig. 13.1D).

We will first solve the two-dimensional problem of the bearing capacity of a simple plastic. Then we will briefly review the solution for the pressure under a circular punch. Finally, we will return to the problem of transport of boulders and blocks by plastic debris and add effects of buoyancy. One reason that we will follow this rather circuitous path to a solution to the problem is that it provides an introduction to the mathematical theory of plasticity, one of the three fundamental theories of rheology.

Equations Describing Plane Strain of a Rigid-Plastic Substance.— *Equations for stresses.*—According to the one-dimensional model of the rigid-plastic substance, the shear stress within the substance must be less than or equal to the yield strength of the substance in order for equilibrium to be maintained:

$$|\sigma_s| \leq k. \tag{13.5}$$

We need to generalize the rheological model to two dimensions and to define orientations of planes of maximum shear stress. The planes of maximum shear stress are the all-important *slip lines* or *shear-stress trajectories* of plasticity theory.

Figure 13.2 shows the state of stress near a point in a rigid-plastic substance. The yield criterion is satisfied, so that the Mohr Circle representing the stress state is tangent to the yield envelopes, which appear as dashed horizontal lines in the figure and which are a distance k from the normal-stress coordinate. If the diameter of the Mohr Circle were less than $2k$, yielding would not be possible and the rigid

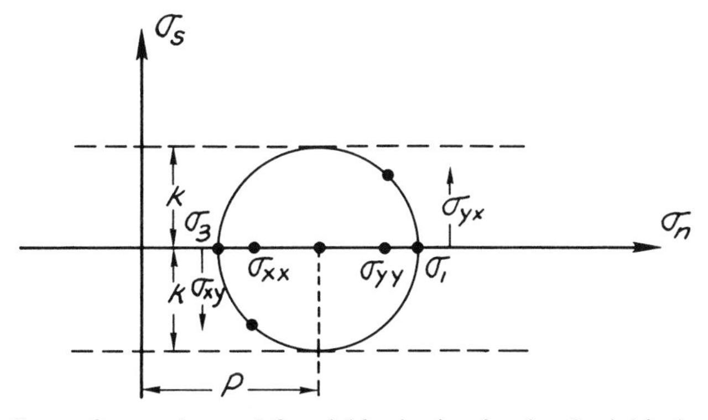

Figure 13.2. State of stress that satisfies yield criterion for simple rigid-plastic substance.

plastic would not deform at all. The rigid plastic, by definition, does not deform unless the yield criterion is satisfied. In effect, we assume that a real material is a rigid plastic if elastic strains in it are so small relative to permanent, plastic strains that we can ignore the elastic strains. Further, the diameter of the Mohr Circle cannot be greater than $2k$ because the shear stress cannot exceed the critical value without causing acceleration.

According to Fig. 13.2,

$$\sigma_1 = p + k,$$

and

$$\sigma_3 = p - k,$$

if failure is impending. Thus,

$$\left.\begin{aligned} |\sigma_1 - p| &\leq k \\ &\text{and} \\ |\sigma_3 - p| &\leq k \end{aligned}\right\} \tag{13.6}$$

are yield criteria for plane strain. Here

$$p = \frac{\sigma_1 + \sigma_3}{2} = \frac{\sigma_{xx} + \sigma_{yy}}{2} \tag{13.7}$$

is the mean normal stress, or pressure, on the substance.

The state of stress near a point in a yielding plastic body could be defined in terms of stress components, σ_{xx}, σ_{yy}, and σ_{xy}. However, the components of stress are interdependent. In fact, we can reduce the number of dependent variables from three to two with the yield criterion. Let p and ρ (rho) be the two variables. They are defined in Fig. 13.3. There, p is the distance from the origin of coordinates to the center of the Mohr Circle. The angle, ρ, is the angle between the x-direction and the α-direction (alpha) in physical space, Fig. 13.3B. The alpha-direction in physical space is one direction of maximum shear stress. The other direction of maximum shear will be designated by beta (β). We showed earlier that the directions of maximum shear stress bisect the angles between the directions of the principal normal stresses.

We derived the following relations between stress space and physical space in Chapter 9. The angle between the point in stress space, representing the stresses acting on the plane parallel to the x-direction in physical space, and the point representing the stresses acting on the plane parallel to the β-axis in physical space is 2ρ. We will call ρ positive in physical space if it is a counterclockwise angle and we will call 2ρ positive in stress space if it is a clockwise angle. Accordingly, the angle 2ρ in Fig. 13.3A is measured along the circumference of the Mohr Circle from point x to point α. It is twice the angle between the x-direction and the α-direction in physical

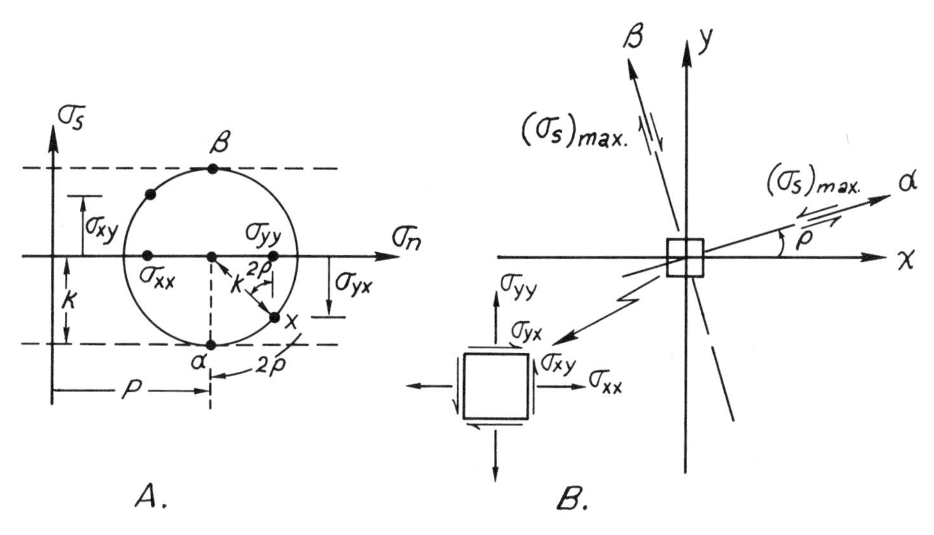

Figure 13.3. Definitions of two new variables, p and ρ.

A. Stress space. The angle 2ρ is twice the angle between the x- and α-planes. The variable p is the mean normal stress.

B. Physical space. Alpha- and beta-lines in physical space.

space, Fig. 13.3B. These relations must be understood in order to verify the derivations in following pages.

Now we can derive expressions for the stress components, σ_{xx}, σ_{yy}, and σ_{xy}, in terms of the new variables, p and ρ. According to Fig. 13.3A,

$$\left.\begin{aligned}\sigma_{xx} &= p - k\sin(2\rho),\\ \sigma_{yy} &= p + k\sin(2\rho),\\ \text{and}\quad \sigma_{xy} &= k\cos(2\rho).\end{aligned}\right\} \qquad (13.8)$$

Equations of equilibrium.—The stress components must satisfy the differential equations of equilibrium, which we derived in Chapter 7:

$$\left.\begin{aligned}\frac{\partial\sigma_{xx}}{\partial x} + \frac{\partial\sigma_{xy}}{\partial y} + X &= 0,\\ \text{and}\quad \frac{\partial\sigma_{xy}}{\partial x} + \frac{\partial\sigma_{yy}}{\partial y} + Y &= 0.\end{aligned}\right\} \qquad (13.9)$$

We can reduce the number of dependent variables from three to two in eqs. (13.9) by substituting eqs. (13.8) into eqs. (13.9):

$$\left.\begin{aligned} &\frac{\partial p}{\partial x} - 2k\cos(2\rho)\frac{\partial \rho}{\partial x} + 2k\sin(2\rho)\frac{\partial \rho}{\partial y} + X = 0, \\ &\frac{\partial p}{\partial y} + 2k\sin(2\rho)\frac{\partial \rho}{\partial x} + 2k\cos(2\rho)\frac{\partial \rho}{\partial y} + Y = 0. \end{aligned}\right\} \quad (13.10)$$

The y- and x-axes are vertical and horizontal, respectively, in Fig. 13.3B, so that the body force, X, in the x-direction would be zero and the body force, Y, in the y-direction would equal the unit weight of the plastic. However, we want to consider a more general situation where there are components of body force in both directions. Therefore, let the x-axis be rotated clockwise by an angle, δ (delta), from the horizontal (Fig. 13.4). In this case eqs. (13.10) become

$$\frac{\partial p}{\partial x} - 2k\cos(2\rho)\frac{\partial \rho}{\partial x} + 2k\sin(2\rho)\frac{\partial \rho}{\partial y} - \gamma\sin\delta = 0,$$

and

$$\frac{\partial p}{\partial y} + 2k\sin(2\rho)\frac{\partial \rho}{\partial x} + 2k\cos(2\rho)\frac{\partial \rho}{\partial y} - \gamma\cos\delta = 0. \quad (13.11)$$

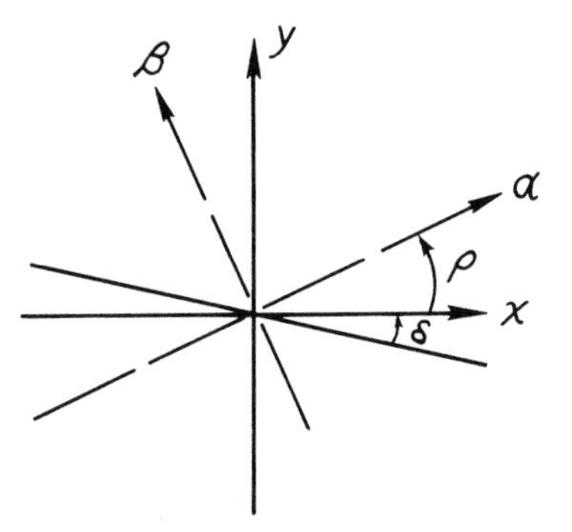

Figure 13.4. Cartesian coordinate system rotated by an angle, δ, with respect to horizon.

These differential equations can be greatly simplified if we convert them to differential equations for the new variables along the slip lines, α and β. The formal procedure of deriving the differential equations for the new variables along the slip lines is extremely tedious. It is necessary, unfortunately, for the Coulomb rigid plastic, the slip lines of which do not intersect at right angles to each other. I refer you to refs. *3,6* for details of the manipulations. Reference *3* includes the best discussion I have found of the physical and mathematical significance of slip lines. For the present problem, fortunately, we can use the following ploy to avoid the more lengthy procedure (*9,11*): Somewhere within a plastic region the x- and the α-directions will coincide. Let us examine differential equations (13.11) where this happens. If x and α coincide, $\rho = 0$, so that eqs. (13.11) reduce to total differential equations. The first is independent of y and the second is independent of x (*14*):

$$\left.\begin{aligned} &\frac{dp}{dx} - 2k\frac{d\rho}{dx} - \gamma \sin \delta = 0, \qquad \text{on an alpha-line,} \\ &\text{and} \\ &\frac{dp}{dy} + 2k\frac{d\rho}{dy} - \gamma \cos \delta = 0, \qquad \text{on a beta-line.} \end{aligned}\right\} \tag{13.12}$$

For the problem of indentation of a plastic by a punch, we can assume that the body forces are negligible compared to the surface stresses, in which case eqs. (13.12) further simplify to (*11*)

$$\frac{dp}{dx} - 2k\frac{d\rho}{dx} = 0,$$

or

$$\left.\begin{aligned} &\frac{dp}{d\rho} - 2k = 0, \qquad \text{on an alpha-line,} \\ &\text{and} \\ &\frac{dp}{d\rho} + 2k = 0, \qquad \text{on a beta-line.} \end{aligned}\right\} \tag{13.13}$$

Thus the differential equations become independent of y and x. The first of eqs. (13.13) relates the angle between the x-direction and the alpha-line, angle ρ, and the mean normal stress, p, along alpha-lines. The second equation gives analogous relations along beta-lines.

Differential equations (13.13) can be integrated immediately to give

$$\left.\begin{aligned} &p - 2k = C_\alpha, \qquad \text{a constant on an alpha-line,} \\ &p + 2k = C_\beta, \qquad \text{a constant on a beta-line.} \end{aligned}\right\} \tag{13.14}$$

We have shown, albeit slightly mysteriously, that the partial differential equations of equilibrium of a rigid-plastic substance reduce to total differential equations along slip lines α and β. The total differential equations can be integrated immediately, if body forces are negligible, to relate the two independent variables, p and ρ, along the slip lines. If a rigid plastic is in critical equilibrium, therefore, the conditions that must be satisfied along the slip lines are stated by eqs. (13.14). The slope of the slip lines can be derived from Fig. 13.3B,

$$\left.\begin{aligned} &\left(\frac{dy}{dx}\right)_{\text{alpha}} = \tan \rho, \\ &\text{and} \\ &\left(\frac{dy}{dx}\right)_{\text{beta}} = \tan (90^\circ + \rho) = -\text{ctn}\ \rho. \end{aligned}\right\} \tag{13.15}$$

The slip lines defined by eqs. (13.14) and (13.15) are similar to the principal stress trajectories that Odé derived to represent the dike pattern at Spanish Peaks (see Chapter 11). The slip lines are inclined at 45 degrees to the principal stress trajectories, however, because the slip lines are directions of maximum shear. If eqs. (13.14) and (13.15) are solved for a plastic body, the variables, p and ρ, are known at every point in the body and the stresses σ_{xx}, σ_{yy}, and σ_{xy} can be calculated at any point by means of eqs. (13.8).

Equations for velocities within rigid plastic.—The equations of equilibrium and the yield criterion are insufficient to characterize completely the mechanical behavior of a rigid-plastic substance. We also need to be able to relate stress and strain or strain rate. We can derive relations between stresses and rates of strain by making two assumptions about the behavior of the ideal substance:

1. The density of the rigid plastic is constant. We assume that there is no volume change during deformation of a rigid-plastic substance. For plane strain, this requires that

$$\dot{\epsilon}_{xx} + \dot{\epsilon}_{yy} = 0 = \dot{\epsilon}_1 + \dot{\epsilon}_3.$$

2. The directions of the principal strain rates coincide with the directions of the principal stresses. Thus, we assume that the slip lines for the stress field are parallel to the slip lines for the strain-rate field. Deformation is the result of shear along the slip lines.

The second assumption requires that the figure of the Mohr Circle of strain rates is similar to the figure of the Mohr Circle of stresses. Let θ be the angle between the y-direction and the direction of the minimum principal stress in physical space. Then, according to Fig. 13.5A,

$$\tan 2\theta = \frac{2\sigma_{xy}}{\sigma_{xx} - \sigma_{yy}}. \tag{13.16}$$

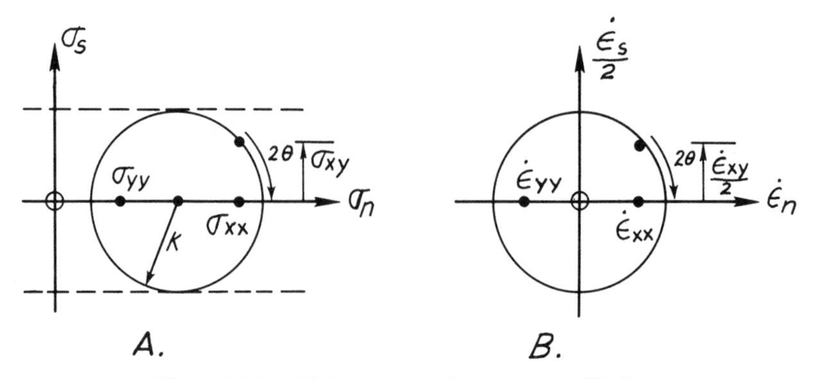

Figure 13.5. Mohr stress and strain-rate Circles.

According to that figure,

$$\tan 2\theta = \frac{\dot{\epsilon}_{xy}}{\dot{\epsilon}_{xx} - \dot{\epsilon}_{yy}}. \tag{13.17}$$

Equations (13.16) and (13.17) are equal, so that the relation between stresses and rates of strain for a rigid-plastic substance is:

$$\frac{2\sigma_{xy}}{\sigma_{xx} - \sigma_{yy}} = \frac{\dot{\epsilon}_{xy}}{\dot{\epsilon}_{xx} - \dot{\epsilon}_{yy}}. \tag{13.18}$$

Equation (13.18) does not imply viscous effects of a rigid-plastic substance. The stress required to cause a certain amount of strain does not depend upon the time taken to produce that strain.

With eq. (13.18) and Fig. (13.3) we can show that the relations between stress and rate of strain are (e.g., see Prager and Hodge, ref. *11*, p. 31):

$$\left.\begin{aligned} \sigma_{xx} &= p + \frac{k\dot{\epsilon}_{xx}}{\sqrt{I}}, \\ \sigma_{yy} &= p + \frac{k\dot{\epsilon}_{yy}}{\sqrt{I}}, \\ \sigma_{xy} &= \frac{k}{2}\frac{\dot{\epsilon}_{xy}}{\sqrt{I}}, \end{aligned}\right\} \tag{13.19}$$

where

$$I = \frac{\dot{\epsilon}_{xx}^2 + \dot{\epsilon}_{yy}^2}{2} + \left(\frac{\dot{\epsilon}_{xy}}{2}\right)^2. \tag{13.20}$$

We can see that eqs. (13.19) furnish a unique expression for the stresses if the strain rates are known. However, if the stresses are given we can only calculate ratios of the rates of strain. It should be reiterated that eqs. (13.18) and (13.19) are a result of our assumption that *directions* of maximum shear stresses and of maximum rates of shear strain coincide. The rigid-plastic model is significantly different from the viscous model. For the latter we assume that rates of strain are proportional to magnitudes of the differential stresses.

With these two assumptions we can formulate velocity equations for a rigid-plastic substance (*15*). Let $\dot{u}$ and $\dot{v}$ be the components of velocity along the α- and β-lines, respectively (Fig. 13.6). The α-line is inclined at an angle, ρ, with respect to the x-axis. Let $\dot{U}$ and $\dot{V}$ be components of velocity in the x- and y-directions, respectively. The latter components of velocity can be related to the former components through the following equations, which can be derived with the aid of Fig. 13.6A.

$$\begin{aligned} \dot{U} &= \dot{u}\cos\rho - \dot{v}\sin\rho, \\ \dot{V} &= \dot{u}\sin\rho + \dot{v}\cos\rho. \end{aligned} \tag{13.21}$$

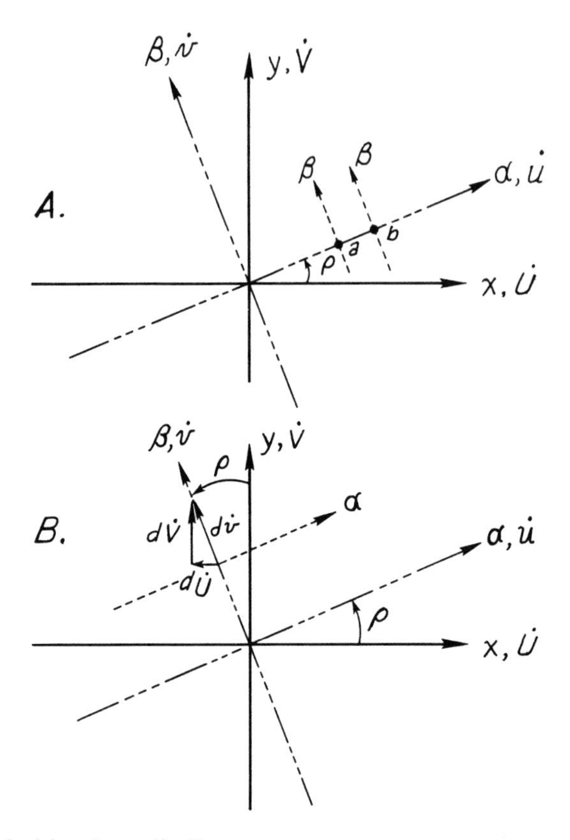

Figure 13.6. Velocities along slip lines.

A. The rigid plastic is incompressible so that there is no relative movement between two points, such as a and b, in the direction of the alpha-line, if the points lie on the same alpha-line. The relative displacement of points a and b is caused solely by shearing along the beta-lines that pass through the two points.

B. The change in velocity in the beta-direction at any point on an alpha-line can be related to changes in velocity in the x- and y-directions.

Now, consider two points, a and b, on the same α-line (Fig. 13.6A). By assumption (1), the rigid plastic is incompressible so that there is no relative displacement of points a and b *in the direction of* the α-line. The relative motion of points a and b is due solely to shearing along the β-lines that pass through the two points (Fig. 13.6A).

The change in velocity in the β direction, at any point on an α-line, can be derived from Fig. 13.6B:

$$(d\dot{v})^2 = (d\dot{V})^2 + (d\dot{U})^2,$$

anywhere on an alpha-line.

According to Fig. 13.6B,

$$\frac{d\dot{V}}{d\dot{U}} = -\tan(90° - \rho) = -\text{ctn } \rho,$$

which can be rewritten in the form

$$d\dot{V} \sin \rho + d\dot{U} \cos \rho = 0. \tag{13.22}$$

But according to eqs. (13.21),

$$d\dot{V} = d\dot{u} \sin \rho + \dot{u} \cos \rho \, d\rho + d\dot{v} \cos \rho - \dot{v} \sin \rho \, d\rho,$$

and

$$d\dot{U} = d\dot{u} \cos \rho - \dot{u} \sin \rho \, d\rho - d\dot{v} \sin \rho - \dot{v} \cos \rho \, d\rho,$$

so that eq. (13.22) becomes

$$d\dot{u} - \dot{v} \, d\rho = 0, \qquad \text{on alpha-lines.} \tag{13.23}$$

Similarly, we could show that

$$d\dot{v} - \dot{u} \, d\rho = 0, \qquad \text{on beta-lines.} \tag{13.24}$$

Now we have the complete set of equations for a rigid-plastic substance. If we have suitable boundary conditions we can solve problems involving the deformation of such a substance.

For some problems, called *statically determinant*, the boundary conditions are such that the stress equations [eqs. (13.14) and (13.15)] can be solved without reference to the stress-strain law outlined above. However, in many practical problems of plane plastic flow, some of the boundary conditions involve velocities. In such problems, the determination of the stresses cannot be separated from that of the velocities. In any case, the stress solution should be checked with the velocity solution to make sure that the slip-line pattern satisfies the velocity boundary conditions.

The solutions to problems in plasticity are similar to those in elasticity and viscosity, because a solution generally is arrived at by a combination of experience and intuition. Solutions to problems normally are explained logically, but the solutions usually are not obtained in that manner. The usual practice is to propose a solution and then show that it fits all the boundary conditions. For example, in plasticity it is not uncommon to find that slip-line fields are first obtained experimentally and then derived theoretically.

Bearing Capacity of a Rigid Plastic.—The problem of computing the bearing capacity of material with a simple yield criterion is instructive because it is simple, yet it demonstrates several properties of slip-line patterns and it provides an example of the importance of the velocity solution in determining the correct slip-line pattern. The problem will be shown to be statically indeterminant in the sense that the correct slip-line pattern must be selected on the basis of velocity boundary conditions.

The problem of calculating the bearing capacity of the rigid plastic beneath the building in Fig. 13.7 will be solved as a problem of plane strain. Thus, the length of the building in Fig. 13.7 is assumed to be very large relative to its width, *AB*. Further, we will assume that the foundation of the building is a perfectly rigid slab and that failure of the plastic is impending. If failure were not impending, the stresses would be undefined.

Figure 13.7A shows plastic regions that would begin to form in real soil as soon as the structure is built. However, for a rigid-plastic material, there can be no subsidence of the building until the plastic field extends all the way across its base, as is shown in Figure 13.7B. The material beneath the slab remains undeformed until the yield criterion is satisfied all along the contact between the slab and the material.

The present discussion will be restricted to incipient flow of the plastic material. At advanced stages of plastic flow, material will be pushed laterally and upward along the sides of the building. The boundary conditions on the deformed surface would be difficult to analyze.

Now we need to examine the boundary conditions in detail to understand why the slip-line patterns shown in Fig. 13.7 were selected. For region *AB* below the building it is assumed that there is no friction between the base of the building and the plastic. The principal normal stresses must be normal and tangential to the surface of the plastic, so that the slip lines must intersect this surface, *AB*, at angles of ± 45 degrees. The slip lines are maximum shear-stress trajectories, which we have shown previously to be normal to one another and at 45 degrees to the principal stress directions. It is apparent, then, that the stress boundary conditions below the building are satisfied by the slip lines shown within *AGC* and *CIB* in Figure 13.7B.

On either side of the building the vertical normal stress, σ_{yy}, and the shearing stress, σ_{xy}, must vanish on the ground surface, *DA* and *BE*, because there are no external loads applied there. The magnitude of σ_{xx} is known on part of the ground surface because we know that the yield criterion must be satisfied for critical equilibrium. One would guess that σ_{xx} is compressive, so that Figure 13.8 represents the probable state of stress on surfaces *DA* and *BE*. The dashed line shows the other possible interpretation, where σ_{xx} is tensile.

Because $\sigma_{xx} = -2k$, and $\sigma_{yy} = \sigma_{xy} = 0$, eqs. (13.14) can be solved for ρ:

$$\sigma_{yy} = 0 = p + k \sin (2\rho),$$

so that

$$\rho = \tfrac{1}{2} \sin^{-1} (-p/k).$$

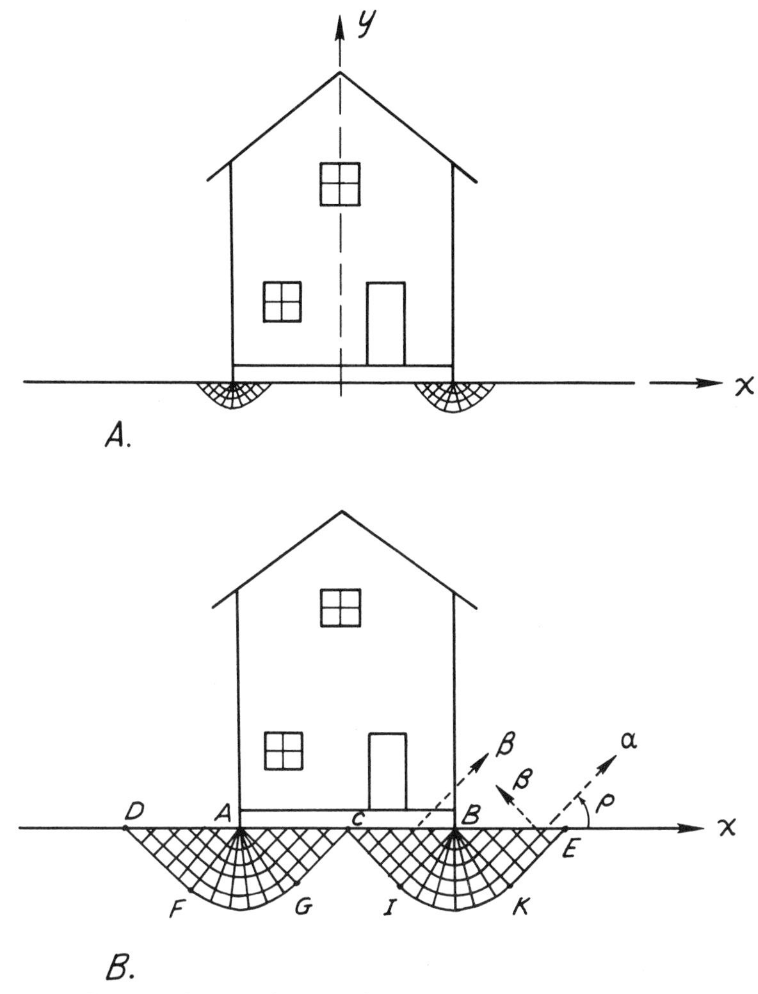

Figure 13.7. Growth of zones of plastic flow beneath a building on a rigid slab.

A. Plastic zones presumably begin to form as the structure is built.

B. Plastic zones have spread across the entire base of the building by the time the building begins to subside.

But

$$p = \frac{\sigma_{yy} + \sigma_{xx}}{2} = \frac{-2k}{2} = -k, \tag{13.25}$$

so that

$$\rho = \tfrac{1}{2}(\pi/2) = \pi/4 \tag{13.26}$$

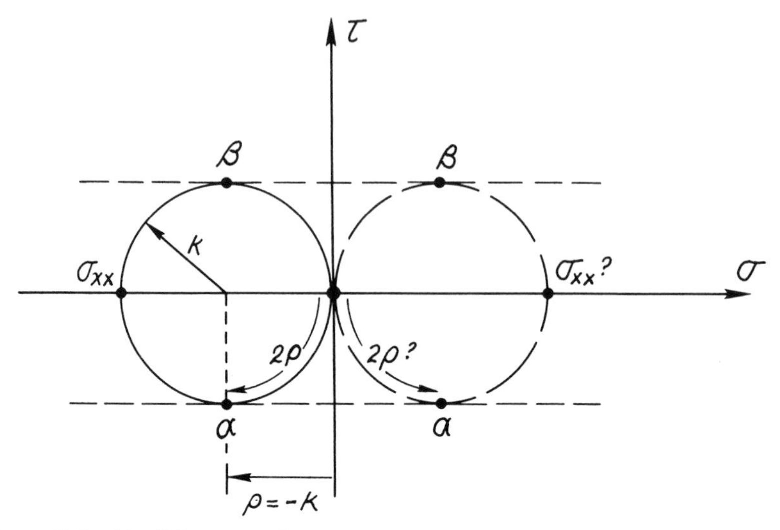

Figure 13.8. Possible states of stress at ground surface at either side of building.

along surfaces DA and BE in Fig. 13.7B.

Thus the shear lines sketched below surfaces DA and BE satisfy the boundary conditions and line EK is an α-line, according to our definition.

The shear trajectories in the region below the surfaces DA and BE necessarily form two families of straight slip lines because p and ρ are constant on these surfaces, *which are not slip lines*. This probably can be illustrated best with an example, which also illustrates how a pattern of slip lines can be calculated approximately, say with a computer.

Figure 13.9 shows a stress-free surface, *ae*. The problem is to calculate the slip-line pattern for the region to the left of this surface. Points $a, b, \ldots, e$ represent

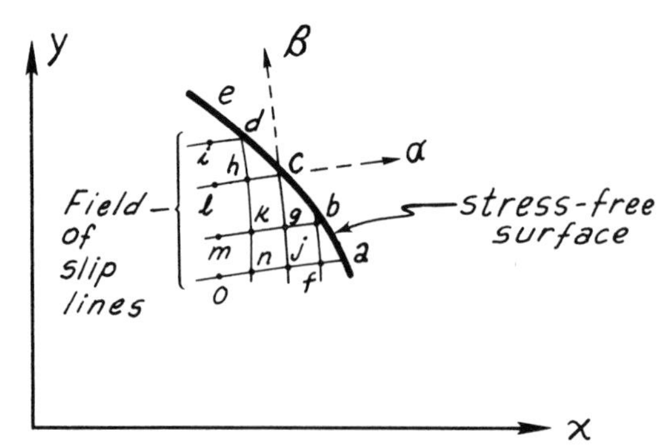

Figure 13.9. Method of constructing pattern of slip lines beneath stress-free surface that is in critical equilibrium.

arbitrarily selected positions on the stress-free surface, where values of p and ρ are known from the boundary conditions.

The values of p and ρ at point g, for example, can be calculated with the data from points c and b. Because $p - 2k\rho$ is constant on an α-line and $p + 2k\rho$ is constant on a β-line [eqs. (13.14)],

$$p_g - 2k\rho_g = p_b - 2k\rho_b,$$

and

$$p_g + 2k\rho_g = p_c + 2k\rho_c.$$

With these two equations, p_g and ρ_g can be found in terms of the known values of p and ρ at points b and c on the boundary:

$$2p_g = (p_b + p_c) + 2k(\rho_c - \rho_b),$$

so that

$$p_g = \tfrac{1}{2}(p_b + p_c) + k(\rho_c - \rho_b). \tag{13.27}$$

Similarly,

$$\rho_g = \tfrac{1}{2}(\rho_c + \rho_b) + \frac{1}{4k}(p_c - p_b). \tag{13.28}$$

Thus the values of p and ρ could be calculated point by point throughout the problem region.

The slip lines in regions *ADF* and *BEK* of Figure 13.7B must be straight, because eqs. (13.27) and (13.28) show that if $\rho_c = \rho_b$ and $p_b = p_c$, then $p_g = p_c = p_b$, and $\rho_g = \rho_c = \rho_b$. Thus, the total slip-line pattern consists of straight lines and the state of stress is constant throughout regions *ADF* and *BEK* of Fig. 13.7B. The slip-line pattern and the state of stress can be examined in a similar manner for regions *AGC* and *CIB*. The results are identical to those above.

The derivation of the slip-line pattern shown in region *BIK* can be explained by means of a theorem due to Hencky (e.g., see ref. *11*, p. 131). The slip lines, *BK* and *BI*, are necessarily straight, because they are parts of slip-line fields, *BKE* and *CIB*, which are composed entirely of straight lines. According to a theorem by Hencky, if one slip line of a family is straight, all the slip lines in *that* family are straight. Now, if a family of straight slip lines radiates from a point, as is shown in Fig. 13.7B, the second family of slip lines must be circular arcs, because the two families of slip lines are normal to each other. The slip-line fields *AFG* and *BKI* are called *centered fans*.

The transition between two regions of constant state, *ADF* and *ACG*, is, therefore, provided by a fan with a center at *A*. Also, the straight slip line, *CG*, is continued by the circular slip line *GF*, which, in turn, is continued by the straight slip line *DF*. This demonstrates that the plastic region, *DA*, at the left-hand side of the building, has a width equal to AC. Also, because EK is an α-line, the centered fan, BIK, is formed by straight β-lines. Therefore, BI is a β-line and

$$\rho = -\pi/4, \text{ in } CBI. \tag{13.29}$$

Now the problem is to determine the value of p for region CBI, so that the normal stress, σ_{yy}, on CB can be calculated. To do this, however, we must first calculate the value of p in the centered fan, using the boundary conditions on one side of it, given by eqs. (13.25) and (13.26). The angle, ρ, is constant along any β-line in the centered fan, so that p must also be constant [eqs. (13.14)]. This checks out with the conclusion that p is constant in region BKE. The constant, C_α, for the centered fan can be found with the first of eqs. (13.14),

$$C_\alpha = -2k\rho + p, \tag{13.30}$$

which holds along the α-lines and, therefore, throughout the problem region, BKI. But ρ and p in region BKE are given by eqs. (13.25) and (13.26),

$$p = -k, \tag{13.25}$$

and

$$\rho = \pi/4, \tag{13.26}$$

so that eq. (13.30) becomes

$$C_\alpha = p - 2k\rho = -k(1 + \pi/2). \tag{13.31}$$

Substituting eqs. (13.31) for C_α and eq. (13.29) for ρ, along surface BI of the centered fan, into the second of eqs. (13.14),

$$p = 2k - k(1 + \pi/2) = -k\pi/2 - k(1 + \pi/2) = -k(1 + \pi), \tag{13.32}$$

which is the expression for p along surface BI and, therefore, throughout region CBI. σ_{yy} on the base, CB, of the foundation in Fig. 13.7B can be found, therefore, by substituting eqs. (13.32) and (13.29) into the second of eqs. (13.8):

$$\sigma_{yy} = p + k \sin(2\rho) = -(2 + \pi)k. \tag{13.33}$$

The solution of this problem required a great deal of discussion, so that the steps of the solution are summarized below:

on BE:

$$\sigma_{xx} = -2k, \sigma_{yy} = 0,$$

that is, we assumed that failure is impending. Thus,

$$\rho = \pi/4 \quad \text{and} \quad p = -k$$

throughout region BEK.

on BK:

$$p = -k, \qquad \rho = \pi/4.$$

Now IK is an α-line, so that

$$p - 2k = C_\alpha,$$

or

$$C_\alpha = -k(1 + \pi/2)$$

in region BIK.

on BI:

$$\rho = -\pi/4,$$

$$C_\alpha = -k(1 + \pi/2).$$

Therefore, in region CIB,

$$p = C_\alpha + 2k\rho = -k(1 + \pi/2) - k\pi/2 = -k(1 + \pi).$$

But,

$$\sigma_{yy} = p + k \sin(2\rho),$$

so that

$$\sigma_{yy} = -k(1 + \pi) + k \sin(-\pi/2),$$

or

$$\sigma_{yy} = -k(2 + \pi),$$

on the footing of the building.

Our solution indicates that, during incipient plastic flow of material below the slab of the building, the foundation exerts a uniform pressure of magnitude $(2 + \pi)k$ on the rigid-plastic material beneath it. The maximum force the plastic can accept from the building, without failure, is

$$(\text{length})(\text{width})(2 + \pi)k.$$

Before we can be sure that the solution based on stress analysis is correct, the velocity field associated with the stress solution must be investigated to see if it makes sense (*11*). The velocities in a plastic substance are proportional to the loading rates, so that the rate of loading of the building, for simplification, can be taken to be unity. Thus, let $\dot{V}$ be -1 along AB in Fig. 13.7B. The base of the building was assumed to be frictionless, so that no statement can be made about $\dot{U}$ from the boundary conditions.

The plastic region below the building is bounded by the surface $CIKE$. Below this surface the material is at rest, so that the velocity *normal* to this surface must be zero. Thus,

$$\dot{v} = 0 \text{ along } CIKE, \tag{13.34}$$

where $\dot{v}$ was defined as the velocity along a β-line.

Eq. (13.24),

$$d\dot{v} - \dot{u}\, d\rho = 0,$$

is the differential equation for the change of velocities with distance along a β-line. In the two regions with straight slip lines, $d\rho$ is zero, so that eq. (13.24) reduces to

$$d\dot{v} = 0.$$

According to eq. (13.34), $\dot{v}$ is zero at the boundaries of the two regions consisting of straight slip lines. Therefore it must be zero throughout those regions. Flow takes place, then, along the α-lines.

Throughout region *CBI*, the velocity is in the direction of *CI* and the magnitude of this velocity is $(2)^{1/2}$, because $\dot{V} = -1$. *IK* is an α-line, so that eq. (13.23) describes the velocities along the circular arcs:

$$d\dot{u} - \dot{v}\, d\rho = 0, \tag{13.23}$$

where $\dot{v}$ is the velocity in the β-direction, and u is the velocity in the α-direction. Thus, because $\dot{v} = 0$, $\dot{u} = C_\alpha$ along *IK*, and $|\dot{u}| = (2)^{1/2}$.

Finally, region *BEK* must move with a velocity of $(2)^{1/2}$ in the direction of *KE*. Thus, according to the pattern of slip lines shown in Fig. 11.5B, the material on each side of the building should move upward at the same rate as the building moves downward.

The foregoing solution is due to Hill (*5*), who derived it by assuming that plastic flow begins at the corners, *A* and *B*, of the foundation and spreads laterally. However, the stress solution is not unique. Prandtl (*12*) earlier derived a slip-line pattern that is shown on the left-hand side of Fig. 13.10. *ACG* and *AFD* are regions of constant state, and *AFG* is a centered fan. The stresses in these regions are the same as those in the corresponding regions in Figure 13.7B, but the velocities are different. Region *ACG* moves as a rigid body with a velocity of -1 in the direction *CG*. Thus, *ACG* behaves as though it were part of the foundation of the building. In region *AFG* the slip lines are circular with the center at *A* and the velocity along the circles has a magnitude of $(2)^{-1/2}$. The region *DAF* moves with a velocity of $(2)^{-1/2}$ in the direction of *DF*. This solution is generally assumed in soil mechanics (e.g., see Terzaghi, ref. *16*, p. 121).

The right-hand half of Fig. 13.10 illustrates a compromise proposed by Prager and Hodge (*11*). The stresses and velocities in region *HKMP* are the same as those in Hill's solution, whereas in region *CHKMPENLI* the velocities are the same as those in Prandtl's solution.

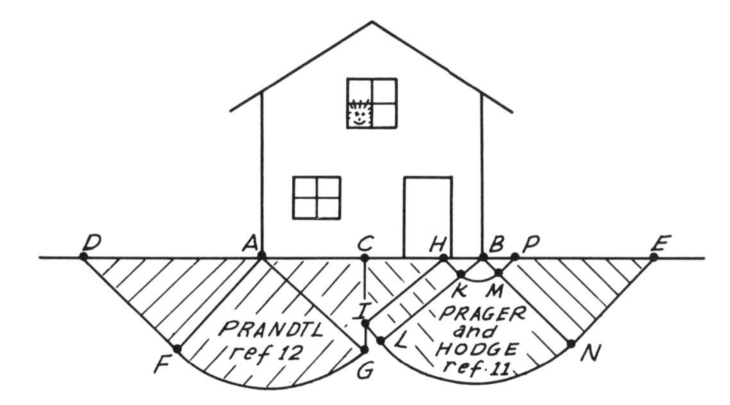

Figure 13.10. Alternative slip-line fields in plastic zone beneath building.

Thus, there are at least three possible solutions for the problem. The correct solution for a particular problem can be found by determining, experimentally, the velocity boundary conditions on surfaces *DA* and *BE* in Fig. 13.10. If the velocity boundary conditions were known when the problem was analyzed, presumably there would be no question about which of the solutions for the slip patterns is correct. This is an excellent illustration of the necessity of satisfying both the velocity and the stress boundary conditions in the solution of a problem in plasticity. However, if the question were only "what is the bearing capacity of the soil?" beneath the building sketched in Fig. 13.8, then any one of at least three slip-line patterns results in the same answer. The latter conclusion is not necessarily true in general. A stress solution of a problem in plasticity involves consideration of both the velocity and the stress boundary conditions.

Relation Between Depth of Penetration of Sphere and Strength of Plastic.— *Modifications to account for three-dimensional conditions and buoyancy of plastic.*—According to our solution for the two-dimensional problem of the resistance of a long, flat punch or the slab of a building, the pressure exerted by the plastic is

$$\sigma_{yy} = -(2 + \pi)k \approx -5.14k. \tag{13.35}$$

The problem of determining how deeply a sphere will penetrate into a plastic, however, also involves buoyancy. The effect of buoyancy can be calculated readily because, according to Archimedes' principle, the force of buoyancy on an object is equal to the weight of the fluid displaced by the object. Thus, the plastic with a unit weight of γ_d applies a force to the underside of the sphere tending to float the sphere.

Another problem is the fact that a sphere is definitely a three-dimensional body, not a two-dimensional one. However, according to Shield (*13*) and Levin (*10*), the upward pressure exerted by a simple plastic on the base of a flat, smooth, circular punch is

$$\sigma_{yy} = -4.83k, \tag{13.36}$$

if the plastic is very thick. If the thickness of the plastic is approximately equal to the radius of the punch, the pressure is

$$\sigma_{yy} = -5.2k \text{ to } -5.7k, \text{ say } -5.5k. \tag{13.37}$$

Apparently our two-dimensional solution is not seriously in error.

Now we will combine the resistances of the plastic—its buoyancy and its strength—in order to determine, approximately, the relation between the depth of penetration of a sphere and the strength and density of the plastic (*7*).

Let a be the radius of the assumed flat side of the sphere, that is, the radius of the circular punch (Fig. 13.11A). For a sphere,

$$a^2 = R^2 - y^2 \qquad [y \geq 0],$$

where y is the distance from the center to the flat side of the sphere (Fig. 13.11B).

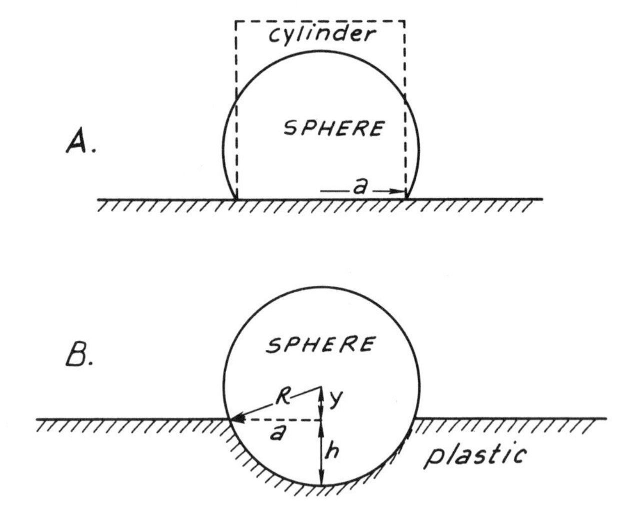

Figure 13.11. Sphere supported by dense, rigid-plastic substance.

The area of the flat side is $\pi a^2 = \pi(R^2 - y^2)$, so that the upward force exerted by the plastic on the sphere is

$$\pi(R^2 - y^2)\sigma_{yy}.$$

The effective force exerted downward by the weight of the sphere is equal to the total volume of the sphere times its unit weight, minus the submerged volume of the sphere times the unit weight of the plastic. Thus,

$$\text{total volume of sphere} = \tfrac{4}{3}\pi R^3,$$

$$\text{submerged volume of sphere} = \frac{h^2}{3}(2R - h)\pi,$$

so that

$$\left[\gamma_b(\tfrac{4}{3})\pi R^3 - \gamma_d\left(\frac{h^2}{3}\right)\pi(3R - h)\right]$$

is the total driving force. Here γ_b is the unit weight of the sphere and γ_d is the unit weight of the plastic.

For equilibrium, the upward force must equal the downward force, so that

$$-\sigma_{yy}\pi(R^2 - y^2) = \left[\gamma_b(\tfrac{4}{3})\pi R^3 - \gamma_d\pi\left(\frac{h^2}{3}\right)(3R - h)\right]. \tag{13.38}$$

But $h = R - y$, according to Fig. 13.11B, so that

$$\frac{-3}{\gamma_d}\sigma_{yy}\left[1 - \left(1 - \frac{h}{R}\right)^2\right] = 4\frac{\gamma_b}{\gamma_d} - \left(\frac{h}{R}\right)^2\left(3 - \frac{h}{R}\right),$$

or (*5*)

$$K = \frac{(4\gamma_b/\gamma_d) - H^2(3 - H)}{1 - (1 - H)^2} \qquad \left[0 \le H = \frac{h}{R} \le 1\right], \tag{13.39}$$

where

$$\left.\begin{aligned} K &= (-3\sigma_{yy}/R\gamma_d), \\ &\text{and} \\ H &= h/R. \end{aligned}\right\} \tag{13.40}$$

If the sphere has penetrated to a depth greater than the radius of the sphere,

$$-\sigma_{yy}\pi R^2 = \left[\gamma_b(\tfrac{4}{3})\pi R^3 - \pi\gamma_d\left(\frac{h^2}{3}\right)(3R - h)\right],$$

so that

$$K = 4\frac{\gamma_b}{\gamma_d} - H^2(3 - H) \qquad [1 \le H \le 2]. \tag{13.41}$$

where K is defined in eqs. (13.40).

K in these equations is a dimensionless strength variable and H is a dimensionless depth variable.

Equations (13.39) and (13.41) are plotted in Fig. 13.12, where the strength variable, K, is related to the unit-weight or density ratio of the sphere and plastic for different depths of penetration, h/R.

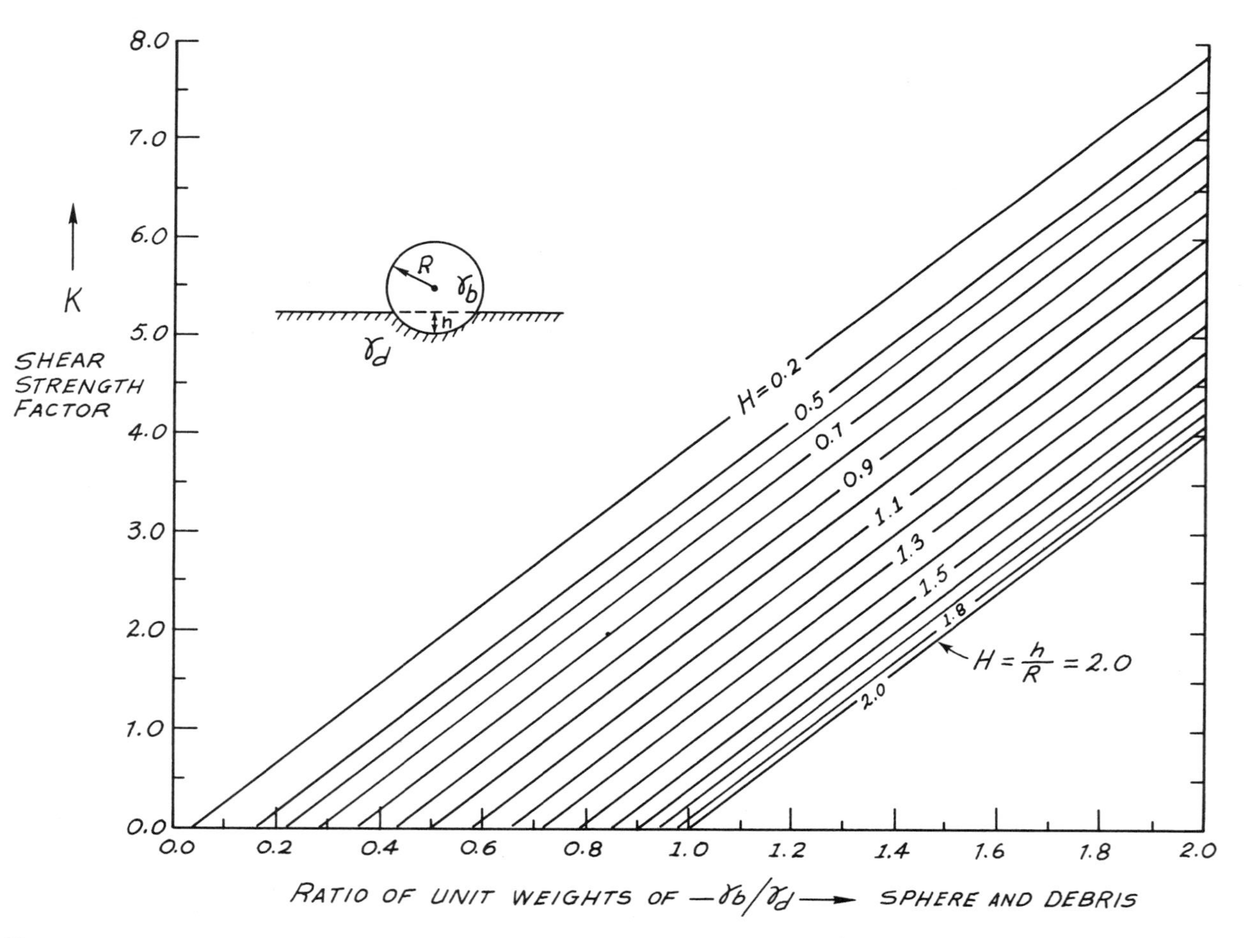

Figure 13.12. Relation between shear-strength factor, K, and ratio of unit weights of sphere, γ_b, and debris, γ_d. Shear-strength factor defined in text.

Comparison of theory and experiment.—Our solution for the depth of penetration of spheres into plastic material ignores the effect of the part of the sphere below the surface of the plastic on the slip-line pattern in the plastic. The magnitude of the error introduced by this simplification is difficult to assess. The existence of the error is recognizable in experimental data (*8*) collected in the laboratory, however. The spheres used in the experiment consisted of four wooden balls, which were drilled and loaded with molten lead to modify their specific gravities. The spheres ranged in radius from 1.62 to 2.27 cm and in specific gravity from 1.195 to 2.160 (Fig. 13.13). Various mixtures of medium sand, kaolin clay, and water were used for the slurries. Fig. 13.13 shows the relation between strength, k, and weight percent of water of three types of slurries. The solids in the slurries consisted of 90, 70, and 30 weight percent clay in the three types of slurries. Water contents ranged from 30 to 62 percent of the combined weight of solids and water.

The spheres were individually lowered into the slurries and their depths of

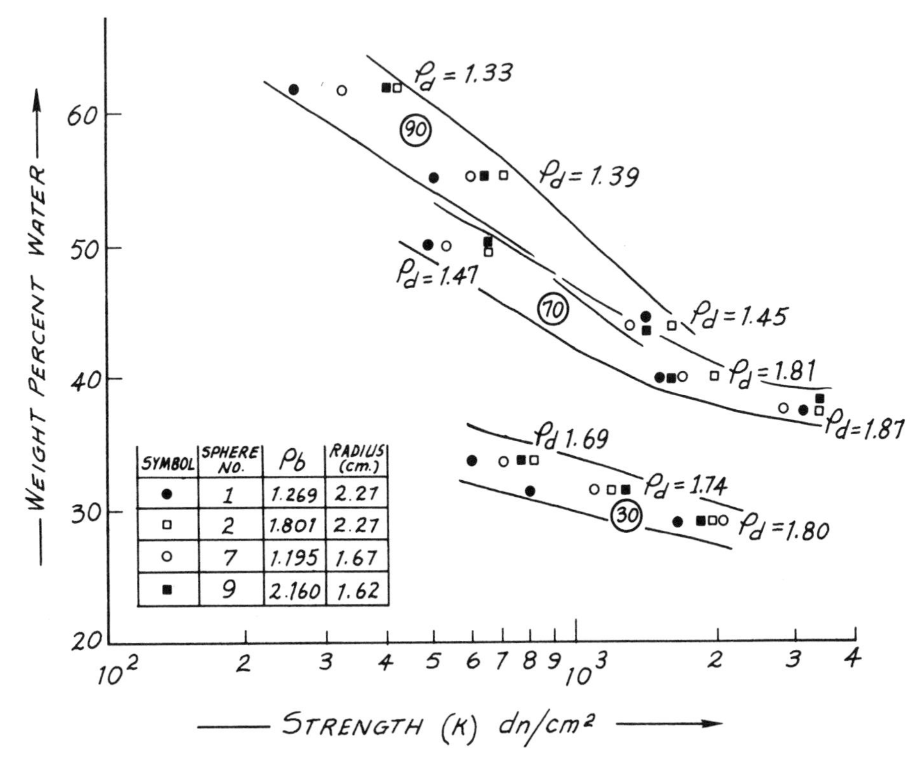

SYMBOL	SPHERE NO.	ρ_b	RADIUS (cm.)
●	1	1.269	2.27
□	2	1.801	2.27
○	7	1.195	1.67
■	9	2.160	1.62

Figure 13.13. Variations in apparent strengths of slurries measured with spheres of different sizes and densities. ρ_d = Density of slurry. 70 = Weight percent of kaolin clay in solid fraction of sand-clay-water slurries. ρ_b = Density of sphere.

penetration were measured with a micrometer. Then the strength of each slurry was calculated with curves similar to those shown in Fig. 13.12.

The data are not perfectly consistent, largely, I think, because of operator error, but they do show that sphere no. 7, with the lowest specific gravity, always indicated a strength lower than the mean of the four estimates. Two spheres with greater specific gravities, spheres 2 and 9, always indicated strengths that were higher than the mean. Thus, the strength estimated by this test depends upon the depth of penetration of the spheres in a way not entirely accounted for by our simple theory.

Transport of boulders and blocks on Surprise Canyon alluvial fan

The debris flow of 1917 preserved on the Surprise Canyon alluvial fan in Panamint Valley, California, carried hundreds of boulders and blocks of rock for distances of up to a kilometer across fan slopes ranging from about eight degrees near the fan apex to two or three degrees at the place where most of them came to rest (*6*). Finer-grained waves of the flow traveled about two kilometers, across the fan and onto the playa at its base. The deposits of this gigantic flow are remarkably well preserved in spite of their age of more than 50 years (*6*).

A block of alaskite, which originated in the narrows of Surprise Canyon, about three kilometers from its present site on the fan, is shown in Fig. 13.14. The block is in the western lateral deposit of the 1917 flow. The view is from the medial channel, toward the west. The debris of the 1917 flow is the coarser-grained unit, vertically exposed from about 15 cm below the base of the block to the horizon in Fig. 13.14. The block is 1.65 m high, 2.74 m wide, and 2.14 m broad, measured normally to the plane of the photograph. The debris-flow deposit is about 1.60 m thick here and its surface is sloping at an angle of six degrees, from right to left in the photo. Note the debris that remained preserved for 50 years on the top of the block; apparently it splashed up on the block as the block was being dragged along by the flow.

Let us assume that the strength of the debris was barely adequate to support the block at the time the block and the debris surrounding it stopped moving. The block is crudely ellipsoidal; therefore, its volume is approximately $\frac{4}{3}\pi(abc)$, where a, b, and c are its three principal axes. The appropriate form of the equilibrium equation is eq. (13.38) if we replace the area $\pi(R^2 - y^2)$ by πab, the volume by $(\frac{4}{3})\pi(abc)$, and the submerged volume by $(\frac{4}{3})\pi(abc)/n$, where n depends upon the depth of penetration of the ellipsoid into the debris. Thus, eq. (13.38) should be rewritten in the form

$$-\sigma_{yy}\pi(ab) = \left[\gamma_b(\tfrac{4}{3})\pi(abc) - \gamma_d\pi\left(\frac{4}{3}\right)\left(\frac{abc}{n}\right)\right].$$

However, according to eq. (13.37),

$$\sigma_{yy} \approx -5.5k,$$

Figure 13.14. Alaskite block in lateral deposit of Surprise Canyon debris-flow deposit of 1917.

so that

$$k \approx \frac{(4/3)c}{5.5}\,[\gamma_b - \gamma_d/n],$$

or

$$k \approx \frac{c}{4}\,[\gamma_b - \gamma_d/n]. \tag{13.42}$$

The height, c, of the boulder is about 165 cm. The specific gravity of the alaskite block is 2.5 and the specific gravity of the debris was about 2.4 (*6*). Approximately $\frac{3}{4}$ of the volume of the block was submerged at the time of transport; thus, $1/n = \frac{3}{4}$ and eq. (13.42) becomes

$$k \approx \frac{165\text{ cm}}{4}\,(980\text{ cm/sec}^2)(2.5 - (\tfrac{3}{4})2.4)\text{ gm/cc},$$

so that $k \approx 3 \times 10^4$ dn/cm^2. (13.43)

The yield strength of the same debris, shown in Fig. 13.14, can be estimated independently by considering the thickness and surface slope of the debris deposit in which the block rests (*6*). If we suppose that the debris-flow deposit shown in Fig. 13.14 was at critical equilibrium at the time it stopped moving, we can calculate the strength by considering the requirements of equilibrium. The thickness of the deposit is small relative to the width of the deposit, so that, as a first approximation to actual conditions, we can represent the debris as an infinite sheet of plastic material. The requirements of equilibrium of any element of an infinite sheet of plastic material on an inclined surface can be visualized with the aid of Fig. 13.15, which shows part of the sheet of plastic material. Each element of material is under the same state of

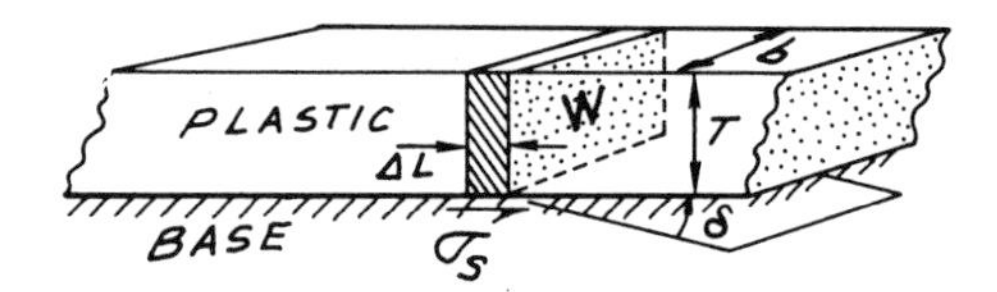

Figure 13.15. Part of infinite sheet of plastic resting on inclined base.

stress in such a sheet, so that, for equilibrium, the resisting force, $\sigma_s \Delta L b$, must equal the driving force. That is,

$$\sigma_s \Delta L b = \gamma_d \Delta L b T \sin \delta,$$

where ΔL is the length of the element, T is the thickness of the plastic, b is the unit of width, and δ is the slope angle of the surface and base of the plastic sheet. Thus,

$$\sigma_s = T \gamma_d \sin \delta. \tag{13.44}$$

The maximum shear stress the simple plastic substance can withstand is k, its yield strength, so that at critical equilibrium, the shear stress on the base of the element is

$$|\sigma_s| = k,$$

and eq. (13.44) becomes

$$k = T_c \gamma_d \sin \delta, \tag{13.45}$$

and the critical thickness for flow is (*6*)

$$T_c = \frac{k}{\gamma_d \sin \delta}.$$

According to eq. (13.44), the shear stress increases as the thickness of the plastic increases, so that, according to the concept we have adopted for the plastic substance, the material above the critical depth, T_c, is rigid. The plastic material apparently slips along its base, where the shear stress equals the yield strength of the plastic.

If we assume that the debris in the lateral deposit shown in Fig. 13.14 was at critical equilibrium at the time it stopped flowing, we can estimate the yield strength of the debris with eq. (13.45).

The thickness of the deposit is about 1.6 m, the unit weight of the debris is (2.4)(980) dn/cc, and the slope angle is six degrees. Thus,

$$k = T_c \gamma_d \sin \delta,$$

$$k \approx (160)(2.4)(980)(0.105) \text{ dn/cm}^2 \approx 4 \times 10^4 \text{ dn/cm}^2, \quad (13.46)$$

which is remarkably close to the value of 3×10^4 dn/cm² estimated by consideration of the block of alaskite in Fig. 13.14.

In another place on Surprise Canyon alluvial fan, a boulder of alaskite with the following dimensions,

$$a = 2.16 \text{ m}$$
$$b = 1.78 \text{ m}$$
$$c = 1.15 \text{ m},$$

is three-fourths submerged in a deposit of 0.89 m thickness with a surface slope of four degrees. Thus,

$$k = T_c \gamma_d \sin \delta \approx 1.7 \times 10^4 \text{ dn/cm}^2, \quad (13.47)$$

using the depth of the flow, and

$$k = \frac{c}{4}\left[\gamma_b - 3\frac{\gamma d}{4}\right] \approx 4 \times 10^4 \text{ dn/cm}^2, \quad (13.48)$$

using the size of the boulder.

In another place on the fan there are two boulders, one of alaskite and one of gneiss. Their dimensions are:

	alaskite	*gneiss*
a	3.8 m	2.4 m
b	0.92 m	1.5 m
c	1.2 m	1.5 m

They are both about half submerged in a lateral deposit with a thickness of about 0.92 m and with a surface slope of five degrees. Thus, using the lateral deposit,

$$k = T_c \gamma_d \sin \delta \approx 2 \times 10^4 \text{ dn/cm}^2. \quad (13.49)$$

Using the alaskite boulder,

$$k = \frac{c}{4}\left[\gamma_b - \frac{\gamma_d}{2}\right] \approx 4 \times 10^4 \text{ dn/cm}^2, \quad (13.50)$$

and using the gneiss boulder,

$$k \approx 5 \times 10^4 \text{ dn/cm}^2. \tag{13.51}$$

These estimates of strength of the debris on the Surprise Canyon alluvial fan, given by eqs. (13.43) through (13.51), range from 1.7 to 5 $\times$ 10^4 dn/cm^2. The estimates made by using the blocks range from 3 to 5 $\times$ 10^4 dn/cm^2, and those made by using the measurements of thickness and slope of the lateral deposit range from 1.7 to 4 $\times$ 10^4 dn/cm^2. It is apparent, therefore, that the estimates are not far apart and that the estimates based on the sizes of blocks are generally slightly higher than the estimates based on the thickness of the lateral deposit. This is quite reasonable, because we have considered only one factor of boulder transport. Other factors (*6,8*), such as dispersive forces, and viscous and plastic drag, have been ignored.

The similarities of estimates of strengths based on the thickness of the lateral deposit with those based on the sizes of the blocks transported by the debris are compatible with the concept, based on many observations of debris flows and debris deposits, that larger blocks in debris flows are not dropped because of incompetency in a turbulent medium, but that, instead, they stop because the debris carrying them, and of which they are a part, stops or thins markedly.

References cited in Chapter 13

1. Bagnold, R. A., 1954, "Experiments on Gravity-free Dispersion of Large Solid Spheres in a Newtonian Fluid under Shear": *Proc. Roy. Soc.*, London, Ser. A., V. 225, p. 49–63.
2. ———, 1968, "Deposition in the Process of Hydraulic Transport": *Sedimentology*, V. 10, p. 45–56.
3. Fox, L., Ed., 1962, *Numerical Solution of Ordinary and Partial Differential Equations*: Pergamon Press, N.Y.
4. Goldsmith, H. L., and Mason, S. G., 1967, "The Microrheology of Dispersions": In, *Rheology*, Ed. by Eirich, F. R., Academic Press, N.Y., V. 4, p. 85–250.
5. Hill, R., 1950, *The Mathematical Theory of Plasticity*: Oxford at the Clarendon Press, London.
6. Johnson, A. M., 1965, "A Model for Debris Flow": Ph.D. dissertation, The Pennsylvania State University, Univ. Park, Penna.
7. ———, and Hampton, M. A., 1968, "Subaerial and Subaqueous Flow of Slurries": Ann. Progress Report to U.S. Geological Survey; Branner Library, Stanford University, Stanford, Calif.
8. ———, and Hampton, M. A., 1969, "Subaerial and Subaqueous Flow of Slurries": Final Report to U.S. Geological Survey; Branner Library, Stanford University, Stanford, California.
9. Johnson, W., and Mellor, P. B., 1962, *Plasticity for Mechanical Engineers*: D. Van Nostrand Co., Ltd., N.Y.
10. Levin, E., 1955, "Indentation Pressure of a Smooth Circular Punch": *Quart. Appl. Math.*, V. 13, p. 133–137.

11. Prager, W., and Hodge, P. G., 1951, *Perfectly Plastic Solids*: John Wiley & Sons, Inc., N.Y.
12. Prandtl, L., 1920, "Ueber die Haerte Plastischer Koerper": *Goettinger Nachr., math.-phys.* Kl. 1920, p. 74–85.
13. Shield, R. T., 1955, "The Plastic Indentation of a Layer by a Flat Punch": *Quart. Appl. Math.*, V. 13, p. 27–46.
14. Sokolovsky, V. V., 1965, *Statics of Granular Media*: Pergamon Press, N.Y.
15. Spencer, A. J. M., 1964, "A Theory of the Kinematics of Ideal Soils under Plane Strain Conditions": *Jour. Mech. Phys. Solids*, V. 12, p. 337–351.
16. Terzaghi, K., 1943, *Theoretical Soil Mechanics*: John Wiley & Sons, Inc., N.Y.

Other references on soil mechanics and plasticity theory

Benari, M. M., 1961, "Rheology of Granular Material II—A Method for the Determination of the Intergranular Cohesion": *Brit. Jour. Appl. Phys.*, V. 12, No. 9.

Bishop, A. W., 1954, "The Use of the Slip Circle in the Stability Analysis of Slopes": *Geotechnique*, V. 5, p. 7–17.

Bishop, J. F. W., 1953, "On the Complete Solution to Problems of Deformation of a Plastic-Rigid Material": *Jour. Mech. Phys. Solids*, V. 2, p. 43–53.

Caquot, Albert, 1934, *Équilibre des massifs a frottement interne*: Gauthier-Villars, Paris.

Collin, Alexandre, 1846 (transl. by Schriever, W. R., 1956), *Landslides in Clays*: Univ. Toronto Press.

Geiringer, Hilda, 1953, "Some Recent Results in the Theory of an Ideal Plastic Body": *Advances in Appl. Mech.*, V. III, p. 199–294.

Giesekus, Hanswalter, 1964, "Statistical Rheology of Suspensions and Solutions with Special Reference to Normal Stress Effects": In *Internat. Union Theor. and Appl. Mech., Symp. at Haifa, Israel*, Macmillan Co., N.Y., p. 553–584.

Goodier, J. N., and Fields, F. A., 1962, "Plastic Energy Dissipation in Crack Propagation": In *Fracture of Solids*, Drucker, D. C., and Gillman, J. J., Eds., Interscience Publishers, N.Y., p. 103–118.

Green, A. P., 1954, "On the Use of Hodographs in Problems of Plane Plastic Strain": *Jour. Mech. Phys. Solids*, V. 2, p. 73–80.

Haythornthwaite, R. M., 1960, "Stress and Strain in Soils": In *Proc. Second Symp. Naval Structural Mech., Brown Univ.*, Pergamon Press, N.Y. p. 185–193.

Hodge, P. G., Jr., 1960, "Boundary Value Problems in Plasticity": *Proc. Second Symp. Naval Structural Mech., Brown Univ.*, 1960, Pergamon Press, N.Y., p. 297–337.

———, 1965, "The Theory of Piecewise Linear Isotropic Plasticity": In "Deformation and Flow of Solids," *Internat. Union of Theor. and Appl. Mechanics*, Colloquium in Madrid, Springer-Verlag, Berlin, p. 147–170.

Jenike, A. W., et al., 1959, "Flow of Bulk Solids Progress Report": *Utah Engrg. Exp. Stat., Bull. 96.*

Johanson, J. R., and Jenike, A. W., 1962, "Stress and Velocity Fields in Gravity Flow of Bulk Solids": *Univ. Utah Engrg. Exp. Stat., Bull. 116.*

Jumikis, A. R., 1962, *Soil Mechanics*: D. Van Nostrand Co., Inc., N.Y.

Kvapil, Rudolf, 1965, "Gravity Flow of Granular Materials in Hoppers and Bins": *Int. Jour. Rock Mech., Mining Sci.*, V. 2, p. 35–41.

———, "Gravity Flow of Granular Materials in Hoppers and Bins and in Mines: Part II: Coarse Material": *Int. Jour. Rock Mech. Mining Sci.*, V. 2, p. 277–304.

Lee, E. H., and Symonds, P. S., Eds., 1960, "Plasticity": *Proc. Second Symp. Naval Struct. Mechanics, Brown Univ.*, Pergamon Press, N.Y.

McClintock, F. A., 1962, "On the Plasticity of the Growth of Fatigue Cracks": In *Fracture of Solids*, Interscience Publ., N.Y., p. 65–102.

Morrison, R. G. K., and Coates, D. F., 1955, "Soil Mechanics Applied to Rock Failure in Mines": *Canadian Mining and Metal. Bull.*, V. 48, p. 701–711.

Muller, Leopold, 1964, "The Stability of Rock Bank Slopes and the Effect of Rock Water on Same": *Internat. Jour. Rock. Mech., and Mining Sci.*, V. 1, p. 475–504.

Nadai, A., 1963, *Theory of Flow and Fracture of Solids*, V. II: McGraw-Hill Book Co., Inc., N.Y.

Nye, J. R., 1951, "The Flow of Glaciers and Ice-Sheets as a Problem in Plasticity": *Proc. Royal Society*, Ser. A., V. 207, p. 554–572.

———, 1952, "The Mechanics of Glacier Flow": *Jour. Glaciology*, V. 2, p. 82–93.

Onat, E. T., 1960, "The Influence of Geometry Changes on the Load-Deformation Behavior of Plastic Solids": In *Proc. Second Symp. Naval Struct. Mech., Brown Univ.*, Pergamon Press, N.Y., p. 225–238.

Scott, R. R., 1963, *Principles of Soil Mechanics*: Addison-Wesley Publ. Co., Inc., Palo Alto.

Shield, R. T., 1955, "Plastic Flow in Converging Conical Channel": *Jour. Mech. Phys. Solids*, V. 3, p. 246–258.

Skempton, A. W., 1964, "Long-Term Stability of Clay Slopes": *Geotechnique*, V. 14, p. 75–102.

Sokolovsky, V. V., 1960, "Axial Plastic Flow Between Non-Circular Cylinders": In *Proc. Second Symp. Naval Struct. Mech., Brown Univ.*, Pergamon Press, N.Y.

Slibar, A., and Paslay, P. R., 1964, "On the Analytical Description of the Flow of Thixotropic Materials": In *Internat. Union Theor. and Appl. Mech., Symp. at Haifa, Israel*, Macmillan Co., N.Y., p. 314–330.

Spencer, A. J. M., 1962, "Perturbation Methods in Plasticity—II. Plane Strain of Slightly Irregular Bodies": *Jour. Mech. Phys. Solids*, V. 10, p. 17–26.

———, 1962, "Perturbation Methods in Plasticity—III. Plane Strain in Ideal Soils and Plastic Solids with Body Forces": *Jour. Mech. Phys. Solids*, V. 10, p. 165–177.

Spurr, R. T., 1958, "The Friction of Mineral Particles": *Brit. Jour. Appl. Phys.*, V. 9, No. 12.

Taylor, D. W., 1948, *Fundamentals of Soil Mechanics*: John Wiley and Sons, Inc., N.Y.

Tetelman, A. S., 1962, "The Plastic Deformation at the Tip of a Moving Crack": In *Fracture of Solids*, Interscience Publishers, N.Y., p. 461–502.

Terzaghi, K., 1960, *From Theory to Practice in Soil Mechanics*; Selections from the Writings of Karl Terzaghi, prepared by C. Bjerrum et al., John Wiley and Sons, Inc., N.Y.

———, 1962, "Stability of Steep Slopes in Hard, Unweathered Rock": *Geotechnique*, V. 13, p. 1–20.

———, and Peck, R. B., 1948, *Soil Mechanics in Engineering Practice*: John Wiley and Sons, Inc., N.Y.

Truesdell, C., 1964, "Second-Order Effects in the Mechanics of Materials": In *Internat. Union Theor. and Appl. Mech., Symp. at Haifa, Israel*, Macmillan Co., N.Y., p. 1–47.

14

Chapter Sections

Debris Flow in Channels
Circular Channel Viscometer
Flow of Bingham Substance in Semicircular Channel
Velocity Profile of Kaolin-Sand Slurry in Semicircular Channel
Velocity Profile for Flow in Infinitely Wide Channel
Simplified Method of Determining Strength and Viscosity of Debris Flows in Noncircular Channels
Properties of Natural Debris Flow at Wrightwood, California
Implications of Solutions and Experiments
Velocity Profiles of Newtonian and Pseudoplastic Substances in Semicircular Channels
Comments on Model Fitting
Viscosity and Strength of Lava in Makaopuhi Crater, Hawaii
Theory of Rotational Viscometer
Application of Solution to Apparatus Used to Measure Properties of Lava
Viscosity and Strength of Basaltic Lava
Glen's Experiments with Ice Cylinders
Velocity and Rheology of Saskatchewan Glacier, Alberta, Canada
Saskatchewan Glacier
Surface Velocities of Ice Tongue
Velocities within Ice
Rheological Model for Ice
Comparison of Properties of Basaltic Lava, Glacial Ice, and Debris
References Cited in Chapter 14

14

Rheological Properties of Debris, Ice, and Lava

Debris flow in channels

We have shown that certain features of debris deposits can be explained in terms of plastic strength of debris. Field and laboratory observations of the movement of debris, however, are not in accord with such a simple picture of its rheology. For example, thicknesses of most debris flows moving in channels must have been greater than the critical values required for flow of a simple plastic substance, because the lateral deposits are topographically above the deposits within the channels. Also, small debris flows produced in the laboratory increase in thickness as their flow rates increase and decrease in thickness as their flow rates decrease, finally reaching a minimum thickness at cessation of flow (*5*).

These observations are not in accord with the behavior of the simple plastic substance, which would accelerate indefinitely if its yield strength were exceeded. On the contrary, it would seem that a debris flow is not at critical thickness until the moment it stops flowing. Then a boundary of a rigid layer of "plug" coincides with a channel boundary, as is shown in Fig. 14.1A. Thus, Fig. 14.1A is an interpretation of debris when it is static and when the bottom of the rigid layer coincides with a channel boundary. An interpretation of the state of flow in moving debris is shown in Fig. 14.1B, where the total thickness of the debris is greater than the critical thickness and where there is a region below the "plug" or "raft" that is in a state of flow, here assumed to be laminar. The debris flow shown in this figure would stop moving when the debris had thinned until its thickness equaled the thickness of the rigid layer, which rides on top of the laminar layer.

This major discrepancy between actual behavior of debris flows and theoretical behavior of plastic substances seems to indicate some form of viscous resistance of

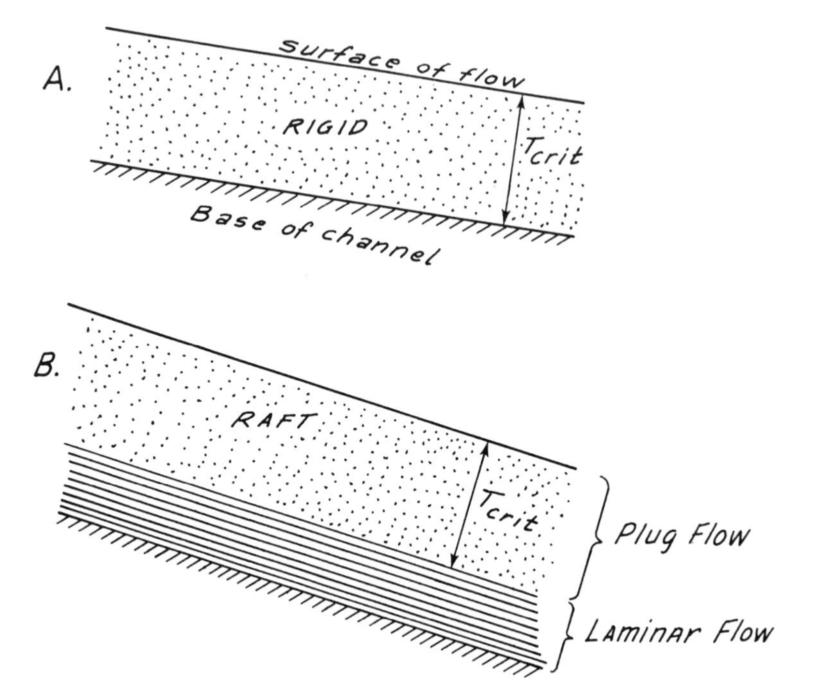

Figure 14.1. Types of flow of debris in wide channel.

A. Entire thickness of debris is rigid. Shear stress at base of channel is less than yield strength of debris. No flow.

B. Top part of flow is rigid, bottom part is shearing. Plug flow at top, laminar flow below.

debris. The simplest correction we can make on our plastic rheological model is to add Newtonian viscosity, giving a Bingham model:

$$|\sigma_s| = k + \eta_b \dot{\epsilon}_s; \qquad [|\sigma_s| \geq k], \tag{14.1}$$

which states that, if the shear stress is greater than the yield strength of the material, the resistance to shear is the sum of the plastic and viscous resistances. Here η_b is coefficient of viscosity, k is yield strength, and $\dot{\epsilon}_s$ is rate of shear strain.

Circular Channel Viscometer.—Natural debris flow, being inherently an episodic process, and occurring generally under most uncomfortable conditions, is rarely observed and extremely rarely studied analytically. In any given drainage basin, normal runoff may be punctuated every 25 to 100 years by a catastrophic debris flow. For example, major debris flows moved down Surprise Canyon in the Panamint Range, California, in the 1880's and in 1917 (*5*). Smaller ones flowed in 1942 and 1968. The one in the 1880's was observed by residents of Panamint City, a silver-mining town, because Panamint City was in the bottom of Surprise Canyon.

The one in 1942 was observed momentarily by the operator of a road grader, working in the narrows of Surprise Canyon. The other two apparently were unobserved.

Thus, the student of debris flow might be an old man before he had the opportunity to analyze a natural debris flow. Fortunately, however, we can learn a great deal about the process of debris flow by producing small flows with natural materials in the laboratory. The behavior of debris seems to be independent of scale, as was suggested by observations reported in Chapter 12.

We can study velocity profiles and properties of debris flowing in channels by adding water to a sample from a natural debris-flow deposit and by pouring the reconstituted debris into channels of various shapes. When I did research for my dissertation, I experimented with the flow of fine-grained debris, from the 1917 Surprise Canyon debris deposit, in channels of triangular and rectangular cross-sections (*5*). The mathematical solutions for flow in triangular and rectangular channels are rather involved, however, so that a simpler means for determining rheological properties is in order. In this regard I would comment that the most important design criterion for an apparatus used for determining rheological properties of a substance is that the solution can be derived for the velocity distribution in the apparatus. Another important criterion is that the observations of the flow behaviors be as direct as possible; the increase in difficulty of interpreting data from an apparatus seems to be exponentially related to the complexity of the apparatus.

I have recently (*6*) adopted the semicircular channel as being most nearly suited to direct determination of properties of slurries. The surface of debris moving in open channels can be observed and surface velocities can be measured accurately. A semicircular channel was selected because the equations describing the flow of complex substances in a semicircular channel are relatively easily derived. There are many other designs of equipment used to measure viscosities and related properties of flowing materials. For example, there are concentric-cylinder viscometers, capillary-tube viscometers, rotating-disc viscometers, and so on (e.g., see ref. *2*). But all these viscometers seem to be unsuited to direct observation of flow behaviors, or else solutions for the velocity distributions of complex substances in them are complicated or even impossible.

Flow of Bingham Substance in Semicircular Channel.—The semicircular channel has a simple geometry, and thus the pattern of steady flow of a Bingham substance in it is relatively simple. The method of solving for the velocity distribution in a semicircular channel, which I will usually refer to as a *circular channel*, for brevity, consists of deriving an equation of equilibrium of forces, substituting a rheological model into the equilibrium equation, and integrating.

When we assume *steady flow* of any material in a channel, we assume that the flowing material maintains the same velocity profile everywhere along the length of the channel. Thus, the velocities may change with respect to distance from the sides of the channel, but at a given distance from the sides, the velocity is constant at different places along the length of the channel. Further, the shear stress acting parallel to concentric cylinders, representing flow laminae in circular-channel flow, is independent of position along the length of the channel.

Suppose a substance is flowing uniformly in a circular channel of radius R (Fig. 14.2A). Each circular element of material is under the same state of stress, so

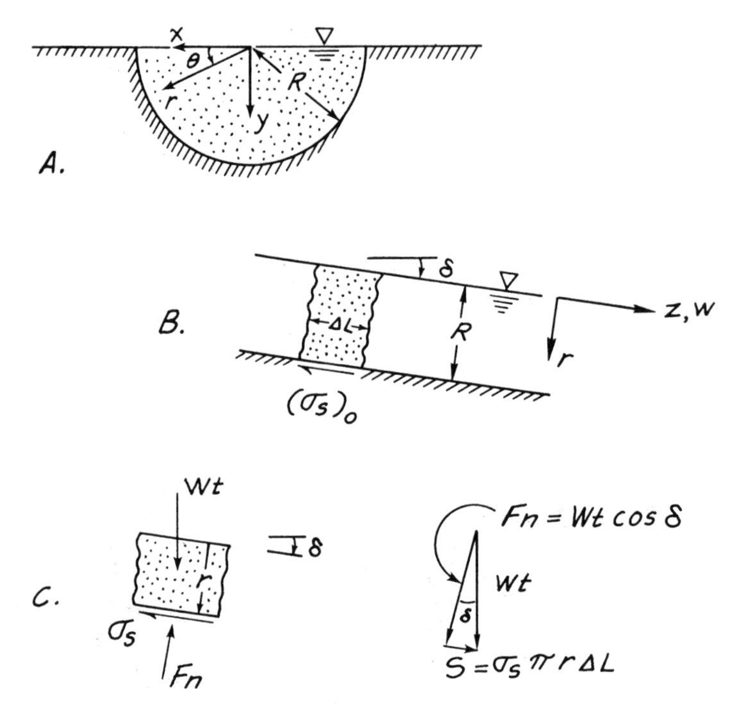

Figure 14.2. Equilibrium of small element of material flowing uniformly in semicircular channel.

A. Cross-section of channel.
B. Longitudinal section of channel.
C. Forces and shear stress acting on small, semicircular element of material.

that we can analyze the equilibrium of any small element of length ΔL (Figs. 14.2B and 14.2C) and with radius r, and our analysis will be representative of any part of the flow. Figure 14.2C shows a free-body diagram of the element, which is acted upon by a force, F_n, normal to the base of the element; a shear force, S, tangential to the base; and a weight, Wt, acting parallel to the direction of gravitational acceleration. For equilibrium,

$$\left[\sum F_z = 0\right] \qquad Wt \sin \delta - S = 0,$$

or

$$S = Wt \sin \delta. \tag{14.2}$$

The shear force, however, is equal to the shear stress, σ_s, times the area on which the shear stress acts, $\pi r\, \Delta L$, and the weight is equal to the volume of the element,

$$\frac{\pi r^2\, \Delta L}{2},$$

times the unit weight of the material, γ, so that eq. (14.2) can be rewritten as

$$\pi \sigma_s r\, \Delta L = \gamma\, \Delta L \frac{\pi r^2}{2} \sin \delta,$$

or

$$\sigma_s = \frac{\gamma r}{2} \sin \delta. \tag{14.3}$$

This latter equation is valid for all possible radii and for all types of materials in the channel. The shear stress vanishes at the center of the channel, where $r = 0$, and it is a maximum at the edge of the channel, where $r = R$.

The rheological model for a Bingham substance flowing in a circular channel is (*12*)

$$|\sigma_s| = k + \eta_b \frac{d\dot{w}}{dr}; \qquad [|\sigma_s| \geq k]. \tag{14.4}$$

Here $d\dot{w}/dr$ is the shear rate and the dot over the w signifies velocity, or differentiation of the displacement with respect to time.

Combining the rheological model, eq. (14.4), and the equilibrium equation, eq. (14.3),

$$k + \eta_b \frac{d\dot{w}}{dr} = \frac{\gamma r}{2} \sin \delta,$$

or

$$\frac{d\dot{w}}{dr} = \frac{1}{\eta_b}\left(\frac{\gamma r}{2} \sin \delta - k\right). \tag{14.5}$$

The velocity gradient, $d\dot{w}/dr$, is negative because the velocity decreases with increasing distance, r, from the center of the channel. Thus, for the coordinate system shown in Fig. 14.2A, eq. (14.5) becomes

$$-\frac{d\dot{w}}{dr} = \frac{1}{\eta_b}\left(\frac{\gamma r}{2} \sin \delta - k\right). \tag{14.6}$$

In order to derive the form of the velocity profile in the channel, we integrate eq. (14.6),

$$\dot{w} = -\frac{1}{\eta_b}\left(\frac{\gamma r^2}{4}\sin\delta - rk\right) + C_1, \tag{14.7}$$

where C_1 is an arbitrary constant.

If there is no slippage between the Bingham substance and the edges of the channel, the velocity of the substance is zero there. Thus, the boundary condition we use to eliminate the arbitrary constant in eq. (14.7) is

$$\dot{w} = 0 \qquad \text{at } r = R.$$

It follows that

$$C_1 = \frac{1}{\eta_b}\left(\frac{\gamma R^2}{4}\sin - Rk\right),$$

and that the velocity distribution in a circular channel is

$$\dot{w} = \frac{1}{\eta_b}\left[\frac{R^2 - r^2}{4}\sin\delta - (R - r)k\right] \qquad [r \geq R_0]. \tag{14.8}$$

The meaning of the restriction on possible values of r, shown in brackets following eq. (14.8), will become clear as we proceed with the analysis.

There is an r^2 term in eq. (14.8) so that the velocity distribution is parabolic in form, as is shown in Fig. 14.3B. However, our solution is incomplete because we have not examined the solution to determine where, if anywhere, flow actually can occur

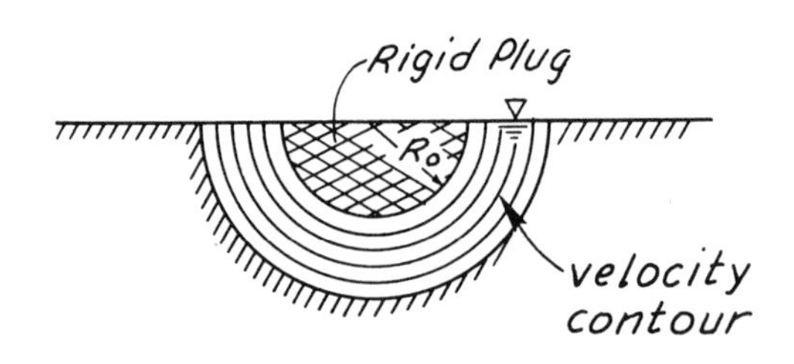

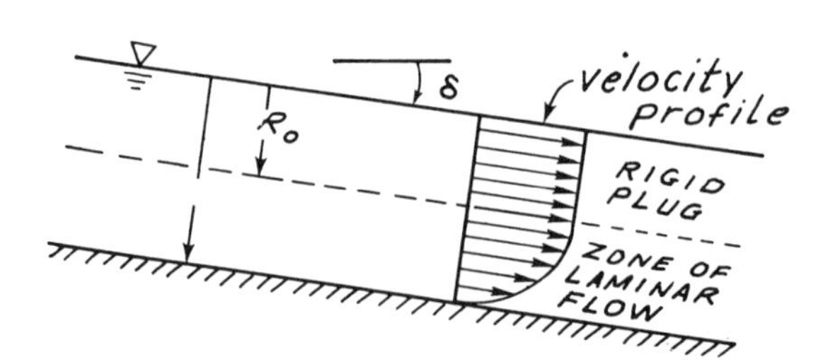

Figure 14.3. Velocity distribution of Bingham substance flowing in circular channel.
A. Cross-section.
B. Longitudinal section.

in the channel. In other words, we have ignored the "greater than or equal to" symbol in the rheological equation, eq. (14.4). Also, if we find that flow is possible, we must determine the radius of the "plug" of material that is rigid. Our solution for the velocity profile, eq. (14.8), is invalid within the plug. This is the reason for the "$[r \geq R_0]$" in eq. (14.8).

A "plug" of material must exist in the channel, because the shear stress vanishes at the center of the channel, where r is zero, and it increases with distance from the center of the channel, in accordance with eq. (14.3),

$$\sigma_s = \frac{\gamma r}{2} \sin \delta.$$

Somewhere, therefore, the shear stress is less than the strength of the material. The radius at which the shear stress equals the strength, k, of the Bingham material is the radius of the "plug." At that critical radius, the velocity gradient, $d\dot{w}/dr$, must vanish, according to eq. (14.4),

$$|\sigma_s| = k + \eta_b \frac{d\dot{w}}{dr},$$

so that we can compute the radius of the "plug" by setting the velocity gradient in eq. (14.6) equal to zero:

At $r = R_0$, the edge of the "plug," $d\dot{w}/dr = 0$,

$$-\frac{d\dot{w}}{dr} = 0 = \frac{1}{\eta_b}\left(\frac{\gamma R_0}{2} \sin \delta - k\right).$$

Solving for the radius, R_0,

$$R_0 = \frac{2k}{\gamma \sin \delta}. \qquad \text{Radius of plug} \qquad (14.9)$$

The velocity of the "plug" can be determined by substituting eq. (14.9) into the equation for the velocity distribution, eq. (14.8).

The velocity profile of a Bingham substance flowing in a circular channel, therefore, consists of a straight part, within the "plug," and a curved part, below the "plug." The curved part is parabolic in form (Fig. 14.3B).

Velocity Profile of Kaolin-Sand Slurry in Semicircular Channel.— According to our solution for the flow of a Bingham substance in a circular channel, we can determine the Bingham material constants of a substance by determining the radius and the velocity of the plug. The plug radius is proportional to the Bingham strength, k, according to eq. (14.9), and the plug velocity is proportional to the Bingham viscosity, η_b, according to eqs. (14.8) and (14.9).

We use this method routinely at the U.S. Geological Survey Slurry Laboratory, in Menlo Park, California (*6,7*), to determine properties of slurries. Slurries are pumped at a uniform rate into a semicircular channel and the velocity profile of the surface of the slurry is determined photographically. For example, Fig. 14.4 is a vertical photograph of a kaolin-sand slurry moving in a circular channel of about 16 cm diameter. The light-toned streaks are paths of flakes of "glitter" spread on the

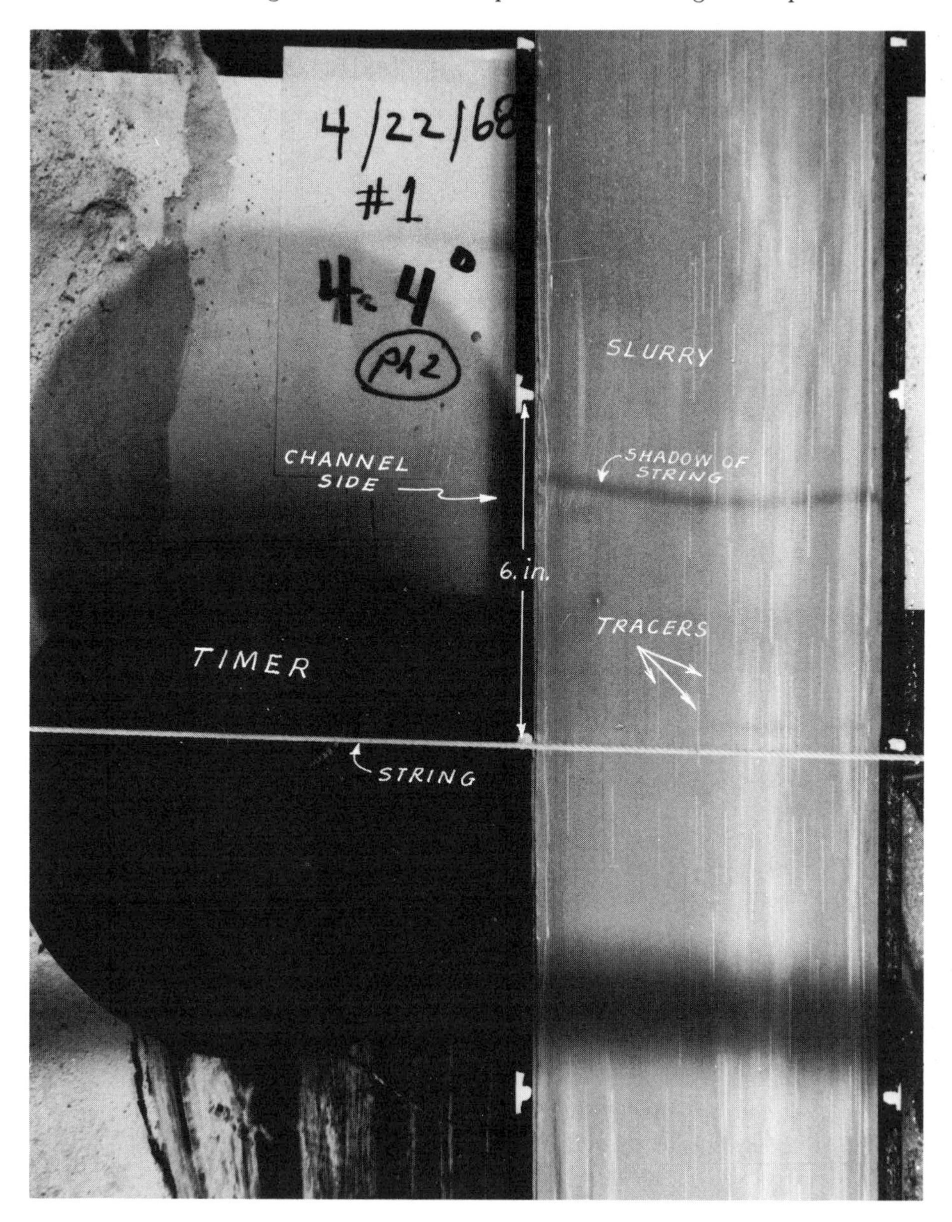

Figure 14.4. Slurry moving in semicircular channel. Exposure time was $\frac{1}{15}$ second. Lengths of light streaks on surface of slurry are proportional to velocities.

surface of the moving slurry. The lengths of the streaks are proportional to the velocities of the surface of the slurry. The lengths, converted into velocities, are shown in Fig. 14.5, where velocity is plotted in the vertical direction and distance from the center of the channel is plotted in the horizontal direction. Each solid circle on either half of Fig. 14.5 represents a measurement of the length of a light streak on the photograph. The same data has been used on each side of the center of the channel, so that the number of data points has been doubled. Several sources of error inherent to this method of determining rheological constants give rise to the wide range of velocities shown in the figure (see refs. *5*, *6*, and *7*). The most significant error is operator error.

One of the three smooth curves fitted to the data in Fig. 14.5 represents the theoretical profile of the Bingham substance. It fits the data moderately well. The data seem to indicate a plug with a radius of about 6.5 cm and a velocity of about 21 cm/sec. The slope angle of the channel was 4.4 degrees and the specific gravity of the slurry was 1.7.

Exercise

Calculate the Bingham strength and viscosity of the slurry shown in Fig. 14.4.

Velocity Profile for Flow in Infinitely Wide Channel.—The solution we have derived for flow of a Bingham substance in a semicircular channel can be slightly modified to give a solution for flow in a wide channel. If y instead of r is used for distance below the surface of the flowing material and if a unit width of material is considered, the equilibrium equation, eq. (14.3), becomes

$$\sigma_s = y\gamma \sin \delta. \qquad \text{Infinitely wide channel} \qquad (14.10)$$

The rheological model also changes slightly to

$$|\sigma_s| = k + \eta_b \frac{d\dot{w}}{dy}, \qquad \text{Infinitely wide channel} \qquad (14.11)$$

and the solution for the velocity profile is

$$\dot{w} = \frac{1}{\eta_b}\left[\frac{H^2 - y^2}{2}\gamma \sin \delta - k(H - y)\right] \qquad [y \geq T_c], \qquad (14.12)$$

where H is the total depth of flow and where

$$T_c = \frac{k}{\gamma \sin \delta} \qquad (14.13)$$

is the thickness of the plug or raft of rigid material at the top of the flow. There is a y^2 term in eq. (14.12) so that the velocity profile has the same form as the profile in a channel of semicircular cross-section (Fig. 14.3B).

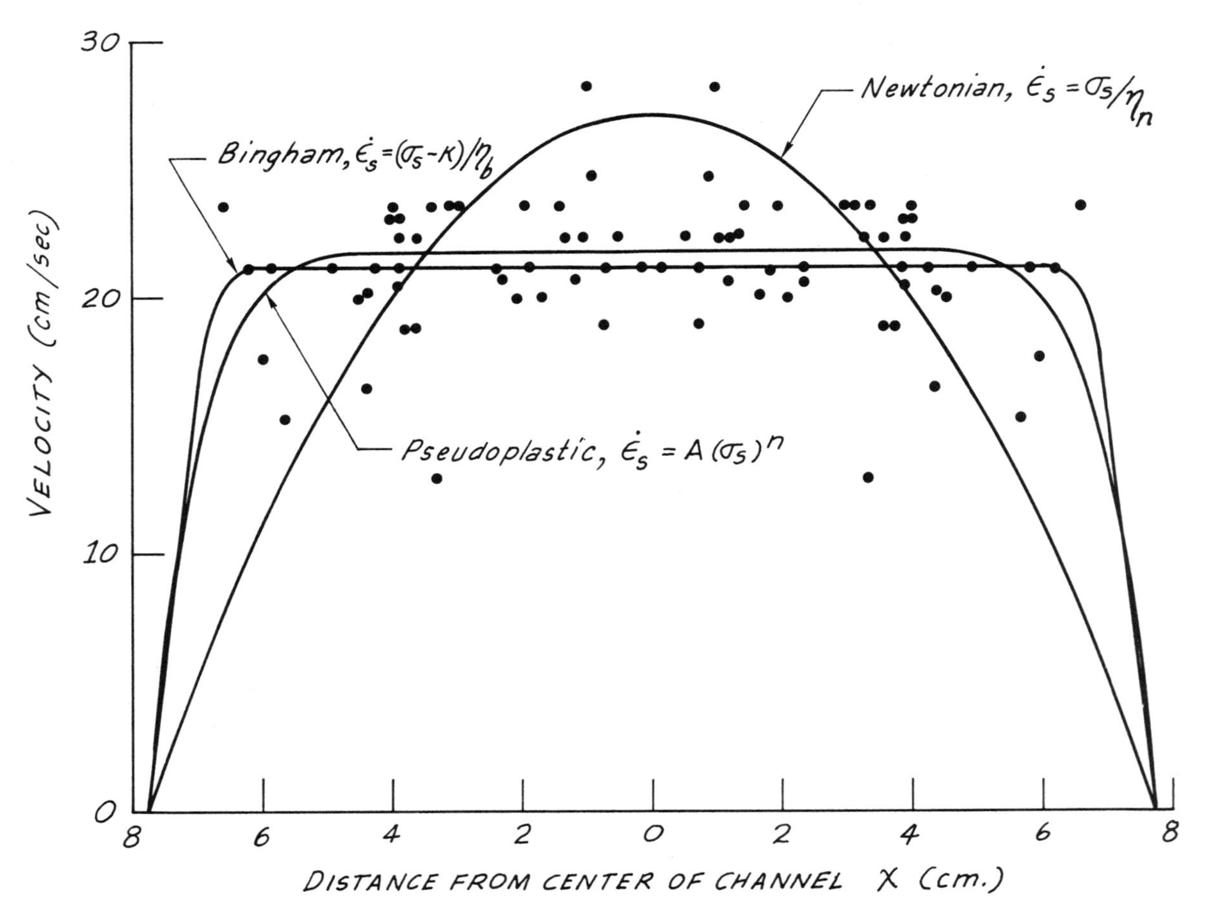

Figure 14.5. Velocities of kaolin-sand slurry moving in semicircular channel. "Best fit" Bingham, pseudoplastic, and Newtonian velocity profiles shown with solid lines.

Simplified Method of Determining Strength and Viscosity of Debris Flows in Noncircular Channels.—Most debris-flow channels are neither circular nor infinitely wide in cross-section. Also, a circular channel may be only partly filled with moving debris so that the cross-section of the flow may be noncircular. Fortunately, however, we can represent the cross-sectional shape of nearly any channel by half an ellipse. The circular and the infinitely wide channels, of course, are special cases of elliptic channels. In addition, however, flow in rectangular and triangular channels can be represented with sufficient accuracy for most purposes by flow in an imaginary elliptic channel, consisting of an ellipse inscribed in the form of the actual channel. Thus the solution for flow in elliptic channels is of practical importance.

The general rheological equations we derive in Chapter 15, in order to study the flow of Bingham substances in channels of triangular and rectangular cross-sections, can be solved for channels of elliptic cross-sections (*7*). Indeed, this problem is assigned as homework there.

Figure 14.6 shows the velocity distribution of a kaolin slurry flowing in a partly-filled circular channel. The slurry contained 73 weight percent of water and had a specific gravity of 1.196. The depth of the flow was 4.4 cm and the half-width of the flow was 4.5 cm. The width of the plug, x_0, was 3.8 cm and the slope angle of the flow surface was 4.1°. The problem is to calculate the Bingham strength and viscosity of the slurry.

The half-width of the flow is slightly different from the depth of the flow, so that

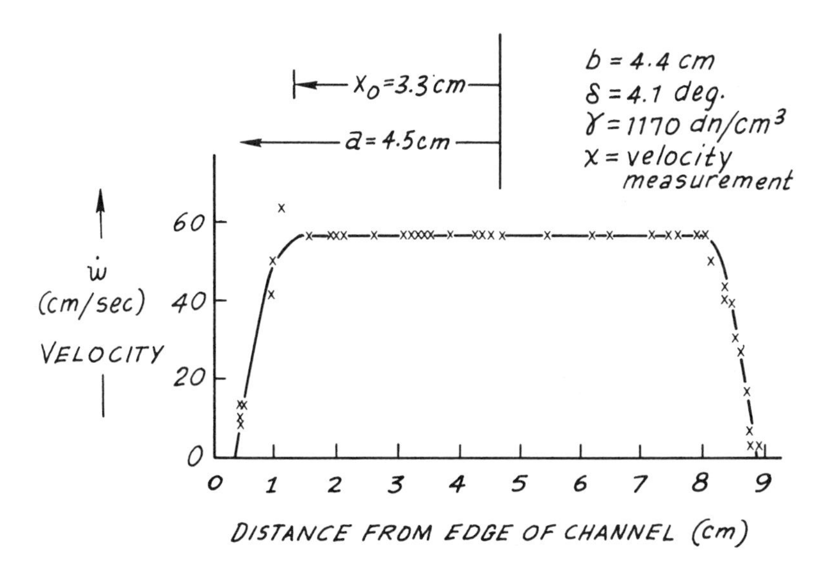

Figure 14.6. Velocity distribution of kaolin slurry flowing in nearly semicircular channel.

our solution for flow in a circular channel is not applicable. As you will show when you solve a problem in Chapter 15, the equation for the half-width of a plug in a semi-elliptic channel is

$$x_0 = \frac{k}{\gamma \sin \delta}\left[\left(\frac{a}{b}\right)^2 + 1\right], \tag{14.14}$$

where

k = debris strength,

γ = unit weight (density times gravity) of debris,

δ = slope angle (degrees) of flow surface,

a = half-width of flow,

and

b = depth of flow.

The equation for the thickness of the plug is

$$y_0 = \frac{k}{\gamma \sin \delta}\left[\left(\frac{b}{a}\right)^2 + 1\right]. \tag{14.15}$$

Usually we will be able to measure only the width of the plug; therefore, eq. (14.14) is the pertinent one for channelized flow. We can put it in dimensionless form by dividing each side of eq. (14.14) by a:

$$\frac{x_0}{a} = \frac{k}{a\gamma \sin \delta}\left[\left(\frac{a}{b}\right)^2 + 1\right], \tag{14.16}$$

or

$$\frac{x_0}{a} = k'\left[\left(\frac{a}{b}\right)^2 + 1\right], \tag{14.17}$$

if we use, for shorthand,

$$k' = \frac{k}{a\gamma \sin \delta}.$$

Equation (14.17) is plotted in Fig. 14.7A so that calculations can be simplified. For example, if the ratio of the plug width to the total width of the channel is 0.4, and if the ratio of the half-width to the depth of a rectangular or elliptic channel is 3, the strength variable, k', is about 0.04, according to the scale at the bottom of Fig. 14.7A. Thus,

$$k' = 0.04 = \frac{k}{a\gamma \sin \delta},$$

so that the strength of the debris is

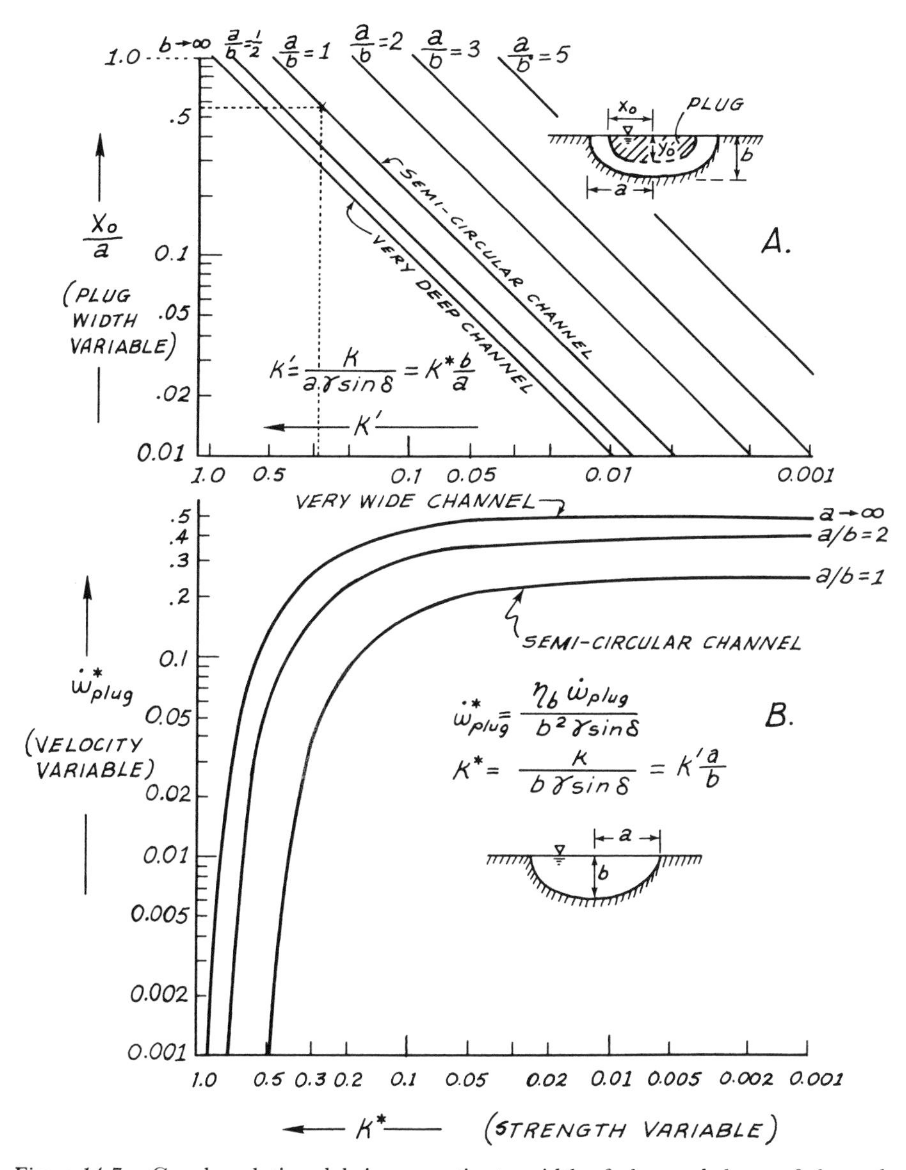

Figure 14.7. Graphs relating debris properties to width of plug and shape of channel.

$$k = (0.04)(a)(\gamma \sin \delta)$$

in any consistent system of units—feet, meters, centimeters, and so forth. To calculate strength, therefore, we would simply substitute appropriate values for channel half-width, a, unit weight, γ, and slope angle, δ, into the last equation, above.

According to the experimental data for the kaolin slurry, shown in Fig. 14.6,

$$\left(\frac{a}{b}\right) \approx 1.02,$$

and

$$\left(\frac{x_0}{a}\right) \approx 0.84,$$

so that, according to Fig. 14.7A, the dimensionless strength variable for the kaolin slurry is

$$k' \approx 0.4.$$

It follows that the shear strength of the slurry is

$$k = k'a\gamma \sin \delta \approx (0.4)(4.5)(1170)(0.072),$$

or

$$k \approx 150 \text{ dn/cm}^2.$$

In order to calculate the Bingham viscosity of the slurry, we need to consider the velocity of the plug in an elliptic channel. It is (*7*)

$$\dot{w}_{\text{plug}} = \frac{1}{\eta_b}\left[\frac{\gamma \sin \delta}{2[1 + (a/b)^2]}\left\{b^2\left[\frac{k(1 + (a/b)^2}{\gamma \sin \delta}\right]\right\} - kb + \frac{k^2[1 + (a/b)^2]}{\gamma \sin \delta}\right], \quad (14.18)$$

which obviously is rather cumbersome. A graphical solution of the equation is easily employed, however. In order to put the equation in dimensionless form, let

$$\dot{w}^*_{\text{plug}} = \frac{\dot{w}_{\text{plug}}}{(\gamma \sin \delta)(b^2)}, \quad (14.19)$$

and

$$k^* = \frac{k}{b\gamma \sin \delta} = k'\left(\frac{a}{b}\right). \quad (14.20)$$

Then the equation for the velocity of the plug, eq. (14.18), simplifies to:

$$\dot{w}^*_{\text{plug}} = \frac{1}{2[1 + (a/b)^2]} + \frac{k^{*2}}{2}[1 + (a/b)^2] - k^*. \quad (14.21)$$

Equation (14.21) is plotted in Fig. 14.7B for three ratios of channel half-width to depth.

According to the experimental data shown in Fig. 14.6,

$$\left(\frac{a}{b}\right) \approx 1.02,$$

and

$$\dot{w}_{\text{plug}} \approx 57 \text{ cm/sec}.$$

Thus, according to Fig. 14.7B, if

$$k^* = k'\left(\frac{a}{b}\right) \approx 0.41,$$

$$\dot{w}^*_{\text{plug}} \approx 0.009,$$

and

$$\eta_b = \frac{\dot{w}^*_{\text{plug}}}{\dot{w}_{\text{plug}}} b^2 \gamma \sin \delta \approx \frac{0.009}{57} (4.4)^2 (1170)(0.072),$$

or, the Bingham viscosity of the kaolin slurry is

$$\eta_b \approx 0.3 \text{ poise},$$

which is about thirty times the Newtonian viscosity of water.

Properties of Natural Debris Flow at Wrightwood, California.—I have believed for a long time that the photographic method used to measure debris properties in laboratory flows would be quite useful for measuring properties of lava flows and of debris flows in the field. Glowing spots on surfaces of lava flows and reflective spots on surfaces of debris flows provide light-toned streaks similar to those produced in the laboratory by sprinkling reflective flakes on surfaces of experimental flows.

On 20 May 1969 several graduate students and I were presented with a unique opportunity to try this idea, when we observed dozens of debris flows moving through channels near Wrightwood, in southern California (*7*). The flows began as small slides mobilized by melting snow in the source area and occurred for a period of about ten days. (See Chapter 12 for a more nearly complete description.)

Figure 14.8 is a photograph taken with the camera axis essentially normal to the surface of a relatively fluid phase of one of the last large debris flows at Wrightwood on 20 May. The photograph was taken at 11:15 AM so that the sun was nearly directly overhead. The exposure time was 0.062 second. The flow was about two meters wide and one meter deep at the time the photograph was taken. The slope of the channel was about seven degrees. The right-hand side of the channel was slightly undercut; for that reason about 15 to 20 cm of the width of the flow was masked in the photograph. The right-hand edge is visible in the lower, right-hand part of the photograph.

The light-toned streaks on the surface of the flow are reflections from mud-coated pebbles, such as we see in the wet mud on the left-hand side of the photograph.

Lengths of streaks were measured in the central third of the photograph. Distances were measured from the left-hand edge of the flow, which is clearly visible in the photograph, and traces were measured to the nearest $\frac{1}{50}$ inch, photo scale, or

Figure 14.8. Time-exposure photograph of moving debris flow at Wrightwood, California. Flow is about two meters wide. Light streaks on flow are traces of reflective spots.

about one cm, actual scale. Traces were ignored in two areas where flow was disturbed by boulders beneath the surface of the flow. The results are shown in Fig. 14.9.

The width of the plug was about 100 cm, as is indicated by the rapid decrease in velocity to the left of an area about 40 cm from the left-hand side of the channel (Fig. 14.9). In order to calculate strength of the debris, we can approximate the shape of the channel cross-section by a semicircle; the width of the flow was about twice the depth. Thus, we use the graph in Fig. 14.7A, which applies to $a/b = 1$.

The following are the measurements:

x_0 = half-width of plug = 50 cm,

γ = unit weight of debris = (2 gm/cc)(980 cm/sec^2) = 1960 dn/cm^3,

δ = slope angle = 7 deg. ($\sin \delta = 0.122$),

a = half-width of flow = 90 cm,

and

b = depth of flow = 90 cm.

According to Fig. 14.7A, we need a value of x_0/a. Thus,

$$x_0/a \approx 50 \text{ cm}/90 \text{ cm} \approx 0.56.$$

This value is plotted as an "x" in Fig. 14.7A. By reading the horizontal scale, we find that

$$k' = 0.28 \frac{k}{a\gamma \sin \delta},$$

so that the strength of the debris was, approximately,

$$k = (0.28)(90 \text{ cm})(1960 \text{ dn/cm}^3)(0.122) = 6020 \text{ dn/cm}^2.$$

Accordingly, we estimate the strength of the debris in the Wrightwood flow to have been about 6000 dn/cm^2 at the time the photograph was taken. Presumably, preceding debris in the drier, frontal portion of the flow and following debris in the more fluid terminus of the flow had different strengths.

The Bingham viscosity of the debris can be estimated as follows. According to our calculations, the strength variable, k', is 0.28. Thus, because $k^* = k'(a/b)$, $k^* = 0.28$, also. However, it is important to remember that k' and k^* are usually different. We can determine the value of the velocity variable, $\dot{w}^*_{\text{plug}}$, by reading the graph for $a/b = 1$ at a value of $k^* = 0.28$. The pertinent value of $\dot{w}^*_{\text{plug}}$ is about 0.044, so that

$$\dot{w}^*_{\text{plug}} = (0.044) = \frac{\eta_b \text{ plug}}{b^2 \gamma \sin \delta}.$$

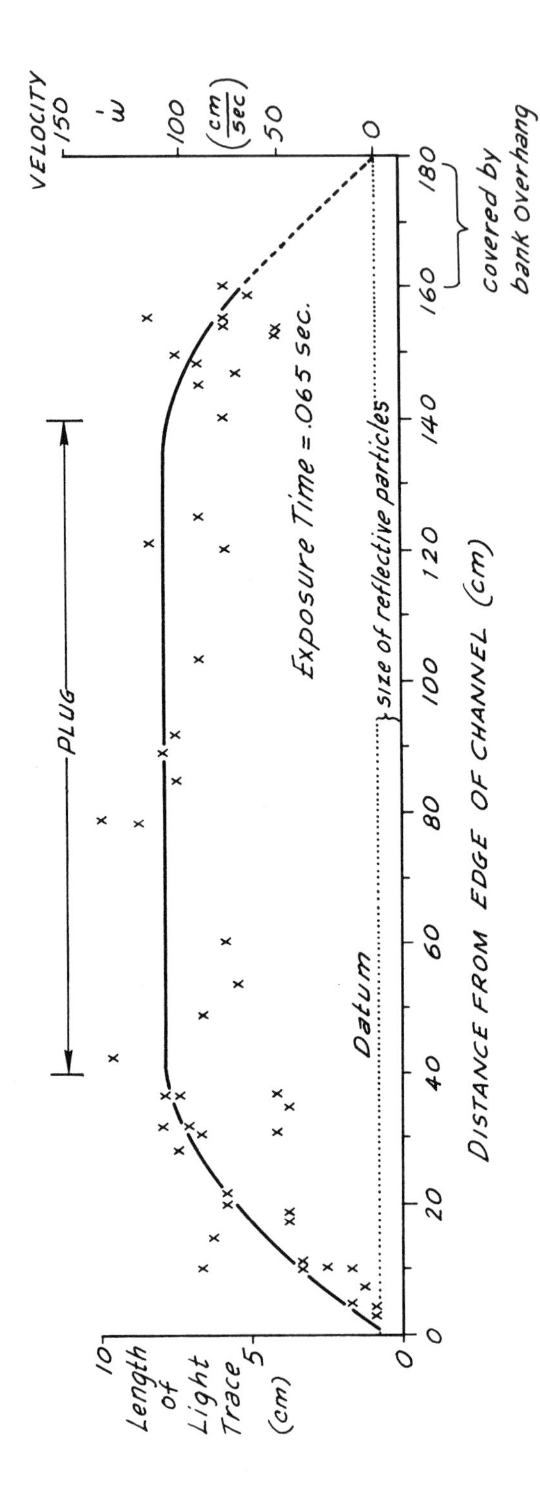

Figure 14.9. Velocities of surface of flow shown in Fig. 14.8. Each "x" represents a measurement.

The Bingham viscosity is, therefore,

$$\eta_b = \frac{(0.044)b^2\gamma \sin \delta}{\dot{w}_{\text{plug}}}.$$

The velocity of the plug was 110 cm/sec, according to Fig. 14.9, so that

$$\eta_b \approx \frac{(0.044)(90 \text{ cm})^2(1960 \text{ dn/cm}^2)(0.122)}{(110 \text{ cm/sec})} \approx 7.6 \times 10^2 \text{ poise},$$

which is about 80 thousand times the Newtonian viscosity of water.

Most investigators have determined the Newtonian viscosity rather than the Bingham constants of debris flows. The Newtonian viscosity of debris, unfortunately, depends as much on channel size as on actual debris properties. The Newtonian viscosity of the debris at Wrightwood was 4.5×10^3 poise, which is six times greater than the Bingham viscosity of the same debris. Sharp and Nobles (*13*) estimated about the same value of Newtonian viscosity for the Wrightwood flows of 1941.

Implications of Solutions and Experiments.—The theoretical analysis of debris flow in channels provides explanations for several peculiar phenomena of debris flow which were described in Chapter 12. For example, the shear stresses within the rigid plug of a channelized flow are so small that the Bingham substance does not deform there. This theoretical behavior seems to account for the field observation that clasts such as fractured boulders, tree stumps, tin cans, and numerous fragile objects can retain their respective identities even though they have been carried for miles within the central part of a debris flow. Another result of plug flow is that large clasts transported by debris flows characteristically are unabraded and angular in shape (*5*).

Recognition of the fact that the Bingham model describes quite well the flow of debris allows us to explain behaviors of debris under a variety of conditions. The model predicts the plug, which we see in experimental as well as natural flows. The solution for the radius of the plug,

$$R_0 = \frac{2k}{\gamma \sin \delta},$$

helps us to explain the observation that the surface of a lobe or tongue of debris usually slopes more steeply than the surface over which the debris moves. According to the solution, the radius of the plug becomes extremely large as the slope angle, δ, of the surface of the debris approaches zero. Thus, plastico-viscous substances such as debris must create their own surface slopes if they are to flow on flat underlying surfaces. In accord with this conclusion, the lobes and cones of debris and damp sand described in Chapter 12 built their surface slopes by upward thrusting in their snout regions.

The "smoothing effect," another characteristic of debris flow, also can be explained in terms of the Bingham model. Many people have noted that, after a debris flow has passed through a canyon, the channel bottom and sides have been swept clean of loose debris. But after the flow has reached the fan, they note, the regimen is largely depositional. Intense erosion by relatively underloaded flood waters may follow the major debris waves, but study of recent debris deposits indicates that the most characteristic phenomena reflect deposition and topographic smoothing by debris flow. For example, debris flows commonly move over road pavement without damaging the pavement at all, and topographic irregularities visible on fan surfaces adjacent to recent debris-flow deposits are not reflected in the smooth surfaces of the deposits (*5*).

This smoothing seems to be a result of the raft or plug of relatively rigid material in the upper parts of flows. In terms of the equation for the radius of the plug, we see that, unless pre-existing channels on the fan surface are large, at least equal to

$$\frac{2k}{\gamma \sin \delta},$$

the debris will not flow along the axes of the pre-existing channels. The sizes of channels that would be reflected in the surface of a debris flow moving over the fan surface can be estimated with equations we have already derived. If the thickness of a lobe of debris is small relative to its width, the thickness, T_c, of material behaving rigidly is given, approximately, by eq. (14.13),

$$T_c = \frac{k}{\gamma \sin \delta}.$$

In order for the same material, with strength, k, and unit weight, γ, to flow in a channel that is approximately semicircular, the radius of the channel must be at least

$$R_0 \geq \frac{2k}{\gamma \sin \delta},$$

which is precisely twice the critical thickness of the lobe of debris. Therefore, if a channel on a fan has a radius that is less than twice the thickness of the rigid layer of debris moving over the fan surface, the channel will not be reflected in the surface of the debris-flow deposit. One variable we have ignored is the slope angle, δ, of the channel and the fan surface. If the slopes are the same, our conclusions are valid. If they are different, our conclusions need to be slightly modified.

Figure 14.10 shows the lateral deposit of a small debris flow that was moving from left to right down a channel that was about two feet wide and one foot deep. The strength of the debris was so great that the debris was unable to flow laterally from the main channel into the smaller, subsidiary channel in the foreground, as explained by our analysis.

Figure 14.10. Smoothing effect. A recent debris-flow deposit near Cris Wicht Camp in Surprise Canyon, Panamint Valley, California. Debris flowed across the rill, but the form of the rill is not reflected in the surface of the debris-flow deposit.

Velocity Profiles of Newtonian and Pseudoplastic Substances in Semi-circular Channels.—Thus far we have restricted our attention to channelized flow of a Bingham substance. Most early investigators assumed that ice, lava, and debris behave as viscous or Newtonian substances (*1,8,9,12*). Several recent studies have indicated that ice deforms according to a pseudoplastic law, which is a flow law of the form (*3,4,9*)

$$\dot{\epsilon}_s = A\sigma_s^n, \tag{14.22}$$

which states that the rate of shear is proportional to the applied shear stress, σ_s, raised to some power, n.

We will derive equations for flow of Newtonian and pseudoplastic substances. The equilibrium equation, eq. (14.3), is valid for steady flow of any material in a circular channel, so that

$$\sigma_s = \frac{\gamma r}{2} \sin \delta.$$

For a Newtonian substance,

$$\sigma_s = \eta \frac{d\dot{w}}{dr},$$

so that the velocity profile in a circular channel is

$$\dot{w} = \frac{R^2 - r^2}{4\eta} \gamma \sin \delta. \qquad \text{Newtonian} \tag{14.23}$$

For a pseudoplastic, the rheological model is given by eq. (14.22), so that

$$\frac{d\dot{w}}{dr} = A\sigma_s^n = A\left(\frac{\gamma r}{2} \sin \delta\right)^n,$$

and the velocity profile is given by

$$\dot{w} = \frac{A[(\gamma \sin \delta/2)]^n}{n+1} [R^{(n+1)} - r^{(n+1)}]. \qquad \text{Pseudoplastic} \tag{14.24}$$

If $n = 1$ and $A = 1/\eta$, eq. (14.24) reduces to the equation for velocities of a Newtonian substance, eq. (14.23). The Newtonian model is a special case of the pseudoplastic model. Figure 14.11 shows velocity profiles for pseudoplastic substances in which the exponent, n, ranges from 0.1 to 5.0.

The theoretical velocity profiles for Newtonian, pseudoplastic, and Bingham models that best fit the experimental data for kaolin-sand slurry flowing in a circular channel are shown as solid lines in Fig. 14.5. The Bingham and pseudoplastic sub-

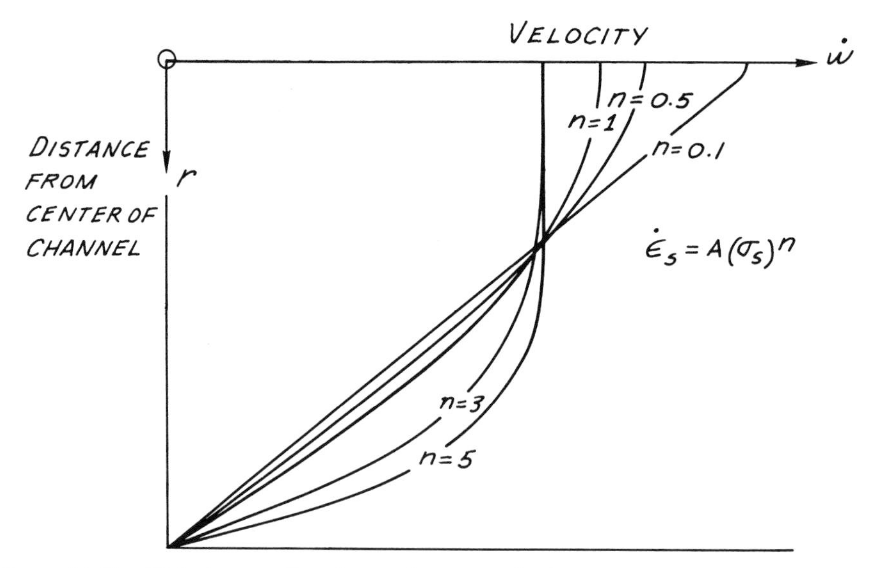

Figure 14.11. Velocity profiles for various pseudoplastic substances in semicircular channel.

stances seem to fit the experimental data most closely. The Newtonian substance deviates markedly from the actual behavior of the slurry.

Comments on Model Fitting.—The fact that the pseudoplastic substance and the Bingham substance fit the experimental data in Fig. 14.5 equally well may be confusing, but it is definitely instructive. A possible question is, which model *really* better fits the data? After all, the Bingham model involves yield strength and viscosity, whereas the pseudoplastic model involves high sensitivity of shear rate on shear stress, so that we might think that we should be able to distinguish between the two behaviors. For example, if the exponent, n, in the equation of the pseudoplastic,

$$\dot{\epsilon}_s = A\sigma_s^n,$$

is 10, a slight increase of σ_s above unity will greatly increase the flow rate, $\dot{\epsilon}_s$, whereas a slight decrease of σ_s below unity will greatly decrease the flow rate. Thus, if σ_s is 10, σ_s^{10} is 10^{10}, a very large number; whereas if σ_s is 0.1, σ_s^{10} is 10^{-10}, an extremely small number. Partly due to the range of data, however, this behavior might be indistinguishable from that of a Bingham substance, as we noted when we fitted curves to the experimental flow data, Fig. 14.5.

Thus we might not be able to differentiate Bingham and pseudoplastic behaviors with one experiment. Suppose, then, that we try two experiments. The first experiment allowed us to reject the Newtonian model for slurry so, then, another experiment might allow us to determine whether Bingham or pseudoplastic behavior more closely describes the channelized flow of slurry. Thus, we will design a new experiment in which we will cause slurries that are identical in composition to flow in channels of two different sizes. Then we will determine which pseudoplastic and Bingham substances best fit the flow profiles across the two channels. That is, we will independently determine the constants, A and n of the pseudoplastic model and k and η_b of the Bingham model. If we actually did this, we would probably find that the "constants" depend upon channel size, so that they are not truly material constants at all. We might find that the pseudoplastic "constants" change more than the Bingham "constants," so that on this basis we could accept the Bingham model and reject the pseudoplastic model for slurries. The first experiment allowed us to reject the Newtonian model and the second probably would allow us to reject the pseudoplastic model. As a result, we might proudly proclaim that we have discovered the rheological model for slurries: The slurries are Bingham substances. A skeptic might well ask us, however, why we did not try different models, such as a combination of plasticity and pseudoplasticity:

$$\dot{\epsilon}_s = B(\sigma_s - k)^m$$

or

$$\sigma_s = k + \left(\frac{\dot{\epsilon}_s}{B}\right)^{1/m},$$

or perhaps a combination of Newtonian and pseudoplastic models,

$$\sigma_s = D\dot{\epsilon}_s + E(\dot{\epsilon}_s)^{1/n},$$

or a model of the form

$$\dot{\epsilon}_s = F\sigma_s + G\sigma_s^n,$$

and so on, where k, b, m, n, D, E, F, and G are constants. The skeptic might argue that some model other than Bingham's might better fit the data. If so, he would be absolutely correct. No matter how complicated a rheological model we might select to describe empirical data, there is always another, perhaps more complicated model that better describes the data. This is because there probably is an infinite number of combinations of rheological models.

So, how do we select a model for slurry, or for any other substance? And when do we know we have selected a good rheological model for a substance? We select a rheological model by a complicated process of trial and error, presumably. We can discover how certain rheological models behave under different conditions, and we know how the actual substance has behaved under a few conditions, so that we might select the simplest model we know will behave approximately like the actual substance under the conditions with which we are familiar. Our selection should be guided, as well, by consideration of the structure of the material. For example, slurries are largely granular solids; *thus*, we might study the interaction of theoretical grains in order to deduce a rational model for slurries. In this way we could show that the Coulomb model is theoretically a good representation of slowly flowing granular solids. I used the Coulomb model to explain the transport of large clasts, the shape of debris-flow snouts, and the existence of a raft or plug in moving debris flows when I wrote my dissertation on debris flows (*5*). The essential elements of the Coulomb model are the same as those of the simple plastic model; *therefore*, I adopted the simpler model for most of the derivations here.

The selection of the combination of a Coulomb or simple plastic model and a viscous model to explain certain features of moving debris flows, however, was dictated more by convenience than by consideration of the fundamental properties of granular solids or water. I would conclude, therefore, that we select a rheological model for a material on the basis of our understanding of the internal mechanisms of deformation of the material, on the basis of convenience, and on the basis of knowing the consequences of assuming certain models or combinations of models.

An equally important question is, when do we know we have selected a good rheological model for a substance? One criterion of a good rheological model is that it accounts for many of the phenomena that affect the material. For example, if we assume a Coulomb-viscous model,

$$|\sigma_s| = C + \sigma_n \tan\phi + \eta_c\dot{\epsilon}_s,$$

where C is cohesive strength, $\sigma_n \tan\phi$ is frictional strength, and η_c is viscous resistance, we can explain much about the behavior of debris with a wide range of compositions. We can explain the longitudinal profiles of many landslide masses, very small debris flows, and very large debris flows; we can explain the ability of debris to transport extremely large clasts; we can explain the smoothing effect of debris flows; and we can explain the raft or plug of rigid debris in a moving debris flow. Using our criterion, then, the Coulomb-viscous model is a good one for debris flow.

Another criterion of a good rheological model for a material is that it allows us to predict and discover behaviors that we had not noticed before. For example, after tentatively adopting the Coulomb-viscous model or the Bingham model to describe debris flow, we predict lateral deposits parked high on the gently sloping banks of debris channels, and we can find them in the field and in experimental debris flows. In addition, we can predict that debris channels will tend to be U-shaped in cross-section and we can provide a reasonable explanation for the characteristic U-shaped cross-sections of glacial valleys (Chapter 15).

If we select a rheological model that does not allow us to predict certain behaviors that we observe, we need to modify the model. Thus the selection of rheological models is a process of trial and error.

Viscosity and strength of lava in Makaopuhi Crater, Hawaii

I know of no measurements of surface velocity distributions of moving lava flows. The speculation that properties of lava are similar to properties of debris, based on the observation that debris flows and lava flows have similar morphologies, is, however, corroborated by measurements of rheological properties of lava in Makaopuhi Crater, Hawaii (*14*). At the time the Kilauea volcano erupted in 1965, a lake of tholeitic magma about 80 meters deep and 400 meters in diameter formed in the Makaohpui pit crater (*14*). Measurements of properties of the lava were begun when the solid crust on the lake was about four meters thick. The measurements were made by inserting a shaft into a borehole through the crust and into the lava and then measuring rates of rotation of the shaft when various torques were applied to the shaft. Attached to the end of the shaft was a stainless-steel spindle, consisting of four long and narrow blades radiating from the shaft. The spindle was about 0.32 meters long and 3 cm in diameter (*14*).

During each test, the spindle was first placed at a depth of about 2.5 meters below the base of the solid crust and measurements of rate of rotation were made for four different torque values. Then the spindle was lowered to about 3.2 meters below the crust and the measurements were repeated (*14*).

The temperature of the lava was about 1130°C where the measurements were made. Crystals comprised about 25 percent, by volume, and gas bubbles probably comprised about 5 percent, by volume, of the lava, so that the lava was a three-phase substance: liquid melt, solid crystals, and gas bubbles (*14*).

Theory of Rotational Viscometer.—The rotation of the spindle in the lava is similar to the rotation of a liquid cylinder in a large body of the same liquid, because the spindle and the liquid trapped between its blades are essentially a cylinder of liquid. The rotation of a cylinder in an infinite body of liquid is a special case of rotation of one cylinder inside another, coaxial cylinder. Therefore, our theory of the deformation of the lava around the spindle will assume that the lava is contained in the annulus between a small cylinder of radius a and a larger cylinder of radius b (Fig. 14.12).

We will assume tentatively that lava is a Bingham substance so that it deforms according to the flow rule,

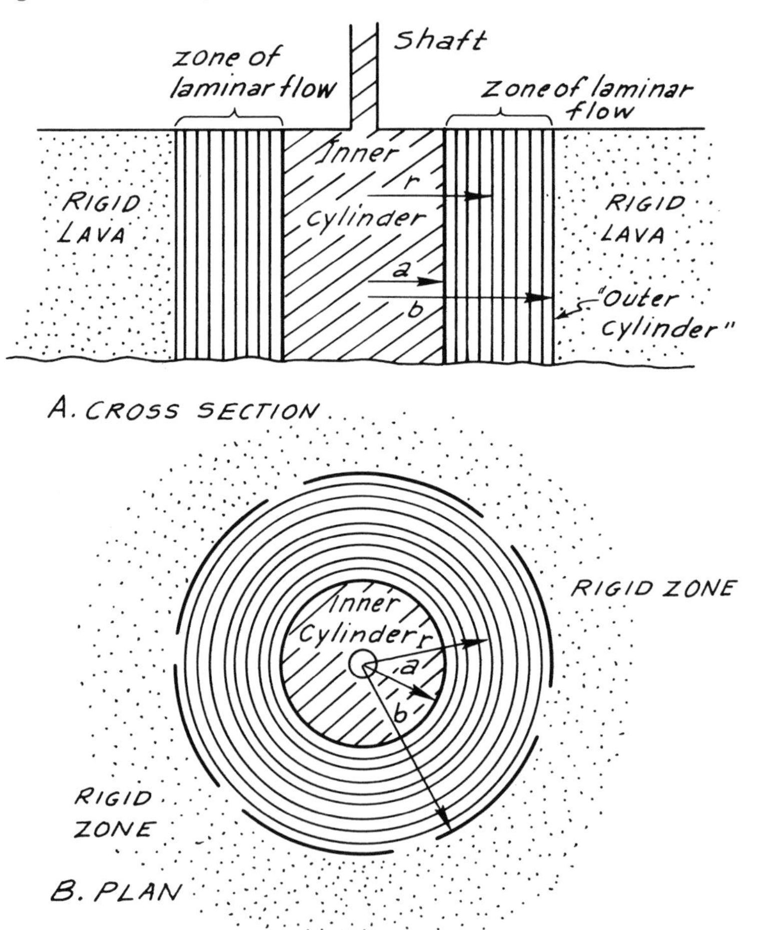

Figure 14.12. Zones of flow within large body of lava during rotation of cylinder. Problem is same as flow within coaxial-cylinder viscometer.

$$|\sigma_s| = k + \eta_b \dot{\epsilon}_s; \qquad [|\sigma_s| \geq k], \tag{14.25}$$

where k is yield strength, η_b is coefficient of viscosity, and $\dot{\epsilon}_s$ is rate of shear strain. In order to measure the constants, k and η_b, of the Bingham substance in a rotational viscometer, we must combine eq. (14.25) with expressions for the geometry of the viscometer and the torques applied to it.

For equilibrium, the torque, M, must be the same for the inner and outer cylinders and for any cylindrical section of the lava between them. That is (*5*),

$$M_{r=b} = M_{r=a} = M_{r=r}. \tag{14.26}$$

However, the torque is proportional to the shear stress between adjacent cylindrical laminae of the liquid being sheared, so that

$$M = (\sigma_s)(\text{radius})(\text{surface area of cylinder}).$$

Now the torque, M, remains constant, according to eq. (14.26), and the surface area of a cylindrical lamina of liquid increases as the radius increases, so that the shear stress, σ_s, must decrease with increasing radii. That is,

$$\sigma_{s_{r=h}} < \sigma_{s_{r=r}} < \sigma_{s_{r=a}}. \tag{14.27}$$

Thus the greatest shear stress in the lava will be tangential to the surface of the inner cylinder and the yield strength, k, of the lava could be determined by measuring the torque required to begin rotation of the inner cylinder, that is, the spindle. The rheological model for impending shear of the lava is [eq. (14.25)]

$$|\sigma_s| = k, \tag{14.28}$$

so that the resisting torque applied by the lava on the inner cylinder is

$$\sigma_s(\text{radius})(\text{surface area of inner cylinder})$$

or

$$k(a)(2\pi a H),$$

where a is the radius and H is the height of the inner cylinder. If M_0 is the torque required to begin rotation of the inner cylinder, the strength is

$$k = \frac{M_0}{2\pi a^2 H}. \tag{14.29}$$

We have an expression for the strength of the lava in terms of the torque required to start rotation of the inner cylinder within the lava. However, if the torque is greater than the critical value for rotation, M_0, we have no expression for how much lava is being sheared. The radius of the interface between lava being sheared and static lava, beyond the interface, can be calculated with the aid of the following principle: Whereas the torque applied by the inner, rotating cylinder to any coaxial cylinder of lava is independent of the distance, r, of the coaxial cylinder from the center of rotation, the shear stress transferred from one coaxial cylinder of lava to another decreases as the inverse square of the distances of the cylinders from the center of rotation. Thus, in a cylindrical viscometer consisting of two coaxial cylinders, part of the substance between the cylinders will not deform unless the torque is large enough to cause the shear stress along the outer cylinder to equal the shear strength of the substance. And in a very large body of a material, such as a lava lake, there will always be some cylinder of the material across which the shear stress is less than the yield strength of the material.

The radius of the lava cylinder where the shear stress equals the shear strength of the lava can be calculated as follows: If M is the torque applied to the inner cylinder, and b is the radius to the interface between the moving and static lava, then

$$M = 2\pi b^2 Hk,$$

or

$$b = \sqrt{\frac{M}{2\pi Hk}}; \qquad [M > M_0]. \tag{14.30}$$

The radius of the outer cylinder shown in Fig. 14.12 is variable; it depends upon the amount of torque applied by the inner cylinder, that is, the spindle, to the large body of lava. All the lava between the inner cylinder and the distance, b, is being sheared. Lava beyond the distance, b, is undisturbed by the rotation of the spindle. The lava beyond a distance b is rigid and corresponds to the plug or raft in channelized flow of the Bingham substance.

We have now derived equations for calculating strength by means of measurements of torques applied to the spindle of the viscometer. The remaining problem is to calculate the Bingham viscosity, η_b. According to the rheological equation, eq. (14.25),

$$|\sigma_s| = k + \eta_b \dot{\epsilon}_s.$$

The rate of shear strain, $\dot{\epsilon}_s$, for rotation of a cylinder in a fluid is

$$r\frac{d\dot{\theta}}{dr}$$

where $\dot{\theta}$ is the angular velocity of a lamina of material. Therefore, the rheological equation becomes

$$|\sigma_s| = k + \eta_b r \frac{d\dot{\theta}}{dr}. \tag{14.31}$$

The torque applied by the lava to a cylinder of lava inside it is

$$M = \sigma_s(\text{radius})(\text{surface area of cylinder}),$$

$$= \sigma_s(r)(2\pi rH) = 2\pi Hr^2\sigma_s,$$

or

$$M = \left(k + \eta_b r \frac{d\dot{\theta}}{dr}\right) 2\pi H r^2. \tag{14.32}$$

Solving eq. (14.32) for $d\dot{\theta}/dr$ and integrating from the outer cylinder to the inner cylinder,

$$\int_0^{\Omega} d\dot{\theta} = -\int_a^b \left[\frac{M}{2\pi H}\frac{1}{r^3} - \frac{k}{r}\right] dr,$$

where Ω is the angular velocity of the inner cylinder.

Integrating,

$$\Omega\eta_b = -\left[-\frac{M}{4H\pi}\frac{1}{r^2} - k\ln(r)\right]_a^b$$

$$= \frac{-M}{4H\pi}\left[\frac{b^2 - a^2}{a^2 b^2}\right] - k\ln\left(\frac{b}{a}\right),$$

so that the Bingham viscosity is

$$\eta_b = -\frac{1}{\Omega}\left[\frac{M}{4H\pi}\left(\frac{b^2 - a^2}{a^2 b^2}\right) - k\ln\left(\frac{b}{a}\right)\right]. \tag{14.33}$$

Here b is the distance to the interface between the moving and static Bingham substance in a very large body of the substance. The negative sign in front of the angular velocity, Ω, is a result of the coordinate system we selected. The change of angular velocity with radial distance, $d\dot{\theta}/dr$ in eq. (14.31), should be negative for the problem here, because the angular velocity decreases with increasing distance, r, from the center of rotation.

Application of Solution to Apparatus Used to Measure Properties of Lava.—In actual viscometers, eq. (14.33) for the Bingham viscosity must be corrected for conditions not accounted for in our theoretical analysis. For example, the shaft that rotates the spindle, used by Shaw et al. (*14*) to measure properties of lava, is in the lava so that the shear it transmits to the lava must be accounted for. In addition, the solution we derived assumes that there is no drag on the distal end of the spindle. If the diameter of the spindle is small relative to its length, however, the correction for the end drag is small. Finally, friction within the viscometer apparatus itself must be accounted for. Thus the total torque, M_T, is the sum of the torque caused by friction within the apparatus, plus the drag of the end of the spindle and along the sides of the shaft, M_f, plus the torque delivered to the lava, M. That is,

$$M = M_T - M_f.$$

Viscometers are calibrated, in order to determine M_f, using fluids with known properties. Shaw et al. used asphalt liquids for calibration of their apparatus.

Two independent measurements of torque and angular velocity are required in order to determine the strength and viscosity of a Bingham substance. We see in eq. (14.33) that there are three unknowns in any test: viscosity, η_b, strength, k, and radius of the flowing material, b. However, we can eliminate either b or k by substituting eq. (14.30) into eq. (14.33).

If practical, the most desirable method of measuring k is to determine the torque at which the spindle begins to rotate. Then we could use eq. (14.29) to calculate k, because when the spindle begins to rotate, the radius of the flowing lava coincides with the radius of the spindle itself. Shaw et al. (*14*), however, used an apparatus in which the torque was not readily adjustable so they resorted to another method to calculate the constants. They determined angular velocities corresponding to two different torques and solved simultaneously for the Bingham constants.

Viscosity and Strength of Basaltic Lava.—Fig. 14.13 shows relations between torque and angular velocity of the spindle used by Shaw et al. to measure properties of lava in the lava lake in Makaopuhi Crater, Hawaii. The solid line represents measurements taken at depths of 6.8 meters and the dashed line represents measurements taken at depths of 7.5 meters below the top of the lava crust. The numbers next to the data points refer to the sequence of measurement. Thus the torque was increased between points 1, 2, and 3 and decreased between points 3 and 4 for the measurements made at 6.8 meters depth. Points 1, 2, and 3 define a

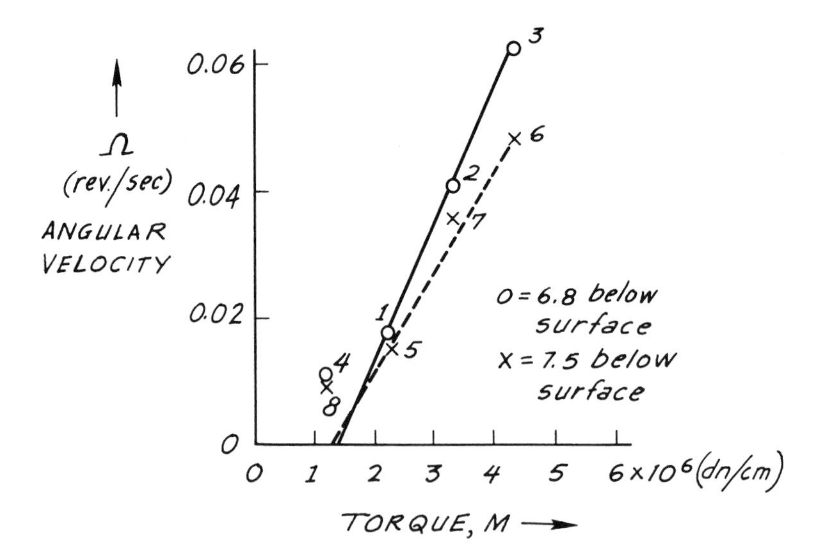

Figure 14.13. Relation between torque and angular velocity of rotational viscometer used to measure properties of lava in Makaopuhi lava lake, Hawaii. Sequence of measurements is indicated by numbers (after ref. *14*, p. 245).

straight line relation between torque and angular velocity, so that they seem to indicate that the lava behaves approximately like the Bingham model. Also, a straight line through the points intercepts the torque axis at a value of about 1.5×10^6 dn/cm, so that there is a yield strength, as is required by Bingham's model. However, when the torque was reduced, for the fourth test, the shear rate was higher than expected. According to the line defined by points 1, 2, and 3, the torque applied during the fourth test should have caused no rotation of the spindle in the lava. The torque applied during test 4 was below the initial yield torque of the lava. This behavior suggests that the lava was slightly thixotropic, which is just another way of saying that the strength of the lava decreased as it was sheared.

Shaw et al. (*14*) calculated the coefficient of Bingham viscosity and the strength for the two positions of the spindle:

Depth below top of lava crust	Bingham viscosity (η_b)	Strength (k)
6.8 meters	6500 poises	1200 dn/cm^2
7.5 meters	7500 poises	700 dn/cm^2

Glen's experiments with ice cylinders

J. W. Glen deformed cylinders of polycrystalline ice in order to propose a rheological equation for glacial ice (*3*). He prepared cylindrical samples of ice by filling a cylindrical mold with air-free, distilled ice water and by seeding the water with tiny crystals scraped from the pipes of a refrigerator. The specimens were then placed in a refrigerator and loaded with weights. Their deformation was measured with a dial gauge.

Glen found that his ice specimens deformed rapidly when the weights were first applied but then deformed more slowly. It is this latter deformation that is of primary interest in glacier flow, because it allows glaciers to flow continuously under uniform loading (*3*).

The results of one of his tests is shown in Fig. 14.14, which shows a relation between time and shortening of an ice cylinder under a constant load (*3*). The strain rate could be calculated from the curve in Fig. 14.14 by dividing the shortening by the original length of the cylinder and by the time required to attain the shortening. If the ratio of the shortening and the original length are plotted with respect to time, the rate of strain would be the slope of the resulting strain-time curve. After about 60 hours, the curves relating shortening and time become nearly straight (see Fig. 14.14A), so that the strain rate reaches a constant value. If the value of strain rate is plotted with respect to the load applied to the specimen, divided by the cross-sectional area of the specimen—that is, if the strain rate is plotted with respect to the average axial stress—the resulting curve is an estimate of the stress-strain-rate curve for steady flow of ice. Such a curve is shown in Fig. 14.14B.

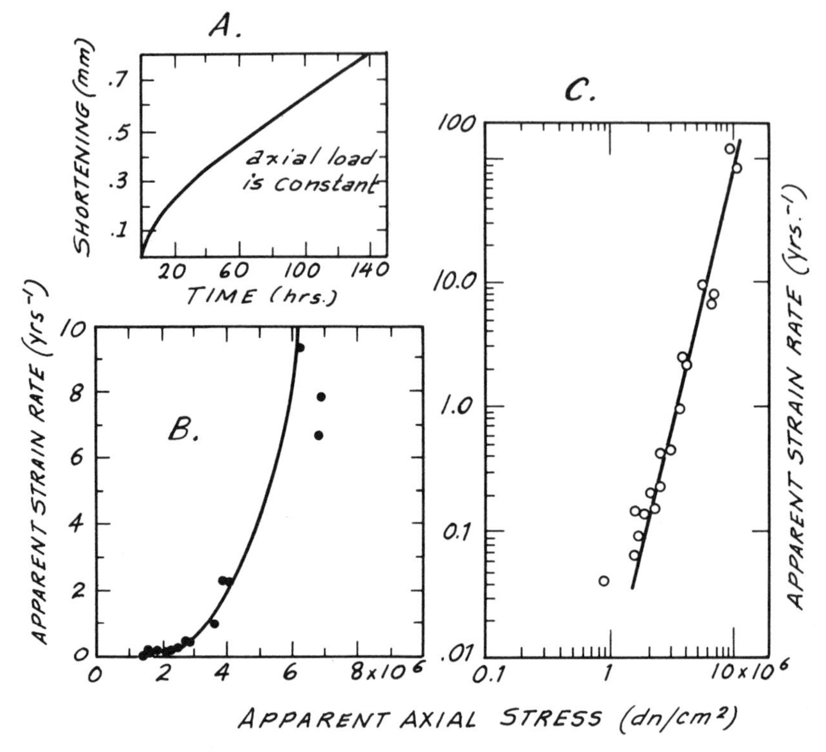

Figure 14.14. Behavior of polycrystalline ice under uniaxial loading (after ref. *3*).

If ice were a simple Newtonian fluid, the resulting curve, Fig. 14.14B, would be a straight line through the origin. The actual curve, however, is quite nonlinear, so that a Newtonian rheological model poorly describes the steady deformation of the ice tested by Glen and, presumably, the deformation of a glacier.

Glen proposed a nonlinear rheological model to describe the deformation of his ice specimens (*3*). If the points shown in Fig. 14.14B are replotted on log-log graph paper, a straight line closely describes the relation between the logarithm of normal strain rate and the logarithm of average normal axial stress. That is,

$$\ln \dot{\epsilon} = n \ln (k_0 \sigma).$$

This equation is identical in form to

$$\dot{\epsilon} = k_0 \sigma^n. \tag{14.34}$$

For the ice Glen tested, $n \approx 4$ and $k_0 \approx 0.0074$ if σ is in bars and $\dot{\epsilon}$ is in years^{-1}.

Velocity and rheology of Saskatchewan Glacier, Alberta, Canada

Mark Meier's study (*9*) of Saskatchewan Glacier, in Alberta, Canada, is one of the most detailed analyses of velocities within and on the surface of a glacier (*9,11*).

The purpose of Meier's work was to measure velocity fields within a glacier and to relate the measurements to results of theoretical studies and laboratory tests. He selected Saskatchewan Glacier because its shape is relatively simple, simplifying mathematical analysis.

Saskatchewan Glacier.—Saskatchewan Glacier is a tongue of ice extending about five miles eastward from Columbia Icefield. The total drop in altitude between the source and the terminus of the tongue is about 2400 feet and the width of the tongue ranges from about 5000 feet at the source to 2000 feet near the terminus. The ice tongue has an average surface slope of about 3°23′, ranging up to about 8.5° near the source and up to 12° near the terminus (*9*).

The shape of the channel of the glacier was determined by seismic reflection (Fig. 14.15). The channel is roughly U-shaped, and can be approximated by a circular cylinder with a diameter of about one mile and with an inclination of about 0°46′ eastward. The surface of the ice is inclined at an angle of about 3.5° eastward, so that the semicircular channel is only partly filled with ice (*9*).

Surface Velocities of Ice Tongue.—Surface movement of the glacier was measured by resurveying stations on the glacier. The stations consisted of stakes placed in holes bored eight to nine feet into the ice. The positions of most of the stakes were determined once or twice each summer for three years. Some of the stakes were resurveyed several times during a summer, however, in order to measure the magnitudes of velocity fluctuations.

Meier (*9*) found that there are marked short-term velocity fluctuations so that the flow is comprised of a large number of pulses of short wavelength. The velocity fluctuations decrease with increasing time intervals between measurements so that the variations of velocity from day to day commonly are greater than the variations from year to year. The maximum surface velocity was 383 feet per year in the source area and it decreased to about 12 feet per year near the terminus.

The directions of the velocities of different parts of the surface of the glacier generally are not parallel to the bottom or sides of the channel or to the top of the glacier. Unlike a lava flow or a debris flow, the quantity of ice in a glacier increases or decreases by addition or subtraction of ice in the direction normal to the ice surface. In the source area of a glacier, ice is added to the glacier each year and elsewhere ice is being removed by wastage, largely surface melting, of the ice.

The vertical component of velocity of the measuring stakes placed near the center of the glacier surface was downward, about ten feet per year, in the sourceward end of the ice tongue and it was upward, about two feet per year, in the terminal end of the tongue. The horizontal component along the center line was about 200 to 250 feet per year in the sourceward part and about 100 feet per year in the terminal part of the tongue. Near the edges of the sourceward part of the tongue the velocities are not parallel to the channel sides, rather they deviate slightly toward the channel sides (Fig. 14.16) (*9*).

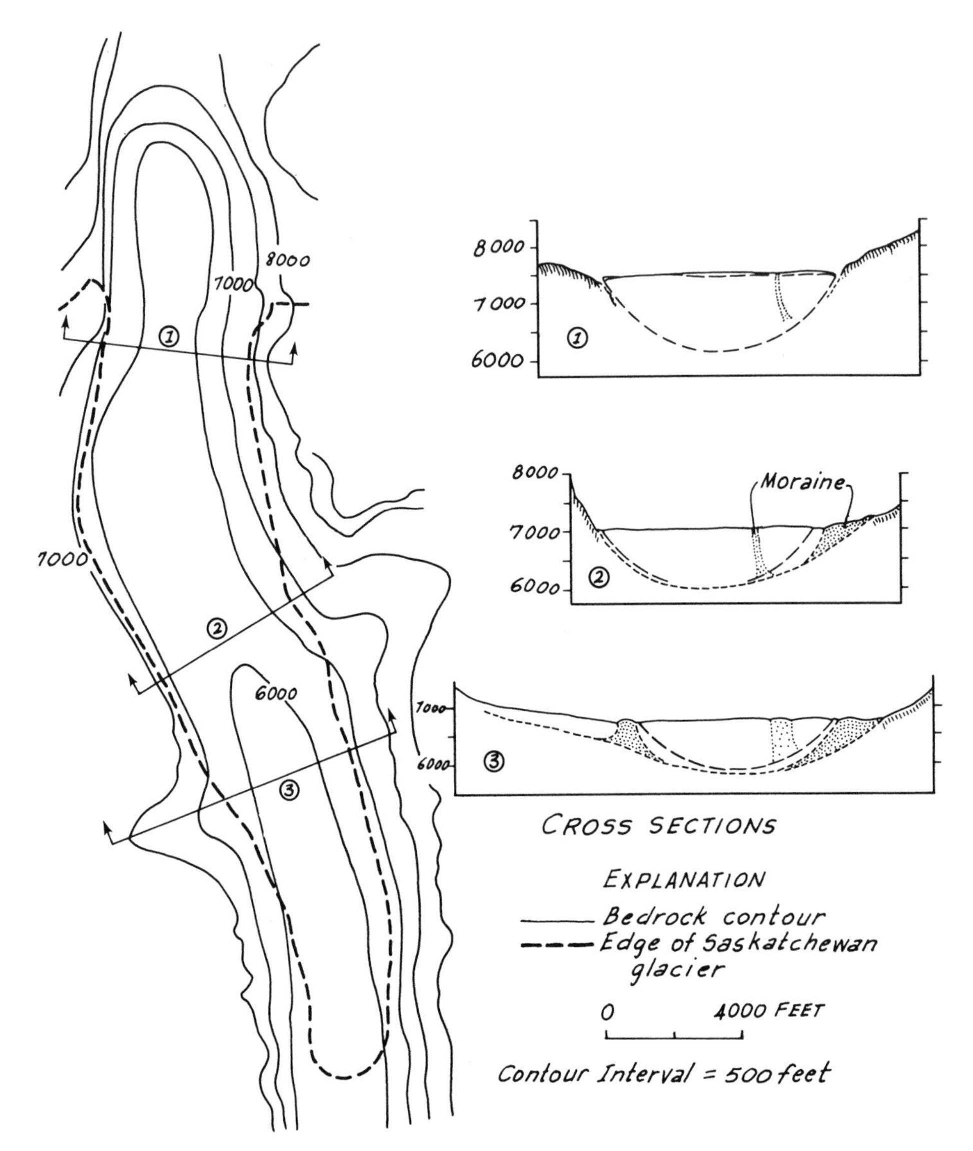

Figure 14.15. Shape of Saskatchewan glacier (after ref. *9*, p. 9).

Figure 14.17 shows profiles of surface velocities at several places in the western end of the tongue. The velocity is nearly constant across the central half of the tongue and it rapidly decreases near each edge of the tongue. Meier estimated that the slippage between the ice and the channel is less than five percent along the southern edge of the tongue (see Fig. 14.17).

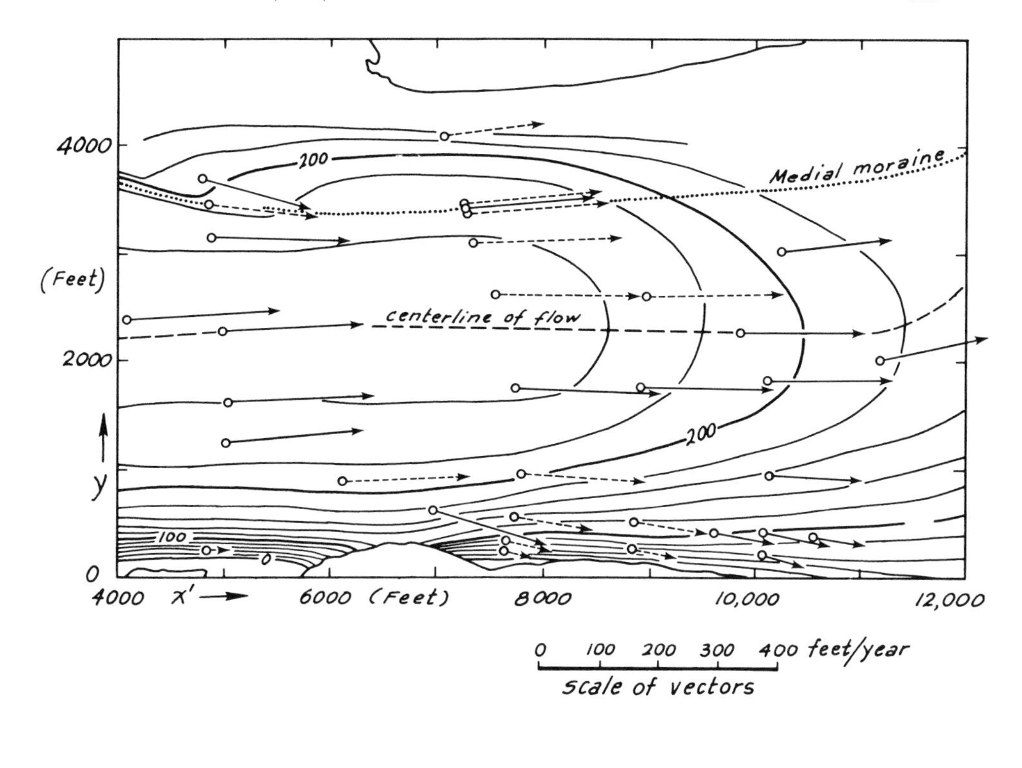

Figure 14.16. Horizontal velocities in part of Saskatchewan glacier (after ref. *9*, p. 21).

Velocities within Ice.—The velocity distribution within the ice tongue was measured at three places along the center line of the western end of Saskatchewan Glacier. An aluminum pipe with a heated point was sunk vertically into the ice and displacements were calculated using measurements of subsequent inclinations of the pipe. The orientations of the pipe were determined at 25-foot intervals of depth with an inclinometer.

One pipe was successfully emplaced to a depth of 150 feet in 1952. Between 1952 and 1954 the pipe was slightly tilted and bent, resulting in 0.71 foot of displacement of the top relative to the base of the pipe (Fig. 14.18). During the two-year period the base of the pipe was displaced 428 feet down the glacier. Figure 14.18 shows that even the upper part of Saskatchewan Glacier is sheared and that a rigid plug such as we seem to see in debris flows is absent in the glacier. The glacier is about

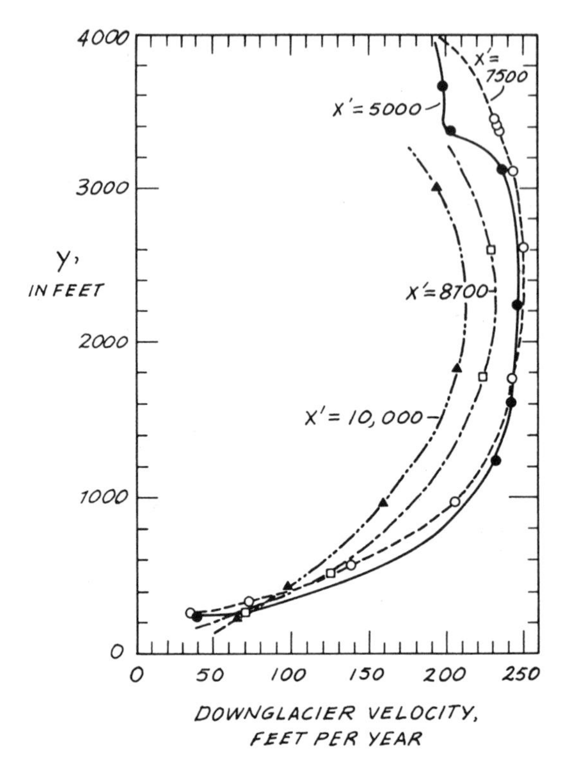

Figure 14.17. Distributions of surface velocities at several places on Saskatchewan glacier (after ref. *9*). Positions, x', of profiles are shown in Fig. 14.16.

1500 feet thick at the place where the measurements were made so that the pipe penetrated about one-tenth of the thickness of the glacier. The amount of shear is small, but measureable, in the upper few tens of feet of the glacier (*9*).

Rheological Model for Ice.—Meier (*9*) proposed a rheological model for the ice in the Saskatchewan Glacier by analyzing transverse profiles of the surface velocities (Fig. 14.17) and vertical profiles of internal velocities of the glacier (Fig. 14.18). Rate of shear is directly related to the slope of the velocity profile at any point along the profile. Maximum shear stress, the other variable required in deriving a rheological equation, was calculated at various points along the surface and subsurface profiles by making certain assumptions about the conditions of glacier flow. One important assumption in Meier's analysis is that the velocities are parallel to the sides of the channel in which the ice moves. We saw in Fig. 14.16 that there are components of velocity normal to the channel sides in the actual glacier, but, to a first approximation, these components can be neglected. Another important assumption is that flow is steady. However, the ice of Saskatchewan Glacier definitely pulsates. The magnitude of the error caused by assuming steady flow is unknown.

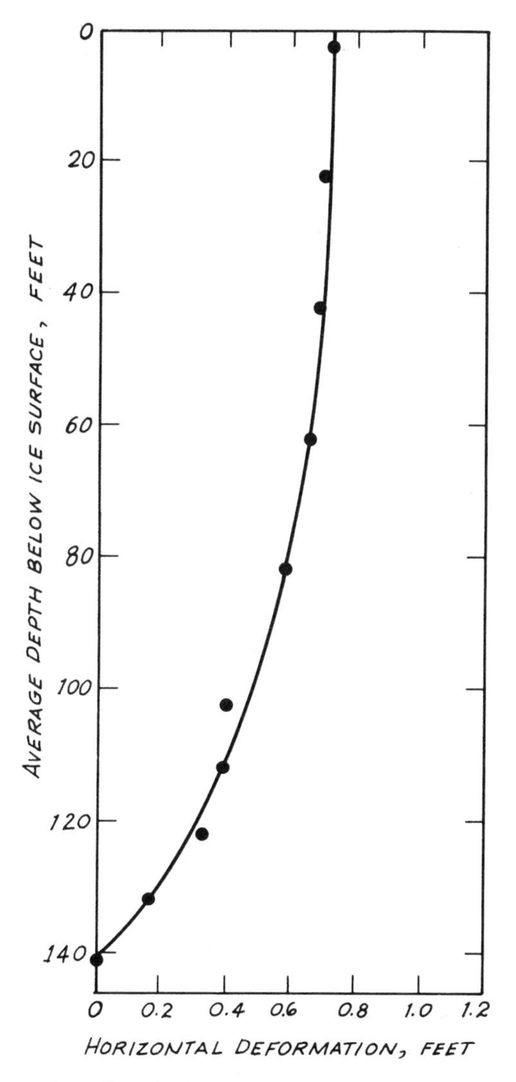

Figure 14.18. Deformation of vertical borehole in Saskatchewan glacier between 6 Aug. 1952 and 5 Aug. 1954 (after ref. *9*).

Assuming that flow is steady and is parallel to the channel sides, the shear stress on planes parallel to the channel sides is

$$\sigma_s = \left(\frac{\gamma r}{2}\right) \sin \delta, \tag{14.35}$$

where γ is the unit weight of ice, r is the distance from the center of a circular channel, and δ is the slope angle of the surface and the base of the glacier. Using an equation similar to this one and using the velocity profiles measured on and in Saskatchewan Glacier, Meier estimated that a rheological model that best fits his data is of the form (Fig. 14.19)

$$\dot{\epsilon}_s = C_1\sigma_s + C_2\sigma_s^n. \quad (14.36)$$

The value of the exponent, n, in eq. (14.36), is about 4.5, according to Meier (*9*).

Exercise

Solve eq. (14.36) for flow in infinitely wide channels and in semicircular channels.

Ans. (infinitely wide channels):

$$\dot{w} = C_2\left(\frac{\gamma}{2}\right) \sin \delta[H^2 - y^2] + C_3 \frac{(\gamma \sin \delta)^n}{n+1} [H^{(n+1)} - y^{(n+1)}].$$

Comparison of properties of basaltic lava, glacial ice, and debris

We showed in Chapters 12 and 13 that certain forms of depositional features of glaciers, lava flows, and debris flows are strikingly similar, although the sizes of the features may be different. Now we have shown that the properties of flowing lava, ice, and debris can be approximated with the same or similar rheological models, so that we can better understand the similarities in depositional features.

The strengths of slurries, consisting of mixtures of quartz sand, kaolin clay, and water, are a few hundred dynes per square centimeter, whereas the strength of basaltic lava at Makaopuhi Crater was a few thousand dynes per square centimeter. The viscosities of the slurries are about three orders of magnitude greater than the viscosity of warm water, which is about 10^{-2} poise, whereas the viscosity of the lava was about five orders of magnitude greater than that of water.

A nonlinear or pseudoplastic model has been suggested as a good approximation to the behavior of glacial ice. Thus, Glen (*3*) has suggested a rheological model of the form

$$\dot{\epsilon}_s = A\sigma_s^n, \quad (14.37)$$

where n is about 4; Meier has suggested a model of the form

$$\dot{\epsilon}_s = C_1\sigma_s + C_2\sigma_s^n, \quad (14.38)$$

where n is about 4.5. The same model might be adopted for kaolin-sand slurries, but for them the exponent, n, is probably on the order of 8 to 12.

One aspect of the velocity measurements recorded in Fig. 14.17 indicates that ice flows similarly to debris and lava. The velocities are nearly constant across large regions on each side of the center line of the glacier surface. Nearly one-half the total

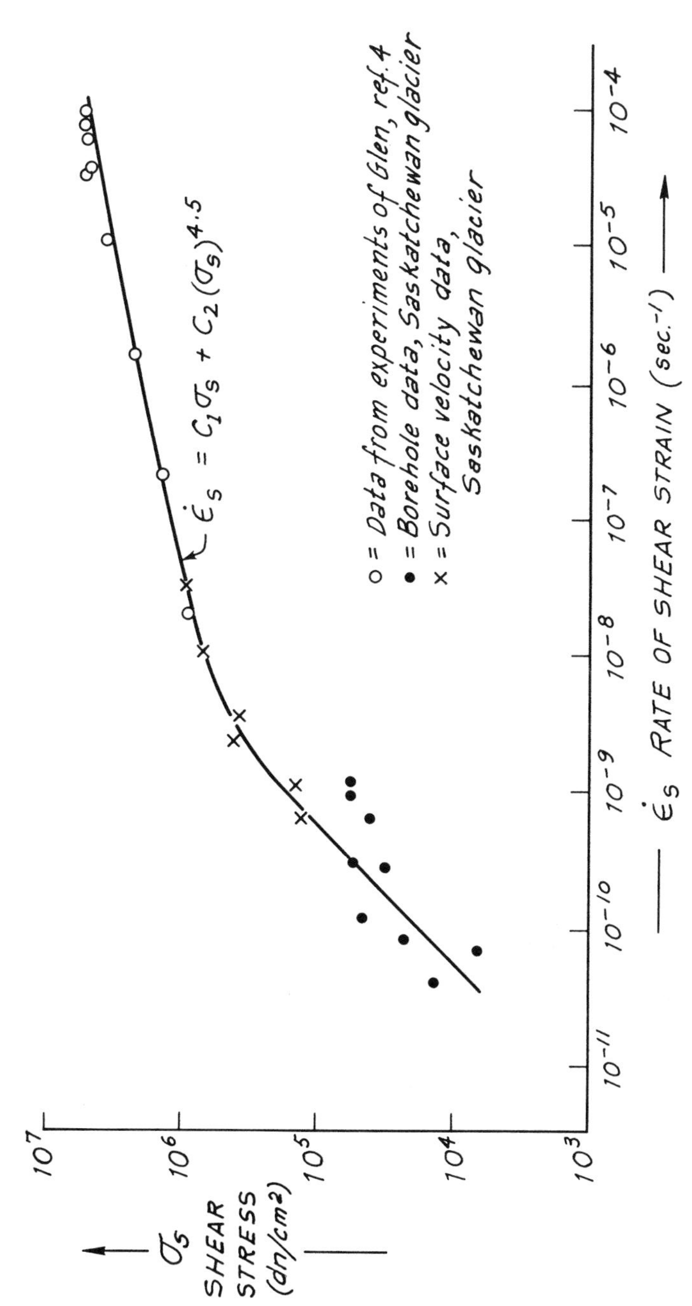

Figure 14.19. Relation between shear stress and rate of shear strain for polycrystalline ice (after ref. 9).

width of the glacier is moving at about the same velocity, so that the center of the glacier is essentially in a state of "plug" flow. This behavior means that glacial ice is essentially nondeforming, and may even be static, in areas of low shear. It is this behavior which will allow us to speculate about the reason why glacial valleys tend to be U-shaped in cross-section, as we will in the next chapter.

I think that we can safely conclude, on the basis of the measurement reported in this chapter, that ice, lava, and debris behave qualitatively similarly but quantitatively differently. That is, they have similar types of rheological constants but they have different magnitudes of rheological constants.

References cited in Chapter 14

1. Curry, R. R., 1966, "Observation of Alpine Mudflows in the Tenmile Range, Central Colorado": *Geol. Soc. America Bull.*, V. 77, p. 771–776.
2. Eirich, F. R., Ed., 1967, *Rheology*: Academic Press, N.Y.
3. Glen, J. W., 1952, "Experiments on the Deformation of Ice": *Jour. Glaciology*, V. 2, p. 111–114.
4. ———, 1955, "The Creep of Polycrystalline Ice": *Royal Soc. London, Proc.*, Ser. A., V. 228, p. 519–538.
5. Johnson, A. M., 1965, "A Model for Debris Flow: Ph.D. dissertation, The Pennsylvania State Univ., Univ. Park, Penna.
6. ———, and Hampton, M. A., 1968, "Subaerial and Subaqueous Flow of Slurries": Progress Report, U.S. Geological Survey; Branner Library, Stanford University, Stanford, Calif.
7. ———, and Hampton, M. A., 1969, "Subaerial and Subaqueous Flow of Slurries": Final Report, U.S. Geological Survey; Branner Library, Stanford University, Stanford, Calif.
8. Krauskopf, K. B., 1948, "Lava Movement at Paricutín Volcano, Mexico": *Geol. Soc. America Bull.*, V. 59, p. 1267–1284.
9. Meier, M. F., 1960, "Mode of Flow of Saskatchewan Glacier, Alberta, Canada": U.S. Geological Survey Prof. Paper 351.
10. Nye, J., 1952, "The Mechanics of Glacier Flow": *Jour. Glaciology*, V. 2, p. 82–93.
11. Patterson, W. S. B., and Savage, J. C., 1963, "Measurements on Athabasca Glacier Relating to the Flow Law of Ice": *Jour. Geophys. Research*, V. 68, p. 4537–4543 (see also, p. 4513–4536).
12. Reiner, M., 1960, *Deformation, Strain and Flow*: Interscience Publishers, N.Y.
13. Sharp, R. P., and Nobles, L. H., 1953, "Mudflow of 1941 at Wrightwood, Southern California": *Geol. Soc. America Bull.*, V. 64, p. 547–560.
14. Shaw, H. R., Wright, T. L., Peck, D. L., and Okamura, R., 1968, "The Viscosity of Basaltic Magma: Analysis of Field Measurements in Makaopuhi Lava Lake, Hawaii": *Amer. Jour. Science*, V. 266, p. 225–264.

15

Chapter Sections

Introduction
Flow of Newtonian Substance in Rectangular Channel
Navier-Stokes Equations
Methods of Computing Flow
Poisson and Laplace equations
Separation of variables
Method of reflection
Fourier series
Calculation of Fourier coefficients
Velocity distribution
Flow of Bingham Substance in Rectangular Channel
Two-Dimensional Bingham Substance
Shear Stress Distribution
Dead Regions
Velocity Distribution
Flow of Debris in Channel of Triangular Cross-Section
Formation of Lateral Deposits
U-Shape of Glacial Valleys
References Cited in Chapter 15

15

U-Shape of Valleys

Introduction

Nearly every budding geologist has learned in a course in physical geology that one of the distinct differences between valleys carved by valley glaciers and valleys worn down by streams is their cross-sectional profiles. Usually, it is explained, streams attack the bottoms of their deep valleys, undercutting slopes and causing mass wasting of the valley sides so that the valleys in time become crudely V-shaped. Glaciers, the story continues, carve the sides as well as the bottoms of their valleys, so that valleys they have visited are characteristically U-shaped in cross-section (Fig. 15.1). Further, valley glaciers usually move down valleys that have been formed by stream erosion, so that glacial erosion usually is superimposed upon topography already sculptured by streams. For example, whereas streams can easily move around overlapping spurs projecting into mountain canyons, glaciers tend to remove the ends of the spurs, straightening the canyons, because ice is more "viscous" than water and, therefore, ice has difficulty negotiating sharp corners (*10*).

One clue to the origin of the peculiar shape of glacial valleys probably is the mechanisms by which glaciers carve valleys. If we read on in our text in physical geology, we learn that glaciers erode their valleys principally by two mechanisms: abrasion and plucking, which is a process of incorporation of joint blocks into the moving mass. Rocks of assorted sizes and shapes are embedded in the bottom and sides of a glacier so that, as the glacier advances, it acts as a geologic rasp, scratching, gouging, and generally abrading the surfaces over which it rides. Rock underlying glaciers is eroded by plucking.

We might well ask, however, whether we have thus explained why glacial valleys are U-shaped. Why, for example, does not a glacier erode a pre-existing, V-shaped valley uniformly, all along the parts of the valley sides actually in contact with the glacial rasp, causing the valley to increase in depth and width while it

Figure 15.1. Glacial valley on eastern side of Sierra Nevada Mountains, California (photograph by U.S. Geological Survey).

maintains the same V-shape? Why, in short, *are* typical glacial valleys smoothly curved in transverse profile, rather than V-shaped?

The explanation we will examine later on for the U-shape of glacial valleys is based on the assumption that the flow of glacial ice is similar to the flow of debris. Channels that have been visited by debris flows tend to be U-shaped, as shown in Fig. 15.2, for example, and I think we can explain why. Perhaps the same explanation applies to glacial valleys.

In order to explain the U-shape of debris-flow channels we will study the flow of a Bingham substance in rectangular and triangular channels. The pattern of flow is the key to the explanation.

It is interesting to note, in passing, that lava tubes, channels, and caves tend to be elliptic in cross-section. This observation becomes significant when we examine it in the light of the similarities between the rheological properties of lava and debris, as discussed in Chapter 14.

Figure 15.2. U-shaped debris-flow channel near Cris Wicht Camp, Surprise Canyon, Panamint Range, California.

Flow of Newtonian substance in rectangular channel

The equations describing the velocity distribution of a Bingham substance in channels with sharp corners, such as rectangles or triangles, are rather difficult to derive, since they necessitate reference to many concepts we were able to avoid when discussing flow patterns in circular and infinitely wide channels. A method of deriving a solution for the flow of a Bingham substance in a rectangular channel arises from study of the solution for the flow of a Newtonian substance in a rectangular channel. We will consider first a Newtonian substance.

Navier-Stokes Equations.—When we derived solutions for the flow of a Newtonian substance in a circular channel we separately employed equations of equilibrium, rheological model, and simple boundary conditions. The more complex boundary conditions needed to describe flow in channels of more complex shapes require that we immediately combine the rheological model and the equilibrium equations. The resulting equations are called the *Navier-Stokes equations* (*7,8*). They were explained in Chapter 7. The rheological equations for three-dimensional flow of a Newtonian substance are [eqs. (7.89) and (7.97)]

$$\left.\begin{aligned} \sigma_{xx} &= p + 2\eta\frac{\partial \dot{u}}{\partial x}, \\ \sigma_{yy} &= p + 2\eta\frac{\partial \dot{v}}{\partial y}, \\ \sigma_{zz} &= p + 2\eta\frac{\partial \dot{w}}{\partial z}, \end{aligned}\right\} \tag{15.1}$$

where

$$p = \frac{\sigma_{xx} + \sigma_{yy} + \sigma_{zz}}{3},$$

and

$$\left.\begin{aligned} \sigma_{xy} &= \eta\left(\frac{\partial \dot{v}}{\partial x} + \frac{\partial \dot{u}}{\partial y}\right), \\ \sigma_{yz} &= \eta\left(\frac{\partial \dot{v}}{\partial z} + \frac{\partial \dot{w}}{\partial y}\right), \\ \sigma_{xz} &= \eta\left(\frac{\partial \dot{u}}{\partial z} + \frac{\partial \dot{w}}{\partial x}\right). \end{aligned}\right\} \tag{15.2}$$

The equations of equilibrium are [eqs. (7.5) to (7.7)]

$$\frac{\partial \sigma_{xx}}{\partial x} + \frac{\partial \sigma_{xy}}{\partial y} + \frac{\partial \sigma_{xz}}{\partial z} + X = 0, \tag{15.3}$$

$$\frac{\partial \sigma_{yy}}{\partial y} + \frac{\partial \sigma_{xy}}{\partial x} + \frac{\partial \sigma_{yz}}{\partial z} + Y = 0, \tag{15.4}$$

$$\frac{\partial \sigma_{zz}}{\partial z} + \frac{\partial \sigma_{xz}}{\partial x} + \frac{\partial \sigma_{yz}}{\partial y} + Z = 0, \tag{15.5}$$

where X, Y, and Z are body forces in the x-, y-, and z-directions, respectively.

If the rheological equations, eqs. (15.1) and (15.2), are substituted into the equilibrium equations, eqs. (15.3) to (15.5), the following form of the Navier-Stokes equations can be derived:

$$\eta\nabla^2\dot{u} + \eta\frac{\partial \xi}{\partial x} + \frac{\partial p}{\partial x} + X = 0,$$

$$\eta\nabla^2\dot{v} + \eta\frac{\partial \xi}{\partial y} + \frac{\partial p}{\partial y} + Y = 0,$$

$$\eta\nabla^2\dot{w} + \eta\frac{\partial \xi}{\partial z} + \frac{\partial p}{\partial z} + Z = 0,$$

where ∇^2 is the Laplacian operator,

$$\nabla^2 = \left(\frac{\partial^2}{\partial x^2} + \frac{\partial^2}{\partial y^2} + \frac{\partial^2}{\partial z^2}\right),$$

and where

$$\xi = \frac{\partial \dot{u}}{\partial x} + \frac{\partial \dot{v}}{\partial y} + \frac{\partial \dot{w}}{\partial z}$$

is the rate of volumetric expansion, which we assumed to be zero in deriving eqs. (15.1). Thus the pertinent form of the Navier-Stokes equation is, for an incompressible viscous substance (*7,8*),

$$\left.\begin{aligned} &\eta\nabla^2\dot{u} + \frac{\partial p}{\partial x} + X = 0, \\ &\eta\nabla^2\dot{v} + \frac{\partial p}{\partial y} + Y = 0, \\ \text{and} \quad &\\ &\eta\nabla^2\dot{w} + \frac{\partial p}{\partial z} + Z = 0. \end{aligned}\right\} \tag{15.6}$$

The right-hand sides of eqs. (15.6) would contain accelerative terms if motion were non-uniform (e.g., see refs. *7* and *8*).

Exercise

Solve eqs. (15.6) for one-dimensional flow, in which $\dot{u} = 0 = \dot{v} = X$.

Ans.: $p = -\gamma y \cos \delta,$

$$\dot{w} = \frac{H^2 - y^2}{2}\frac{\gamma}{\eta}\sin \delta,$$

where H is the depth of flow and δ is the slope angle.

Methods of Computing Flow.—Suppose that a Newtonian fluid is flowing uniformly through a long channel of rectangular cross-section (Fig. 15.3). The fluid moves parallel to the axis of the channel so that the velocities normal to the channel sides are zero. That is, $\dot{u} = \dot{v} = 0$, if z is parallel to the channel axis. Under these flow conditions, the Navier-Stokes equations, eqs. (15.6), become

$$\eta\nabla^2\dot{u} = 0 = \frac{\partial p}{\partial x} + X,$$

$$\eta\nabla^2\dot{v} = 0 = \frac{\partial p}{\partial y} + Y, \tag{15.7}$$

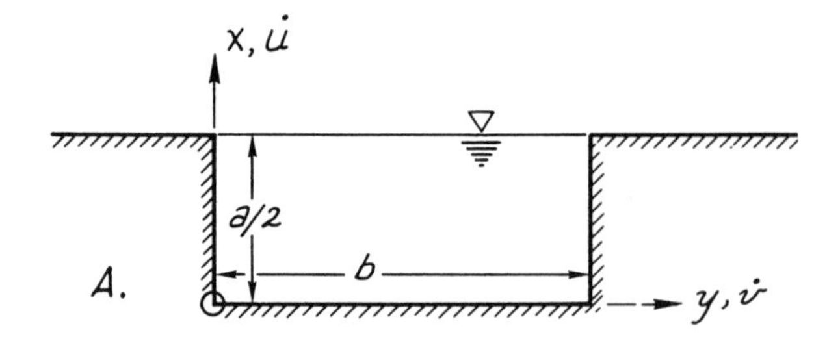

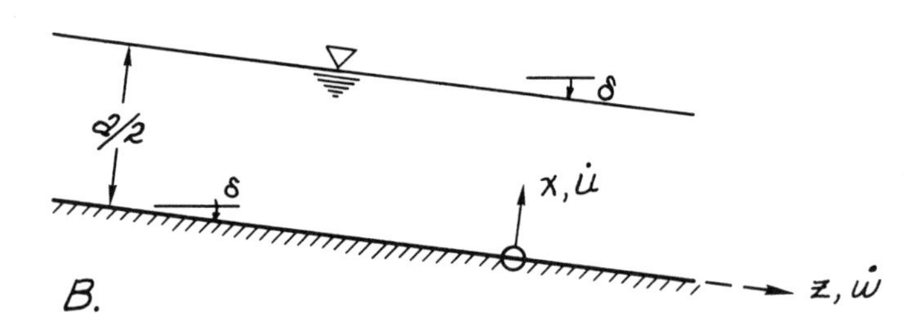

Figure 15.3. Parameters of rectangular channel.
A. Cross-section.
B. Longitudinal section.

and

$$\eta \nabla^2 \dot{w} + \frac{\partial p}{\partial z} + Z = 0, \tag{15.8}$$

where

$$\nabla^2 \dot{w} = \frac{\partial^2 \dot{w}}{\partial x^2} + \frac{\partial^2 \dot{w}}{\partial y^2} + \frac{\partial^2 \dot{w}}{\partial z^2}. \tag{15.9}$$

For steady flow, however, the down-channel velocity is the same for any channel cross-section, so that

$$\frac{\partial^2 \dot{w}}{\partial z^2} = 0,$$

and eqs. (15.8) and (15.9) become, when combined,

$$\left(\frac{\partial^2 \dot{w}}{\partial x^2} + \frac{\partial^2 \dot{w}}{\partial y^2}\right) + \frac{\partial p}{\partial z} + Z = 0. \tag{15.10}$$

Thus, the velocity in the z-direction, $\dot{w}$, varies with respect to the x- and y-directions, but not with respect to the z-direction.

The change of pressure within the channel can be evaluated as follows: The upper surface of the fluid is under atmospheric pressure, so that the pressure is independent of position along the surface. That is, $\partial p/\partial y = \partial p/\partial z = 0$ there. We could show that the pressure is independent of z and y everywhere within the channel. It depends only upon the vertical distance below the surface of the fluid, and thus the change of pressure with respect to the x-direction, according to eq. (15.7), is

$$\frac{\partial p}{\partial x} = -X = -\gamma \cos \delta,$$

and

$$p = -x\gamma \cos \delta + C_1 = -\left(\frac{a}{2} - x\right)\gamma \cos \delta.$$

The component of body force in the z-direction is $-\gamma \sin \delta$, if γ is the unit weight of the fluid and δ is the slope angle of the surface of the fluid and the base of the channel. Therefore, eq. (15.10) simplifies to

$$\frac{\partial^2 \dot{w}}{\partial x^2} + \frac{\partial^2 \dot{w}}{\partial y^2} = \frac{\gamma}{\eta} \sin \delta. \tag{15.11}$$

Exercises

1. Show that $\dot{w} = C_0[(x/a)^2 + (y/b)^2 - 1]$ is a solution to eq. (15.11). Here a, b, and C_0 are constants. What is the value of C_0?
2. The answer to problem 1 is the solution for the velocity distribution of a viscous material in an elliptic channel with minor axis $2a$ and major axis $2b$. Show that three special cases are solutions for velocities in circular channels, infinitely wide channels, and infinitely deep channels.

Poisson and Laplace equations.—Equation (15.11) is a kind of Poisson equation (*11*), which is of the form

$$\nabla^2 \dot{w} = f(x, y), \tag{15.12}$$

where $f(x, y)$ is some function of x and y or a constant. The function in eq. (15.11) is a constant, so that we can write eq. (15.11) as

$$\nabla^2 \dot{w} = k_0, \tag{15.13}$$

where $k_0 = \gamma/\eta \sin \delta$.

We can solve eq. (15.13) by finding a general solution to the homogeneous part, the part to the left of the equal sign, called Laplace's equation (*11*),

$$\nabla^2 \dot{w} = 0, \tag{15.14}$$

and then by adding a particular solution that satisfies the Poisson equation.

Separation of variables.—The general solution to the Laplace equation can be derived by separating variables (*11*). We saw an example of the method of separation of variables in Chapter 6; here we will skip several details of the derivations. Let

$$\dot{w} = f(x)g(y), \tag{15.15}$$

where f is an unknown function of x and g is an unknown function of y. If we substitute eq. (15.15) into eq. (15.14), we produce two total differential equations,

$$\frac{d^2f}{dx^2} - fC^2 = 0, \tag{15.16}$$

and

$$\frac{d^2g}{dy^2} + gC^2 = 0, \tag{15.17}$$

where C is a constant.

The solutions to eqs. (15.16) and (15.17) are

$$f = C_1 \exp(Cx) + C_2 \exp(-Cx),$$

and

$$g = C_3 \exp(Ciy) + C_4 \exp(-Ciy),$$

as we have shown several times in previous chapters.

The expression for f can be rewritten in terms of hyperbolic cosines and sines and the expression for g can be rewritten in terms of trigonometric sines and cosines, as we have shown before (see Chapters 3 and 7). Therefore, the general solution, eq. (15.15), to the Laplace equation is

$$\dot{w} = [C_5 \cosh(Cx) + C_6 \sinh(Cx)][C_7 \cos(Cy) + C_8 \sin(Cy)], \tag{15.18}$$

where $C_5 \cdots C_8$ are arbitrary constants.

A particular solution to the Poisson equation can be guessed by inspection of eq. (15.13). Try

$$\dot{w} = k_0\left(\frac{y^2}{2} + C_{11}y\right) = \int k_0(y + C_{11})\, dy,$$

where C_{11} is a constant. There is no slippage between the Newtonian substance and the edges of the channel, so that (Fig. 15.3)

$$\dot{w} = 0 \qquad \text{at } y = b,$$

$$C_{11} = -b/2,$$

and a particular solution is

$$\dot{w} = k_0\left(\frac{y^2}{2} - \frac{by}{2}\right). \tag{15.19}$$

By combining eqs. (15.18) and (15.19), we have the general solution to the Poisson equation, eq. (15.13):

$$\dot{w} = \frac{k_0}{2}(y^2 - by) + [C_5 \cosh (Cx) + C_6 \sinh (Cx)][C_7 \sin (Cy) + C_8 \cos (Cy)] \tag{15.20}$$

Method of reflection.—In order to proceed further with the solution, we need to examine the boundary conditions in detail. According to Fig. 15.4, the boundary conditions are

$$\dot{w} = 0 \qquad \text{at } y = 0, \text{ at } y = b, \text{ and at } x = 0. \tag{15.21}$$

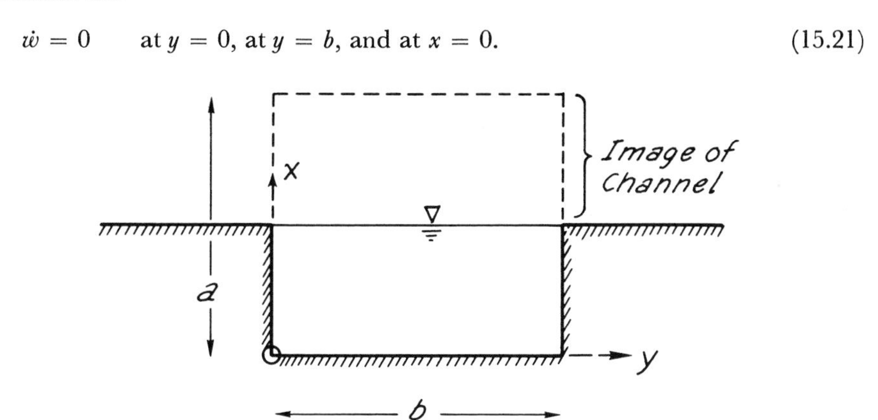

Figure 15.4. Reflection of channel across upper surface of flow. The problem of calculating velocities in a closed rectangular conduit is the same as the problem of calculating velocities in an open rectangular channel.

Thus, there are four arbitrary constants, $C_5 \cdots C_8$, but only three equations involving boundary conditions, eqs. (15.21). In order to obtain another boundary condition, we can reflect* the channel, Fig. 15.3, across the axis, $x = a/2$, Fig. 15.4. This procedure does not affect the theoretical velocities because the velocities are independent of hydrostatic pressures. The velocity distribution will be symmetric across the surface, $x = a/2$, so that there will be no shear on this surface.

By reflection we have added another boundary condition, at $x = a$,

$$\dot{w} = 0.$$

Now we have five unknowns, $C_5 \cdots C_8$, and C, and four independent equations. But we can eliminate two of the constants by inspection. According to the second boundary condition, eqs. (15.21),

$$\dot{w} = 0 \qquad \text{at } y = b.$$

* Eq. (15.13) is classified as elliptic (e.g., see ref. *11*) because it is similar to the equation of an ellipse in cartesian coordinates. For the solution of an elliptic equation, the boundaries must be closed surfaces. By reflection we have closed the bounding surfaces of the fluid.

Substituting these values into eq. (15.20),

$$\dot{w} = 0 - 0 + [C_5 \cosh(Cx) + C_6 \sinh(Cx)][C_7 \cos(Cy) + C_8 \sin(Cy)].$$

In order for the right-hand side of this equation to be zero, either

$$C_7 = 0 \qquad \text{if } C = m\pi/b,$$

where m is an integer, because $\sin(m\pi) = 0$, or

$$C_8 = 0 \qquad \text{if } C = m\pi/2b, \text{ because } \cos(m\pi/2) = 0.$$

Either possibility is acceptable so let

$$C = m\pi/b$$

and

$$C_7 = 0.$$

Then eq. (15.20) becomes

$$\dot{w} = \frac{k_0}{2}(y^2 - by) + \sin\left(\frac{m\pi y}{b}\right)\left[C_9 \cosh\left(\frac{m\pi x}{b}\right) + C_{10} \sinh\left(\frac{m\pi x}{b}\right)\right]. \qquad (15.22)$$

The remaining problem is to select values of C_9 and C_{10} so that the boundary conditions

$$\dot{w} = 0 \qquad \text{at } x = 0 \quad \text{and} \quad x = a$$

are satisfied. This turns out to be much more difficult than one might guess. The difficulty is a result of the rectangular shape of the channel. However, we can deal with such a problem because sums of solutions to a differential equation are solutions to the above equation. Also, we can represent complicated forms, such as rectangular, triangular, or even irregular forms, by an infinite series of sine and cosine terms. Putting these two ideas together, we can imagine that the solution for the velocities in a rectangular channel can be written in the form (see ref. *8*; ref. *5*, p. 472; and ref. *11*, pp. 102 and 219):

$$\dot{w} = \frac{k_0}{2}(y^2 - by) + \sum_{m=1}^{\infty} \sin\left(\frac{m\pi y}{b}\right)\left[Am \cosh\left(\frac{m\pi x}{b}\right) + Bm \sinh\left(\frac{m\pi x}{b}\right)\right], \qquad (15.23)$$

where Am and Bm are constants, having different values for each m. Equation (15.23) is in the form of a Fourier series, which we shall examine now in order to learn how to calculate the constants, Am and Bm.

Fourier series.—Apparently any single-valued periodic function, $f(y)$, can be expressed by a Fourier series, which is the name of the sum of an infinite number of sine and cosine waves of successively higher order (*1,5,11*). In order to illustrate the use of a Fourier series to represent a discontinuous function and to illustrate the method of calculating the coefficients, Am and Bm, we will derive the Fourier series that describes a rectangular waveform. This has more than academic interest to us because we are concerned with flow in a rectangular channel.

Fourier's series is

$$f(y) = a_0 + \sum_{m=1}^{\infty} \left[Am \cos\left(\frac{m\pi y}{b}\right) + Bm \sin\left(\frac{m\pi y}{b}\right)\right], \tag{15.24}$$

where a_0 is a constant. Equation (15.24) actually does represent the waveform shown in Fig. 15.5 because the coefficients, $Am = A_1, A_2, \ldots, A_m$ and $Bm = B_1, B_2, \ldots, B_m$, can be determined, first by solving eq. (15.24) point by point for successive values of $f(y)$ and then by solving the resulting equations simultaneously. For the rectangular waveform shown in Fig. 15.5, the successive values of $f(y)$ are (*5*)

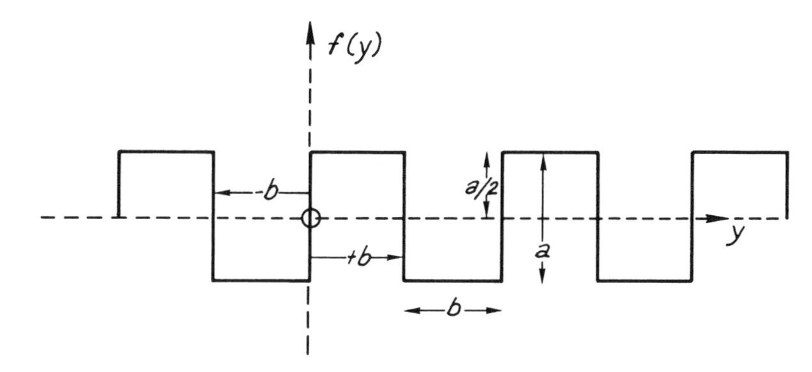

Figure 15.5. Parameters of rectangular waveform.

$$\left.\begin{array}{ll} \text{value of } f(y) & \text{value of } y \\ -a/2 & \text{from } -b \text{ to } 0 \\ a/2 & \text{from } \quad 0 \text{ to } b \\ -a/2 & \text{from } \quad b \text{ to } 2b \\ \vdots & \quad\vdots \\ \text{etc.} & \quad\text{etc.} \end{array}\right\} \tag{15.25}$$

We can eliminate the constant, a_0, and the coefficients of the cosine term, Am, in eq. (15.24) by selecting the origin of coordinates, so that at $y = 0, f(y) = 0$, as is shown in Fig. 15.5. Thus, the pertinent form of eq. (15.24) is

$$\begin{aligned} f(y) &= \sum_{m=1}^{\infty} \left[Bm \sin\left(\frac{m\pi y}{b}\right)\right] \\ &= B_1 \sin(\pi y/b) + B_2 \sin(2\pi y/b) + \cdots B_m \sin(m\pi y/b) \cdots \end{aligned} \tag{15.26}$$

Now we will show that the values of Bm can be computed with the *Fourier formula*,

$$Bm = \frac{2}{b}\int_0^b f(y)\sin(m\pi y/b)\,dy. \tag{15.27}$$

In order to evaluate any coefficient, B_m, multiply each side of eq. (15.26) by $\sin(m\pi y/b)$ and integrate from $-b$ to $+b$. For example, if B_2 is to be calculated (*1,11*),

$$\int_{-b}^{b} f(y)\sin(2\pi y/b)\,dy = \int_{-b}^{b} B_1 \sin(\pi y/b)\sin(2\pi y/b)\,dy + \int_{-b}^{b} B_2 \sin^2(2\pi y/b)\,dy. \tag{15.28}$$

In order to evaluate eq. (15.28), let us consider two integrals:

$$\left.\begin{aligned} 2\int_0^b B_m \sin(m\pi y/b)\sin(n\pi y/b)\,dy &= 0,\\ 2\int_0^b B_m \sin^2(m\pi y/b)\,dy &= B_m b. \end{aligned}\right\} \tag{15.29}$$

Therefore, if m and n are unequal integers, in the first of eqs. (15.29), the integral over a full wavelength is zero. It follows that

$$2\int_0^h f(y)\sin\left(\frac{m\pi y}{b}\right)dy = Bmb,$$

which is Fourier's formula, eq. (15.27).

Now we can derive a Fourier series for a rectangular waveform, using eqs. (15.25) and (15.27). According to eqs. (15.25),

$$f(y) = a/2$$

for odd integers, so that, according to eq. (15.27),

$$Bm = (a/b)\int_0^b \sin(m\pi y/b)\,dy = \left(\frac{2a}{m\pi}\right).$$

When we substitute this expression for the coefficients into the Fourier series, eq. (15.26),

$$\begin{aligned} f(y) = (2a/\pi)[&\sin(\pi y/b) + (1/3)\sin(3\pi y/b)\\ &+ (1/5)\sin(5\pi y/b)\cdots(1/m)\sin(m\pi y/b)\cdots], \end{aligned} \tag{15.30}$$

where m is an odd integer.

Figures 15.6A through 15.6C show the approach, through the fifth order of approximation, of the Fourier series to the rectangular waveform. The more terms that are added, the closer the Fourier form approaches the rectangular form.

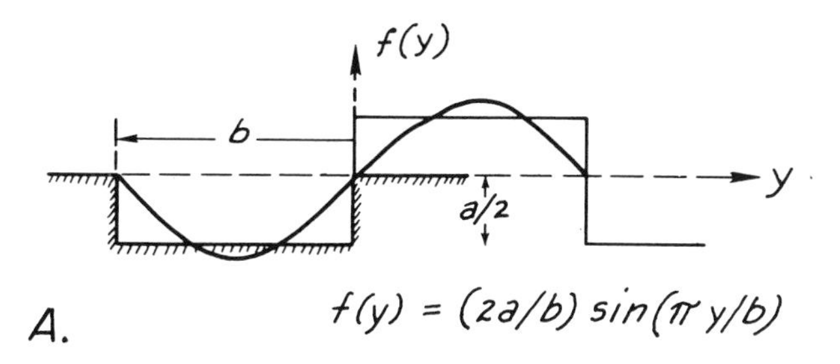

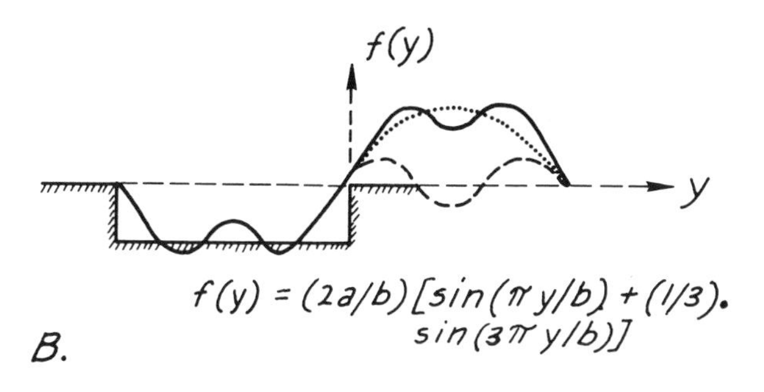

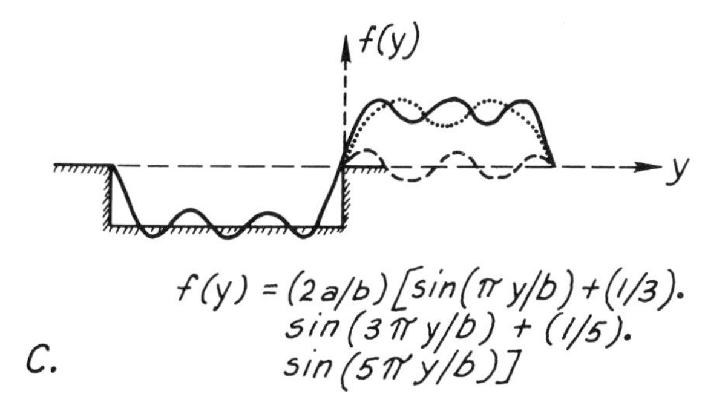

Figure 15.6. Approximation of rectangular waveform by first five terms of Fourier series.

Calculation of Fourier coefficients.—Now we have the background needed to compute values for the coefficients, Am and Bm, in eq. (15.23), our solution for the velocities of a Newtonian substance in a channel of rectangular cross-section.

One of the boundary conditions is that, at $x = 0$, $\dot{w} = 0$, Fig. 15.4, so that eq. (15.23) becomes

$$\dot{w} = 0 = \frac{k_0 y}{2}(y - b) + \sum_{m=1}^{\infty} \sin(m\pi y/b)[Am \cosh(0)]. \tag{15.31}$$

The first term in eq. (15.31) corresponds with $f(y)$ in Fourier's formula, eq. (15.27). Thus,

$$f(y) = \frac{k_0 y}{2}(y - b),$$

so that

$$Am = -\frac{2}{b}\int_0^b \frac{k_0 y}{2}(y - b)[\sin(m\pi y/b)]\,dy,$$

or

$$Am = -\frac{2k_0 b^2}{m^3\pi^3}[\cos(m\pi) - 1]. \tag{15.32}$$

The boundary condition at $x = a$ is $\dot{w} = 0$, Fig. 15.4, so that

$$\dot{w} = 0 = \frac{k_0 y(y - b)}{2} + \sum_{m=1}^{\infty} \sin(m\pi y/b)[Am\cosh(m\pi a/b) + Bm\sinh(m\pi a/b)]. \tag{15.33}$$

Solving eqs. (15.31) and (15.33) simultaneously,

$$Am = Am\cosh(m\pi a/b) + Bm\sinh(m\pi a/b),$$

$$Bm = \frac{Am[1 - \cosh(m\pi a/b)]}{\sinh(m\pi a/b)}. \tag{15.34}$$

Velocity distribution.—Now we have an exact solution* for velocities of a Newtonian substance in a rectangular channel.† The solution is

$$\dot{w} = \frac{\gamma\sin\delta}{2\eta}y(y - b) + \sum_{m=1}^{\infty}\sin(m\pi y/b)[Am\cosh(m\pi x/b) + Bm\sinh(m\pi x/b)]. \tag{15.35}$$

The coefficients, Am and Bm, are given by eqs. (15.32) and (15.34).

Figure 15.7 shows velocity contours for a rectangular channel in which the width-to-depth ratio is four to one, that is, in which $b/a = 2$. The velocity contours are in nondimensional form.

* For numerical methods of determining coefficients and of estimating the error caused by using a finite number of terms, see ref. *5*, p. 494.

† Because of the m^3 term in the denominators of the coefficients, Am and Bm, the solution for the velocities in a rectangular channel, eq. (15.23), converges for all pertinent values of x and y that are within the limits $0 \le x \le a$ and $0 \le y \le b$. The term in brackets in eq. (15.33) becomes arbitrarily small as $m \to \infty$ because in this case Bm approaches $-Am$ and $\cosh(m\pi a/b)$ approaches $\sinh(m\pi a/b)$. The derivatives up to the third order of $\dot{w}$ with respect to x or y also converge, so that eq. (15.23) is differentiable up to the third order (see ref. *2*, p. 28–30).

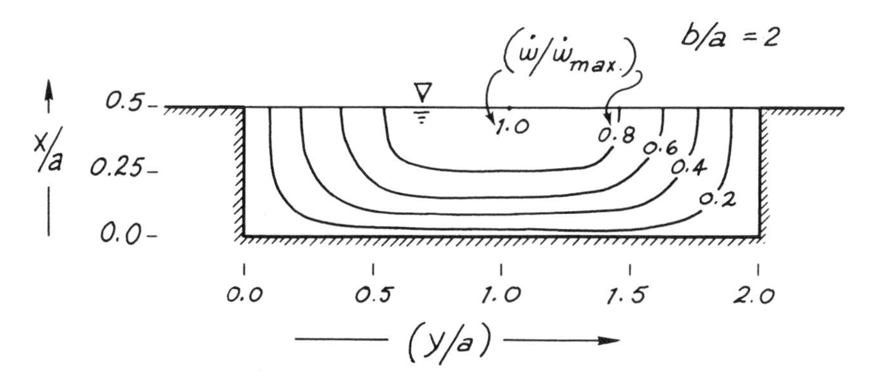

Figure 15.7. Velocity contours for flow of a viscous substance in a rectangular channel. The contours and the distances within the channel are in dimensionless form (after ref. *8*, p. 220).

Flow of Bingham substance in rectangular channel

We will use the methods illustrated in the solution for velocities of a Newtonian substance flowing in a rectangular channel in order to derive a solution for velocities of a Bingham substance in a rectangular channel. The yield strength of the Bingham substance introduces some difficult mathematical problems but the results of the analysis are so interesting that solutions of the problems seem to be worth the effort. The solution derived here is based on a solution that I derived (*3*) earlier, using finite-difference equations and a computer.

Two-Dimensional Bingham Substance.—According to the one-dimensional Bingham model that we have used in previous derivations, the shear stress is equal to the shear strength, k, plus the Bingham viscosity, η_b, times the rate of shear strain, $\dot{\epsilon}_s$,

$$|\sigma_s| = k + \eta_b \dot{\epsilon}_s; \qquad [|\sigma_s| \geq k].$$

We need to modify the rheological model in order to treat two-dimensional problems.

We have already derived the appropriate relations between stress and strain rate for the viscous component of the Bingham model. They are

$$(\sigma_{xy})_\eta = \eta_b\left(\frac{\partial \dot{u}}{\partial z} + \frac{\partial \dot{w}}{\partial x}\right) = \eta_b \frac{\partial \dot{w}}{\partial x},$$
$$(\sigma_{yz})_\eta = \eta_b\left(\frac{\partial \dot{v}}{\partial z} + \frac{\partial \dot{w}}{\partial y}\right) = \eta_b \frac{\partial \dot{w}}{\partial y}, \qquad (15.36)$$

for steady flow in a channel. Also, we know that the appropriate form of the equilibrium equation is

$$\frac{\partial \sigma_{xz}}{\partial x} + \frac{\partial \sigma_{yz}}{\partial y} = -\gamma \sin \delta. \tag{15.37}$$

The remaining problem is to derive relations for the part of the shear stress due to the strength, k, of the Bingham substance.

Consider a cross-section of a rectangular channel, *abcd*, filled with flowing Bingham substance (Fig. 15.8A). On any plane containing the z-axis, inclined to the x- and y-directions, and represented by its projection as line, S, in Fig. 12.8A, there acts a shear stress, σ_s, in the z-direction, due to the strength of the substance. Let θ be defined as the angle between the y-axis and the S-line at any point along the S-line (Fig. 15.8B). Then, according to Fig. 15.8B, which represents an infinitesimal

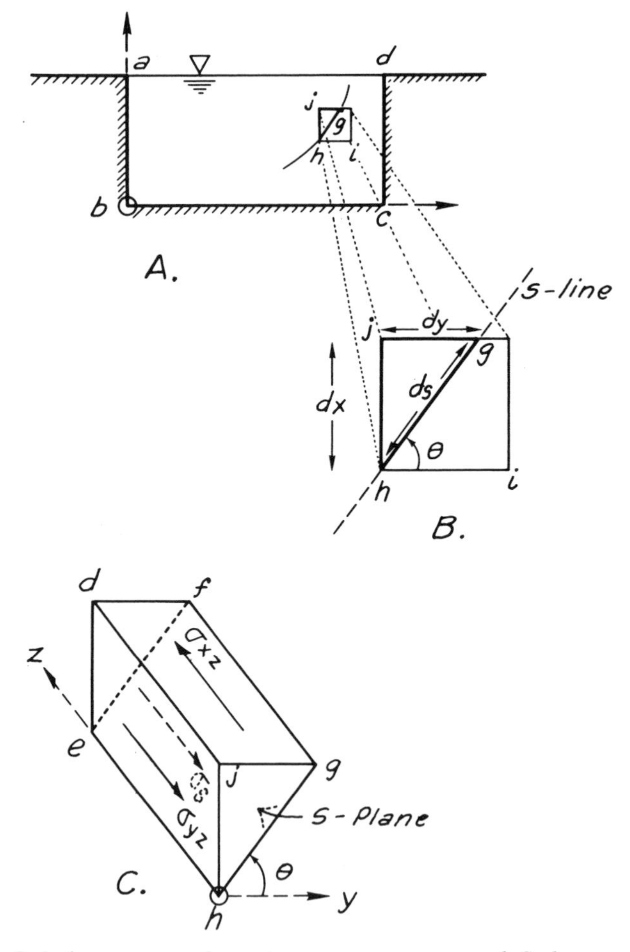

Figure 15.8. Relations among shear stresses on x-z, y-z, and S-planes near a point in channel filled with Bingham substance.

element at some point along the S-line, and Fig. 15.8C, which represents an infinitesimal wedge projected onto the x–y plane of Fig. 15.8B as triangle hjg,

$$\left[\sum F_z = 0\right]; \qquad \sigma_s \, ds - \sigma_{xz} \, dy + \sigma_{yz} \, dx = 0,$$

or

$$\sigma_s = \sigma_{xz} \cos \theta - \sigma_{yz} \sin \theta. \tag{15.38}$$

Equation (15.38) is satisfied if the components of shear stress are given by the equations

$$\sigma_{xz} = \sigma_s \cos \theta,$$

and (15.39)

$$\sigma_{yz} = -\sigma_s \sin \theta.$$

The directions of σ_{xz} and σ_{yz} in Fig. 12.8C are in accord with their positive directions, as defined in Chapter 5.

Now k is the maximum magnitude of shear stress a Bingham substance can withstand without flowing, so that surfaces of maximum shear stress are of interest. We can determine the orientation of the plane of maximum shear by taking the derivative of eq. (15.38) with respect to θ and equating the result to zero. Thus,

$$\frac{d\sigma_s}{d\theta} = 0 = \sigma_{xz} \sin \theta + \sigma_{yz} \cos \theta.$$

If α is the angle for which the shear stress is maximum or minimum,

$$\alpha = \pm \tan^{-1} \left| \frac{\sigma_{yz}}{\sigma_{xz}} \right|. \tag{15.40}$$

The two-dimensional rheological model for the shear strength term of the Bingham model, therefore, is

$$\left.\begin{aligned} \sigma_{xz} &= k \cos \alpha, \\ \sigma_{yz} &= -k \sin \alpha. \end{aligned}\right\} \tag{15.41}$$

Combining eqs. (15.41) and (15.36), we have the two-dimensional Bingham model,

$$\left.\begin{aligned} \sigma_{xz} &= k \cos \alpha + \eta_b \frac{\partial \dot{w}}{\partial x}, \\ \sigma_{yz} &= -k \sin \alpha + \eta_b \frac{\partial \dot{w}}{\partial y}. \end{aligned}\right\} \tag{15.42}$$

The general rheological equations of a Bingham substance can be written in an alternative but equivalent form (see ref. *4*).

Shear Stress Distribution.—If we substitute eqs. (15.42) into the equilibrium equation, eq. (15.37), we have a form of the Poisson equation:

$$\frac{\partial^2 \dot{w}}{\partial x^2} + \frac{\partial^2 \dot{w}}{\partial y^2} = \frac{1}{\eta_b}\left[-\gamma \sin \delta + k\left(\sin \alpha \frac{\partial \alpha}{\partial x} + \cos \alpha \frac{\partial \alpha}{\partial y}\right)\right]. \tag{15.43}$$

We know we can derive a solution to the homogeneous part of this differential equation. Although a particular solution might be difficult to find, the most difficult problem is to define boundary conditions. There will be regions of no flow in the channel corners, where shear stresses are less than the yield strength of the substance. We need to define the edges of these regions before we can solve for velocities because the edges are places where velocities are zero.

Probably the most satisfactory method of solving eq. (15.43) is to define a shear-stress function, Γ (capital gamma), such that (*3*)

$$\left.\begin{aligned} \sigma_{xz} &= \frac{\partial \Gamma}{\partial x} + k \cos \alpha + \eta_b \frac{\partial \dot{w}}{\partial x}, \\ &\text{and} \\ \sigma_{yz} &= \frac{\partial \Gamma}{\partial y} - k \sin \alpha + \eta_b \frac{\partial \dot{w}}{\partial y}. \end{aligned}\right\} \tag{15.44}$$

Then the equation of equilibrium, eq. (15.37), becomes

$$\frac{\partial^2 \Gamma}{\partial x^2} + \frac{\partial^2 \Gamma}{\partial y^2} = -\gamma \sin \delta. \tag{15.45}$$

Equation (15.45) is identical in form to the Navier-Stokes equation for steady flow in a conduit. Only the derivatives of Γ are pertinent to the problem of calculating velocities, eqs. (15.44), so that we can arbitrarily let $\Gamma = 0$ on the boundaries of the channel. Thus the pertinent solution to eq. (15.45) is eq. (15.35),

$$\Gamma = \frac{\gamma \sin \delta}{2} y(y - b) + \sum_{m=1}^{\infty} \sin (m\pi y/b)[Am \cosh (m\pi x/b) + Bm \sinh (m\pi x/b)], \tag{15.46}$$

where [eqs. (15.32) and (15.34)]

$$\left.\begin{aligned} Am &= \frac{-2\gamma \sin \delta b^2}{m^3 \pi^3} [\cos (m\pi) - 1] \\ &\text{and} \\ Bm &= \frac{Am[1 - \cosh (am\pi/b)]}{\sinh (am\pi/b)}. \end{aligned}\right\} \tag{15.47}$$

Dead Regions.—We can calculate the shear stresses within the channel by performing the appropriate differentiations on the solution for the shear-stress function, eq. (15.46),

$$\sigma_{xz} = \frac{\partial \Gamma}{\partial x} = \sum_{m=1}^{\infty} (m\pi/b) \sin (m\pi y/b)[Am \sinh (m\pi x/b) + Bm \cosh (m\pi x/b)] \quad (15.48)$$

$$\sigma_{yz} = \frac{\partial \Gamma}{\partial y} = (\gamma \sin \delta/2)(2y - b) + \sum_{m=1}^{\infty} (m\pi/b) \cos (m\pi y/b)$$
$$[Am \cosh (m\pi x/b) + Bm \sinh (m\pi x/b)]. \quad (15.49)$$

The magnitudes of maximum shear stresses can be calculated at every point within the rectangular channel by using the relation

$$(\sigma_s)_{\max} = \sqrt{(\sigma_{xz}^2 + \sigma_{yz}^2)} \quad (15.50)$$

and eqs. (15.48) and (15.49).

The expressions for the components of shear stress [eqs. (15.48) and (15.49)] involve infinite series; therefore, we cannot exactly determine the values of maximum shear stress. The easiest way to determine shear stresses is to use a computer to calculate magnitudes of shear stresses at several points within and along the edges of rectangular channels. Then we construct contours of equal shear stress by interpolating between the known values. Obviously, we do not use an infinite number of terms in the calculations. The number of terms actually used depends upon the rapidity of the convergence. Thus, we select an arbitrarily small value of variation and continue adding terms in the infinite series until the addition of another term changes the value of the quantity less than does the arbitrarily selected variation. Eight or ten terms usually are sufficient to cause the series to converge closely enough to a fixed value.

Figure 15.9 shows contours of equal shear stress within a rectangular channel with a width-to-depth ratio of 2, that is, with $b/a = 1$. The diagram is in dimensionless form and the values of shear stresses can be calculated with the relation

$$(\sigma_s)_{\max} = (\sigma_s^*)_{\max} \gamma a \sin \delta.$$

Shear stresses are high along the upper left and lower central parts of the channel boundaries and they are low near the corners. They are zero in the corners and in the center of the upper, free surface. The sharp corners of the channel apparently affect the stress distribution for a considerable distance into the flow.

Figure 15.10 shows shear stresses within another rectangular channel, this one with a width-to-depth ratio of 4, that is, $b/a = 2$. We can imagine, by comparing Figs. 15.9 and 15.10, that, as the width of the channel becomes large with respect to its depth, the effects of the corners on the stress distribution become restricted to a small area of the entire channel.

Figure 15.11A illustrates this effect of width-to-depth ratio, b/a, on the shear stresses along the line of symmetry of the channel, from top to base (*4*). Apparently the maximum shear stress on the base of the channel is relatively independent of the cross-sectional shape of the channel if the channel width is greater than about four times the channel depth (i.e., b/a greater than 2). The shear stress in a semicircular

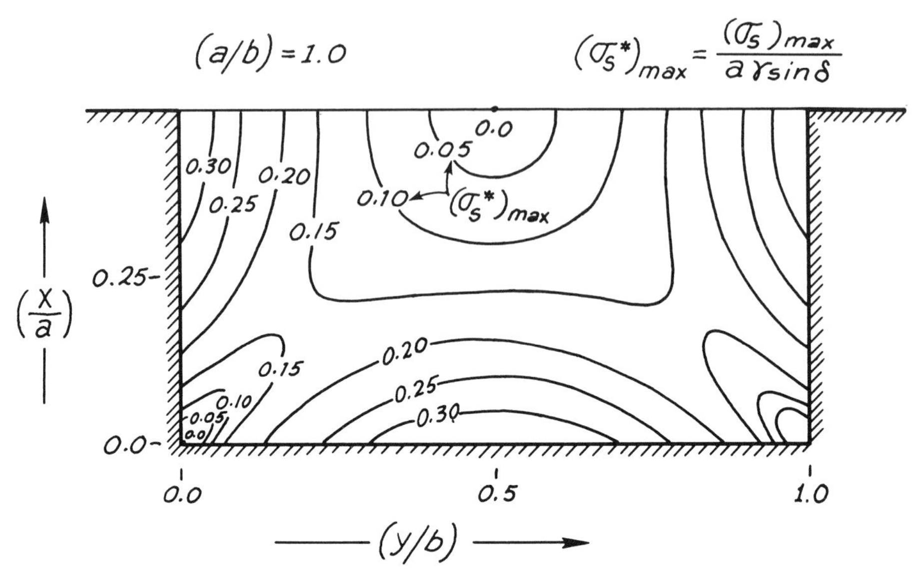

Figure 15.9. Contours of equal shear stress in rectangular channel with a width-to-depth ratio of two, $(a/b) = 1$.

channel also is shown in the figure, and it is shown to be less than that in a channel of semisquare cross-section.

If we consider the distribution of maximum shear stress along the diagonal of a semisquare channel, however, we see a quite different picture (Fig. 15.11B). Whereas in a semicircular channel the shear stress increases linearly from the center to the edge, in a semisquare channel the maximum shear stress first increases with increasing distance from the center of the channel, then goes through a maximum, and

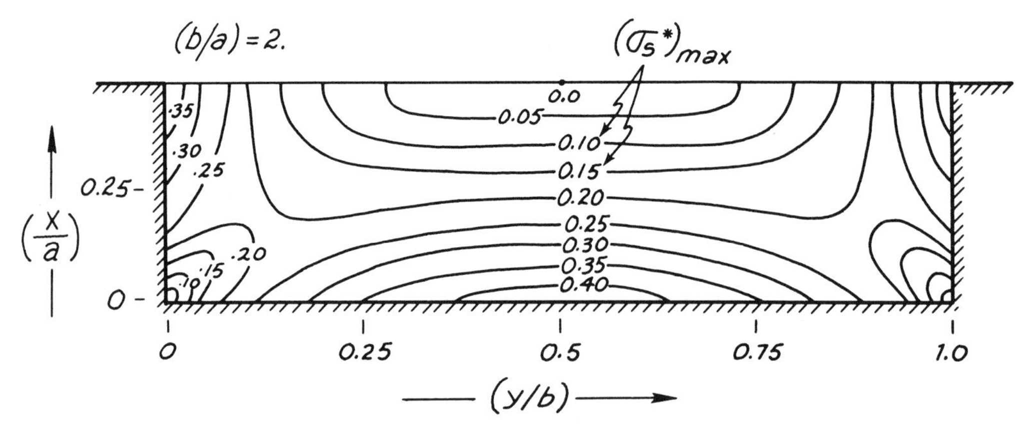

Figure 15.10. Contours of equal shear stress in rectangular channel with a width-to-depth ratio of 4, $(b/a) = 2$.

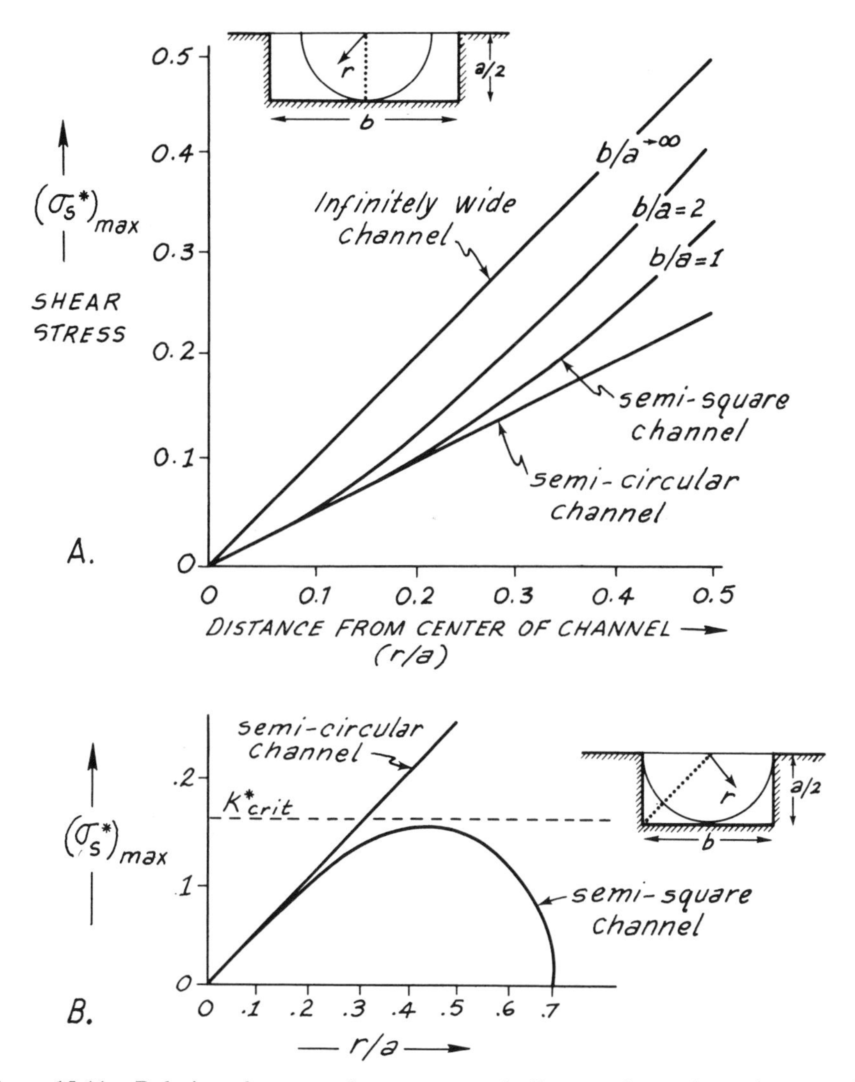

Figure 15.11. Relations between shear stress and distance from channel center for rectangular and circular channels.

A. Relations along line from center to base of channels.

B. Relations along line from center to corner of square channel. k^*_{crit} indicates a critical shear strength. Strengths greater than this value but less than 0.25 should cause flow to hang up in square channel but not in circular channel.

finally decreases to zero at the corner. This is shown clearly in the shear stress contours in Figs. 15.9 and 15.10.

A result of this peculiarity in the distribution of shear stresses in rectangular channels is the occurrence of "dead regions," or regions of no flow at the corners of

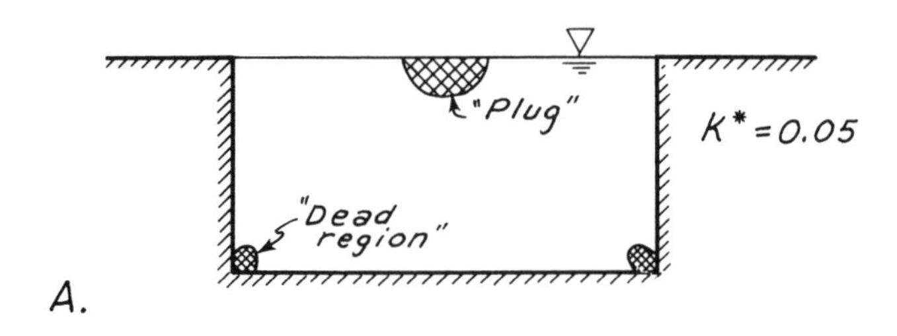

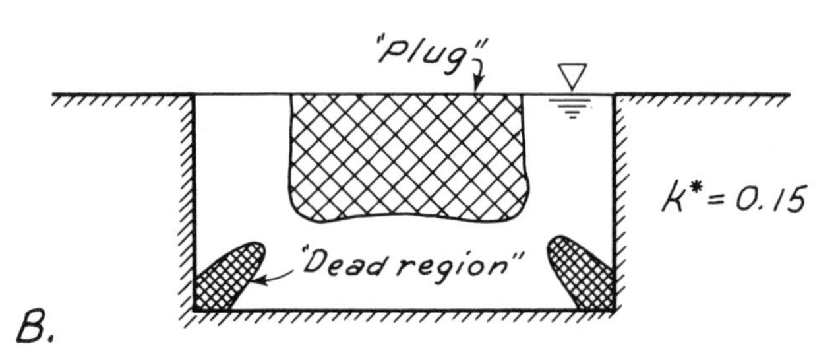

Figure 15.12. Relation between shear strength of Bingham substance and sizes of "plug" and "dead regions."
A. Low shear strength.
B. High shear strength.

the channels, as well as plugs like those we saw in the centers of elliptic and circular channels (*3,4*). Consider, for example, the channel shown in Fig. 15.9, where contours of shear stress variable, $(\sigma_s^*)_{\max}$, range from zero to slightly greater than 0.15. If the dimensionless shear strength variable, k^*, where

$$k^* = \frac{k}{a\gamma \sin \delta}$$

were 0.05, there would be small dead regions at each corner of the channel and a small plug in the center (Fig. 15.12A). If the shear strength variable had a value of 0.15, however, the dead regions would be large and the plug would be large. As the strength variable becomes greater than 0.15, the dead regions would coalesce with the plug and the entire mass of Bingham substance would be dead (Fig. 15.12). There would be no flow.

Velocity Distribution.—Various approximation techniques are available (*9*) for numerically integrating the rheological equations, eqs. (15.44), in order to compute velocities throughout the channels. The computations are tedious; a computer must be used.

We can determine velocities along the upper surface and along the vertical axis of symmetry of the channel relatively easily (*4*). Along the upper surface, the velocity gradient normal to the surface is zero so that the rheological equations, eqs. (15.44), reduce to

$$\frac{\partial \dot{w}}{\partial x} = 0,$$

$$\frac{\partial \dot{w}}{\partial y} = \frac{d\dot{w}}{dy} = -\frac{\sigma_{yz}}{\eta_b} - \frac{k}{\eta_b}\sin\alpha.$$

However, $\alpha = -90$ degrees at the surface, so that

$$\frac{d\dot{w}}{dy} = -\frac{\sigma_{yz}}{\eta_b} + \frac{k}{\eta_b}.$$

Integrating,

$$\dot{w} = -\frac{1}{\eta_b}\int_0^y \sigma_{yz}\,dy + \frac{k}{\eta_b}\int_0^y dy. \tag{15.51}$$

Substituting eq. (15.49) for the shear stress into eq. (15.51), and performing the appropriate integrations, we find that the velocity distribution is

$$\dot{w} = \frac{1}{\eta_b}\left\{\frac{\gamma\sin\delta}{2}(y^2 - yb) - ky + \sum_{m=1}^{\infty}\sin(m\pi y/b)\right.$$

$$\left.\cdot\,[Am\cosh(m\pi a/2b) + Bm\sinh(m\pi a/2b)]\right\}. \qquad [y \leq y_0] \tag{15.52}$$

A similar expression can be derived for velocities along the vertical axis. The solution is valid for all values of y less than or equal to the y-coordinate of the plug, y_0.

Exercises

1. Show that $\Gamma = C_0[(x/a)^2 + (y/b)^2 - 1]$ is a solution to eq. (15.45). Determine the appropriate value for C_0, assuming an elliptic channel.
2. By performing the appropriate differentiations and integrations on the equation in problem 1, derive eqs. (14.14), (14.15), and (14.18), which describe the flow of a Bingham substance in an elliptic channel.

Figure 15.13 shows velocity contours, dead regions, and plug for a Bingham substance flowing in a semisquare channel (*4*). The velocities are expressed in terms of a dimensionless variable,

$$\dot{w}^* = \frac{\eta_b\dot{w}}{a^2\gamma\sin\delta}.$$

Details of the shapes of the contours have been estimated, using values calculated along the upper surface and along the vertical axis of symmetry.

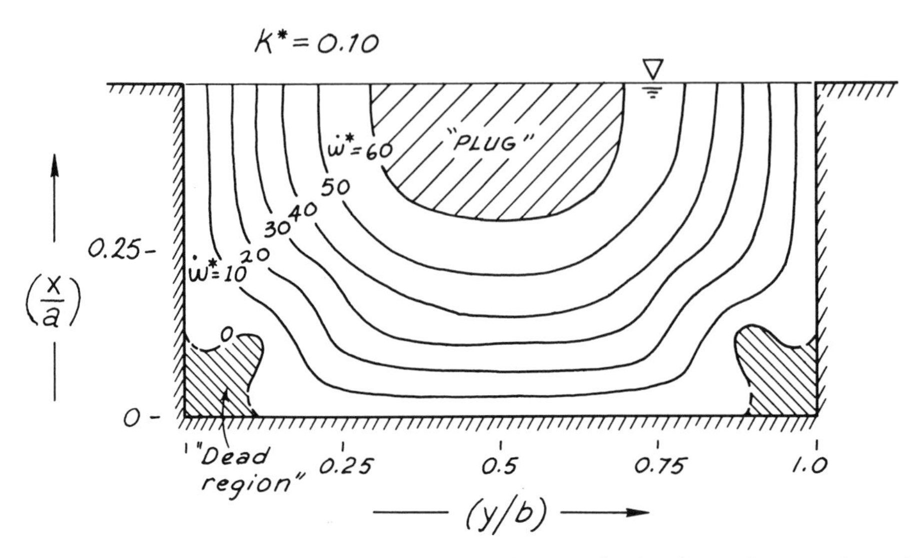

Figure 15.13. Velocity contours of Bingham substance flowing in semisquare channel.

Figure 15.14A shows the theoretical velocity contours and Fig. 15.14B shows an experimental velocity profile for fine-grained slurry flowing in a small, rectangular channel (*3*). The fine-grained debris used for the slurry was collected from a unit of the 1917 deposit on Surprise Canyon alluvial fan in Panamint Valley, California. The solid line in Fig. 15.14B is the theoretical velocity profile, which corresponds remarkably well with the velocity profile defined by the measurements.

Flow of debris in channel of triangular cross-section

The flow of a Bingham substance in channels of triangular cross-section are of interest for several reasons. The dead regions we predict theoretically should occur at the surface of a flow moving in a triangular channel. Thus we should be able to see the dead regions in experimental flows in triangular channels. Also, many natural channels and valleys can be approximated by triangular channels. The V-shaped mountain canyon is visualized as the typical predecessor of the U-shaped glacial valley (*10*). Therefore, an understanding of the distribution of velocities and dead regions in a triangular channel might have an important bearing on our understanding of the evolution of mountain canyons during episodes of glaciation.

We could solve Poisson's equation for flow in channels of triangular cross-section, but we can save ourselves a great deal of work if we simply reflect for a moment about the solutions we have already derived. We have derived the solution for the flow of a Bingham substance in rectangular channels, and a special case of rectangular channel is a square channel. We recall that we derived the solution for

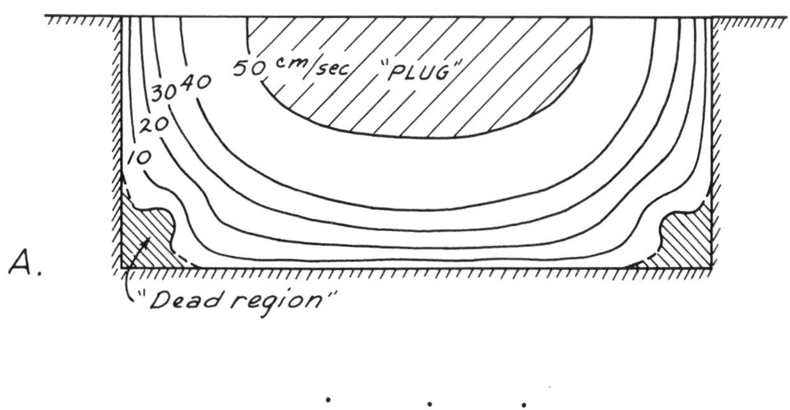

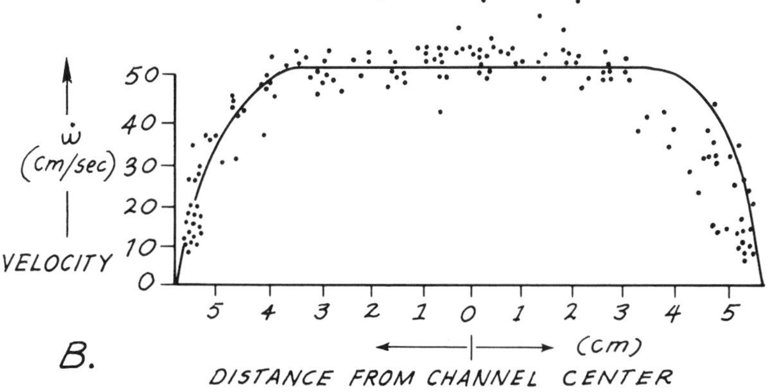

Figure 15.14. Velocity distribution of small debris flow in rectangular channel.
A. Theoretical velocity contours plotted on cross-section of channel.
B. Surficial velocities of debris. Each dot represents one measurement.

the contours of equal shear stress in a rectangular channel by reflecting the channel across its upper surface, forming a closed conduit (Fig. 5.15A). Thus our solutions are valid for open, rectangular channels and for closed, rectangular conduits; the solutions are identical. If we rotate the closed conduit about its axis, that is, about a line representing the direction of flow (Fig. 15.15B), we change nothing in the solution except the coordinate system. If we rotate the coordinate system also, we change nothing in the solution. Suppose, then, that we rotate the square conduit through an angle of 45 degrees (Fig. 15.15B). In this new position, a different line of symmetry is horizontal. The part of the conduit above the line of symmetry is still an image of the part below the line of symmetry. Indeed, this is what we mean by "a line of symmetry." If we agree that the flow pattern is unchanged by rotating the conduit 45 degrees, then we agree that the dead regions and velocity contours are unchanged by the rotation. And we should agree that the flow pattern in a

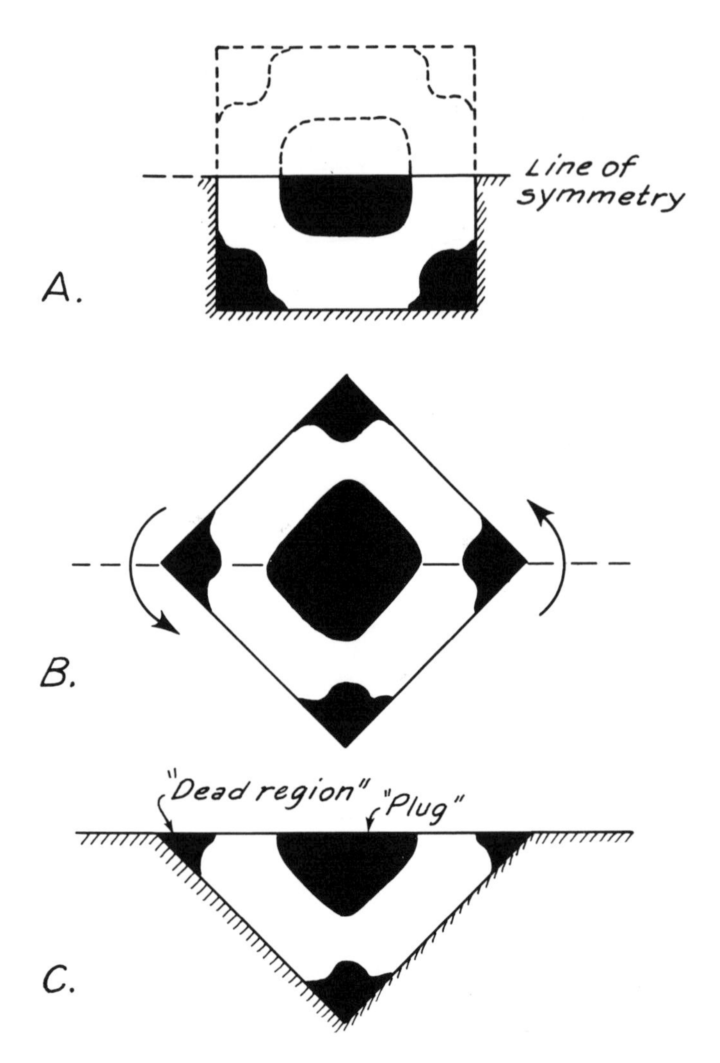

Figure 15.15. Conversion of flow pattern of Bingham substance in square channel to flow pattern in triangular channel.

triangular channel, consisting of one-half of a square conduit cut diagonally from one corner to the other, is the same as the flow pattern in a square conduit (Fig. 15.15C). Thus, we know the solution for velocities and dead regions in one type of triangular channel.

One peculiar feature of the flow pattern of a Bingham substance moving in a triangular channel is that the dead regions appear at the outer edges of the surface of the flow, so that we should be able to observe them experimentally. Indeed we can. Fig. 15.16 shows a time-exposure photograph of the surface of an experimental

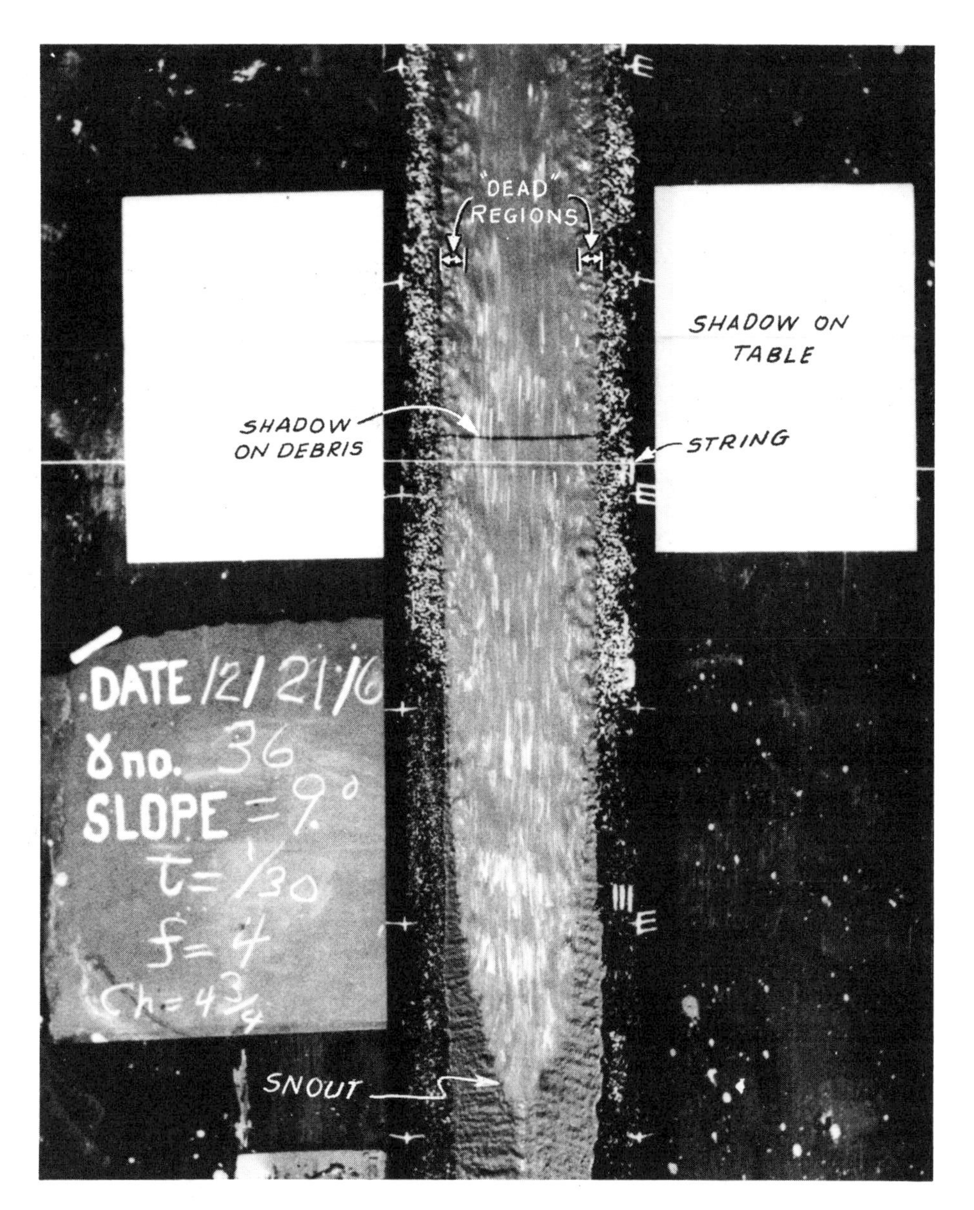

Figure 15.16. Vertical photograph of experimental debris flow in triangular channel. Light streaks on surface of flow are traces of highly reflective flakes.

debris flow moving in direction from the top toward the bottom of the figure (*3*). The channel is triangular and the channel sides slope inward at 45 degrees. The debris used for the experiment is fine-grained material collected from a natural debris-flow deposit in Surprise Canyon alluvial fan. The surface of the moving debris was sprinkled with highly reflective flakes. The photograph shown in Fig. 15.16 was taken normally to the surface of the debris flow, so that the light-toned streaks indicate distances the reflective flakes traveled during the exposure time of the photograph, $\frac{1}{30}$ second. The surficial velocity distribution of the debris flow, therefore, can be determined by measuring the lengths of the light traces.

The distribution of surficial velocities of the debris flow shown in Fig. 15.16 has been plotted in Fig. 15.17B. Each measurement of velocity is represented by a point in the figure. The solid line is the theoretical velocity profile, assuming that the debris behaves like a Bingham substance with a Bingham viscosity of about 7 poises and with a strength of 300 dn/cm^2. The theoretical widths of the dead regions are shown by crosshatching in the figure. The theoretical velocity profile fits the

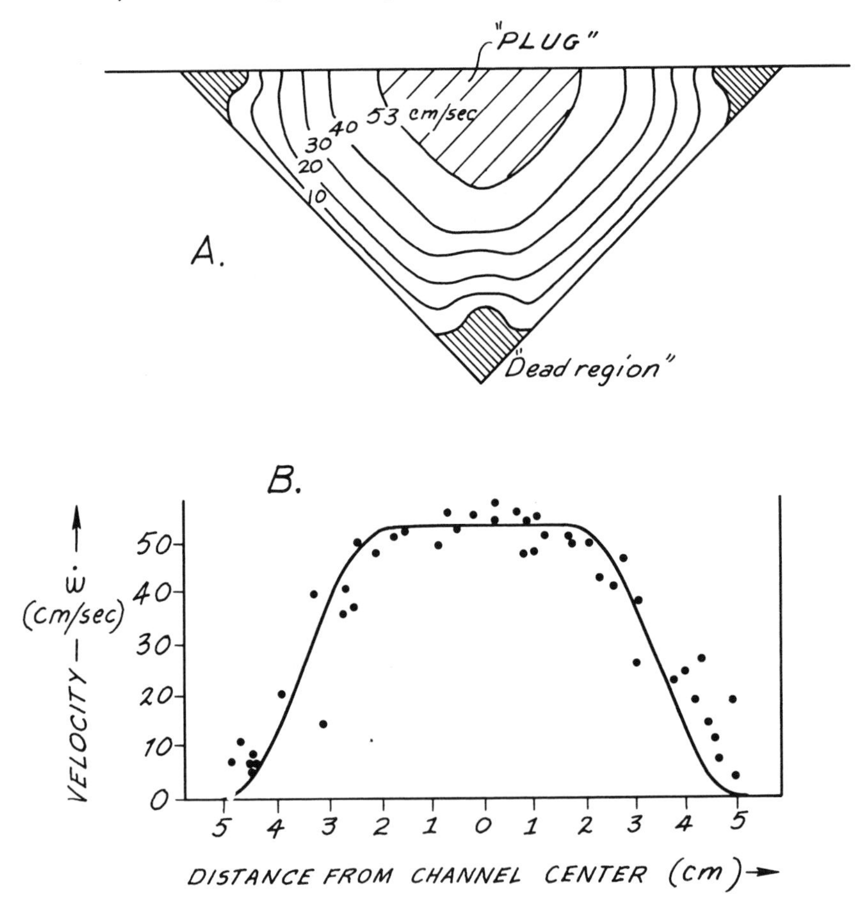

Figure 15.17. Velocity distribution of debris flowing in channel of triangular cross-section.

A. Theoretical velocity contours on cross-section.

B. Velocities of surface of small debris flow in channel of triangular cross-section.

velocity data quite well, so that the debris seems to behave approximately as specified by the Bingham model.

Fig. 15.17A shows the theoretical velocity distribution in transverse section through the channel shown in Fig. 15.16. There are small dead regions at each side and a plug in the center of the flow. The dead regions and the plug are areas where the shear stress is less than or equal to the shear strength of the debris (*3*).

Formation of lateral deposits

The occurrence of lateral deposits in natural debris-flow deposits and in lava flows indicates that part of the debris or lava either became immobilized within the channel, forming a linear ridge, or else overflowed the channel banks, becoming immobilized in a sheet-like form. We will restrict our analysis to lateral deposits consisting of linear ridges on channel banks. We considered the other type in Chapter 12.

We showed that dead regions form in a Bingham material on either side of a triangular channel (Fig. 15.17B), where the yield strength of the material is not exceeded. Perhaps one result of the dead regions is the formation of linear deposits that border debris and lava channels. For example, just above the piles of debris in the foreground of Fig. 15.18 are marked lateral ridges (see arrows) that strikingly resemble the dead regions shown in Fig. 15.17A. There are two lateral ridges on the right-hand side of the channel shown in Fig. 15.18. Multiple lateral ridges would be preserved if succeeding waves of debris moving in the channel were smaller. Each wave would leave behind part of the material in its dead regions (Fig. 15.19).

Figure 15.20 shows lateral deposits forming on the banks of a rectangular experimental channel. The figure is a vertical photograph of a flow of fine-grained debris moving from left to right in the channel. The channel sides or banks slope toward the right in the figure, and where the debris flow is forming a marked lobe, the debris is flowing on the surface of a flat table. The white streaks are traces of highly reflective flakes, some of which moved short distances during the exposure of the film. The flakes appear as dots on the sides of the flow, in areas where the debris has flowed onto the channel banks. When the supply of debris has waned and most of the debris in the channel has flowed onto the table surface, making a large lobe, the debris on the channel sides remains as lateral deposits. Natural analogs of these deposits are conspicuous on many alluvial fans; one excellent example is on the south side of the 1917 debris-flow deposits on the Surprise Canyon fan in Panamint Valley, California (*3*).

U-shape of glacial valleys

One of the characteristics of glacial, lava, and debris valleys is a U-shaped transverse profile. The U-shape of valleys through which lava or debris has moved

Figure 15.18. Linear ridges, representing "dead regions" of channel flow, near Klare Springs, in Titus Canyon, Death Valley National Monument, California. Arrows mark one lateral ridge on left and two on right.

probably is due largely to the formation and preservation of deposits roughly corresponding to dead regions in the channelized flow of a Bingham substance. Perhaps lava remains in contact with relatively cool channel sides in the dead regions so that it solidifies there, preserving the dead regions as lateral ridges. Similarly, wet debris remains in contact with dry channel sides in the dead regions, so that the debris loses part of its water and part of its mobility and tends to stick to the channel sides in the vicinity of the dead regions. The U-shape of lava and debris channels, therefore, is probably caused largely by deposition or immobilization.

The U-shape of some glacial valleys may be due also to deposition. Lateral moraines may in part be manifestations of regions of a glacial tongue that are relatively "dead." And the existence of lateral moraines can give a glacial channel a U-shaped transverse profile.

Some U-shaped glacial valleys have bare rock walls, however, so that some must be erosional in origin. The mechanisms of erosion of solid rock by moving glaciers apparently are abrasion by rock-studded bottoms and sides of glaciers and plucking of loose blocks of rock. Both of these erosional mechanisms require that the ice within a few inches or a few feet of the rock channel be moving with respect to the

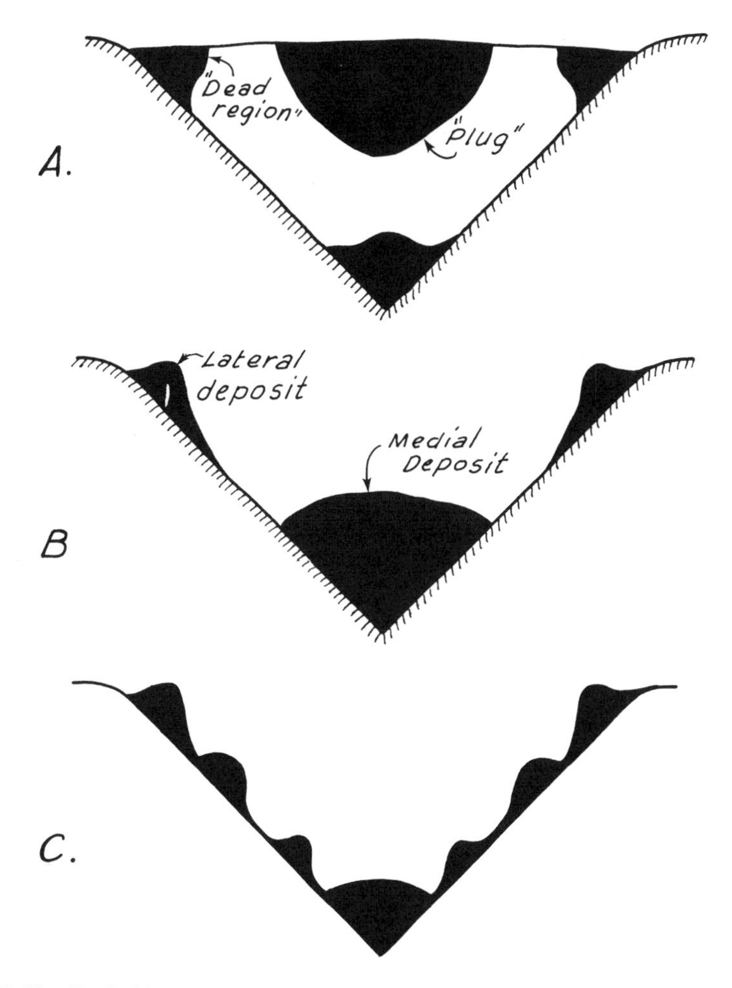

Figure 15.19. Probable sequence of events leading to multiple lateral deposits on sides of channel.

channel; if a rasp is simply held firmly to a piece of wood, the wood is not worn down. Therefore, if there are dead regions in glaciers, we would expect erosion to be minimum where they occur. Suppose that dead regions actually do exist in glaciers. Then, if we imagine that a valley glacier moves into a pre-existing, V-shaped mountain canyon, parts of the glacier, in the vicinity of the juncture of the sides of the canyon and the top of the glacier, and in the bottom of the canyon, will be rigid. Correspondingly, the rate of erosion of the valley sides adjacent to the dead regions is negligible, because the valley sides are protected by a veneer of stagnant ice. But erosion might be appreciable along other parts of the valley sides, where high

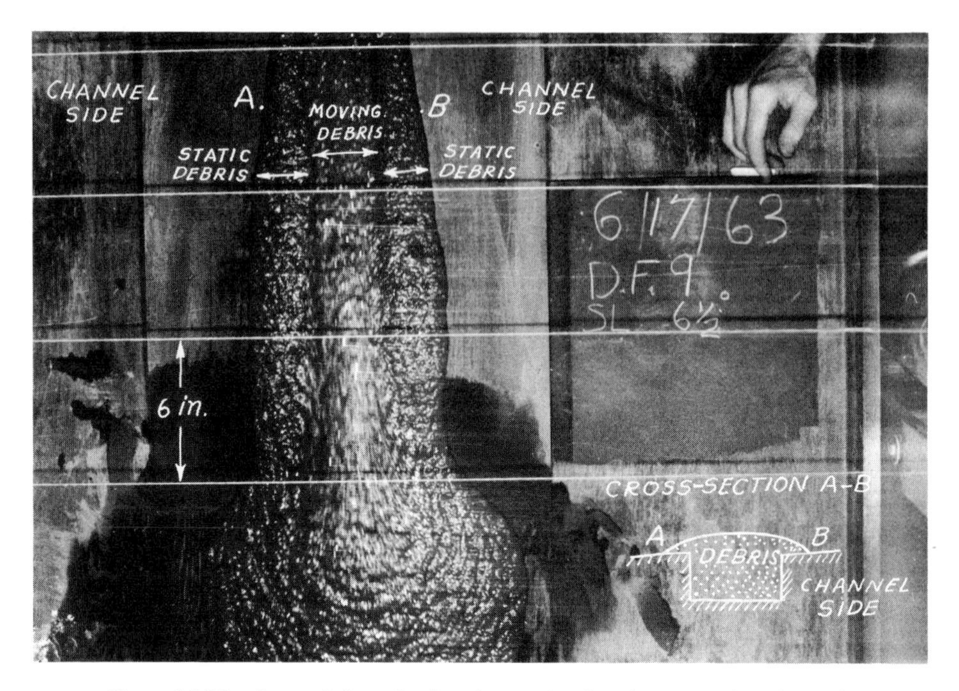

Figure 15.20. Lateral deposits forming on banks of rectangular channel.

shear stress causes a steep velocity gradient. These places will be roughly midway between the top and bottom of the glacier (Fig. 15.9). Erosion along the midpoint of the valley sides will cause the profile of the valley sides to become convex outward, tending toward a U-shape. The more the profile is eroded outward, the smaller the dead regions become and the greater is the erosion, until, in the limit, the glacier is rasping the rock all along its periphery. Fig. 15.21 shows an idealized picture of the evolution of a mountain canyon to a glacial valley. The arrows indicate places of intense erosion along the valley walls.

A rather fundamental assumption of this presentation of valley erosion is that ice behaves as a Bingham substance. The evidence we examined in the preceding chapter indicates that a better rheological model for ice is a pseudoplastic in which the exponent, n [eq. (14.37)], is approximately two to four. Nye (*6*) has solved equations for velocities and shear stresses in channels of elliptic, rectangular, and parabolic cross-sections. I have replotted part of his data for flow of a pseudoplastic in a square conduit. Nye's solution shows that velocities and shear stresses in a pseudoplastic in which $n = 3$ are very low in the corners and bottom of the triangular channel (Fig. 15.22). Therefore, I think we can state that the explanation given here for the U-shape of glacial valleys is reasonable, whether we assume ice behaves as a Bingham or a pseudoplastic substance.

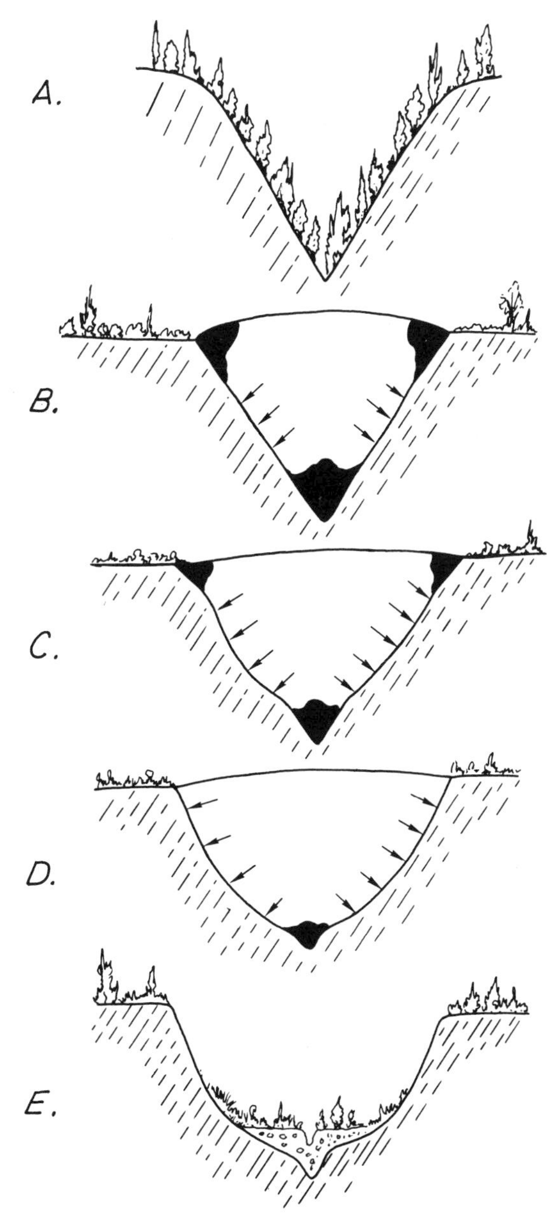

Figure 15.21. Possible sequence of events leading from V-shaped mountain canyon to U-shaped glacial valley.

A. V-shaped mountain canyon.

B. V-shaped canyon visited by glacier.

C. Glacier erodes sides of canyon.

D. "Dead" regions disappear and entire side of canyon is rasped by rock-studded glacier.

E. Glacier disappears, leaving U-shaped valley.

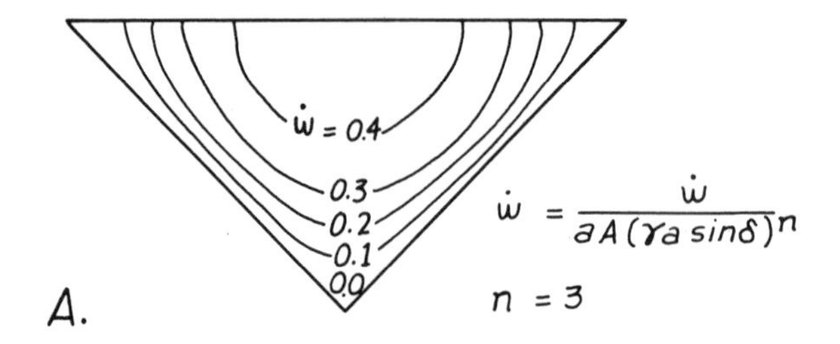

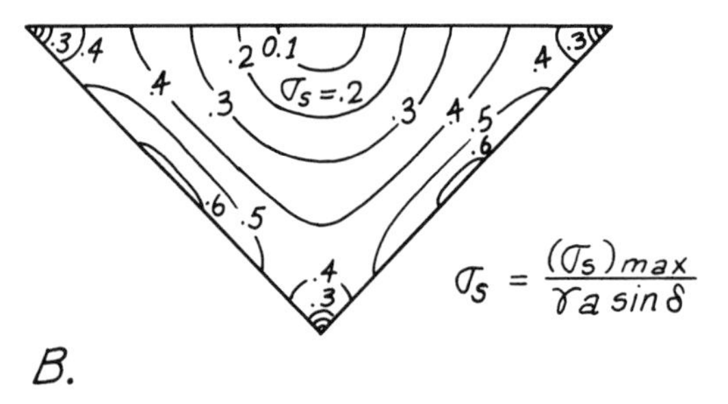

Figure 15.22. Contours of equal velocity and equal shear stress in a pseudoplastic flowing in triangular channel (after ref. *6*, p. 672, 673).
A. Contours of equal velocity.
B. Contours of equal shear stress.

References cited in Chapter 15

1. Alger, P. L., 1957, *Mathematics for Science and Engineering*: McGraw-Hill Book Co., Inc., N.Y.
2. Churchill, R. V., 1963, *Fourier Series and Boundary Value Problems*: McGraw-Hill Book Co., Inc., N.Y.
3. Johnson, A. M., 1965, "A Model for Debris Flow": Ph.D. dissertation, The Pennsylvania State University, University Park, Penna.
4. Johnson, A. M., and Hampton, M. A., 1969, "Subaerial and Subaqueous Flow of Slurries": Final Report, U.S. Geological Survey; Branner Library, Stanford University, Stanford, Calif.
5. Kreyszig, E., 1962, *Advanced Engineering Mathematics*: John Wiley & Sons, Inc., N.Y.
6. Nye, J. F., 1965, "The Flow of a Glacier in a Channel of Rectangular, Elliptic or Parabolic Cross-section": *Jour. Glaciology*, V. 5, p. 661–690.
7. Rouse, H., 1938, *Fluid Mechanics for Hydraulic Engineers*: Dover Publications, Inc., N.Y.
8. ———, Ed., 1959, *Advanced Mechanics of Fluids*: John Wiley & Sons, Inc., N.Y.

9. Shaw, F. S., 1953, *Relaxation Methods*: Dover Publications, Inc., N.Y.
10. Shelton, J. S., 1966, *Geology Illustrated*: W. H. Freeman and Company, San Francisco.
11. Sokolnikoff, I. S., and Redheffer, R. M., 1958, *Mathematics of Physics and Modern Engineering*: McGraw-Hill Book Co., Inc., N.Y.

Closing Comments

Closing Comments

We have touched upon subjects from a variety of fields of geology and a wide range of topics in applied mechanics. We have probed volcanology, geomorphology, glaciology, structural geology, engineering geology, sedimentology, and igneous petrology. We have developed parts of the theories of elasticity, hydrodynamics, plasticity, and fracture mechanics. Indeed, we have covered a great deal of ground, but the methods we have developed are applicable to a much broader class of problems—the problems of describing the physical processes of geology.

The basic method of approach presented here is Gilbert's. Following Gilbert, we have compared the responses of theoretical analogs under various conditions with the responses of natural materials under conditions which we deduce by field study. The use of Gilbert's method sharpens our field observations. We begin to realize which imagined behaviors actually are possible. We begin to discard some preconceptions, such as certain supposed relations between stresses and orientations of faults and joints, and replace them with analogies that, for the moment anyway, appear to be soundly based on physical principles. For example, the supposed analogy between ice flow and debris flow seems to allow us to explain the geological observation that cross-sections of glacial valleys are U-shaped.

I hope that from now on you will ask yourself *how* the features you see in an outcrop were formed. I hope that the next time you see a termination of a sill or a dike you ask yourself how the intruding magma and the intruded country rock behaved at the moment of intrusion. Next time you see a fold or a fault, I hope you examine the bent or fractured rocks and ask yourself how the rocks behaved. If you do this, and, especially, if you are able to answer your question, I congratulate you.

After you have answered the question of how a material behaved when it formed a geologic feature, you are in a position to ask yourself why the material behaved the way it did. For example, adoption of the Bingham model for debris flow naturally led to the question of the cause of the strength of debris. It seems that the strength of debris is largely controlled by the clay fraction, as long as the clay fraction comprises at least thirty percent of the solids. The strength of the clay-water slurry in debris, in turn, seems to be controlled by the distances between double layers surrounding clay particles. The lower the water content of a clay slurry, the more the double layers approach each other and interact, and the higher is the strength of the slurry. Just as the answer to the question of how a certain type of debris behaved allowed us to predict behaviors of that debris under new conditions, the answer to the question of why that debris behaved the way it did allows us to predict behaviors of other types of debris.

Now that you have read this book I recommend that you set it aside on your shelf. It is intended as a workbook to be studied and understood, not as a reference work to be cited. The books mentioned in the preface are appropriate reference works. As I stated in the Preface, my intent has been to provide you with background and perspective, by means of working out solutions to a variety of problems. Now you can understand and assess the theoretical developments presented in more nearly comprehensive books in applied mechanics. Thus I recommend that you begin to study books that deal with the fundamentals of the types of mechanics you need to understand in order to solve your problems that relate to physical processes.

I wish that I could tell you about an easy way to solve problems involving geologic processes. I know of none. I can only recommend more work and more searching.

Index

Airy's stress function 256
Archemede's principle 481

Bearing capacity
  of long, rectangular punch 473
  of circular punch 481
Bending moment
  meaning 44
  sign convention 43
  within elastic member 57
  within viscous member 281
  within multilayer 230
Bending resistance, effect of layering on 81, 150
Bingham model 18, 496, 552
Buckling
  granite sheets 385
  isolated member 84
  single member in elastic medium 91, 136
  single member in viscous medium 279
Boundary conditions 22

Cantilever beam 58, 59
Centered fan, slip-line pattern 477
Circular hole, stresses around
  in plate with edge loads 352, 371
  in body with pressurized hole 413
Coefficient of viscosity 273
Compatibility equation for strains
  meaning 253
  plane strain of elastic body 255
  plane strain of viscous body 278
  plane stress of elastic body 255
  polar coordinates 406
Competency of debris flows 486
Competent structure, law of 110
Complex functions, trigonometric 87
Concentric folds 327
Critical axial load 88, 117, 140, 157
Critical stress for spontaneous crack growth 379
Critical depth of debris flow 488
Curvature of beam
  meaning 55, 57
  of cantilever beam 59
  large deflections 165

Dead regions 554
Debris flow
  shape of longitudinal profile 445,453
  height of snout 456
  plug flow 501
  ability to transport boulders 461
  process of 433, 443
  viscosity 433
  evidence of laminar flow 442
  similarities with lava flow and glacier flow 444
Del operator 257
Differential equations
  integration 59, 63
  homogeneous 86, 93, 139
  particular solution 115, 139
  solution by separation of variables 237, 260, 285
  solution by Fourier series 546
  solution by substitution of $r = e^t$ 410
Dike patterns
  Dike Mountain 402
  Spanish Peaks 414

Dilation 191
Distributed load 52
Domains of folds 311
Drag flexures 103

Elastica 162
Elastic body
  stress-strain relations 199
  Lamé's constants 203
  Young's modulus 200
  Poisson's ratio 200
  shear modulus 209
Elastic constants
  of some rocks 202
  of multilayer 228, 234
Elasticity theory
  textbooks ix
  Hookean model 14, 199
Elliptic hole
  stresses around, in elastic plate 374
Elliptic integral 168
Elliptic partial differential equation 545
Equilibrium
  of forces and bending moments 10
  in terms of stresses, cartesian coordinates 250, 258
  in terms of stresses, polar coordinates 406
Euler wavelength 91
Experiments
  value of 7, 8, 110, 130
  of folding 98, 151

Fault patterns
  Timber Mountain 335
  associated with folds 350
  Anderson's theory 345
Flow
  in semicircular channel
    Newtonian 515
    Pseudoplastic 515
    Bingham 497
  in infinitely wide channel
    Bingham 503
    Newtonian 541
  in elliptic channel
    Bingham 559
    Newtonian 543
  in rectangular channel
    Newtonian 550
    Bingham 559
  in triangular channel
    Bingham 560
    Pseudoplastic 568
Folding
  references 126
  comparison of elastic and viscous folds 288
  of quartz veins 140
Folding experiments
  by Bailey Willis 98
  rubber strips in gelatin 145
  paper cards 322
  rubber and cardboard in clay 314
Fourier coefficients 548
Fourier series 119, 547
Free-body diagram 46
Fiber stresses 56

Gilbert's method of research 3, 129
Griffith criterion of fracture 370
Griffith substance
  meaning 390
  failure 392

Hencky's theorem 477
Hinge lines 313
Homogeneous differential equation 86
Hyperbolic sine and cosine 148, 268, 269

Ice, rheological properties
  ice cylinders 525
  glacial ice 530
Images, method of 267, 417
Imaginary number, *i* 86
Initial dips 110
Initial deflections
  simple 110, 112
  complex 119
Initial stress 323, 380

Kelvin model 18
Kink bands
  orientation 314, 321
  In Franciscan Formation 311

Laccolith
  Gilbert's concept of limital diameter 39
  Idealized form 34
  Relation between depth and diameter 39, 69
Lamé's constants 203
Laplace equation 543
Lateral deposits of debris flows

description 435
theory of formation 565
Lava, basaltic, strength and viscosity of 525
Levees (see lateral deposits)

Maxwell model 18
Method of physical processes 3, 7
Microfractures
in Chelmsford granite 366
possible origin 386
Mode of buckling 88
Modulus, B 62, 217
Modulus of rigidity 205
Mohr Circle
use 205
derivation 340
Moment of inertia of cross sections 57
Multilayer
frictionless contacts 150
Biot model 222
interlayered clay and rubber 314

Navier-Stokes equations 540
Newton's laws 10
Nonlinear-elastic material, buckling 308
Nonlinear rheological models 19

Plane stress and plane strain 210
Plasticity theory
St. Venant model 16
bending of plastic member 298
penetration of elongate punch 464
textbooks ix
references 491
equations of equilibrium 467
equations for velocities along slip lines 473
slip lines 465
Plug flow of debris 500
Plug width
semicircular channel 501
elliptic channel 506
Poisson's equation 543
Poisson's ratio 200, 202
Prandtl model 18
Pressure 275, 278
Principal stresses 338, 352
Pseudoplastic
buckling of member 304
flow in channel 515
model for ice 526
Ptygmatic folds 161

Radius of curvature 55, 57, 162
Reflection, method of 267, 417, 545
Rheology
meaning 13, 21
selection of model 517
criterion of good model 518

Scale models 130
Separation of variables 237, 260, 285, 544
Settling of sphere in Bingham substance 481
Shear force
definition 43
relation to distributed load 53
relation to bending moment 53
deflection of member 108
Shear rate
semicircular channel 499
infinitely wide channel 503
Sheet fractures
equation for 382
spacing of 386
Sheet structure, origin of 357, 368
Shear-stress function 554
Slip-line patterns
constant state 477
centered fan 477
beneath punch 480
Sign conventions
forces and bending moments 43
stresses 184
Smoothing effect of debris flow 513
Soil mechanics
references 491
bearing capacity of foundation 479
Statically determinate problem 473
Steady flow 497
Strain energy 377
Strain, finite
meaning 195, 197
relation to displacements 197
Strain gauge 203
Strain, infinitesimal
meaning 13, 186
types 189, 195
dilatation 191
relation to displacements 191, 194
shear 191
normal 191
tensor components 193
Strain rate 272
Strength

of debris 456
of basaltic lava 525
of some rocks 202
concept of 14
Stress
definition 12, 175
as vector quantity 178
on various planes 179
difference between force and stress 180
reason for nine components 181
notation used in book 182
other notations 184
sign conventions 184
initial 323
Stress-strain relations
elastic, Hookean body 14
initially-stressed body 381
Strut member concept 80, 133
Superposition
of strains 197
of stresses 267, 417
Surface energy of cracks 374

Thickness of debris flows 488
Trajectories
of principal stresses 413, 420
of shear stresses 465
shear stresses in plastic 465
Transport of boulders by debris
theory 463, 486
description 439
gentle handling 513

U-shape of glacial valleys
of Saskatchewan glacier 527
possible explanation 566

Velocity distributions (see flow)
Velocity, measurement
by using glitter 501
in natural flows 509
in glaciers 529
Viscometer
circular-channel type 496
rotational cylinder type 520
other types 497
Viscosity coefficient 273
Viscosity theory
Newtonian model 14, 272
textbooks x
stress—strain-rate relations 275
Viscous-plastic member, buckling of 304

Wavelengths of folds
elastic member in infinite elastic medium 141
multilayer, frictionless contacts 152
effect of gravity 155
multilayer 243, 244
viscous member in infinite viscous medium 288
Wave train of folds 96

Yield strength 298
Young's modulus 199, 202

Some Commonly Used Trigonometric Identities

Sine Law: $\dfrac{a}{\sin A} = \dfrac{b}{\sin B} = \dfrac{c}{\sin C}$

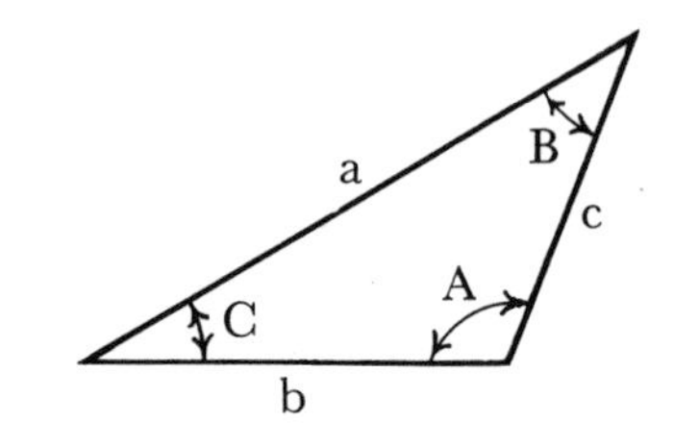

Cosine Law: $b^2 = a^2 + c^2 - 2ac \cos B.$

$a^2 = b^2 + c^2 - 2bc \cos A.$

$c^2 = a^2 + b^2 - 2ab \cos C.$

$$\frac{1 + \sin\phi}{\cos\phi} = \frac{\cos\phi}{1 - \sin\phi} = \tan\left(45° + \frac{\phi}{2}\right) = \text{ctn}\left(45° - \frac{\phi}{2}\right)$$

$\sin^2\phi + \cos^2\phi = 1$

$\sec^2\phi - \tan^2\phi = 1$

$\sin(90° - \phi) = \cos\phi$ $\quad$ $\sin(90° + \phi) = \cos\phi$

$\cos(90° - \phi) = \sin\phi$ $\quad$ $\cos(90° + \phi) = -\sin\phi$

$\tan(90° - \phi) = \text{ctn}\,\phi$ $\quad$ $\tan(90° + \phi) = -\text{ctn}\,\phi$

$\sin(180° - \phi) = \sin\phi$ $\quad$ $\sin(180° + \phi) = -\sin\phi$

$\cos(180° - \phi) = -\cos\phi$ $\quad$ $\cos(180° + \phi) = -\cos\phi$

$\tan(180° - \phi) = -\tan\phi$ $\quad$ $\tan(180° + \phi) = \tan\phi$

$\sin(-\phi) = -\sin\phi$

$\cos(-\phi) = \cos\phi$

$\tan(-\phi) = -\tan\phi$

$\sin(A + B) = \sin A \cos B + \cos A \sin B$ $\quad$ $\sin(A - B) = \sin A \cos B - \cos A \sin B$

$\cos(A + B) = \cos A \cos B - \sin A \sin B$ $\quad$ $\cos(A - B) = \cos A \cos B + \sin A \sin B$

$$\tan(A + B) = \frac{\tan A + \tan B}{1 - \tan A \tan B} \qquad \tan(A - B) = \frac{\tan A - \tan B}{1 + \tan A \tan B}$$